T0181027

THE FRONTIERS COLLECTION

THE FRONTIERS COLLECTION

The books in this collection are devoted to challenging and open problems at the forefront of modern science, including related philosophical debates. In contrast to typical research monographs, however, they strive to present their topics in a manner accessible also to scientifically literate non-specialists wishing to gain insight into the deeper implications and fascinating questions involved. Taken as a whole, the series reflects the need for a fundamental and interdisciplinary approach to modern science. Furthermore, it is intended to encourage active scientists in all areas to ponder over important and perhaps controversial issues beyond their own speciality. Extending from quantum physics and relativity to entropy, consciousness and complex systems—the Frontiers Collection will inspire readers to push back the frontiers of their own knowledge.

More information about this series at http://www.springer.com/series/5342

Robert Ayres

ENERGY, COMPLEXITY AND WEALTH MAXIMIZATION

 Springer

Robert Ayres
INSEAD
Fountainebleau, France

ISSN 1612-3018   ISSN 2197-6619   (electronic)
The Frontiers Collection
ISBN 978-3-319-80835-2   ISBN 978-3-319-30545-5   (eBook)
DOI 10.1007/978-3-319-30545-5

Printed on acid-free paper

This Springer imprint is published by Springer Nature
The registered company is Springer International Publishing AG Switzerland

Praise for *Energy, Complexity and Wealth Maximization*

"Economists and physicists, like oil and water, resist mixing, sadly to the detriment of useful human knowledge. Bob Ayres is the rare combination of a physicist and a resource economist, giving him a unique understanding of the importance of useful energy services to all of life. This unique understanding is critical to the massive challenge human kind now faces – how to 'power' continued wealth creation without destroying the planet we call home. This book will almost certainly alter the way we approach this great challenge."

Thomas R. Casten, Chair, Recycled Energy Development LLC

"This is a must read for those who wish to understand what we've got wrong in our contemporary development paradigm and how we can fix it. By far the most important book in years that will reshape physics the way Darwin and Einstein have done, and will hopefully reshape economics too!"

Dr. Stefanos Fotiou, Director of the Environment and Development Division, UNEP

"The fact that the world's nominal GDP shrank last year by 4.9%, while the planet experienced no financial crash, no earthquake, and no sovereign default, remains impossible to understand unless one acknowledges the dependency of our economies on finite natural resources. Bob Ayres is among the pioneers of this biophysical approach to economics, which may prove to be the most fruitful innovation in economics since Keynes. This extraordinary book crosses disciplinary boundaries to takes a broad, evolutionary perspective on human societies as thermodynamical dissipative structures. As natural resources become scarce and quality declines, knowledge is the one ingredient that may save us from following a path analogous to supernovae explosions. At a time when most economists confine themselves to partial and local micro-explanations, Ayres provides a big-picture understanding of the forces that underlie our current economic paradoxes."

Gaël Giraud, Professor of Economics, Ecole Normale Superieur (Paris), and chief economist, Agence Francais pour Developpement

"The energy system of the world has gone through several changes in recent decades, and it is expected that with current global developments and the Paris agreement on climate change, major changes in global energy developments and wealth need to be assessed in depth with the revolutionary changes that are likely to occur. No one better than Prof. Robert Ayres, who understands the industrial metabolism of the world and the role of energy globally, could attempt the analysis presented in this book. This analytical study combines science, economics and technology issues in a remarkable manner to provide rare insights into where the world is heading and why. The book is a must read for concerned citizens and decision makers across the globe."

R K Pachauri, Founder and Executive Vice Chairman, The Energy and Resources Institute (TERI) and ex-chair, International Panel on Climate Change (IPCC)

"This magisterial synthesis traces the evolution of order and complexity from the Big Bang to Big Data to Big Dangers ahead. The book delineates the urgent collective challenge of making the 'great transition' from an economy that squanders nature's wealth to a new paradigm rooted in a knowledge-based wealth."

Dr. Paul Raskin, Founder and President Tellus Institute

"Robert Ayres' new book is a historic, a contemporary, and a future oriented work of immense depth of thought, written by an author of incredible knowledge and wisdom, and encompassing views and concepts of both social and natural sciences. It is theoretically interesting, empirically relevant and timely regarding integrated assessments of social and natural systems. I think the work is a seminal contribution to looking at the co-evolution of human (economic and social) development and the Earth system, and will especially help to comprehend the new geological era – the 'Anthropocene.'"

Udo E. Simonis, Professor emeritus for Environmental Policy at the Berlin Social Science Center (WZB)

"Recommending this book is done best by stating a fact and making a wish. The fact: most people who run the modern world (politicians, economists and lawyers) have a very poor grasp of how it really works because they do not understand the fundamentals of energy, exergy and entropy. The wish: to change the ways of thinking of all those decision-makers, who would greatly benefit from reading this book. But so would scientists and engineers who may be familiar with its basic messages. They would profit from Bob's life-long examination of fundamental ideas and from their lucid distillation and synthesis: an important book, indeed."

Vaclav Smil, Distinguished Professor Emeritus, University of Manitoba, Canada

"Bob Ayres, the doyen of the intellectual universe encompassing physics and economics, has hit again. Beginning, of course, at the origin of the physical universe, he traverses galaxies, stars, planets and then our own planet's history. He then concentrates on human history and offers explanations for the dynamics of natural wealth creation that must become our new paradigm after conventional 'progress' has destroyed so much of natural wealth. And it is knowledge, rather than Gigabytes of 'information,' that can lead humanity into a better future. A grandiose design; impressive; worth reading and reflecting!"

Dr. Professor **Ernst Ulrich von Weizäcker**, Founder of Wuppertal Institute; Co-President of the Club of Rome, Former Member of the German Bundestag, co-chair of the UN's Resource Panel.

"In an age of sustainable development goals, there is no more urgent need for the policy makers and the public alike than to have a clear understanding of the complex linkages among energy, innovation, and wealth. Bob Ayres' book has done a superb job, weaving back and forth between physics and economics seamlessly, in illuminating the history of wealth creation in the past through the conversion of materials into 'useful things' based on the consumption of energy, and providing insights into the future when wealth will be created by knowledge accumulation, de-materialization and institutional innovation. It is a must read for all of us who wish for a sustainable future for humanity."

Lan Xue, Dean of School of Public Policy and Management, Tsinghua University, and Co-chair, UN Sustainable Development Solution Network

Preface

This book has had a long gestation. It is, in effect, a follow-on of a book I wrote in 1994, entitled *Information, Entropy, and Progress* (Ayres 1994). That book was an attempt to explain evolution in terms of accumulation of "useful information," as distinguished from just information. I was reminded of this yesterday when I read a surprisingly favorable review of a new book entitled *Why Information Grows: The Evolution of Order, from Atoms to Economies* by Cesar Hidalgo (Hidalgo 2015). I could have used that title for my 1994 book or for this one.

There is only one problem, really. Information, in the proper sense of the word (as in information theory), is not wealth. In fact, it is mostly junk. At any rate, too much can be as harmful as too little. Information technology may have "progressed" by leaps and bounds, and it has made a lot of people wealthy in Silicon Valley. But there is little or no evidence that the rest of us have prospered thanks to smartphones or Facebook (or even Google, which I couldn't live without). A better word than "information" would be "knowledge." Economists do use the term "knowledge economy," where "knowledge" is intended to convey something like the "essence of information."

But knowledge is not well defined, and its role in driving growth is very unclear. I'm afraid Hidalgo—like many in the "commentariat"—has put the cart before the horse. While the rich countries have more information processing and denser information flows, that is not necessarily why they are rich. Having better universities would be a better explanation of relative wealth, but having a lot of oil in the ground probably helps even more. The real connection between economic growth (useful) information and knowledge is much subtler. It is what the latter part of this book is about.

So why the long delay between from 1994 and 2016? That is partly because a group of us with backgrounds in physics or other sciences have been arguing with mainstream economists (but not being heard) for many years. The topic of the argument is the proper role of energy in economic science. (The role of entropy in economics is not being discussed at all, so far as I am aware.) This is not the place to summarize arguments (which still continue) except to say that progress is agonizingly slow because there is a widespread conviction among supposedly well-educated people, including business leaders and decision-makers, that they don't need to know anything about basic science to make good decisions.

I went to the University of Chicago to study physics in 1954 at the time when its president, Robert Hutchins, and his sidekick, Mortimer Adler, were famously promoting the *Great Books of the Western World* (Adler et al. 1990). The original 54-volume set included only two on economics (Adam Smith Vol. 39; Marx and Engels Vol. 50), plus a scrap of J.S. Mill. Science was covered only slightly better (Ptolemy, Copernicus, and Kepler Vol. 16; Gilbert, Galileo, and Harvey Vol. 28; Newton and Huygens Vol. 34; Lavoisier, Fourier, and Faraday Vol. 45; Darwin Vol. 49). In the second edition (Adler et al. 1990), Volume 56 was added. It included Einstein, Eddington, Planck, Bohr, Heisenberg, and Schrödinger.

The fact that the choices of who to include, or not, were not made by physicists is clear from some of the obvious omissions in physical science: Boltzmann, Carnot, Clausius, Dirac, Fermi (who was at the University of Chicago at the time), Feynman, Gell-Mann, Gibbs, Leibnitz, Maxwell, Mayer, Mendeleev, Pauli, Prigogine, and so on (to the end of the alphabet). In economics, the absence of Arrow, Jevons, Keynes, Malthus, Marx, J.S. Mill, Ricardo, Samuelson, J-B Say, Schumpeter, Solow, Veblen, von Neumann, Walras, and Max Weber makes the same point.

Nothing in the first edition of the *Great Books* mentions the most important laws of nature, namely, the first and second laws of thermodynamics. The first law is conservation of energy and the second (entropy) law says that all spontaneous processes in nature go in one direction ("time's arrow"). Whether Volume 56 in the second edition mentions either of these laws, I do not know. But the fact that the non-scientists who compiled that list of "great books," and the "great ideas" in them, were unaware of those laws—and a lot else—is shocking.

After leaving Chicago, I spent 2 years (1956–1958) at King's College of the University of London, working on a Ph.D. in theoretical physics. It was impossible not to notice that the higher levels of the socioeconomic strata in Britain at the time were heavily recruited from students with honors degrees in *Literae Humaniores*, known as "The Greats" at Oxford University. That course was (and is) focused on reading the Greek and Roman classics (Homer, Virgil) in the original languages and writing weekly essays on a variety of topics. The ability to quote appropriate passages in Latin was one of the criteria for being "one of us" at the top levels of British society.

The prevailing attitude, as conveyed by the media, was that scientists were "boffins in the back room" where they were paid very modestly to discover or invent things for the rest of society, which the rest of society didn't necessarily want or need. The 1951 Ealing comedy "The Man in the White Suit," starring Alec Guinness, made that point very clearly. The fact that those clever "boffins" had also invented jet engines and radar, decrypted the German codes, and created the atomic bomb was very disconcerting. The nerdy people who made such a huge contribution to winning the war were ignored or (in one notorious case) actively persecuted.

This gap—a chasm—was central to the novels of C.P. Snow and his famous "Two Cultures" lecture at Harvard in 1959. But that didn't open the doors of the elite clubs on Pall Mall to boffins, nor did the great companies bring scientists into the executive suite or onto their boards of directors.

Back in the USA, working in my chosen field, I could not help but notice the rise of the Harvard Business School. (My sister-in-law, in the 1970s, divorced my brother in order to go to HBS. Her great ambition was to become the first female VP of Generous Electric, Inc.) But the point here is that HBS and its upcoming rivals were teaching smart young people that "management" is a science and that to be a good—or great—CEO of a company like GE it is not necessary to understand what they produce or how they produce it. All of that detailed stuff can be left to the "boffins in the back room." What CEOs do is grand strategy, which turns out to be about some combination of finance, law, stockholder relations, labor relations, and lobbying the government regulators. In other words, HBS thought that it is possible to run General Electric Co. without having a clue about how electric power is generated and distributed, or how it is used to do work, still less about the laws of thermodynamics.

Sadly, most of the people who run the world now have a grossly inadequate grasp of important ideas that are fundamental to how the world (and the economy) works. That degree of ignorance among the powerful is dangerous. Energy and entropy are among the fundamental ideas that cannot be safely ignored. But thermodynamics is inadequately understood because it is badly taught, or not taught at all (except in specialized science courses), in schools and universities. This book started as an ambitious—probably overambitious—attempt to explain energy and entropy to otherwise educated people who thought that energy is the secret ingredient of "Red Bull," or the reason for drinking coffee in the morning, or is just a topic for nerds with calculators. (This book has evolved somewhat *en route*.)

As for exergy and entropy, the words are scary and unfamiliar, but they should not be. I use the word "should" in the normative sense. *Exergy* is that part of energy that can do work. *Exergy* is what gets "consumed" and "used up." Engineers say that energy is "destroyed" when it does work, but that is a little overdramatic. *Anergy* is the useless part of energy that cannot do any work. *Entropy* is a measure of the state of the world that increases after every spontaneous change and whenever exergy is consumed.

Entropy is invisible and intangible. It is not a substance. There are no "entropy meters." It was originally defined by a relationship, much as positrons and neutrinos (and the former planet Pluto) were discovered: because they were missing pieces of a puzzle needed to satisfy a law of nature. For the record, the relationship is simple: the difference between total energy E and exergy B (in a chemical system) is the product TS of temperature T times entropy S. Of course T is measurable on a thermometer. Does that help? Probably not, if you didn't study science. I won't mention it again in this book.

Recently I realized that there is a deeper connection between the origin of the universe and the reality of today. This book is my best attempt to explain it. In brief, the second law of thermodynamics isn't only about irreversibility, the "arrow of time," or the "heat death" of the universe. It is far from it. The keyword in the title of this book is "complexity." I could have used the words "order" or "structure" or even "resilience."

Yet, the universe is 13.77 billion years old (according to the Big Bang theory), and apart from being very large, it is extremely diverse. There are many billions of galaxies containing millions of trillions of stars, many of which have planets, some of which probably have carbon-based life. Where there is life, proliferation and organization occur, resulting in increasing complexity. When the complexity reaches a certain level, intelligence emerges. Intelligence creates more complexity and, ultimately, knowledge.

Speaking of our own planet Earth, the variety of life forms—past and present—is astonishing. And within our own species, the variety of social organizations, religious beliefs, business plans, scientific theories, chemicals, products, artworks—and book titles—is also very large. My point is that the cooling and aging of the universe have been accompanied by an explosion of different, increasingly orderly, configurations of matter on all scales, from the microscopic to the cosmic.

In fact, the increasing complexity of the universe is causally related to the second law of thermodynamics and irreversibility. This book will explain some of the reasoning behind that statement. I hasten to point out that the underlying idea that biological evolution, in particular, is a consequence of the entropy law has been stated before, by others. I will cite the sources in due course.

This brings me to "wealth," the word in the title of this book. The first definition in a typical dictionary is "A great quantity or store of money, valuable possessions, property, or other riches." Is that what you thought the title of this book was about? Well, it is but only up to a point. It was the second definition that I had in mind: "A rich abundance or profusion of valuable resources, or valuable material possessions." In particular, I stress the notion of profusion or diversity. The reason a lot of money is called wealth is that it offers a lot of different choices. The more choices you have, the greater your wealth. If there is nothing for sale in the shops, as in Zimbabwe a few years ago or in Venezuela today, money is worthless. When the Berlin Wall came down in 1989, it was the range of choice—including bananas and oranges—in the shops of West Berlin that was so attractive to the people who had been trapped for so long behind the Wall.

The idea that increasing wealth is a consequence of information flow is being bandied about. There is undoubtedly some truth in that proposition. The Internet does seem to promote social organization. It can also destroy it. But I would emphasize the importance of knowledge, rather than information as such. We are all surrounded by a flux of useless information, much of which is counterproductive if not toxic. (Think about "cyber-wars" and all the complex and wasteful efforts to secure "privacy" and protect personal information of little value.) Information is not a source of wealth, except insofar as information contributes to knowledge. Knowledge is hard to define and hard to measure, but increasing knowledge surely explains why "produced" wealth keeps increasing while natural wealth is being dissipated.

This book concludes with several chapters on economic theory, as regards energy flow, economic growth, and wealth accumulation. For a rigorous discussion of those relationships, I recommend *The Second Law of Economics* by Reiner Kümmel of Würzburg University (Kuemmel 2011). The present book is much less mathematical (and less rigorous) than his but considerably broader in scope. There is no need for me (or anyone) to recapitulate the mathematical derivations in that book. They constitute a permanent contribution to economic growth theory. Instead I have tried to write for a larger but less mathematically sophisticated audience. I believe there is room for both books and that they should be viewed as complementary rather than competitive.

However, I think there is more to be said that less specialized readers—especially people interested in science—may find interesting. Darwinian natural selection plays an important role in economics, of course. But the role of complexity, as a precursor of selection, is rarely mentioned in the academic literature.

And here I should say for whom I am writing this book. One group consists of people who read books like Weinberg's *The First Three Minutes* (Weinberg 1977) or Lederman's *The God Particle* (Lederman 1993) or *What is Life?* (Schrödinger 1945) or *From Being to Becoming* (Prigogine 1980) or *Into the Cool* (Schneider and Sagan 2005) or the books by Carl Sagan or Jared Diamond or *Scientific American* and other comparable science publications. But I also want to speak to people who read popular economics books, like *The Constitution of Liberty* (Hayek 1960), *Capitalism and Freedom* (Friedman 1962), *More Heat than Light* (Mirowski 1989), *Debunking Economics* (Keen 2011a), or *The Global Minotaur* (Varoufakis 2011). In short, I will present some ideas relevant to both camps of C.P. Snow's *Two Cultures*, and I hope to convey some new ideas to both groups.

Fontainebleau, France   Robert Ayres

Acknowledgments

'

Most books like this start (or end) by acknowledging the lifetime support of a spouse or partner. In some cases, this may be *pro forma*. In my case, it is definitely not so. My wife, Leslie, has contributed to all my books for more than 60 years, in every possible way except the actual writing. Without her I could not function. Enough said.

Once or twice in the past, I have made the mistake of carelessly overlooking someone who deserved acknowledgment. In the present case, I find myself in a dilemma. Some of the people who have influenced my ideas did so long ago. I could also list some authors whom I admire but have never physically met, but to keep this list from excessive length, I forebear to do so. To avoid any suggestion of rank-ordering, the list is alphabetical. It is long because I have been active for a long time and I have worked professionally in several fields of science. A fair number of those listed below are now deceased, but I am no less thankful for the times we spent together. A few are people I disagree with, but disagreement sharpens the argument. Here goes:

David Allen, Julian Allwood, Ralph D'Arge (dec.), Ken Arrow, Brian Arthur, Nick Ashford, Bob Aten, Rob Axtell, Jeroen van den Bergh, Alan Berman, Eric Britton, Colin Campbell, Michael Carroll, Tom Casten, Mike Chadwick, Xavier Chavanne, Cutler Cleveland, Joe (dec.) and Vary Coates, Bob Costanza, David Criswell, Paul Crutzen, Jim Cummings-Saxton (dec.), Otto Davis (dec.), Tiago Domingos, Cornelis van Dorsser, Faye Duchin, Nina Eisenmenger, Marina Fischer-Kowalski.

Paulo Frankl, Bob Frosch, Jeff Funk, Murray Gell-Mann, Gael Giraud, Fred Goldstein, Tom Graedel, Arnulf Grübler, Charles Hall, Bruce Hannon, Miriam Heller, Ludo van der Heyden, Buzz Holling, Leen Hordijk, Paul Horne, Jean-Charles Hourcade, J-M Jancovici, Michael Jefferson, Dale Jorgenson, Herman Kahn (dec.), Astrid Kander, Felix Kaufman, Yuichi Kaya, Steve Keen, Ashok Khosla, Allen Kneese (dec.), Andrey Krasovski, Michael Kumhof, Reiner Kuemmel.

Jean Laherrere, John (Skip) Laitner, Xue Lan (as he is known in China), Tom Lee (dec.), Tim Lenton, Dietmar Lindenberger, Hal Linstone, Amory Lovins, Ralph (Skip) Luken, Katalin Martinàs, Andrea Masini, Fran McMichael, Steve Miller, John Molburg, Elliot Montroll (dec.), Granger Morgan, Shunsuke Mori, Nebojsa Nakicenovic, Mancur Olsen (dec.), Philippe Pichat, Vladimir Pokrovski, Ilya Prigogine (dec.), Paul Raskin, Bob Repetto, Ronald Ridker.

Sam Rod (dec.), Don Rogich, Hans-Holger Rogner, Pradeep Rohatgi, Rui Rosa, Adam Rose, Ed Rubin, Tom Saaty, Warren Sanderson, Heinz Schandl, Bio Schmidt-Bleek, Uwe Schulte, Andre Serrenho, Gerry Silverberg, Herb Simon (dec.), Udo Simonis, Mitchell Small, Vaclav Smil, Kerry Smith, Rob Socolow, Gus Speth, Martin Stern, Jim Sweeney, Laura Talens, Joel Tarr, Ted Taylor, Iouri Tchijov, John Tilton, Michael Toman, Richard Tredgold (dec.).

Gara Villalba, Genevieve Verbrugge, Sylvestre Voisin, Vlasios Voudouris, Ingo Walter, Benjamin Warr, David Wasdell, Luk van Wassenhove, Chihiro Watanabe, Helga Weisz, Ernst von Weizsäcker, Eric Williams, Ernst Worrell, Philip Wyatt, Huseyin Yilmaz (dec.).

In the text, I have occasionally used the word "we" where "I" would normally suffice. You may think of this usage as a kind of "royal we," but it often has a narrower sense of referring to several of my most active collaborators in recent years, especially Reiner Kümmel, but also Jeroen van den Bergh, Marina Fischer-Kowalski, Paul Horne, Astrid Kander, Steve Keen, Michael Kumhof, Dietmar Lindenberger, Kati Martinàs, Uwe Schulte, Andre Serrenho, Gerry Silverberg, Udo Simonis, Vlasios Voudouris, and Benjamin Warr.

Contents

List of Figures

List of Tables

Chapter 1
Introduction

What, then, is the connection between energy, complexity and wealth? That will take some explaining, because the idea of complexity may be simply confusing (too complex?) while energy is neither money nor wealth, at least, not in any simple sense. Here I think it is appropriate to refer to a book I have found interesting, though I disagree profoundly with its key message. That book is *More Heat Than Light* by Philip Mirowski (Mirowski 1989). What I disagree is with his interpretation of the history of science. In his own words on p. 99:

> The discipline of physics owes its coherence and unity to the rise of the energy concept in the middle of the nineteenth century. However as soon as the discipline was consolidated, further elaboration and scrutiny of the energy concept began to undermine its original content and intent ...

He goes on for another 20 lines of print to explain that energy does not exist. In the book itself, he mentions the vexing problem of "renormalization" (adding and subtracting infinities) in quantum field theory, the lack of energy conservation in Einstein's general theory of relativity, and the fundamental question of whether or not it makes sense to imagine that the universe was created out of nothing.

I agree that those questions, including the last one, are still vexing because we do not know what actually happened and what "causation" can possibly mean before time began. But some modern versions of quantum field theory have been formulated in a way that avoids renormalization while Einstein's theory of gravity has been challenged and (I think) superseded by another theory that does satisfy the energy conservation condition. (See Chap. 4). Frankly, I am sure, as Eddington was, that the laws of thermodynamics are fundamental laws of nature, and that any theory implying the contrary is wrong. There are "free lunches" in economics, but they do not contradict the first or second laws of thermodynamics.

Wealth, unlike energy, is a human concept. It did not exist until humans appeared on this planet. Wealth is a word that captures the notion of material possessions with value to other humans. Material possessions imply ownership, and ownership implies rights of use and rights to allow, or prohibit, rights of use by

© Springer International Publishing Switzerland 2016
R. Ayres, *Energy, Complexity and Wealth Maximization*, The Frontiers Collection,
DOI 10.1007/978-3-319-30545-5_1

others. Owners may exchange these rights for money by selling the possession for
money. But money is only valuable if there is a choice of goods or services
available to purchase.

All wealth until very recently had a material base. A few thousand years ago
cattle or slaves were wealth. Hundreds of years ago jewels, or bars of gold, or gold
coins were wealth. Now a number on an account, or on a printed paper with a
picture of a president, or a title to land or a house or a mining claim or a share of
stock, or a financial derivative, or a bitcoin, can be wealth—again, assuming the
availability of other goods or services to buy. The monetary value of a "good", such
as a chicken or a ship, is usually determined in a marketplace. Or, it may be
exchanged for another material object, such as a sack of grain, a tank full of oil,
or a diamond ring.

Energy is the essence of every substance. Everything—including mass and every
material thing—is a form of energy. The "theory of everything"—the *ne plus ultra*
of physics—is therefore a theory of energy. Not only that, but energy flux is the
driver of change. Nothing happens without a flow of energy. Not in the natural
world and not in the human world. Thus, it is perfectly true that energy—not
money—makes the world go round.

I'll try to keep it as simple as possible, but I have to begin at the beginning. The
thesis of the book is summarized in Chap. 2 which follows. Part I of the main text
begins with Chap. 3 about the history of human thought about energy and thermodynamics. Its purpose is to explain that the concepts of "energy" and its cousin
"entropy", have changed meaning greatly over time. Even now they are understood
differently (or even completely misunderstood) by most people. Chapter 4 focuses
on the history of the cosmos, the sun, the origins of the elements and the "terra-
forming" of Earth. Chapter 5 deals with the origins of organic chemicals, the origin
of life, DNA, the oxygen catastrophe, the Cambrian explosion, the carboniferous
era, and evolution since the asteroid that killed off the dinosaurs. Chapter 6 is all
about long-term chemical changes in the Earth's atmosphere, hydrosphere and
biosphere. Climate change is discussed there.

In Part II, Chap. 7 is about how our species *H. sapiens* came to dominate the
Earth during the last several glacial episodes, starting half a million years ago with
the taming of fire, the domestication of animals, language, social organization,
slavery, money, and pre-industrial technology. Chapter 8 is about technology as it
evolved after 1500 CE, from printing to coking, iron smelting, steam power, and the
industrial revolution. Chapter 9 then focuses on new materials and new forms of
energy, electric power, petroleum, the internal combustion engine, our current
dependence on fossil fuels, and the demographic transition. Chapter 10 is about
the coming shift from fossil fuels laid down during the Carboniferous Era to "Peak
oil", renewables and energy efficiency technologies for the future.

Part III, starting with Chap. 11 discusses core ideas of economics. It explains
why economic growth is an aspect of Darwinian evolution, by exploiting natural
resource discoveries and innovations that made the resources useful and created
new products and markets. Yet, mainstream economic theory today still neglects
energy and complexity as the primary sources of economic surplus, and thus the

drivers of growth. Current economic theory is inconsistent with the laws of thermodynamics. Chapter 12 discusses what economics has to say (and needs to say) about a world in which material resources are no longer unlimited, where the "cowboy economy" is in transition to the circular "spaceship economy", and where knowledge is the only new resource. Yet the "circular economy" remains a figure of speech. Growth cannot continue indefinitely. Perpetual motion and perfect recycling are not possible in this universe. Appendix A provides the details of a theory of energy as a driver of economic growth.

Chapter 2
Thesis

In the very beginning, there was only pure energy—neither particles nor photons—and the laws of physics. As the energy has cooled and dissipated, an immense diversity of particles, elements, chemicals, organisms and structures, has been created (and also destroyed) by the blind functioning of those laws. Wealth in nature consists of complex structures of condensed ("frozen") energy, as long-lived mass. Wealth in human society is the result of conscious and deliberate reformulation and dissipation of energy and materials, consisting of frozen energy, for human purposes. This book is about both natural and human wealth creation, preservation and maximization. Knowledge is a new sort of immaterial wealth that enables us to dissipate—and utilize—that natural wealth more and more effectively for human purposes. Can the new immaterial wealth of ideas and knowledge ultimately compensate for the dissipation of natural wealth? This is the question.

During the first expansion (and cooling) of the universe, mass was distinguished from radiation by an interaction not yet well understood, but thought to be driven by the so-called "Higgs field" which (supposedly) permeates everything. All of the (several dozen) "known" elementary particles were created by what physicists call "symmetry breaking", which cannot be explained in a paragraph or even a whole chapter. (But if you are interested, look at the "Afterword" of Steven Weinberg's marvelous book, especially pp. 158–160 (Weinberg 1977). However it is clear that most particles were annihilated by anti-particles as quickly as they emerged from the "vacuum" (physics-speak for "nothingness"). So the analog of Darwinian "fitness" for elementary particles was stability and long lifetime. But, for a very, very short time (called "inflation") the baby universe expanded so fast—much faster than the speed of light—that causal linkages between particle-antiparticle pairs were broken. A few elementary particles– the electrons and protons (and the neutrons were unbound) constituting ordinary matter as we know it—survived. They are the building blocks of everything.

When the universe was about 700,000 years old it consisted of a hot, homogeneous "plasma" (~3000 K) consisting of photons, electrons, protons, neutrons and neutrinos (Weinberg 1977). That plasma was the origin of the microwave

© Springer International Publishing Switzerland 2016
R. Ayres, *Energy, Complexity and Wealth Maximization*, The Frontiers Collection,
DOI 10.1007/978-3-319-30545-5_2

"background" radiation, discovered in 1965, that provided the first real evidence of the Big Bang (BB). Quantum fluctuations appeared as infinitesimal density and temperature variations in the plasma. In fact, that plasma was very smooth and uniform, homogeneous and isotropic. But those quantum fluctuations grew over time.

The next phase (as expansion continued) started with the synthesis of hydrogen and helium atoms in the hot plasma. Hydrogen atoms consist of one proton and one electron, dancing together (but at some distance) and held by electromagnetic forces. As temperatures cooled, more and more of the free electrons and free protons decided to "get hitched", as it were. But some of the free protons met and were attracted by neutrons, resulting in deuterons. The deuterons also grabbed electrons, becoming deuterium ("heavy hydrogen"). Some deuterons decided to merge with hydrogen atoms, becoming helium 3. And some of the helium 3 grabbed another neutron while other pairs of deuterons also got married, as it were, creating helium 4. *Each of these mergers took place in order to increase their "binding energy"—a kind of measure of love among the elementary particles.*

As the expansion continued and the temperature continued to drop, the free electrons and protons were "used up" and the substance of the universe became a cloud of atomic hydrogen and atomic helium. Tiny density and temperature fluctuations in the cloud, were gravitationally unstable. The symmetry of homogeneity was broken as the denser regions attracted each other gravitationally. The dense regions got still denser (as the rich nowadays get richer). This "cloud condensation" process resulted in the creation of the stars and galaxies.

As the densest central cores of the proto-stars heated up under extreme pressure, they became nuclear fusion machines. The same process of binding particles into nuclei (by irreversibly converting mass into energy) that made hydrogen atoms and atoms, carried on to form helium, boron, neon, carbon, nitrogen, chlorine, oxygen, silicon and other elements—up to iron, with atomic weight 56. All this took place in young stars. That "nucleo-synthesis" process produced light and made the stars "shine". *But it also resulted in increasing the complexity and diversity of matter. Complexity is a form of natural wealth.*

As the stars used up their fuel (i.e. hydrogen) they also got cooler and denser. The delicate balance between gravity and radiation pressure broke, and the smaller stars became very dense "white dwarves". The bigger stars, especially the ones several times as massive as our sun, collapsed into neutron stars or (maybe) "black holes". The most violent collapses of the biggest stars resulted in "supernovae", which emitted huge amounts of energy almost all at once, by galactic standards. These explosions created heavier-than iron elements by endothermic fusion, and scattered their mass all over their galaxies. Our galaxy (the "Milky Way"), which incorporates about a 100 billion stars, experiences a supernova once or twice every 100 years. When small stars die, they become white, brown and finally black dwarfs Nothing happens to them after that. *It was the catastrophic collapse—call it creative destruction—of big stars, resulting in supernovae, that made life itself and everything we care about possible.*

Some of that scattered mass of nuclei and atoms (together with fresh hydrogen) formed into new "second generation" stars. Our sun is one of those. The inner planets, including Earth, have iron cores and large stocks of all the elements, including carbon, oxygen, nitrogen and others that are essential to life. That endowment was left over from a supernova explosion that occurred about 5 billion years ago. The entire evolutionary process from the Big Bang to the creation of our sun was driven by *exergy destruction* (i.e. the second law of thermodynamics). Yet the cooling also resulted in the creation of physical structures with compartments separated by boundaries and gradients. *The process can be characterized roughly as the "condensation" of useful energy (exergy) into useful mass and massive planets, such as Earth.*

In the Earth the elements that were created in stars and supernovae have undergone gravitational condensation and separation by weight, which is why iron and nickel are predominant in the core while aluminum, silicon are concentrated in the mantle while and calcium, carbon, oxygen and other light elements are more concentrated in the crust. There are also endothermal chemical reactions powered by heat from radioactive decay of the heaviest elements. Those processes have created a variety of mineral concentrations ("ores") that can be differentiated by composition (Ringwood 1969). Our planet has also acquired a lot of water since the beginning, mostly from space, via comets and meteorites (Frank 1990). (The original endowment would have mostly boiled off, when the Earth was hot, as it did from Mars and our Moon.)

Water has an extra-ordinarily useful property: unlike most other substances, the solid phase (ice) is less dense than the liquid phase. So ice floats on top. If it sank to the bottom, the pressure would keep it solid forever. Hence, our blue planet remains liquid on the surface. water is also the "universal solvent". Not quite universal, of course, but it has been essential for the creation and spread of life.

Chemical "monomers" (small molecules), were formed as temperatures dropped, starting with molecular hydrogen, H_2. This was followed by combinations of atomic H atoms with other light elements such as C, O, N. These small molecules were synthesized, both in space (e.g. dust clouds or comets) and on Earth. It was the same process—maximization of stability as binding energy (love among the elementary particles) that accounted for the nucleosynthesis of heavier elements in the stars. As time went on, some of the light elements combined to make small molecules like CO, CO_2, HCN, H_2O, NH_3, CH_2—and so on. And these molecules began to combine with each other. This took place, mostly on the surfaces of silica dust particles in space, but partly in the early oceans.

The combinations of small molecules (monomers) was assisted by catalytic properties of the dust or rock surfaces. At a later stage, more complex molecules were able to assist in the formation of others like themselves (this is called autocatalysis). Finally polymers, consisting of long chains of simple monomers appeared. Thanks to the propensity of carbon atoms to attach to hydrogen atoms, and to each other, the most common polymers were formed from hydrocarbons and carbohydrates. Among the most important early polymers were "lipids" (fats and fatty acids). Some of these attached to phosphate (PO_4) monomers. (There was more reactive phosphorus in the oceans at the time than there is now.) Stable phospho-lipid polymers—formed by as yet unknown micro-processes—enabled the creation of protective cells with "skin". Life was on the verge.

Protected by cell walls, more complex chemical species, such as amino acids and nucleotide bases were able to survive. One of those autocatalytic chains was ribonucleic acid (RNA). Auto-catalysis made chemical replication possible. The first living (metabolizing) cells appeared on Earth around 3.5 billion years ago. At first they were energized by chemical gradients, especially involving the oxidation of iron and sulfur. There wasn't much "free" (atomic or molecular) oxygen around in the early oceans, but there must have been a little, perhaps due to decomposition of some oxides by ultraviolet radiation. Darwinian evolution of species began. It seems to have been promoted near environmental gradients, such as undersea volcanic vents.

The next stage was the "invention"—excuse the word—of *oxygen photosynthesis*, and the protein molecule (chlorophyll) that does the job. This was almost miraculous, since it involved combining two distinct metabolic process into a single one (Lenton and Watson 2011). The result was the emergence of "blue-green algae". Those algae used energetic photons from sunlight to convert carbon dioxide and water molecules into glucose (a simple sugar) plus oxygen. Oxygen was the waste product of photosynthesis. It is also very reactive, chemically. Consequently oxygen is toxic to anaerobic organisms.

The blue-green algae soon took over the oceans, spewing oxygen as they spread. As oxidizable compounds were used up, toxic oxygen accumulated in the atmosphere. That buildup was a time-bomb. Had it continued, life on Earth would have died out due to poisoning by its own waste products.

A lucky mutation (*respiration*) saved the day. It did two things. First, it provided metabolic energy no less than 18 times more efficiently than the prior fermentation process (still used by yeasts). And secondly, it used up some of the oxygen. To us, oxygen is necessary. In fact, it is probably a precondition for the development of intelligent life (Lenton and Watson 2011, pp. 296 et seq). But to the first living cells it was poison.[1] Evidently maximization of life on earth is inconsistent with maximization of chlorophyll. A certain amount of that green stuff needs to be metabolized by oxygen breathers, to keep the plants from dying of their own waste product. And, conversely, the oxygen breathers can't survive without the glucose produced by the plants. That balance is critical.

Thanks to the more efficient energy metabolism enabled by respiration, living organisms became mobile. Some single-celled organisms attached themselves symbiotically to others. Eventually some of those symbiotes "merged" with each

---

[1] People who worry about cancer will have heard of "free radicals" and the dietary components of some fruits and vegetables called "anti-oxidants" that combine with and neutralize those free radicals. Free radicals are chemical compounds that have an attached oxygen atom with a spare "hook" (excuse the metaphor) eager to combine with something—almost anything—else. When that happens, functions essential to life are degraded. Aging is thought to be largely attributable to free radicals.

other, as a cell might become the nucleus of another cell. DNA arose. Multicellularity appeared. Species competed for resources (light, space, nutrients). Some of them began to attack others for food. "Fight or flight" appeared. New territories were invaded (especially land). New niches were invaded, and new species appeared. Complexity and diversity increased.

While the cooling of the universe is irreversible, it led to the formation of the different elements and later, of chemicals. After life appeared, the process of differentiation and complexification accelerated. Darwinian evolution is also irreversible in the sense of continuously increasing complexity and "fitness". Because all organisms depend on "food" in the sense of exergy flux, Darwinian evolution can be characterized as increasing exergetic efficiency at the individual and species levels.

Biomass and complexity increased, as did the exergy "consumption" of the survivors. Living organisms became more and more diverse. There were several "great extinctions" followed by recoveries that enabled new "winners" to dominate—for a while. Atmospheric oxygen increased as plant life flourished in the "Carboniferous" age. Bones appeared, followed by spines and skulls. Teeth appeared. Fish appeared in the oceans. Biomass on land was buried and converted into coal and oil. Dinosaurs came and went. Homeostasis (warm blood) provided an advantage. Birds appeared. Sensory organs evolved. Central nervous systems and brains evolved. Mammals with four legs and tails occupied many "niches" (along with six-legged insects and eight-legged arthropods).

Some mammals climbed trees and developed hands with thumbs for gripping. They came down from the trees and became bipedal. They communicated with each other and used tools, and weapons. Their brains got bigger, allowing greater intelligence. They tamed fire and animals. They organized. They learned to transmit knowledge. They took over the world. They may destroy it.

The "wealth of the world" 10,000 years ago—at the end of the last ice age—consisted of natural resources. There were rich soils, great forests and grasslands, streams and springs delivering clean water, useful and tame-able animals (such as horses, oxen, sheep, cattle and dogs) and visible and extractable concentrations of metals, clay and stone. The environment had other beneficial characteristics that were not noticed or needed 10,000 years ago, notably a benign climate and the innate ability to absorb and recycle wastes.

Humans started multiplying and using up those stored natural resources—both organic and mineral—partly for useful material properties and partly for their stored exergy. Humans are spending our inheritance, like children with no idea of saving or investment. This happened slowly at first, but faster and faster. As natural resources have been used, and abused, much of that original endowment have been used up or damaged. Resource exhaustion in human civilization bears a certain resemblance to the process that led to supernovae explosions. The explosion creates a bright but brief light, and what follows is devastation.

Or, is there another way to go? Luckily another "resource" has emerged. It is knowledge, embodied in brains and books and (more importantly) in organization and societal institutions. Those "new" resources have enabled humans to greatly extend the life of our original material resource inheritance. We can find and mine metal ores far less concentrated than those our ancestors exploited. We can see better and dig much deeper wells and mines. We can plow, plant and fertilize soils that our ancestors could not. We can multiply the muscle power of animals—and ourselves—by enormous factors. We can fly. We have, thanks to fossil fuels and flowing water—in the words of Reiner Kümmel—"energy slaves" of great flexibility and power (Kuemmel 2011). Not only that, our energy slaves are "smarter" and less material-intensive as our knowledge base grows.

The evolution of human civilization has been Darwinian, in the sense that it was based on competition for resources (including mates). It has also been irreversible in the sense that the "winners" in every niche are able to capture and exploit more exergy and use it more efficiently than the "losers". The winners survive, the losers disappear. (The vast majority of species that once thrived on Earth are long gone. Even if they could be re-created, dinosaurs could not survive in the wild today. Their eggs would never even hatch.) The rule applies to societies: there is no way a society of primitive hunters or self-sufficient peasant farmers can compete with an industrial society.

This irreversible "progress" depends essentially on selection within increasing diversity and complexity. It may, or may not, be a direct consequence of the second law of thermodynamics (the "entropy law"). After all, one's intuitive understanding of the entropy law is that things degrade, wear out, fall apart, and gradients disappear. But there seems to be a general "law of irreversibility" affecting all dynamical interactions between entities, from atoms to molecules, to living organisms. Each successful evolutionary innovation increases complexity and organization, increases information content of structures, increases both exergy consumption for maintenance and also increases exergy efficiency. It applies to nuclear reactions and chemical reactions. It applies to biological interactions. It also applies to competition between human individuals (as they grow and learn), and to societies, corporations and nations. The attractive "binding energy" applies to humans, not just in marriage or families, but in tribes, enterprises, and nations.

Evolutionary innovations often occur as a result of "creative destruction" (Schumpeter's phrase). Examples range from the collapse of a star that has burned all its fuel, to a planetary collision (such as the one that gave us our Moon), the "snowball Earth" episodes, and the asteroid strike that killed off the dinosaurs. Those were episodes of creative destruction. Glaciation also qualifies. Others were "Noah's flood" (probably due to the post-glacial rise in sea level that re-connected the Mediterranean Sea with the Black Sea), the volcanic explosion of Santoro that ended the Cretan dominance of the Aegean Sea, another (unnamed) volcanic outburst that occurred in Indonesia in 550 AD. That eruption may have shifted a balance of power in the steppes of Asia and indirectly kicked off a series of westward migrations, from Attila the Hun to Genghis Khan (Keys 1999). A more recent example was the spread of the "black death" in the fourteenth century. It caused a labor scarcity that shifted the power balance between towns and castles.

More pertinent to the problems of today was the more gradual, but equally important consequence of deforestation of England in the sixteenth and seventeenth centuries. That deforestation caused the price of charcoal to rise dramatically and accelerated the use of coal. But, of course, the great human innovation was the use of coking as a way to utilize coal (instead of charcoal) for iron-smelting. That discovery-innovation, in the early eighteenth century, arguably maximized wealth (by making coal into a substitute for charcoal, thus increasing available exergy reserves) and kicked off the industrial revolution. The need to dig deeper coal mines, in turn, led to mine flooding. The creative response to mine floods was the development of the steam engine, by Newcomen and Watt, followed by steam-powered railways and much else. During that industrial development, which fed on coal, exergy consumption, *per capita*, rose enormously.

Another resource-related problem was the near extinction of sperm whales (whose spermaceti was the source of whale-oil for lamps) in the nineteenth century. That scarcity—signaled by rising prices—triggered the search for "rock oil" in Pennsylvania and its active exploitation where it was already well-known, in Azerbaijan. The petroleum industry and its "children"—automobiles, aircraft and plastics—followed quickly. Moreover, the profits (derived from economic surplus) of that resource discovery financed a great deal of the industrialization of Europe and the USA. It also drove exergy consumption and resource destruction, *per capita*, still higher.

The acceleration of material resource consumption and wealth creation has been accompanied—and arguably caused, at least in part—by social changes. The formalization of coinage and the rise of markets and long-distance trading were important. The spread of literacy, numeracy, and education were important. The end of feudalism, slavery and the "divine rights" of hereditary monarchs, replaced by ideas of free association, freedom of religion, free speech and other freedoms, were critical. The Protestant Reformation and the rise of capitalism and fractional reserve banking were crucial. All of this history is recounted, very briefly, in Part II of this book.

The unprecedented natural resource destruction, in the form of deforestation, environmental pollution, fossil fuel combustion and atmospheric buildup of green-house gases (GHGs), now in progress, may also be the stimulus for a new round of technological and social innovation. We must hope that it will be so.

Today an increasing fraction (albeit still only a fraction) of all competitive human interactions occur non-violently, whether in families or in markets. The days of competition for land by fighting or military conquest are largely (if not quite entirely) past. Capitalism and the production and exchange of goods and services in markets have proven to be more efficient ways of acquiring—and creating—wealth. That is the good news.

There is also bad news. Part of it is the fact that our economy is now "addicted" to economic growth, whereas the natural resources that enabled that growth since the eighteenth century are becoming harder to find and utilize. Moreover, economic growth in recent years is much less beneficial to society as a whole than it was two centuries ago. It is increasingly a sequence of "bubbles" that leave devastation in their wake. Another part of the bad news is the growing inequality between those with access to natural resources, or the capital created by earlier access, and those without it. A conflict is already brewing between inconsistent "rights": notably the property rights of those few who own the material and financial assets of our planet versus the supposed right to equal opportunity for the rest of the population.

In short, *wealth maximization, as a strategy, is not yet what human governments do, or know how do*. That needs to change. Economics is the science that attempts to explain these dynamical interactions and help to manipulate their outcomes. The creation and preservation of wealth—meaning both productive and consumptive assets—is an explicit subject within the domain of economics. But so far, it is being applied only in the realm of finance and technology.

Physical capital (machines, houses, infrastructure) are useless—unproductive—unless they are activated by exergy flows. Motors need electricity. Engines need fuel. Horses must be fed. The same is true of human workers. Exergy is what "makes things go" but knowledge is needed to optimize and control exergy flows and material transformations, as well as social activities and institutions. Knowledge enables us to do more with less. Knowledge in a book is not wealth—or is it?—but applied knowledge certainly creates wealth. Economic growth, per capita, is not driven by capital accumulation *per se*, but by exergy availability and knowledge. That is what Part III of this book is about.

In summary, the history of the universe until humans appeared on Earth was the history of material differentiation and increasing diversity and complexity. The history of Man until now, has been the history of converting materials into "things". The history of the future may be a history of wealth creation by knowledge accumulation, de-materialization and institutional innovation. That history will be subject, of course, to the laws of physics—especially the laws of thermodynamics.

Part I

Chapter 3
A Brief History of Ideas: Energy, Entropy and Evolution

3.1 Aristotle, Descartes, Newton and Leibnitz

The modern usage of the term *energy* in physics and thermodynamics is the end result of a very long evolution of ideas that began with the Greeks and began to accelerate in the sixteenth century. The starting point is sometimes attributed to **Aristotle**, who used the term "energeia" (in his *Ethics*) to mean power and strength to be mentally and physically active, and in his *Rhetoric* to mean a vigorous style. Elizabethan usage of the word was largely based on the rhetorical sense, as vigor of utterances, force of expression and even the quality of personal presence. As recently as 1842, the *Encyclopedia Britannica* defined *energy* as "the power, virtue or efficacy of a thing, emphasis of speech". Clearly none of these definitions is what we mean today.

A more plausible starting point for the modern conception would be *Le Monde (The World)* by the French philosopher **René Descartes** (Descartes 1633 [2000]). This book was withdrawn—because of its defense of forbidden Copernican ideas—after the Holy Office (better known as "the inquisition') formally prosecuted **Galileo Galilei** for having dangerous thoughts, in 1636. Unlike **Giordano Bruno** in 1600, Gallileo escaped with his life by recanting. Luckily the Inquisition could not suppress his discoveries. (I have been told that Galileo was protected from the Inquisition by the Jesuits. He also had allies in the Vatican, including Ascanio Piccolomini, Archbishop of Siena.)

However Descartes's energy-related arguments were included in his later book, *Principles of Philosophy* (Descartes 1644). Descartes viewed the universe as a giant machine. He sought mechanical explanations for all observable phenomena. He accounted for the effect of gravitation by postulating a vortex of something invisible and undetectable, called "aether" (a notion that originated with Aristotle) moving at high speed around the Earth and pushing slower particles or "corpuscles" of mass toward the center. In 1638 he also tried to explain light as a pressure wave through the aether, and he postulated properties of this mysterious fluid to explain

© Springer International Publishing Switzerland 2016

R. Ayres, *Energy, Complexity and Wealth Maximization*, The Frontiers Collection,
DOI 10.1007/978-3-319-30545-5_3

refraction and reflection of light, as well as magnetic fields. The aether theory became popular, possibly because there was no other theory at the time and because it was difficult to prove, or disprove.

Descartes asserted that all bodies with mass possessed a "force of motion" the product of mass times velocity, which we now call momentum (**mv**). (Modern critics have denied this on the grounds that Descartes did not have a proper definition of mass.) He thought that this quantity (momentum) should be *conserved* in all collisions and other reactions. Descartes favored contact forces and rejected the idea of "action at a distance" which some philosophers, since Plato have postulated to explain certain phenomena. Action-at-a-distance as conceived in that era is not equivalent to the action of force fields, such as electric fields or magnetic fields, first postulated by Faraday and Maxwell. (The fact that his hypothetical "aether" acted very much like a force field did not trouble Descartes.) He formulated three "laws of nature", which anticipated **Isaac Newton's** three laws of motion, but energy in the modern sense was not included.

One might think that something like the modern conception of energy must have been in the mind of Isaac Newton, when he formulated his theory of gravity and planetary motion. That is not the case. Newton followed Descartes in this choice of the product of mass and velocity (**mv**) as the appropriate measure of force. (To be sure, Newton's force law—that force is the time rate of change of momentum—is one of the most fundamental laws of physics.) Surprisingly, Cartesian "aether" hung around in post-Newtonian physics, exemplifying the power of belief in established authority, until it was finally banished by the Michaelson-Morley experiments that were essential to **Einstein's** special theory of relativity published in 1905.

Descartes' and Newton's mistake was engendered by the fact that they concerned themselves with planetary motion, where collisions do not (or rarely) occur. However, physicists like **Christiaan Huygens**, inventor of the pendulum clock, as well as the wave theory of light, (which was rejected by his contemporaries but rediscovered later), couldn't help but notice that when inelastic collisions do occur on the micro-scale, momenta appear to cancel out, due to the force of friction. Friction is a real phenomenon, even at the planetary scale, whence, sooner or later, everything in the universe would cease to move. Newton actually believed that the motive force to keep things (planets and stars) going was some sort of divine intervention. It was his view (according to Huygens student, **Gottfried Leibnitz**) that:

> God Almighty wants to wind up his watch from time to time. Otherwise it would cease to move. He had not, it seems, sufficient foresight to make it a perpetual motion.[1]

Leibnitz disagreed with Newton and Descartes on the definition of force as well as much else. In fact he developed his own philosophical system, invented the

---

[1] The possibility of perpetual motion was taken seriously at the time. It was officially rejected by the Paris Academy in 1775, paving the way for the Second Law of thermodynamics, although the second law was not formulated (by Clausius) until nearly a century later.

differential calculus to describe it, and initiated the fundamental idea of *conservation laws* ("economy of nature"). Based on some earlier work by Christian Huygens, Leibnitz proposed (1695) that force should be dual in character, consisting of *vis mortua* (dead force) roughly corresponding to potential energy, and *vis viva* or "living force" corresponding roughly to *kinetic energy*. He thought that *vis viva* should be proportional to the mass times the *square* of the velocity ($\mathbf{mv^2}$). This was still wrong, but only by a factor of 1/2.

Leibnitz also believed that this *vis viva* must be a conserved quantity in the universe. (Like Newton, he was misled by his focus on planets and other heavenly bodies.) In his "Essay on Dynamics" Leibnitz firmly rejected the idea of *"perpetual motion"*. He evidently understood the basic notion of *entropy*, long before thermodynamics was formulated in terms of state variables (temperature, pressure, density, etc.)

As it happened, both Newton and Leibnitz shared the same French translator and commentator, the remarkable Marquise **Emilie du Chatelet**. She was one of the few mathematically educated women in the world at the time, and one of the few persons of either sex who could understand the works of either Newton or Leibnitz. Her lover, **François-Marie Arouet** (who published under the name "Voltaire"), thought that Leibnitz' idea of conservation of *vis viva* implied changelessness "of this best of all possible worlds".

Voltaire supported Newton's position in the dispute. He wrote the novel, *Candide,* to satirize Leibnitz' views, which he had misunderstood. However, du Chatelet decided in favor of Leibnitz and looked for experimental evidence. She found it in the work of a Dutchman, **Willem Jacob's Gravesande**, who was dropping brass balls at different velocities into soft clay and measuring the depth of the resulting craters. These experiments proved the square law for energy, postulated by Leibnitz, except for that pesky factor of 1/2.

Leibnitz maintained his view that $\mathbf{mv^2}$ should be a conserved quantity against fierce opposition from both Newtonians and Cartesians, as well as against apparently contradictory evidence of inelastic collisions resulting in friction. To explain (apparent) inelasticity, he relied on the idea that the *vis viva* must be broken up into many fragments (now called atoms) whose individual motions would be significant but almost undetectable in the aggregate. In the eighteenth century there was extensive debate about the lost *vis viva*. The suspicion grew that it was being converted into heat, which was true. Leibnitz was very close to understanding the mechanical equivalent of heat, as well as entropy. However, conventional wisdom was in his way.

The foregoing section and significant parts of the rest of this chapter are based, in large part, on Mirowski's scholarly book *More Heat than Light* (Mirowski 1989). His history of ideas related to energy is much more thorough than the foregoing few paragraphs, and anyone interested in digging deeper should find it and read it. Lest there be any misunderstanding, I must say that, of the two, Newton was probably the deeper thinker and more influential scientist. But Leibnitz was not far behind.

3.2   Heat: Caloric vs Phlogiston

The prevailing "phlogiston" theory of combustion was introduced in 1667 by **Johan Joachim Becher**, an alchemist and physician. His concept spread, so to speak, like wildfire. The idea was that combustible substances contain a massless, colorless, odorless, and tasteless substance (phlogiston) which was released by combustion. (This invisible, undetectable substance had characteristics in common with Descartes "aether".) However experimental science was in it's infancy at the time. Nearly a century later, **Joseph Black**, a professor of chemistry in Edinburgh, discovered carbon dioxide. More important, Black and his students began to perform quantitative experiments on the relationship between heat flow, latent heat and temperature (calorimetry). His research led to the notion of *specific heat* and the formula for *heat capacity*, though he did not use the term. Black and his student **Daniel Rutherford**, discovered nitrogen in 1772. They interpreted the mixture of nitrogen and carbon dioxide, following combustion, as "phlogisticated air".

Finally, in the 1770s the chemist **Antoine Lavoisier** (who was also a tax collector, which was why he lost his head during the French Revolution) discovered oxygen and the role of oxygen in combustion. He interpreted oxygen, at first, as "dephlogisticated air". But later he realized that his results were inconsistent with the phlogiston theory. In 1783 Lavoisier, together with **Pierre-Simon Laplace**, wrote a review of the competing energy theories, *vis viva* vs. caloric. Lavoisier published a book "*Réflexions sur la phlogistique*" which put forward an alternative, which he called the "caloric theory" (Lavoisier 1783). It postulated that heat was a kind of fluid, which always flowed from warmer to cooler bodies. This observation was rather obvious, but its formal statement was of great importance for the later development of thermodynamics. In fact, simplified discussions of the Second Law often explain it using this example of one-way flows. The caloric theory implied that this pseudo-fluid (caloric) *must also be conserved*, which raised further questions. The caloric theory had a number of successes in terms of explaining observed behavior, not least an explanation of Joseph Black's results on heat capacity and Laplace's correction of Newton's "pulse" equation, which is now known as the adiabatic index of a gas. All of this took place in the eighteenth century.

In 1798 an English engineer **Benjamin Thompson, *Count Rumford*** was boring out cannons for the King of Bavaria, but also recording data and thinking about its implications. Rumford published his results as *An Experimental Enquiry Concerning the Source of the Heat which is Excited by Friction* (Thompson 1798). This work strongly supported the view that mechanical work could be converted into heat by friction alone. This result meant that *caloric was not conserved after all*. Yet the notion of mechanical work, as we understand it today, was not really formulated until three decades later. Most physicists at the beginning of the nineteenth century sought "caloric" explanations of Count Rumford's results, because of the many successful applications of that earlier theory. Consequently Rumford's crucial work on the mechanical equivalent of heat was not cited much until after 1870. By that time, physics had evolved significantly, both empirically and theoretically.

3.3   The Birth of Thermodynamics and Statistical Mechanics

Thomas Young, an Englishman, lectured on mechanics before the British Royal Society in 1807. He remarked that "the term energy may be applied with great propriety to the product of mass or weight of a body into the square of the number expressing its velocity." Of course he was only talking about what we now call "kinetic energy" and, as Leibnitz had done before him, he still omitted the factor 1/2. In short, Young borrowed Leibnitz' idea of *vis viva* without attribution, and claimed it for his own. For many years after, the idea of energy was associated with Young's name.

From Rumford's work on cannons and other engineer's work on boring out the cylinders for guns and steam engines, it was known that mechanical work could be converted into heat. Meanwhile **James Watt's** steam engine itself was a clear demonstration that heat could perform mechanical work. Engineers—the people who designed and built engines—speculated about the maximum amount of work that could be obtained from a heat engine. These speculations introduced the notion of *"thermal efficiency"*. As the mechanical equivalency of heat and work became established, at least qualitatively, after Rumford, it became clear that heat engines of the time were very inefficient.

Because of steam engines, the idea of *"mechanical work"* in the sense of motion to overcome inertia or against gravity (like lifting a weight) was understood and quite well established long before it was articulated in formal, mathematical terms (in 1829). The dangling unsolved efficiency question, for a cyclic heat engine, was solved by the French engineer and genius **Nicolas Sadi-Carnot**, in his *"Reflections on the motive power of heat"* (Carnot 1826 [1878]). Carnot's work—entirely theoretical—was based on Antoine Lavoisier's "caloric theory" and assumed that heat (caloric) is neither created nor destroyed (i.e. it is conserved), *hence cannot account for work done by an engine*. Carnot's conclusion, based on the further assumption that perpetual motion is impossible (Leibnitz), was that only temperature differences between heat source and heat sink could account for the quantity of work done. This became known as Carnot's theorem. Actually, Carnot was wrong because the caloric theory was wrong. The mechanical work done by a heat engine is, indeed, due to the difference between the quantity of heat input at a high temperature and the quantity of heat rejected at a lower temperature. But it was Carnot who, for the first time, clearly distinguished between *reversible* and *irreversible* processes in thermodynamics. Carnot's neglect of frictional losses does not detract from the importance of his insight.

The modern concept of *work*, as the product of force exerted over distance, was also introduced in France, also in 1829. The independent authors were **Gustave-Gaspard de Coriolis** in *Du Calcul de l'éffet des Machines* and **Jean-Victor Poncelet** in his book *Industrial Mechanics*. Poncelet had seen Coriolis' work earlier and acknowledged its primacy, but he was the more influential of the two in his lifetime. Coriolis also introduced the notion of *kinetic energy* for the first time, in his 1829 paper, with its modern meaning and definition, including the previously missing factor 1/2.

The German physician **Robert-Julius Mayer** is now credited with being the first to articulate the mechanical equivalent of heat (*Annalen der Chemie und Pharmacie 43*, 233; 1842) although James Joule, Hermann von Helmholtz and Ludwig Colding (an unknown Dane) came close, as did Count Rumford in his 1798 publication. Mayer's calculation, that the force ("Kraft")—meaning heat energy—required to lift a kilogram 365 m would also raise the temperature of a liter (kg) of water by 10 °C, was off by more than a factor of ten, quantitatively.[2] In 1845 Mayer privately published a pamphlet *"Organic motion in its connection with nutrition"* which stated the first law of thermodynamics as follows: "A force (Kraft), once in existence cannot be annihilated." Even so, he confused force (Kraft) with energy.[3] It is not surprising that he did not receive immediate recognition, although Rudolph Clausius did eventually acknowledge Mayer's claims of priority.

Justus von Liebig, known today as the father of German organic and agricultural chemistry, bravely attempted to explain the heat produced in animal bodies in terms of chemical reactions and muscular work. However Liebig took the experimental approach and could not prove his hypothesis. But starting in the mid-1840s a number of scientists explored the implications of the mechanical equivalence of heat. James Joule, working with **William Thomson** (later **Lord Kelvin**), carried out more accurate experiments and formulated the equivalence principle more coherently than Mayer had done. In 1847 Thomson, introduced the absolute temperature Kelvin scale, using Carnot's theorem, and further elaborated the Carnot theory in another paper in 1849. Everybody was getting into the act.

Another physician (as well as physicist) **Hermann von Helmholtz** undertook a more general and mathematical follow-up to Liebig's work on animal heat. He published an influential paper in 1847 *Die Erhaltung der Kraft* ("On the conservation of force") still using the word "Kraft" instead of "energy". This paper was really an argument against the so-called "vitalists" who still believed in a non-physical source of energy unique to living organisms. Helmholtz pointed out that if such a source existed it would imply the possibility of perpetual motion, which had already been rejected on philosophical grounds by Leibnitz in his "Essay on Dynamics". Helmholtz stated, in conclusion, that *the sum total of all the energies*

---

[2] The energy required to raise a kg by 365 m is 3580 J. The energy required to raise the temperature of a kg (liter) of water by 10 °C. would actually be 41,900 J. Mayer's point was that they were both forms of energy (Kraft).

[3] The word Kraft also means "strong" or "strength" as in Kraft paper.

of the universe, in whatever form, must remain a constant. This conservation principle is now known as the first law of thermodynamics.

Rudolf Clausius rediscovered the work of Carnot and took it further. In 1850 he noted that Carnot's theorem can be derived from the "first law" of thermodynamics (conservation of energy, rather than caloric) together with the independent principle that heat in a solid or fluid always flows from higher to lower temperatures, previously discovered by Lavoisier. This was the first, rather casual, (and slightly incorrect) statement of the famous second law of thermodynamics,[4] at least since Leibnitz rejected the possibility of perpetual motion. More precisely, Clausius' showed (correctly) that *it is impossible for a self-acting cyclic heat engine, unaided by any external agency, to convey heat from a body at one temperature to a body at a higher temperature.* This is another way of saying that heat engines convert heat to work irreversibly.

In a very important paper published in 1854 Clausius focused on the two inverse transformations that occur in a heat engine, namely *heat to mechanical work*, where the working fluid experiences a drop in temperature, and *work to heat*. Clausius noted (as Carnot had pointed out long before) that *not all the heat in the reservoir can be converted to work*. The unavailable difference he termed "sensible heat" (a term still used by engineers).

In that paper, he introduced the concept of *entropy* as a state function depending on temperature and pressure. He also made use of Carnot's important distinction between reversible and irreversible processes, and noted that *entropy never decreases but increases in all irreversible processes.* He stated this in mathematical form as an inequality: that the integral of $dS = -dQ/T$ over any irreversible cyclic process trajectory must always be less than zero, where dS is an increment of entropy, dQ refers to an increment of heat and T is the temperature. He chose the word *entropy* from the Greek word for "transformation". This law of nature, now known as the second law of thermodynamics, does not imply the uni-directionality of heat flow, as once thought, because radiation can reverse this directionality. However it does say that all heat-to-work processes involve losses due to irreversibility.

William Thomson (Lord Kelvin) formulated the two laws of thermodynamics very carefully between 1851 and 1853, attributing the first law to his collaborator, Joule (although Helmholtz had stated it earlier), and the second law to Carnot and Clausius:

> The whole theory of the motive power of heat is based on the following two propositions ... due respectively to James Joule and to Carnot and Clausius:

---

[4] Heat moves by conduction or convection from higher to lower temperatures, but this does not apply to radiation, which can carry heat from lower temperatures to higher temperatures, as for instance in the "greenhouse" effect in the Earth's atmosphere where cold layers of the atmosphere re-radiate thermal (infra-red) radiation back to the Earth. Some of the critics of climate change have misunderstood this point. See Sect. 4.1.

Prop I (Joule)—When equal quantities of mechanical effect are produced by any means whatever from purely thermal sources, or lost in purely thermal effects, equal quantities of heat are put out of existence, or are generated.

Prop II (Carnot and Clausius)—If an engine be such that, when it is worked backwards, the physical and mechanical agencies in every part of its motions are all reversed, it produces as much mechanical effect as can be produced by any thermodynamic engine, with the same temperatures of source and refrigerator, from a given quantity of heat

Lord Kelvin (Thomson) emphasized the irreversibility; he never used the term *entropy* (as introduced by Clausius) but he was certainly the first to introduce the idea of *available energy* (now called *exergy*). That was in 1855. The terminology was finally clarified by **William Rankine**, in papers presented at the Glasgow Philosophical Society during 1853–1855. (See Rankine 1881). He asserted in 1853 that "the term "energy" comprehends every state of a substance which constitutes a capacity for performing work. Quantities of energy are measured by the quantities of work which they constitute the means of performing."

Note that Rankine's definition of energy is essentially the modern definition of exergy, or available work. He also clarified the meanings of *force* and *state*, replacing the German word "Kraft" by "energy." Rankine defined energy as follows:

Energy, or the capacity to effect changes, is the common characteristic of the various states of matter to which the several branches of physics relate: If, then, there be general laws respecting energy, such laws must be applicable mutatis mutandis to every branch of physics, and must express a body of principles to physical phenomena in general.

Rankine also introduced the important distinction between *actual energy* and *potential energy*. He restated the law of the conservation of energy: "*the sum of actual and potential energies in the universe is unchangeable.*" For him *actual energy*

comprehends those kinds of capacity for performing work which consists of particular states of each part of a substance ... such as heat, light, electric current, *vis viva*.

Actual energy is essentially positive. In contrast, he said, *potential energy*

comprehends those kinds of capacity for performing work which consists in relations between substances, or parts of substances; that is in relative accidents. To constitute potential energy there must be a passive accident capable of variation, and an effort tending to produce such variation; the integral of this effort, with respect to the possible motion of the passive accident, is potential energy, which differs in work from this—that in work the change has been effected, which, in potential energy, is capable of being effected.

You may have to read that paragraph several times to make sense of it. He is saying that potential energy is that which *could* be realized in certain situations. The obvious example is that of a pendulum, where the kinetic energy of motion is converted to potential energy at each end of its swing, and conversely the potential energy is converted back to kinetic energy in between. Rankine was trying for a definition sufficiently general to take into account other kinds of potential energy, such as chemical or nuclear energy.

The point is that physicists in the middle of the nineteenth century were still trying to grapple with a very difficult question, namely how to fit the mechanical and thermal aspects of energy—heat, pressure, motion—into a simple comprehensible picture. Actually, they couldn't do it except in a relatively narrow range of processes near thermodynamic equilibrium, because those variables (pressure, temperature, density, entropy) are not defined except at or near equilibrium. Moreover, chemical reactions and electro-magnetism, not to mention nuclear reactions, were not yet well enough understood. But chemistry was rapidly developing in the second half of the nineteenth century, especially thanks to the exploding demand for synthetic dyes for the textile industry.

One other major step forward in the nineteenth century was the development of *statistical mechanics*, between 1865 and 1900, mainly by **Ludwig Boltzmann, James Clerk Maxwell** and **Josiah Willard Gibbs**. The stimulus for their work was a bitter controversy arising from Clausius' statement of the second law: namely that pesky *irreversibility*. The problem, for many physicists, was that no such implication follows from Newton's laws of motion, all of which are reversible. (In other words, they are unchanged if time **t** is replaced by −**t**). This is an example of *symmetry with respect to time*, and irreversibility is an example of *symmetry-breaking*, which will come up later.

Boltzmann's innovation was to combine Newtonian mechanics with statistics. His important innovation was to think of gases as clouds of identical point-particles moving independently of each other, such that the probabilistic distribution function describing all of the microstates of the system as a whole is the product of the probabilistic distribution functions of each particle independently of the others.

The notion of "order"—often used in discussions of thermodynamics—is best understood in terms of correlation over distance. A perfect solid crystal at absolute zero temperature has a high degree of order because knowledge of the positions of a few atoms (molecules) suffices to determine the parameters of the structure and consequently the locations of all the rest of the atoms. At higher temperatures the positional correlations are still relevant but over shorter distances. The phase-transition from a solid to a liquid is characterized by the disappearance of long-range correlation. But liquids still retain some short-range correlation, due to inter-molecular (*Van der Waals*) forces. The ideal gas imagined by Ludwig Boltzmann is the extreme case, of *zero correlation*, i.e. no forces between particles at a distance.

The particles in this "ideal gas" (at equilibrium) are distributed among energy levels—micro-states—according to a normal (Gaussian) distribution, known as the Maxwell-Boltzmann distribution. External forces or constraints can and do change the distribution. Changes can be characterized probabilistically. The entropy of a macro-state can then be defined in terms of the number of micro-states (energy levels) corresponding to a given macro-state. Boltzmann's celebrated formula (which is inscribed on his tomb) is as follows:

$$S = k_B \log W$$

where S is the entropy, W is the number of possible microstates and k_B is the Boltzmann constant, which is 1.38062×10^{-23} J/K.

Boltzmann used this statistical model to derive his famous "H-theorem", which says that entropy will increase, irreversibly, with extremely high probability. The theorem was challenged immediately, both by people who did not believe in atoms (e.g. followers of Ernst Mach), and by those who doubted the independence assumption. The main argument by opponents was that if, after a microsecond of random motion all trajectories were suddenly reversed ("by the Hand of God") then logic says that the original configuration of positions and motions would be regained because the Newtonian equations of motion are symmetric with respect to time. But the H-theorem says that entropy would increase in both micro-time-segments.

This appeared to be a logical contradiction. Boltzmann's answer was that the original configuration of microstates might conceivably happen, but only after many lifetimes of the universe. This didn't satisfy mathematicians like Poincaré and Zermelo who continued to insist on Newtonian reversibility. Boltzmann was very depressed by this opposition (which got personal at times). But his statistical view seems to have prevailed over time, possibly because real gases are not composed of point particles, and real collisions between atoms occasionally involve irreversible chemical reactions.

The probabilistic formula of entropy, from Gibbs, is the (negative) sum over all microstates i of the probabilities $p_i \, ln \, p_i$ of being in that micro-state (Gibbs 1948). This formula was later restated in terms of fractions f_i of all particles or molecules being in the ith microstate. I mention it only because the same formula has become very important in information theory.

The extension of statistical mechanics to deal with non-ideal gases interacting with real molecular forces has been based mostly on Gibbs' formulation. However—like classical thermodynamics—statistical mechanics had little to say about the most interesting cases, as when a gas or liquid approaches a phase transition (gas to liquid, liquid to solid). This is because phase transitions depend upon inter-molecular forces and do not occur in the ideal gas.

The second law of thermodynamics has actually been restated and reformulated several more times since the nineteenth century. A summary of the later revisions (and the reasons for them) is provided later in Sect. 3.9.

3.4   Chemistry: From Lavoisier to Gibbs

The whole point of the *phlogiston theory* (finally put to rest by Antoine Lavoisier) and its successor, the caloric theory, was to explain the heat energy of combustion, measured for the first time by Joseph Black and his students. The tendency of some non-combustion (*exothermic*) reactions to generate heat while other (*endothermic*) reactions absorb heat was, for a long time, undetectable. It was disguised by the fact that pure elements like sulfur and phosphorus (not to mention hydrogen and oxygen) and simple compounds like ethanol or saltpeter (potassium nitrate) were rarely found in nature, and had to be separated or manufactured by tedious processes requiring a lot of heat from fuel combustion.

Gradually, as more and more elements and compounds were purified, so that reactions between them could be studied, and inputs and outputs measured carefully, it became evident that some reactions (like combustion) tend to proceed spontaneously, while others (such as the reduction of metal ores) have to be driven by an external source of energy. The first type is therefore exothermic, or heat releasing, while the second type is therefore endothermic, or heat absorbing.

Endothermic reactions cannot take place spontaneously, because they require an extra input of energy from another source. Quite possibly that other source is an associated combustion reaction. The most familiar exothermic reaction in daily life, if you are not a smoker, is probably the gas burner on your stove or the gas heater in your house. The most familiar endothermic reaction is photosynthesis, the reaction in green leaves of plants, driven by sunlight, that eventually converts carbon dioxide from the atmosphere into carbohydrates like glucose and other natural sugars (lactose, fructose, etc.) Those sugars are the fuel source of every animal body.

A somewhat less familiar example would be the blast furnace that converts coke at high temperatures into carbon monoxide, which in turn reduces iron ore (ferric oxide) to produce pig iron and carbon dioxide as a waste. Another important endothermic reaction starting from natural gas "fixes" nitrogen from the air and produces ammonia, which is the source of all nitrate fertilizers (and most explosives). The most common electrolytic reactions occur in batteries, which convert chemical energy to electric energy. Rechargeable batteries also make use of the reverse reaction which converts electric energy back into chemical energy.

A crucial phenomenon in chemistry is catalysis, by means of which a substance (the catalyst) can enable the reaction of two other reactants yet remain unchanged. It was first observed and described (in the context of oxidation/reduction reactions in water) by Elizabeth Fulhame, who published a book about it in 1794. Humphrey Davy discovered the catalytic properties of platinum (1817) and **J. J. Berzelius** named the phenomenon. Inorganic catalysts are usually planar metal surfaces or crystal edges with imperfect metal valences (or combinations). **Wilhelm Ostwald** carried out extensive studies of the catalytic properties of acids and bases. Organic catalysts in biochemistry, which are essential for metabolism and catabolism, are mostly enzymes (which are proteins).

Probably the most important single contribution to chemistry in history was **Dmitri Mendeleev**'s formulation of the periodic table of elements in 1869 (see Fig. 4.13). This has no direct relevance to the central topic of this book—Energy—but it is very relevant to the second topic, i.e. complexity of matter (and material wealth). The periodicity in the sequence of atomic weights had been noticed previously, especially by Alexandre-Emile Beguyer de Chancourtois in France and John Newlands in Britain, but Mendeleev was the first to "complete" the system, including some gaps, and to correctly predict the weights and properties of the missing elements. Another chemist, Lothar Meyer, arrived at almost the same scheme at almost the same time, but with a few mis-classifications. Mendeleev's table was so good that he was also able to correct the atomic weights of several known elements from knowledge of their chemical properties. But, the Mendeleev system was not quite complete.

In 1894 William Ramsay and **John William Strutt** (Lord Rayleigh) discovered argon, and Ramsay went on to discover four other "noble" gases: helium, neon, krypton and xenon. (Radon, the last, was discovered by a German chemist, F. Dorn, in 1900.) These elements do not form compounds, so some chemists thought, at first, that they must be "outside" the periodic table. It took years to figure out that they constitute a column between the halogens (fluorine, chlorine, bromine, iodine . . .) and the alkali metals (Li, Na, K, etc.).

In 1904 **J.J. Thomson**—discoverer of the electron—postulated his "plum pudding model" in which electrons moved around in a positively charged "pudding" organized themselves into "rings" rotating around inside the atomic nucleus. **Niels Bohr** offered a better explanation of this "ring" structure, in terms of quantum mechanics, in 1913. Bohr envisioned the electrons as moving in orbits around the nucleus, roughly as we do today.

Yet, Bohr did not actually derive his key results on spectroscopic energy levels from quantum theory; he used chemistry. The final explanation of how many electrons can fit into each ring was provided by **Wolfgang Pauli** in 1924. Pauli's solution was a new and very fundamental law of nature, the "Pauli exclusion principle". This rule has become crucial for understanding modern high-energy physics, but an adequate explanation does not belong here.

Only by mid-nineteenth century were the measurement techniques (and instruments) accurate enough to compile data that allowed theorists to begin to explain chemical reactions, and reaction rates, in terms of something called "chemical potential" or "available energy", now called *exergy*. As mentioned already, the basic idea was introduced by William Thomson, Lord Kelvin, in 1855. The word "available" was shorthand for "available to do work". It was rediscovered independently by **Josiah Willard Gibbs** at Yale University. In two papers published in 1873 he developed the graphical methods used today by chemists and chemical engineers. Between 1876 and 1878 Gibbs published his most famous work, on the equilibria of heterogeneous substances, where chemical potential was first defined. It was Gibbs who really extended thermodynamics into chemistry.

On the basis of current understanding, we know that the atoms in a molecule are normally held together by an electromagnetic attraction between the positively charged atomic nuclei and the negatively charged electrons in complex orbits around several or all of the nuclei in the molecule. Of course, this "binding" force can be overcome by high-energy collisions (i.e. high temperature heat) or by external electric charges, or by a solvent (usually water, but sometimes other fluids) capable of breaking weak molecular bonds. One or the other of these mechanisms breaks the molecules apart creating ionized molecular fragments that can recombine in different combinations. If the new combinations result in tighter, more stable bonds than the old ones (the exothermic case), the recombination may go to completion with release of *free energy* (in the form of heat or pressure) that can do work on the surroundings. Free energy is therefore the negative of binding energy, which explains why the term "free" is used.

Exergy is defined as *the maximum work that can be done by a closed chemical system approaching equilibrium reversibly.* Actually there are three versions depending on the definition of equilibrium (or reference state) and on assumptions about the path toward equilibrium. If the equilibrium of a subsystem is defined as the state of an external reference system (such as the atmosphere or the ocean) the available energy to do work is (now) called exergy. If the equilibrium is defined as the final state of a chemical reaction, the maximum work output or chemical potential is called *free energy*. If the volume and pressure are held constant, the chemical potential is called *Gibbs free energy*, named after Josiah Willard Gibbs.[5] However, in the case where volume and temperature are held constant but pressure is not (as in explosions), it is called *Helmholtz free energy* after Hermann von Helmholtz.

In the case of endothermic reactions, there is no output of free or available energy. On the contrary, an external input of free energy is needed to compensate for the binding energy, or negative free energy, of the reactants. As noted above, it can be provided in the form of high temperature heat (in a blast furnace), light (in the case of photosynthesis) or electric power (in the case of aluminum production from aluminum oxide). The binding energy of a very stable compound, like CO_2 or H_2O, can be defined as the negative of the Gibbs free energy that would be released by a reaction between the pure elements. The *exergy* of pure carbon dioxide is the amount of work needed to purify it from atmospheric air. Similarly the exergy of pure water is the amount of work needed to purify it from ocean water.

The key point is that not all energy is available to do work. Unavailable energy (such as most of the heat in the oceans) is now called *anergy*, whence the total energy of a system (which is conserved) is the sum of *exergy* (which can do work) and *anergy* (which cannot). As a consequence of the second law of thermodynamics, exergy is "destroyed" in every process or action, as the total energy in the system becomes less available. Thus actions decrease exergy while increasing anergy (since the sum of the two is conserved), and also increasing entropy. N.B. Do not confuse the two terms: anergy is not entropy, the two simply move in the same direction.

An important methodological tool known as "pinch technology", was introduced in the 1980s (Linnhof and Vredevelt 1984; Linnhof and Ahmad 1989). The modern authorities on exergy analysis are van Gool (1987), Szargut et al. (1988), and Dincer and Rosen (2007). The exergy concept was first applied to resource analysis at the national level by Wall (1986) and in a number of more recent international

---

[5] Later Gibbs developed vector calculus, worked on the electromagnetic theory of light, statistical mechanics, and helped create the mathematical framework now used in quantum mechanics. Meanwhile electricity and magnetism—and electromagnetic radiation—were also being investigated by many scientists, culminating with the theoretical masterwork of James Clerk Maxwell, published in 1873. I discuss that topic in the next section.

comparisons. It was first used to measure waste products in the chemical industry by Ayres and Ayres (1999, Chap. 2.6).

3.5   Electricity and Electromagnetism

According to legend, in about 900 BCE a Greek shepherd named Magnus walked across a field of black stones, which pulled the iron nails from his sandals. (Did the Greeks use iron nails? Never mind, it is a legend.) The black stones became known as "magnesia" and, more recently, *lodestones*. At any rate, the Chinese used lodestones to determine the direction of north. The first European to do so was an Italian, Petrus Peregrinus (1269). The discovery that the Earth itself is a giant magnet is attributed to the English physician William Gilbert (1600). Gilbert also seems to have discovered that a static electric charge can be generated by rubbing two objects together. The possibility of transferring the charge over a wire was discovered over a century later (1729) by Stephen Gray.

A theory that electricity consists of two kinds, resinous (negative) and vitreous (positive) was first put forward by Charles F. du Fay (1733). In 1749 the Abbé Jean-Antoine Nollet based his "two-fluid" theory of electricity on du Fay's observations. But by 1747, **Benjamin Franklin** had already set forth the "one-fluid" theory of electricity, based in part on his famous experiments with lightning (which led to a very practical invention: lightning rods).

The first statement of a "**force law**" involving electro-magnetism seems to have been due to Thomas Le Seur and Francis Jacquier, in a note they added to a French edition of Newton's *Principia* published by them (1742). They said that the force between two magnetic dipoles is proportional to the inverse cube of the distance between them. They got it wrong. (It is to the fourth power). A few years later (1750) John Mitchell showed that the force between two magnetic poles is proportional to the inverse square of the distance between them. In 1766 **Joseph Priestley** (based on a suggestion from Ben Franklin) showed that the force between electric charges is also proportional to an inverse square of the distance. **Charles Coulomb** confirmed this law in 1785, using more precise instruments.

The first clue as to an electro-chemical potential arose from the experiments by **Luigi Galvani** in 1780 using the legs of dead frogs. High school students, even today, can observe the characteristic twitch (even though the frog is dead) due to a static charge, and later by contact with dissimilar metals. Followers developed this into a theory of "animal electricity". But the first electric battery by Alessandro Volta (1793) explained the Galvanic effect as due to ordinary electricity. Volta's first practical electric pile (a stack in which dissimilar metals are separated by wet cardboard) appeared in 1800. The standard measure of electrical potential, the volt, was named after him, while the standard measure of quantity of electric charge was named after Coulomb.

The Voltaic pile was applied in the same year (1800) by William Nicholson and Anthony Carlisle to separate the hydrogen and oxygen components of water. It was the birth of electro-chemistry. **Humphrey Davy** definitely proved (1807) that the electrical effect of the pile was due to chemical action. Modern storage batteries, which store energy as chemical potential, are developments and elaborations of the voltaic pile. The first practical application of storage batteries, in the 1840s, was telegraphy.

Incidentally, Ben Franklin was also the first to notice a connection between electricity and magnetism (1751). He proved by experiment that an electric current could magnetize iron. That discovery was forgotten until 1820, when **Hans Christian Ørsted** showed that an electric current in a wire causes a compass needle to align perpendicularly to it. Magnetic fields are measured in Oersted units today. A week later **Andre Marie Ampère** showed that parallel current-carrying wires repel each other, and wires with opposite currents attract each other.[6] Electric current flow is measured in amperes today. Biot and Savart also did their experiments with wires and magnets in 1820. A year later, **Michael Faraday** repeated Ørsted's experiments and began to take them much further, as noted below. **Georg Simon Ohm** was the first to recognize that voltage drives current (Ohm's law), in 1826. Electrical resistance is now measured in ohms, named for him.

The magnetic induction phenomenon, based on the work of Biot and Savart mentioned above, was first demonstrated by Michael Faraday in England (1831) and by **Joseph Henry** in the US (1832). They proved (independently) the converse of Ørsted's result: that a wire moving through magnetic field will generate a current. To understand the phenomenon better, Faraday envisioned a magnetic field as a bundle of curved lines in space. That was the first conception of a "field", in the modern sense (as in quantum field theory). This notion that the field filled space was quite radical in its time, when most scientists though space was empty, except for the *luminiferous aether* which was supposed to transmit light.

Faraday's realization that electricity and magnetism are two aspects of the same force had huge consequences. The major breakthrough (1832) was the discovery that kinetic energy (mechanical work) could be used to generate electricity, and conversely. Dynamos, motors and generators (Jedlik, Wheatstone, Siemens, Pixii, Pacinotti, Gramme, Edison, Sprague and Tesla) followed in the next half century. These were the electrical counterparts of the heat engine. It also meant that a motion could produce an electric current that could, in turn, produce another motion far away. That understanding led to the telegraph, which was invented barely a decade later (Wheatstone, Morse) and was the basis for a major business enterprise by 1860. Incandescent electric lights (Swan, Edison) followed soon after.

The mathematical descriptions of electrical phenomena were developed during the last decades of the eighteenth century, and subsequently, by **Joseph Louis Lagrange** (1762), Pierre Simon Laplace (1782, 1813), Simeon Louis Poisson (1812, 1821) and **Karl Friedrich Gauss** (1813). Gauss rediscovered Lagrange's

---

[6] Ampere's memoir (1825) seems to have been by far the most sophisticated description of the phenomena, both of the experiments and the theory. James Clerk Maxwell later called him "The Newton of electricity".

1762 result but got his own name on it.[7] The so-called "divergence theorem" (or Gauss' theorem) was rediscovered at least twice more, by George Green (1828) and Mikhail Ostrogradsky (1831). It expresses in vector calculus form the idea that the sum of all sources, minus sinks, of a vector field—or a fluid—inside a boundary is equal to the fluid flow across the boundary. This theorem explains the inverse square law as applied to electrostatics, by introducing the notion of a field.

Meanwhile, during the years after 1850 the Scottish physicist **James Clerk Maxwell** was creating an integrated theory of all electromagnetic phenomena, based largely on Faraday's and Henry's discoveries of induction together with a major component of sheer genius. Despite the earlier contributions of Fresnel, only at the end of the century was the true importance of electromagnetic radiation understood. In 1860 **Gustav Kirchhoff**, working on telegraphy, noticed that the signal propagation in a co-axial cable is close to the speed of light.[8]

About that time, Kirchhoff also noted that the temperature of a hot "black body" was related to its color (Kirchhoff's law of thermal radiation). As the temperature rises, we see first a dull red at 1000 K, then orange, then yellow and finally the light is pure white. We see this relationship in blacksmith's shops, glass-blowing, and magma from volcanic eruptions, as well as the color of stars. The product of the black-body temperature, and the wave-length at which thermal emission is maximum, turns out to be a constant. This is known as Wien's displacement law.

The observations about light color as a measure of temperature also suggested that visible light must also be an electromagnetic phenomenon. The wave theory of light had been suggested in the eighteenth century by **Christiaan Huygens** (who disagreed with Newton on this, but was told he was wrong by the Royal Academy). Huygens work on wave theory was later extended by **Augustin-Jean Fresnel**, based partly on experimental work by Thomas Young in England. Fresnel's work in the years from 1814 to 1827 explained diffraction, interference, and polarization among other things. He demonstrated circular polarization. He is best remembered for the Fresnel lens which replaced mirrors in lighthouses. People started to try to measure the speed of light during those years.

In 1861 Maxwell published his "mechanical" model of the electromagnetic field in which he concluded that *light consists of electromagnetic waves in the aether*. His model expressed in mathematical language the following "laws":

- A time-varying electric field generates an orthogonal magnetic field
- A time-varying magnetic field generates an orthogonal electric field

---

[7] Known today as the divergence theorem or Gauss' Law.

[8] Galileo was the first to try to measure the speed of light (between two hilltops) but he was only able to determine that it was large. The first astronomical measurements were made by Ole Christensen Romer (1676). Later measurements were made by Isaac Newton and Christiaan Huygens leading to estimates of the roughly right magnitude, but about 26 % too low. In 1849 Hippolite Fizeau repeated the Gallileo experiment using a rotating cogwheel to intercept the return beam and calculated $c = 3.13 \times 10^5$ km/s. which is slightly too high. His experiment was later repeated by Leon Foucault (1862) who obtained $c = 2.98 \times 10^5$ km/s. with a rotating mirror in place of the cogwheel. The currently accepted value in a vacuum is 2.99782×10^5 km/s.

- An isolated stationary electric charge repels other stationary charges of the same sign but attracts charges of the opposite sign, with a force that decreases as the square of the distance.
- There are no isolated magnetic poles, only dipoles. (This is still disputed).

He completed his equations by introducing something he called the "displacement current" which had no observable effect but was necessary to conserve energy and thus complete the scheme. Maxwell argued that the displacement current must have real physical significance. This enabled Maxwell to conclude that the speed of light is determined by the ratio of electromagnetic and electrostatic units of charge. That ratio was previously determined with field measurements by Weber and Kohlrausch in 1856 by discharging a Leyden Jar (a form of capacitor).

In 1864 Maxwell calculated the speed of light from this ratio, and found that his calculation agreed quite well with direct time-of-flight measurements by the experiment of Hippolyte Fizeau, as improved by Leon Foucault in 1862. Maxwell's complete theory of electro-magnetism was published in 1873. It explained much more than magnetic induction. It predicted electromagnetic waves and, by implication, that light must consist of electromagnetic waves of high frequency.

During the next 40 years there was a series of experiments to determine the motion of the Earth through the supposedly stationary "*luminiferous aether*". (Yes, they were still assuming the existence of such a medium). The famous Michaelson-Morley experiments (1887) were unable to detect any such relative motion. To save the *aether* theory, Hendrik Lorentz proposed that the measuring apparatus would contract along the direction of motion, and that the local time variable as measured by a clock moving at the apparatus would be "dilated" as compared to a stationary clock. This scheme, extended by Henri Poincaré, became known as the *Lorentz transformation*. It later became the core of **Albert Einstein's** special theory of relativity.

Between 1886 and 1889 **Heinrich Hertz** (von Helmholtz's student) put Maxwell's theory to the test in another way: by developing practical experimental techniques for producing and detecting such waves. He was able to produce standing waves and to measure the velocity, electric field intensity, polarity and reflection. In effect, he finally proved that Maxwell's predicted electromagnetic (radio) waves are real. Quite a few other physicists subsequently worked on radio-telegraphy, which later led to broadcast radio, TV and electronics. They included most notably Thomas Edison, Nikola Tesla (who didn't believe that the radio waves were being propagated through the air) and Guglielmo Marconi, who made radio telegraphy practical and created an international business.

In 1895 Kristian Birkeland used Maxwell's theory to explain the aurora borealis ("northern lights"). He hypothesized the electrical currents in space, created by the magnetic field of the sun ("sunspots") would be captured by the Earth's magnetic field, and concentrated where the field lines enter the atmosphere near the poles. In effect, he created a new branch of science by integrating Maxwell's theory of electromagnetism with the theory of fluid motion. He built equipment to test his theory and, extended it to apply to other astronomical phenomena, including sunspots, Saturn's rings and even the formation of the galaxy. (Birkeland, incidentally, invented an electromagnetic cannon.) From a scientific point of view, he was the father of magneto hydrodynamics (MHD) and plasma science. But after his death in 1920, his theories were forgotten until they were renovated 30 years later by Hannes **Alfvén**, in the context of cosmology.

According to classical statistical mechanics, also originally developed by Maxwell in *The Theory of Heat* (Maxwell 1871) and especially the Maxwell-Boltzmann distribution, developed further by Ludwig Boltzmann (Boltzmann 1872), the energy in each wavelength interval of thermal radiation should be distributed equally in thermal equilibrium. This result was the so-called equi-partition theorem. It has many applications. One of them is known as the Rayleigh-Jeans (R-J) law (published in 1905). Based on classical assumptions, including the equipartition law, R-J implies that the energy $\mathbf{E}$ emitted by a black body[9] should be proportional to the temperature divided by the wavelength λ to the fourth power. The equation is

$$\mathbf{E} = 2\mathbf{ckT}/\lambda^4$$

where $\mathbf{c}$ is the velocity of light and $\mathbf{k}$ is the Boltzmann constant. This function approaches infinity at very high frequencies, as wavelength λ goes to zero. This behavior by the R-J function was later called the "ultra-violet catastrophe" by Ehrenfest (1911).

Max Planck became interested in the relationship between radiation and temperature in 1894 (because the new electric light industry wanted to maximize light output from incandescent light-bulbs). He started with Wien's law, which gave correct results at high frequencies but failed at low frequencies. On the other hand, the equi-partition theorem (later named for Rayleigh-Jeans) gave good results at low frequencies, but not at high frequencies. Finally, Planck derived a different formula by assuming that light was not emitted in continuous frequencies, but only in discrete "quanta". On this basis he derived the famous formula

$$\mathbf{E} = 2\mathbf{hc}^2/\lambda^5 [1/\exp(\mathbf{hc}/\lambda\mathbf{kT} - 1]$$

where $\mathbf{h}$ is the Planck constant, $\mathbf{c}$ is the velocity of light and $\mathbf{k}$ is the Boltzmann constant. This equation, published in 1901, matched the observed black-body radiation spectrum. Planck's formula reduces to the Rayleigh-Jeans law as the wavelength λ gets very large (low frequency). But it reaches a peak and declines again as λ gets very small while the frequency ν gets large (the UV case).

Planck's equation had already solved the ultra-violet catastrophe problem in principle, and set the stage for quantum mechanics. Yet Planck was reluctant to assert the obvious implication (now it is obvious) i.e. that light was not emitted in continuous frequencies, but in discrete "quanta". He did so "in an act of despair"

---

[9] Black bodies were simulated by an artificial cavity penetrated by a small hole, such that almost all of the incoming light is not reflected out but is absorbed inside the cavity. The thermal radiation from the hole is then a very good approximation of black body radiation.

(his words) by consciously giving up the ancient supposition in natural philosophy *Natura non facit saltus* (Nature does not make jumps). Einstein made that leap.

Several scientists, starting with Johann Balmer (1885) combined to work out a formula that accurately explained the spectral series of the hydrogen atom. However nobody at the time could explain that observational fact of nature. It was explained later by Neils Bohr.

Meanwhile **Henri Becquerel** discovered radioactivity (1896). It was studied independently by **Pierre** and **Marie Curie**, **Ernest Rutherford** and **Frederick Soddy**. Rutherford first recognized that there are several types of radiation, which became known as alpha, beta, and gamma. Of those, the first two consist of charged particles and the third does not. The alpha particles turned out to be positively charged helium nuclei, while beta particles are now known to be positively charged electrons (positrons). Rutherford (with Soddy) explained radioactive decay as an atomic, not molecular, process (1902). He went on to discover the proton (1911), and the atomic nucleus. He explained the hydrogen atom as an electron orbiting around a nucleus, as a planet orbits around a star. He later was the first to transmute an element into another by radioactive decay (1919).

Today it is obvious that electricity is one form of energy that is also essentially a form of useful work. Useful work may be thermal (as in a heat engine), mechanical (as by a motor) or chemical (as in smelting or electrolysis). People speak of "electric energy". Today electricity can be converted to heat with nearly 100 % efficiency in a toaster or electric arc furnace. Electricity can also perform mechanical work by means of a motor, which—in its most highly developed forms—also often exceeds 90 % efficiency. Generators are also very efficient nowadays, at around 80 %. The least efficient forms of energy conversion, up to now, are the solid-state devices like thermo-electric or photo-voltaic (PV) systems, which rarely approach 30 %.

3.6 Geology and Earth Science

The first key idea of geology was the recognition by Aristotle, and later by the eleventh century Persian Avicenna (Ibn Sina) and the thirteenth century Albertus Magnus, that fossils were the remains of animals that had once been alive. The next step was the recognition (also by Avicenna) of rock "strata" of different ages, and the fact that later strata are always superposed on earlier ones. In the seventeenth century Nicholas Steno took it further by characterizing the strata as "slices of time". One popular theory of strata at the time, called "Neptunist", was that the strata had been laid down by a giant flood. But in 1785, James Hutton introduced his "Theory of the Earth" (named a "Plutonist" theory to contrast with the Neptunist theory) i.e. that the interior of the Earth is hot.

In the second half of the nineteenth century, a number of physicists calculated the age of the Earth, based on different models. One of the first was Hermann Helmholtz (1856) who calculated the age of the Earth to be 18 million years. Then came William Thompson (Lord Kelvin) in 1862, who estimated it to be between 20 million and 400 million years, based on the rate of cooling of a molten sphere. The geologist Charles Lyell disagreed, as did the biologist Thomas Huxley, both of whom thought the evolutionary process to be much slower (Lyell 1838). But astronomer Simon Newcomb calculated the length of time for the sun to condense from its primordial dust cloud, at 22 million years. George Darwin (the son of Charles) calculated the age of the Earth at 56 million, based on the time needed for the moon to move away from the Earth and slow down to its present orbital period. Lord Kelvin recalculated in 1892 and got 100 million years (still between 20 and 400 million). Then John Joly (1899–1900) calculated it to be between 90 and 100 million, based on the time required for the oceans to "fill up" based on observed runoff and erosion rates.

But when Becquerel discovered radioactivity in 1895, it was soon realized that this could account for the heat source in the Earth's interior. It was also realized about that time that there is a viscous "magma" layer between the solid core and the crust. The existence of the viscous mantle changed the heat flow calculations (Joly 1895). These discoveries undermined Kelvin's theory. In 1895, John Perry recalculated based on data on the amount of lead in the uranium ore, and arrived at 2–3 billion years. During the next decade, Earnest Rutherford and Frederick Soddy worked out the decay rates of various other isotopes and provided the data base for what is now known as *radiometric dating* upon which the whole of geology and paleontology depend.

The first rock dating based on radioactive decay rates was published by **Arthur Holmes** in 1911. His book: *The Age of the Earth: Terra-forming of: An Introduction to Geological Ideas* followed (Holmes 1913). In 1917 Yale geologist Joseph Barrel reconstructed geological history, using Holmes' results, and concluded that strata were laid down in different places at different times—contrary to then-current doctrine (Barrell 1917). Unfortunately, most geologists continued to ignore radiometry until the publication of the report of a meeting of the US National Research Council in 1931 finally established a consensus on this question.

Alfred Wegener was the German geologist who pioneered the theory of Pangea and continental drift, based on fossil similarities between continents (c. 1912). Apart from support by Arthur Holmes and Alex du Toit (who incorporated "Gondwana" into the theory in 1937) Wegener's ideas were ignored until the 1950s. Then other evidence came along, supporting the which finally led to the currently dominant theory of continental rifts and *plate tectonics* e.g. Belousov (1969).

Douglas Mawson was the first to suggest the possibility of global glaciation (later called "snowball earth") (Alderman and Tilley 1960). **Mikhail Budyko** suggested the ice-albedo feedback mechanism for glaciation (Budyko 1988). It was Joseph Kirschvink (1992) who saw the importance of "banded iron" as evidence of global glaciation. He, and many others, (including Dietz, Harland, Heezen, Hess, Mason, Morley and Vine & Mattheson) suggested that accumulations of carbon dioxide from volcanic emissions could explain the "escape" from "snowball Earth" conditions.

One other theorist deserves mention: Charles Hapgood wrote a book entitled "The Earth's Shifting Crust" (Hapgood 1958). The idea is that at some point, possibly 9.5 thousand years BCE, the Earth's solid crust—tectonic plates and all—may have "slid" rather fast (during a few tens of years) over the underlying mantle. If this happened, it was probably due to asymmetric glacial ice buildup, leaving the rotational axis of the Earth unchanged, but moving the crustal positions of the north and south poles several hundred km from their previous locations. This theory is not accepted by the mainstream geologists (Hapgood was an historian, not a geologist), but neither has it been comprehensively disproven.[10] The uncomfortable fact is that it does explain some otherwise inexplicable facts, including the megafauna extinctions that occurred around that time.

3.7   Darwin and Biological Evolution

The modern idea of evolution as a long and irreversible process of increasing organism complexity, lasting billions of years—not the result of instantaneous (biblical) creation by God—has gradually gained acceptance, against considerable theological opposition. This was happening at the same time that the physical sciences were finally shaking off archaic notion of Earth-centrality and *aether*. Otherwise, this section may be considered as background to the later discussion of evolution in Chap. 5, where both exergy dissipation and increasing complexity played an important role.

The idea of evolution scarcely existed before the nineteenth century. It is true that many early Christian and Hebrew theologians regarded the Biblical claims as metaphorical, not literal. The dogma hardened when Bishop Ussher contributed his footnotes to the King James Bible. Religious dogma after Ussher insisted that the Earth was barely 6000 years old, and that all the species of plants and animals on it were put there, in the Garden of Eden, by God. (Some still believe this). The logical starting point for modern biology was the classification work of the eighteenth century Swedish botanist **Carolus Linnaeus**, who created the hierarchical taxonomic system (species. genera, families, orders, classes, phyla, kingdoms and domains) in use today (Linnaeus 1735). The system was based on observed structural differences and similarities.

Jean-Baptiste Lamarck followed Linnaeus as a botanist and taxonomist, especially of invertebrates. He classified the flora of France, among other achievements. He argued that evolution was an irreversible process controlled by natural laws, moving toward greater complexity ("higher species"). The two mechanisms driving

---

[10] The comprehensive 1969 geophysical monograph, "The Earth's Crust and Upper Mantle" (Hart 1969) contains articles by 92 different contributors, but does not cite Hapgood's theory. In fact, the concluding article, by V.V. Belousov, entitled "Interrelations between the Earth's crust and the upper mantle" contains no citations at all, even though it briefly discusses (and effectively dismisses) the possibility of large displacements.

change, in his view, were natural "alchemical motions" of fluids carving out ever more complex organs, and the use (or disuse) of organs in the course of adaptation. Lamarck believed that organisms could adapt to changing conditions during their lifetimes and pass these adaptations on to their offspring ("soft inheritance"). In fairness, the "soft inheritance" idea was not originally his, although it has been used to discredit him. However he made a considerable contribution to the development of biology as a science, and the framing of evolutionary theory.

Another major misstep in the developing theory of evolution came from **Ernst Haeckel** (Haeckel 1866). Haeckel postulated his theory that ontology (embryological development) recapitulates phylogeny (the historical sequence of evolutionary stages). This theory, supported by famous drawings of embryos, was very seductive. However, it has since been thoroughly discredited. As Paul Ehrlich et al. have noted:

> Its shortcomings have been almost universally pointed out by modern authors, but the idea still has a prominent place in biological mythology. The resemblance of early vertebrate embryos is readily explained without resort to mysterious forces compelling each individual to reclimb its phylogenetic tree (Ehrlich et al. 1983, p. 66)

Charles Darwin carried out his active explorations, starting with the voyage of *The Beagle* in the 1830s, but he did not publish his scientific results for a long time because he knew how controversial his conclusions would be (and because his wife believed strongly in the Biblical story). His landmark book *On the Origin of Species by means of Natural Selection; or the Preservation of Favored Races in the Struggle for Life* was finally published in 1859 (Darwin 1859). (In the sixth edition, in 1872, the title was shortened to *On the Origin of Species*). Darwin realized that bio-diversity is a genealogical phenomenon and that differential reproduction (speciation) is a consequence of changing conditions and "natural selection" on the basis of "fitness", meaning capability to survive and reproduce.

Even allowing for ambiguity and confusion of definitions, this was an important scientific insight. Darwin accepted, in part, Lamarck's "use and disuse" mechanism for selection. They agreed on the proposition that all existing organisms had a common ancestor. He differed with Lamarck on the question of soft inheritance,[11] although Darwin had no specific theory to explain the transmission of what we now call genetic information. He disagreed with Haeckel in regard to recapitulation, pointing out that resemblances during embryo development do not imply that adults will also resemble each other.

Thanks to improved microscopes becoming available, the phenomena of cell division and sexual reproduction were well established by the mid-nineteenth century. The missing piece of evolutionary theory, to explain the phenomenon of selective breeding and hybridization, was finally provided by **Gregor Mendel's** famous experiments with peas. His theory of inheritance, first published in 1866, showed that (1) information carriers for specific traits, such as colors (now called "genes") are carried from generation to generation (2) that for each trait there is a

---

[11] The idea was still strongly promoted by Lysenko in the Stalinist USSR.

contribution from each parent and (3) that the parental contribution may not "show up" in every generation, but can reappear in later generations.

The importance of the water and nutrient cycles (carbon, nitrogen, sulfur, phosphorus) in earth science and biology was especially emphasized by **Alfred Lotka** (Lotka and Alfred 1925). Lotka was also one of the first to notice periodic reactions in chemistry (Lotka 1910). He later applied this idea to the predator-prey periodicity, with Vito Volterra (Volterra 1931). The Lotka-Volterra equation is now well-known. The predator-prey cycle (extended) is fundamental to population ecology.

One other important contribution (or digression) was Dollo's "law of irreversibility" proposed in 1893, before the mechanisms of genetics were understood. Paleontologist **Louis Dollo** postulated that *"An organism is unable to return, even partially, to a stage previously realized by its ancestors"*. Dollo's theory is still debated in the literature. However the underlying idea of "progress", of course, was that evolution moves toward some goal (e.g. increasing intelligence) according to a teleological principle, such as increasing complexity or increasing intelligence, e.g. Sagan (1977). Moreover, Dollo's proposed Law has quite a lot of empirical support. It was later reformulated by **Julian Huxley** (Huxley 1956). However, some modern evolutionists, like **Stephen Jay Gould**, have strongly opposed the "iconography of the ladder" at least in the context of popular discussions of the descent of mankind from the apes.

Returning to Mendel's results with peas, they were ignored for a generation (because Mendel himself thought it only applied to particular cases) until their "rediscovery" by **William Bateson** in 1900. Thereafter the theory was widely accepted and finally became the core of **Thomas Hunt Morgan's** "chromosome" theory (Morgan 1913; Morgan et al. 1915). Morgan's work with the fruit fly *Drosophila melanogaster*—beloved of generations of geneticists—explained Mendel's third observation in terms of "dominant" and "recessive" characteristics. Morgan was the first to identify the location of specific genes within chromosomes. The importance of free radicals as a factor in the aging process as well as a cause of mutations, is a relatively recent discovery (Harman 1956, 1972).

Arguably the most important change in evolutionary biology in recent decades is the acceptance of **Lynn Margulis's** *endosymbiotic theory of biogenesis* (new species creation). Her idea is that close symbiotic relationships between species, over many generations, can—and do—result in "DNA mergers" resulting in new species. This was probably how the first multicellular organisms were created. Her theory was first published in 1970 and elaborated ten years later (Margulis 1970, 1981). At first it was regarded as too radical. But the symbiosis mechanism she described has now been demonstrated in the laboratory for mitochondria, centrioles and chloroplasts. Her theory of biogenesis is now accepted.

The other major intellectual dispute within paleontology during the last half century or so has been how to explain the fact that most species, once they appear, do not change much during their existence (*genetic homeostasis*). Yet radical changes in species morphology do occur from time to time, driven by environmental pressures on the whole organism e.g. Mayr (1991, 1997, 2001). Richard

Dawkins has promoted the narrower ultra-Darwinian thesis that evolution is strictly gene-based (Dawkins 1976 [2006], 1982, 1996). The "gene" theory and the "chromosome" theory explain how hybridization can (and does) produce new species from a given collection of genes.

Niles Eldredge and Stephen Jay Gould, both together and independently, have made a very strong case for "punctuated equilibrium", which seems to describe the actual history of evolution pretty well (Eldredge and Gould 1972; Gould and Niles 1972; Gould 2002, 2007). But the specific mechanism that causes a "punctuation" remains controversial. Nobody can explain clearly how "new" genes can appear out of nowhere. That seems to be a consequence of random mutations, triggered by environmental influences ranging from exposure to reactive chemicals such as "free radicals" or "cosmic rays". From a modern biological perspective, mutation and speciation can be regarded as examples of symmetry-breaking. The phrase "punctuated equilibrium" expresses the current paleological view. Energy flux and flow did not play much of a role in biological theory—largely focused on specific species—until very recently. However, Kuuijman's *Dynamic Energy Budget* theory (DEB) deserves mention. (Kuuijman 1993 [2009]; Sousa et al. 2008; Kooijman 2000, 2010). It is an ambitious research program to construct a general theory to explain all metabolic processes, based on five "state variables" viz. structural bio volume, reserves, maturity, reproductive buffer and level of toxics. The theory is currently in the process of being tested and modified, so it is not yet "established", but results are promising.

3.8 Ecology

Ecology is a new—some say 'infant'—science, though the word, preceded by 'the' is commonly misused as a synonym for environment. Ecologists are not necessarily environmental activists. The Greek word *Oikos* means "home" and it is also the root for economics. ("Home economics", in the sense of household management, was where it began back in Aristotle's time). The subject matter of ecology differs from evolutionary biology in that it focuses on inter-species relationships and topics like diversity, species succession, and energy flow. There is an overlap, however, as regards evolutionary theory.

The idea of natural succession (following fire or a clear-cutting) goes back at least to Henry David Thoreau in 1860. He described the sequence starting with fast growing grasses, then shrubs, then sun-loving pine and birch trees, followed by slow-growing long-lived hardwoods. Diversity increases over time, along with biomass. On the other hand, the growth rate is maximum at first, and declines over time and finally falls to zero in a "climax" forest. In fact, a young forest absorbs and "fixes" carbon dioxide, whereas an old one dominated by fallen timber and humus, can be a carbon dioxide generator. This point is worth emphasizing, because it bears on current issues related to climate change. Untouched "old growth" deciduous or conifer forest is not necessarily desirable from a carbon emissions minimization perspective.

Similar patterns are observable in other climatic and soil conditions. Henry Cowles (1889) observed a similar horizontal succession pattern, moving inland from the shore of Lake Michigan cross the dunes. At the shoreline there were no rooted plants, but further away from the wind off the lake there were annual succulents, then grasses, then perennial shrubs (like juniper) and finally trees. Again, diversity and biomass increased with distance from the wind. The succession idea, presented (prematurely) as a "universal law" of sorts, was naturally attacked, notably by Henry Gleason (1926).

The most influential ecologist during the prewar period was G. Evelyn Hutchinson of Yale. He pioneered population studies, the idea of trophic levels and ecosystem modeling, but is best known for his students. The important concept of "trophic levels", starting with *autotrophs* that make their own food (plants), followed by "*heterotrophs*" that do not make their own food was further developed by Raymond Lindeman (Lindeman 1942). Lindeman identified a pyramidal hierarchy of trophic levels above the autotrophs, starting with herbivores (e.g. cattle), carnivores, and so on. Since his famous paper in 1942 the notion of succession has been accepted as a universal.

Lindeman (following Lotka) noted that the efficiency of exergy recovery from level to level is so low that no more than five or so trophic levels are possible. Whereas there is a lot of stored energy (exergy) at the bottom of the pyramid, the exergy available at the top is very limited.

It was Lindeman, followed a generation later by Rachel Carson, who made the point that nutrients and toxic substances (like mercury or DDT) are both stored in fatty tissues of heterotrophs and transmitted around the food web. Carson's powerful "Silent Spring" made that point by noting that the egg-shells of raptors like hawks and eagles were too thin to hatch—because of DDT in the seeds of the plants that were then eaten by the herbivorous birds like sparrows and pigeons that were the prey of the hawks (Carson 1962).

Another foundational concept in ecology is the *niche*, meaning a unique combination of habitat, and predator-prey relationship with other species. The remarkable point is that the number of niches in an island (or other bounded habitat) is actually fixed, as a function of area. The number of species doubles with every tenfold increase in the area of the habitat. More surprising, if a new species succeeds in invading, *another must disappear*. This surprising fact was discovered and documented by R.H. Macarthur and E.O. Wilson (MacArthur and Wilson 1963).

One of the consequences of that fact is that when natural habitat disappears—or is "taken over" by humans for purposes of monoculture agriculture or forestry—species diversity disappears and the number of species on the planet (never actually counted) falls dramatically. The global food web has been—and is still being—drastically simplified. The phrase "Sixth Extinction" has been applied to this situation (Roy et al. 1979).

In recent years exergy flow in ecosystems has been a topic of increasing interest. Ramon Margalef first introduced the use of metabolic rates (as a function of mass) as a measure of the maturity—in the succession sense—of an ecosystem (Margalef 1968). Higher metabolic rates mean more "mature" (whatever that means). His idea was that "progress" along an evolutionary path can be measured by the length of time stored exergy "lingers" in a bio-system. This measure may also be applicable to wealth economic systems (see Part III).

The most influential ecologist of the postwar generation was Eugene Odum, another student of Hutchinson. His textbook *"Fundamentals of Ecology"* (Odum 1953; Odum and Eugene 1971) was the first true synthesis of the field. It emphasized the importance of ecological succession, its predictability and directionality, as well as implications for planetary management. It noted the increasing biomass (along a succession), its exergy content and its increasing information content. It stressed the fundamental importance of energy (exergy) flow through ecosystems. His brother Howard (and students) took the energy flow aspect into economics, especially in his influential book *"Environment, Power and Society"* (Odum 1971) as well as several others (Odum and Odum 1976; Odum 1986).

Another ecologist worth mentioning is **C.S. "Buzz" Holling**, who has generalized Lindeman's "succession" ideas as a kind of cyclic spiral. He is particularly known for an important insight, typically expressed in terms of a "figure eight" curve, that reflects the long term cyclic adaptive behavior of many ecosystems, reflecting the traditional "succession" pattern, but with emphasis on *resilience* (Holling 1973, 1986, 1996). The cycle starts with recovery after a fire or some sort of exploitation and it shows biomass increasing rapidly at first but slowing down as the recovery approaches (or reaches) climax. During this phase internal self-regulation is "fine-tuned" to maximize utility in existing conditions, but, complexity and resilience (species diversity) declines. When a low probability but high impact ("black swan") event such as a great fire or insect outbreak occurs, the system collapses. This "creative destruction" (i.e. the fire) is followed by re-organization, renewal and a new growth. In economics the "creative destruction" is what permits innovation and system change of some sort. The adaptive cycle in economics will be discussed further in Part III.

Others who have focused on energy flow include **Alfred Lotka** "The law of evolution as a maximal principle" (Lotka 1945) (not strictly an ecologist), Jeffrey Wicken *Evolution, Thermodynamics and Information: Extending the Darwinian Program* (Wicken 1987), Robert Ulanowitz *Ecology: The Ascendant Perspective.* (Ulanowicz 1997) and James Kay and Eric Schneider "Life as a manifestation of the second law of thermodynamics" (Schneider and Kay 1994). The most ambitious of them all may be *Evolution as Entropy* by Daniel Brooks and E.O. Wiley (Brooks and Wiley 1986–1988). Their book claims to be a unified theory of biology that "explains a large number of diverse empirical facts and widely held generalizations under a single grand theoretical viewpoint". It unifies the fields of Neo-Darwinism, population genetics, population ecology, and speciation theory.

However, as regards speciation, there is an important insight that relates to the second word in the title of this book, namely complexity. Biologists are still groping for an explanation of the "Cambrian explosion" some 500 million years ago. That was a very sharp increase in biological diversity, starting with multi-cellular organisms and ending with incredibly diverse world as it was a few centuries ago before the "sixth extinction" began (Leakey and Lewin 1995, p 106). A tentative but very plausible explanation was suggested to me a few years ago by the late Franz de Soet, a Dutch ecologist who was consulting for the EEC to design a strategy for protecting biological diversity (de Soet 1985). His observation, based on a lifetime of field-work, was that high bio-diversity is correlated with sharp environmental gradients in altitude, rainfall, wind pressure, temperature, and soil composition. The argument, in brief, is that the greater the diversity of micro-systems located near each other, the greater the likelihood of evolutionary divergence.

The best evidence of this theory can be found in the West Cape region of South Africa. That small region—now a UNESCO World Heritage Site—is one of seven "floral plant kingdoms" in the world (richer than the Amazon rain forest) in plant species diversity, with 2285 plant species, many of which grow nowhere else. The 7750 ha Table Mountain National Park (formerly Cape Peninsula National Park) alone claims 1100 plant species and 250 bird species. The explanation offered locally is that this region is where the cold Benguela current on the west coast meets the warm Agulhas current on the east coast of South Africa. The temperature gradient between the currents is 5 °C. Moreover the coast is mountainous and it combines sharp gradients in altitude, temperature, wind pressure and humidity resulting in a number of micro-climates that are cross-linked by a variety of animal, bird and insect pollinators.

How does this connect to the Cambrian Explosion? Here I am speculating, but I suspect that the symbiotic "mergers" of single-cell organisms—first noted by Lynn Margulis—were the enabling mechanism (Margulis 1970, 1981). Were those mergers triggered by environmental gradients? I don't know. But once the enabling mechanism for symbiotic mergers existed, the gradients must have played a role. The "great melt" following the Huronian glaciation—especially if it was brought about by volcanic eruptions—would have created a large number of locations with sharp gradients, especially along coastlines. These may have been comparable to the gradients existing today in Cape Province in South Africa. The existence of those gradients may have contributed to the following "Cambrian explosion" of speciation. At least I think so. See Sect. 3.10 for more on this.

Here I should also mention the work of multi-disciplinary energy group at the University of Illinois, under Bruce Hannon (a geographer) that focused primarily on economics (see Part III) but also overlapped with ecology (Herendeen 1990; Hannon et al. 1991; Costanza and Daly 1992). Several of the ideas mentioned above also fit into the subject of the next section.

3.9   Entropy, Exergy, Order and Information

Basic knowledge about energy, as understood in classical mechanics, electromagnetic theory, quantum mechanics and relativity, is summarized in standard physics textbooks. However, there is another cross-cutting domain of science where energy is, if anything, even more fundamental. This domain is thermodynamics. As the title suggests, thermodynamics is about the way in which heat, pressure, density and electromagnetic fields affect the properties and behavior of matter in the aggregate.

To recapitulate: *exergy* is the ability to do work. It happens to be what most people really mean when they speak of energy, as if it were a substance that can be consumed or used up. It is exergy, not energy, that can be used up or consumed in a process. Exergy is the energy that is actually available to do useful work in a subsystem, or—equivalently—a measure of the actual distance of a subsystem from its "reference state" or from thermodynamic equilibrium.

When the universe was new, all of the energy from the vacuum was potentially useful—but undifferentiated—exergy. But as the universe has expanded and cooled, much of that primal exergy has been converted into kinetic energy (of galaxies moving outward from the BB) and low temperature radiation. Some of it has been "converted" into cold mass (dust, neutron stars, dark matter). However the stars in the visible universe still contain a lot of exergy capable of doing work.

The non-useful part of energy is called *anergy*, which is a second cousin of *entropy*. (The anergy of a system is essentially the product of entropy S times temperature T.) So when the temperature T is very high, exergy is high—useful work can be done—and anergy is low. Conversely, when T is low, exergy is low and anergy is high. Little or no useful work can be done.

A form of energy that is almost totally available to do useful work, such as natural gas, has high exergy content and a low level of entropy. Conversely, a form of energy that is almost totally unavailable to do useful work—such as the heat content of the land—has a low exergy content and a high level of entropy. To take another example, high temperature steam is available to do useful work, whereas the heat content of water at room temperature is not. Fuels have high exergy, whereas wastes have low exergy and high entropy.[12]

Useful work a term used several times above, is another thermodynamic concept, that (in Newtonian physics) means overcoming inertia or resistance (gravity or friction) in order to generate motion. This kind of work is unrelated to employment or work as performed by humans, in the economic sense of the word. In fact, useful work for human purposes consists of four distinct types. They are as follows: (1) *muscle work* by animals and humans, (2) *mechanical work* performed by machines, including internal combustion engines, such as piston engines and gas

---

[12] For physicists and engineers there is a technical definition that is important for calculations but not worthwhile explaining in detail for a book like this. Two good sources are: Szargut et al. (1988) and Dincer and Rosen (2007).

turbines, or external combustion engines like steam turbines (for electric power generation), (3) *electrical work* done by a voltage overcoming resistance as in lighting, electrolysis, motor drive and (4) *thermal work*, including chemical synthesis, cooking, water heating and space heating. These are all examples of energy (exergy) services that are quantifiable and can be calculated with reasonable accuracy, based on published statistics on the input flows plus some engineering data. The final category (5) *nuclear work*, is the work done by high energy photons (e.g. gamma rays) to induce heavy elements to split (nuclear fission). Physicists can calculate that kind of work, as well.

Power, a more familiar term, is a measure of useful work performed per unit time. However, power requirements (and outputs) for vehicles, for instance, are quite variable. Maximum power is needed only for acceleration or takeoff (e.g. of an airplane) but most engines operate far below maximum power output most of the time. There are statistics on the total rated (maximum) horsepower of prime movers installed in the US. But these data are not easily converted to exergy consumed or useful work performed because of this load variability. Most of the engines are idle most of the time.

The ratio of output to input for any category of work and for any process is a pure number between zero and unity. As it happens there are several inconsistent measures of efficiency in use, depending on definitions and assumptions. To minimize confusion, I reserve the complete explanation for a later Sect. 10.5 where the details can be more fully elaborated. At the aggregate level, this efficiency, however defined, is a quantitative measure of the state of technology, either by function such as transport or space heating, or in the economy as a whole. N.B. thermodynamic efficiency must not be confused with economic efficiency, as that term is used by economists (Chaps. 11 and 12). However, thermodynamic efficiency is a reasonable measure of the state of human knowledge as applied to machines and processes.

There are a number of chemical processes that *convert* chemical energy into heat, or heat into chemical energy, and/or from one form of chemical energy to another. Yet energy is conserved in all of these processes. In every case, it can be shown that every energy gainer is compensated by an energy loser or losers. In other words, energy is neither created nor destroyed by any processor action. This experimental fact has now been formalized as the law of *conservation of energy*, or the *first law of thermodynamics*. It is, arguably, one of the two most important laws of nature. It is also an invaluable tool for analysis, much as the fundamental law of conservation of money—changes in the account balance being equal to income minus expenditures—enables accountants and auditors to determine the financial health of a company or a household.

The second important law of nature, the *second law of thermodynamics*, is not less important than the first law. It says that, while energy is conserved, it becomes less available for purposes of doing useful work, with every process that occurs. The technical term for available energy is *exergy*. In every process, without exception, the universe (or any closed subsystem, for that matter) approaches a little more closely to *thermodynamic equilibrium*. Thermodynamic equilibrium is a state in which there can be no gradients, whether in temperature, pressure, or chemical composition. Without gradients to drive change there can be no spontaneous change of any kind.

During the approach to equilibrium the available component of energy—*exergy*—is destroyed while a certain state-variable called *entropy* increases. Unfortunately, whereas we have instruments like a thermometer to measure temperature and a barometer to measure pressure, there is no direct way to measure entropy as such. This "invisibility" so to speak, gives it an air of mystery that is really undeserved. Equilibrium is defined as a state in which nothing changes, or can change. On the macro-scale this hypothetical—and very distant—changeless state has been called "heat death of the universe". This "heat death" had a certain fascination for scientists who think large thoughts. The pioneers of thermodynamics were among them. But, the Earth will become uninhabitable when our sun becomes a red giant, long before that distant prospect.

Steven Weinberg, author of *The First Three Minutes* was another big thinker: Looking down at the passing scene from an airplane at 30,000 ft, he wrote the final paragraph of his book:

> It is very hard to realize that this present universe has evolved from an unspeakably unfamiliar early condition, and faces a future extinction of endless cold or intolerable heat . . . (Weinberg 1977)

Luckily this end-point is almost inconceivably far in the future. But, because every process takes the universe a little closer to this ultimate state, never away from it, it can be said to define the direction of time. This point was emphasized by Sir **Arthur Eddington** when he introduced the phrase "time's arrow" in reference to the fact that entropy and time increase together (Eddington 1928).

Nicholas Georgescu-Roegen, a Rumanian-American economist was also fascinated by this idea that the universe is on a path to entropic heat-death. He argued that the economic system of us humans is on a parallel path to extinction as we "use up" all exhaustible resources on the finite Earth (Georgescu-Roegen 1971). His gloomy book *The Entropy Law and the Economic Process* inspired many of today's ecologists and ecological economists. I will have more to say about him in Part III. The authors of a recent book called *Thanatia: The Destiny of the Earth's Mineral Resources* also belong to this tradition (Valero Capilla and Valero Delgado 2015).

There is another everyday interpretation of increasing entropy that is worth mentioning. It makes a lot of intuitive sense to think of increasing entropy as a process of increasing *disorder*, or increasing *disorganization*. So a state of low entropy is equivalent to a state of high orderliness, or embodied structural information. And conversely a state of high entropy is a state of disorder, and lack of embodied structural information. A DNA molecule or a computer chip contains a lot of embodied information, and requires a very large amount of exergy to create. An ash heap, on the other hand, is a disordered state containing very little information and corresponding to very high entropy. It is evident that human activity converts low entropy materials (ores, fossil fuels, crops) into high entropy wastes.

As mentioned at the end of Sect. 3.3, the second law of thermodynamics has undergone significant restatement since the nineteenth century. As originally formulated, it refers to what happens in a closed adiabatic system, which is a theoretical construct such that no mass crosses the system boundary, and where—eventually—"*a permanent state is reached in which no observable events occur*" (Schrödinger 1945, p. 70). In this state of thermodynamic equilibrium, there are no gradients of pressure, temperature, density or composition.

A **closed system** permits energy (heat) flow across its boundary, but does not permit mass flows across the boundary. Thus the Earth is almost closed to mass flow *but it is not an isolated* system, since it receives a large exergy flux from the sun. (As it happens, the small mass influx, from comets and meteorites, may also have had a very large cumulative influence on the history of our planet, as will be seen in Chap. 4). The distinction between closed systems, **isolated systems** and **open systems** turns out to be important in the later development of non-equilibrium thermodynamics.

The first important restatement (actually a generalization) of the second law after Clausius and Boltzmann was by **Constantin Caratheodory** in 1908 (Caratheodory 1976). It was already understood that classical thermodynamics in closed systems was only applicable in, or very near, equilibrium. Caratheodory did not use Clausius' word "entropy". What he proved was the following: "*In the neighborhood of any given state, of any closed system, there exist states that are inaccessible from it along any adiabatic path, reversible or irreversible*". What this means, in English, is that there are places *the system cannot go* from any starting point (not just a starting point in the neighborhood of equilibrium). Going to those forbidden places corresponds roughly to **Dollo**'s theory of irreversibility (Dollo 1893) or going backwards in time. Caratheodory's proof (which never mentioned entropy) was the first step towards non-equilibrium thermodynamics.

I should say here that Caratheodory's proof, together with Holling's adaptive cycle ideas from ecology, can almost certainly be applied to economic growth. I will say more about this topic in Part III.

It is tempting to mention information theory at this point, if only because **Claude Shannon** and Warren Weaver, at Bell Telephone Labs, developed a mathematical measure of the information content—or the negative uncertainty—of a message (Shannon 1948; Shannon and Weaver 1949). Their formula was exactly the Gibbs formula for entropy, except for a minus sign, where the probability of being in a given microstate p_i was re-interpreted as information uncertainty f_i. (It seems that the mathematician **John von Neumann** suggested the word "entropy" to Shannon because—according to legend—"*nobody knows what entropy really is; so in a debate you will always have the advantage*" (Tribus and McIrvine 1971).)

The real reason for the adoption of the term "entropy" in information theory was because of **Schrödinger**'s comment that living things "consume negative entropy", not to mention the similarity of the Gibbs and Shannon formulae. Leon Brillouin introduced the term "negentropy" to describe information as such (Brillouin 1953). E.T. Jaynes took it further and showed that the Shannon and Gibbs versions are equivalent—in a theoretical sense—and that each can be derived from the other

(Jaynes 1957a, b). Jaynes introduced a "maximum entropy principle", for information theory.

But Brillouin later pointed out that some irreversible behavior can decrease Shannon-Weaver entropy, whereas this is not possible in the Gibbs formulation of thermodynamics (Brillouin 1962). This suggests that the Shannon-Weaver entropy is not necessarily physically interpretable. The differences between the two entropies have been the subject of considerable discussion in the literature since then. P.W. Atkins, a modern authority, argues that the two are fundamentally different concepts (Atkins 1984).

However, in 1986 Collier emphasized the physical nature of *instructional information* which he distinguished from both the Shannon-Weaver and Brillouin versions (Collier 1986). Using this idea, one team of biologists has developed a version of information theory to explain evolutionary processes in terms of structural information maximization, as indicated by irreversibility ("Dollo's Law"), diversity and complexity and its importance in evolutionary dynamics (Brooks and Wiley 1986–1988).

The next step in the evolution of physical thermodynamics was **Lars Onsager**'s analysis of meta-stable states in near-equilibrium thermodynamics. For instance, he discovered the phenomenon of "thermo-diffusion". This means that, when heat flows in the presence of an imposed temperature gradient, there is a meta-stable state at some distance from equilibrium where entropy production is minimized. In this region, his so-called "reciprocity relations"—where forces and fluxes are coupled—hold (Onsager 1931a, b).[13] Moreover, in the near-equilibrium region he also found a relation similar to **Kirchhoff**'s Law for an electrical circuit, namely that the sum of the (chemical) potentials around a closed circuit (loop) must add up to zero. Onsager's insights reveal why it is possible for entropy production to be minimized in certain circumstances, while being maximized in others.

The next step in the evolution of entropy was the "Law of Stable Equilibrium" viz.

> When an isolated system performs a process, after the removal of a series of external constraints, it will reach a unique state of equilibrium: this state of equilibrium is independent of the order in which the constraints are removed.[14] (Hatsopoulos and Keenan 1965)

Virtually the same "law" was stated differently about the same time by Joseph Kestin as "The unified principle of Thermodynamics" (Kestin 1966). In both cases, the restatement—or corollary—of Caratheodory's theorem does not just assert that all processes are irreversible and that there are places the system cannot go. The new formulation also asserts that *there is an end state where the system will go, and*

---

[13] Examples include Fourier's Law (that heat flow is proportional to temperature gradient), Ohm's Law (that current flow is proportional to voltage), Fick's Law (that diffusion is proportional to the gradient of chemical potential or composition).

[14] For isolated systems in a laboratory, the external constraints are typically externally imposed temperature or imposed pressure, but other possibilities include static electric or magnetic fields, electromagnetic radiation or even gravitational fields.

that the sequence of stages by which it gets there doesn't matter. That end-state happens to be characterized by minimum entropy. However, this form of the Second Law does not actually mention entropy.

One more restatement of Kestin's "unified principle" is worth mentioning. Kestin's proof of the law noted that the minimum entropy equilibrium state must be stable in the Liapounov sense. This means that the system will reorganize itself to resist external forces or change agents such as increasing pressure or temperature. In chemistry this behavior was discovered earlier and independently by Henri Louis Le Chatelier in France ("Chatelier's principle") and by Karl Braun in Germany. The increased resistance to externally forced change takes the form of increasing internal gradients. *This is contrary to what one might intuitively expect in a closed system, where the second law implies that gradients gradually disappear.* It is essentially new physics. Schneider and Kay have proposed the following formal restatement of the second law in open systems (Schneider and Kay 1994, p. 29):

> The thermodynamic principle which governs the behavior of [open] systems is that, as they are moved away from equilibrium they will utilize all avenues available to counter the applied gradients. As the applied gradients increase, so will the system's ability to oppose further movement from equilibrium

An interesting short-cut way of saying this is "nature abhors gradients". The implications of this formulation in evolutionary ecology will be noted from time to time later in this book, see Schneider and Sagan (2005). But nature also creates gradients (by symmetry-breaking).

The relevance of thermodynamic laws to living organisms at the cellular level was, for the first time, examined in the light of (then) modern genetics by **Erwin Schrödinger**, in his lectures at Trinity College, Dublin in February 1943, later published under the title *What is Life?* (Schrödinger 1945). Schrödinger emphasized, that life somehow extracts "order from disorder" (his words) and does so by utilizing "negative entropy". By "negative entropy" he was not referring to information, as some might think nowadays—but to available (free) energy, now known as *exergy* (Schrödinger 1945, pp. 71–72). This idea was developed much further in the 1970s by Ilya Prigogine and others at the Free University of Brussels (Prigogine 1955; Nicolis and Prigogine 1977; Prigogine and Stengers 1984). However, Schrödinger did not suggest an entropy maximization principle. That came later.

In Schrödinger's influential book he emphasized the existence (as previously shown by Mendel) of a "hereditary substance" (the gene), as a chemically stable store of information about the structure and function of some component of the organism. He conceptualized that the gene must be a very large macro-molecule capable of (only) discontinuous changes, consisting of isomeric[15] rearrangements of the component atoms. The isomeric states correspond to different energy levels of the macro-molecule, and a transition from one state to another would be a mutation.

---

[15] An isomer of a molecule is another molecule with exactly the same atoms, hence a re-arrangement.

Here is where quantum mechanics enters the picture: In a "classical" system there is a continuum of isomeric states, but since the discoveries of Niels Bohr (1913) and Arnold Sommerfeld (1916) explaining the observed radiation spectrum of atomic hydrogen, we know that the states of an atom (characterized by energy levels) are discrete, and that transitions between them involve emission or absorption of photons at very precise frequencies. This finding applies also to molecules.

In Schrödinger's words "The energy thresholds separating the actual configuration from any possible isomeric ones, have to be high enough (compared to the average heat energy of an atom) to make the change-over a rare event" (Schrödinger 1945). This stability is attributable to the quantization of molecular energy states. It is the explanation of the observed permanence (resistance to change) of DNA. As he says, "there is no alternative to the molecular explanation of the hereditary substance" (p. 57). Later in his lectures he emphasizes that what distinguishes living organisms is their enigmatic ability to delay or put off the process of decay into thermodynamic equilibrium—maximum disorder (entropy)—*by extracting order from disorder*. In earlier times, he notes "a special, non-physical or supernatural force (*vis viva, entelechy*) was claimed to be operative in the system" (p. 71). Now we know that the genetic stability phenomenon is strictly physical in nature.

Schrödinger's small book contained other insights. One of the most intriguing is this comment:

> A living organism continually increases its entropy—or as you may say, produces positive entropy, and thus tends to approach the dangerous state of maximum entropy, which is death. It can only keep aloof from it (i.e. alive), by continually drawing from the environment negative entropy. ...*What an organism feeds upon is negative entropy* (Schrödinger 1945, p. 72). Italics added.

This is true of life on Earth, as a whole, not only the individual living organisms.

As already mentioned, negative entropy has been thought of by some physicists (and denied by others) as *information* (Brillouin 1950, 1953). Entropy can also be thought of as a measure of disorder W, according to the Boltzmann equation: $S = k\log W$, where S stands for physical entropy. Evidently the reciprocal $1/W$ can be considered a measure of order. It follows that organisms not only "consume" information (as food) to stay alive, they also store structural information in their bodies (partly as genetic material, and partly in somatic material) and they pass it on to their offspring.

The next step was to further clarify the link between order- from-disorder and physical structure. Schrödinger realized that when living cells "feed", in the sense of capturing order from their disorderly environment, it is not only for metabolic purposes. The organism itself is a structure of atoms and molecules. This structure not only controls supportive metabolic maintenance processes like breathing and circulation of the blood. It does so by sending information within the cell by means of the substance known as RNA or (in larger organisms) via nerve pathways or hormones). The structure also "contains" information in the form of genetic material (DNA), which is transmitted by self-replication.

The process of self-replication, which is the core feature of living cells, *cannot occur in closed systems or near thermodynamic equilibrium*, nor can it be understood in terms of standard thermodynamic principles such as are found in textbooks. Indeed the standard variables (like pressure, temperature or density) are not even definable far from equilibrium, where gradients are maximal. Obviously living systems (and most thermodynamic systems of interest) are *open*, not closed or isolated. This means that a "system" of interest can be embedded, with a defined boundary, in a larger system (such as a thermal bath) and that matter and energy fluxes can pass between them. In thermodynamics, this is called an "open system" as opposed to one confined inside a closed boundary.

The next key discovery was the existence of "dissipative structures", far from equilibrium, so-named by their main discoverers, Ilya **Prigogine** and his colleagues at the Free University in Brussels (Nicolis and Prigogine 1971; Prigogine 1976). The word "dissipative" is meant to convey the notion that what the "structure" is doing is maintaining its integrity (or growing and copying itself), in a disequilibrium state, *by utilizing—and dissipating—free energy (exergy) supplied from the outside*. The dissipation can be described as either "exergy destruction" or "entropy production". In fact Prigogine's theory can be labelled "dissipation maximization", meaning that organisms tend to grab (and dissipate) as much of the incoming stream of free energy (exergy) as possible. Prigogine's dissipation theory can be regarded as an extremum principle, comparable to Hamilton's least-action principle in mechanics.

Up to this point, "dissipative structures" in the above sense are just what Alfred Lotka was suggesting, back in 1924 when he wrote about the "law of evolution adumbrated as a law of maximum energy flux" (Lotka 1925, p. 357). As he put it

> ... as long as there is an abundant surplus of available energy running "to waste" over the side of the mill wheel, so to speak, so long will a marked advantage be gained by any species that may develop talents to utilize this "lost portion of the stream". Such a species will therefore, other things being equal, tend to grow in extent (numbers) and this growth will further increase the flux of energy through the system.

In the context of energy conservation, this idea has been recently rediscovered in economics, as the "rebound effect" or the "backfire effect" (Sorrell, Dimitropoulos, and Sommerville 2009).

Similar ideas about evolution have been set forth subsequently by Erich Jantsch (Jantsch 1980), Howard Odum (Odum 1983), Zotin (Zotin 1984), Wicken (Wicken 1987), and others. The idea is that metabolic efficiency at capturing external energy supplies is the key to biological evolution. Prigogine and his colleagues not only described such structures in detailed thermodynamic language, especially in chemical systems. They also emphasized the role of *fluctuations* in non-linear, non-equilibrium systems (with multiple solutions only one of which corresponds to the linearized version), especially near instabilities such as phase-transitions.[16]

---

[16] The best-known phase transitions are melting/freezing or evaporation/condensation. Some of the changes that took place in the first few minutes after the "big bang" can be thought of as phase-transitions.

In such situations local micro-fluctuations can trigger a macro effect (as an avalanche can be triggered by a gunshot or some other loud noise). At such points, an unstable non-equilibrium system can "flip" from a disorganized state with no structure, to another more stable "self-organized" state with a structure maintained by dissipation of exergy from elsewhere. The second law still holds, but low entropy in the structure is compensated by higher entropy outside. Examples in physical systems include Belousov-Zabotinsky (BZ) reactions, self-sustaining vortices (e.g. whirlpools) in turbulent fluids (which can self-replicate), not to mention hurricanes, tornados, the jet stream, sunspots and the red spot on Jupiter (Schneider and Sagan 2005, Chaps. 7 and 8).

Unfortunately Prigogine, et al., could not provide a detailed quantitative thermodynamic explanation of dissipative systems far from equilibrium. Let me be clear: they did everything but the formulation of a complete, quantifiable, theory of non-equilibrium thermodynamics. The problem was that neither classical thermodynamics nor classical statistical mechanics could provide quantitative predictions of behavior in the non-equilibrium region. Even though Prigogine knew, and said repeatedly, that his dissipative systems were far from equilibrium, the theoretical tools available either assumed equilibrium or depended on power-series expansions around equilibrium. The problem, stated another way, was that traditional thermodynamics variables (pressure, temperature, density, etc.) are not well-defined in the regions of interest. Other tools are needed.

Very recent research in statistical physics using Gibbs probabilistic formulation of physical entropy, may have shown a way to overcome the gap. Research by Jarzynski (1997) and Crooks (1999), has proved that the entropy of any system is proportional to a ratio of two probabilities: *the probability of undergoing a given change process divided by the probability of changing in the reverse direction.* This formulation is immediately applicable in chemistry. If the probability of a reaction is low, the entropy will be high, and *vice versa.*

This fact can be used to explore the entropic implications of dissipative processes, as described by Prigogine et al. (Nicolis and Prigogine 1977; Prigogine and Stengers 1984). Jeremy England has derived thermodynamic bounds for such disequilibrium processes as self-replication of molecules (England 2013). He has also shown that, in some cases at least, increasing metabolic efficiency (in utilizing the external exergy supply) provides evolutionary advantages. See also Sect. 3.4.

In the realm of ideas, the "sticking point" about entropy (if I may call it that) is to what extent—if any—the Second Law is a *causal agent*, not just a constraint on reversibility. The fact that living organisms (and some non-living ones) are able to capture useful energy (exergy) from a variety of gradients seems to be crucial (Schneider and Sagan 2005). There are life-scientists, in particular, who argue that an entropy maximization or minimization principle must be involved. (The two are not as incompatible as might be supposed at first sight.)

However, a conclusion of the analysis in this book is that a broader law of irreversible evolutionary "progress" (symmetry-breaking coupled with selection) may be needed. I will discuss it further later (Sect. 5.12), and again in Parts II and III.

3.10 Monism, "Energetics" and Economics

I have no intention of sticking even a toenail in theology, beyond the remark that there was a time when most people believed in a pantheon of Gods, not just one. The ancient Egyptians had a pantheon, as did the Phoenicians, Greeks and Romans and their various rivals; the Hindus still do. It was the Hebrews (Jews) who came up with the idea that there is really only one God (Jehova). This idea was adopted by the early Christians (who were Jews first and Christians later) and later by the Mohammedans. Buddhists, Taoists, Shinto and other oriental religions seem to be comfortable with monotheism, though they do not worry much about the existence or nature of a "supreme being".

Monism is not about money. It is a philosophical doctrine introduced, under that name, in the eighteenth century by **Christian Wolff**. It is the doctrine that the world consists of only one substance, in contrast with "dualism" or "pluralism". The intellectual link to monotheism is obvious. Philosophers differ, of course, as to the nature of that unifying substance. **Baruch Spinoza** took it to be the Deity. **Georg Hegel** thought that it was mind or spirit. This book argues that the substance of everything is energy.

As regards energy in the social sciences, one starting point, so to speak, was a recommendation by George Rankine in his 1855 paper, mentioned already. Across the ocean, Joseph Henry, the co-discoverer of magnetic induction and first Secretary of the Smithsonian Institution, said in 1879 that "the physical labor of man . . .is ameliorated by expending what is called power or energy". Rankine's ideas were promoted in a popular German textbook published in 1887 by a school teacher, **Georg Helm** *Die Lehre von der Energie* (Helm 1887). Helm extended the idea that energy is the central "substance" in the universe, thus creating a link with philosophical "monists", such as Haeckel.

Ernst Haeckel was the most influential of the scientific monists. He was a zoologist, anatomist, promoter of Darwin's theory of evolution, author of the recapitulation theory ("ontogeny recapitulates phylogeny") and "biogenic theory". He promoted monism as a new religion, based on insights from evolutionary biology. He also saw monism as the key to the unification of ethics, economics and politics as "applied biology".

Another nineteenth century visionary from the social sciences, was the Ukrainian socialist, Sergei Podolinsky who tried to reconcile Marx's then-popular labor theory of value with the new developments in thermodynamics (Cleveland 1999). Podolinsky pointed out in articles (and in letters to Friedrich Engels) that the implicit assumption of Marx (that "scientific socialism" could overcome all material scarcities) was inconsistent with the laws of thermodynamics, as developed by Carnot and Clausius. His objections are still pertinent today.

Wilhelm Ostwald was the leading chemist of the late nineteenth century and early twentieth, especially for his work on catalysis. He was also one of Boltzmann's most persistent critics. Ostwald published frequently in *The Monist*, a journal devoted to scientific monism but he personally created "energetics". He championed the work of other scientists who saw the world from an energetics perspective, including Gibbs, Le Chatelier, and Duhem. But, most important, Ostwald pushed the application of energetics into the realms of economics and sociology, just as the intellectual foundations of the movement were being challenged in physics.[17] In the decade after 1900 energetics reached its apogee in terms of Ostwald's fame and public acceptance (Ostwald 1907, 1909). Since then, the subject has faded from the stage.

Interestingly, it was a colleague of Ernest Rutherford, Frederick Soddy who took the energetics program of Ostwald further into the realm of economics. Soddy, who is now called a "neo-energeticist", considered energy to be the basis of all wealth—and let me say at the outset, that I agree with him. Here is an early quote:

> Energy, someone may say is a mere abstraction, a mere term, not a real thing. As you will. In this, as in many other respects, it is like an abstraction no one would deny reality to, and that abstraction is wealth. Wealth is the power of purchasing, as energy is the power of working. I cannot show you energy, only its effects ... Abstractions or not, energy is as real as wealth—I am not sure that they are not two aspects of the same thing. (Soddy 1920, pp. 27–28)

Soddy later published several books on energy, wealth and money, starting with *Cartesian Economics* in 1922. In this book he unambiguously identifies the inanimate sources of life:

> ... life derives the whole of its physical energy or power, not from anything self-contained in living matter, and still less from an external deity, but solely from the inanimate world. (Soddy 1922 p. 9)

So much for physics. Soddy introduced energy to economics, long before any economist. The following quote is taken from a later book by Soddy:

> Debts are subject to the laws of mathematics rather than physics. Unlike wealth, which is subject to the laws of thermodynamics, debts do not rot with old age and are not consumed in the process of living. On the contrary they grow at so much percent per annum, by the well-known laws of simple and compound interest ... as a result of that confusion between wealth and debt we are invited to contemplate a millennium where people live on the interest of their mutual indebtedness. (Soddy 1933, pp. 68–69)

Wall Street's activities in recent years seemed to have brought that curious millennium much closer to reality—until the financial collapse of 2008.

The main point worthy of emphasis here is Soddy's uncompromising identification of wealth creation as being subject to the laws of thermodynamics. Evidently he identified wealth creation with use of available Gibbs free energy (or *exergy* in

---

[17] Ostwald was apparently close to caloric theory, insofar as he conceptualized energy as a kind of substance, rather than as an abstract "state" as it was viewed by his contemporaries. (Even today, the "substance" view of energy is commonplace among non-experts). Planck tried to straighten him out on this misconception, but without success. At a conference in 1895 his views were violently attacked by Boltzmann and Nernst, as well as by Planck. Ostwald also disliked the emerging atomic theory and accepted it only reluctantly, and not until long after it had been well established in physics (Mirowski 1989, pp. 57–58).

modern terminology). The problem, for many economists who don't read carefully, is that Soddy's work is interpreted as an "energy theory of value". This is a red flag to modern neo-classical economists, who view "value" as a market-determined phenomenon, based on "revealed preferences" expressed in the marketplace (Debreu 1959; Alessio 1981).[18] (Here I agree with the economists perspective.) In Part III I discuss the energy theory of value later (without endorsing it). There is no inconsistency between Soddy's insistence on the importance of (free) energy on value-creation, and the understanding of economic value as a consequence of "revealed" preferences.

The last sentence in the third Soddy quote, above, may strike many readers as "over the top", so to speak. But it is time to reconsider, given our current understanding of the role of money as numbers in a computer, in a world where gold has a market price in dollars like any other metal and—since 1971—the gold standard no longer applies. Money today is simply credit and the obverse of debt. The quantity of debt (hence credit) to be created is theoretically determined by a few central banks, especially the US Federal Reserve Bank, but in reality banks can—and do—create debt by creating credit. Debts are liabilities for the borrowers but assets to the lenders, and hence part of the money supply. Increasingly it is lending and borrowing that finance both production and consumption and thus drive economic activity (GDP).

In his last book, *The Role of Money* Soddy made some of these points (Soddy 1935). As an outsider, and a chemist, his work was ignored at first, and later derided, by economists. However, it has received some well-deserved attention in recent years, by at least one economist (Daly and Herman 1980). Since Soddy's energy-based critiques of economics (which were ignored or derogated by mainstream economists), several non-economists have also written about the importance of energy in biology.

An important scientific contemporary of Soddy was Alfred Lotka, whose book *Elements of Mathematical Biology* is now a classic (Lotka 1925). Lotka was the pioneer of nutrient metabolism, including the carbon cycle, the nitrogen cycle, the phosphorus cycle, and a few others. He attempted to explain biological evolution in terms of energetics, more specifically as "a general scrimmage for available energy" and a maximal principle underpinning all bio-physical activity. (He later backed off the maximum principle.)

Lotka postulated that Darwinian natural selection favors species "that maximize the rate of energy flux (power) through their bodies", so Lotka's energetic hypothesis—which he attributes to Boltzmann—is one form of maximization of the rate of entropy production (Lotka 1922, 1945). Lotka's thesis, as applied to natural selection, remained controversial for many years, because the state of thermodynamics was still not far enough advanced. However, since the work of Onsager (Onsager

---

[18] The neo-classical view contrasts strongly with the earlier "classical" theory of value, which viewed value as an inherent property of goods, based on labor inputs. The classical theory is somewhat analogous to the caloric theory of energy.

1931a, b) and, especially, by Prigogine and his colleagues (Prigogine 1955; Nicolis and Prigogine 1971), Lotka's idea—or something like it—is now receiving more attention from ecologists. In his synthesis, published much later than his 1922 paper, Lotka commented as follows:

> One is tempted to see in this one of those maximum laws which are so commonly found to be apt expressions of the course of nature. But historical recollections here bid us to caution; a prematurely enunciated maximum principle is liable to share the fate of Thomsen and Berthelot's chemical 'principle of maximum work. (Lotka 1956)

I concur in that caution, even though I cannot help but notice the wide range of possible applications of such a principle, perhaps in terms of "maximum gradient minimization"—as noted in various places in this book. I reiterate that the phrase "nature abhors a gradient" is from Schneider and Kay (1994).

The central importance of energy flows and metabolism in biology and ecology has been mainstream since Lotka. A short list of ecologists who have ventured into the contentious field of ecological economics was provided above, at the end of Sect. 3.7. More recently, a few non-economists, from geographers and ecologists to physicists and a few heterodox economists with engineering or scientific training, have also challenged the economic mainstream e.g. Hannon (1973, 1975, 2010), Hannon and Joyce (1981), Kümmel (1982, 2011), Costanza and Herendeen (1984), Cleveland et al. (1984), Arrow et al. (1995), Costanza et al. (1997), Kümmel et al. (2010). Other names will be cited later at appropriate points.

I have not attempted to summarize the core ideas of economics in this chapter, primarily because those core ideas from the past two centuries concern human behavior in markets and market dynamics without any causal link to energy (exergy) flows, entropy, or irreversibility. Information flow (via prices), adaptive cycles (based on ecology) and energy flow in ecosystems are the only evident—and very recent—connections between economics and the material discussed in Part I of this book. I will discuss the linkage of those topics to standard mainstream economics and finance in Part III.

Chapter 4
The Cosmos, The Sun and The Earth

4.1 Astronomy and Astrophysics

Parts of this could logically be included in Chap. 3 under *Ideas*. However, the story of ideas underlying current theories of cosmology are too complex to reduce to a few names and dates, although I will start with one. For most modern astronomers, the logical starting point of modern cosmology was Vesto Slipher's discovery (1912) of the "red shift" (Z) of the in absorption lines for hydrogen (and other elements) in starlight from distant galaxies. The red shift, toward the infra-red part of the spectrum, indicates slightly lower frequencies (longer wavelengths). It can be explained in terms of relative motion, like the "Doppler effect". (The Doppler effect is what makes the sound of an approaching siren a little higher in pitch than the sound of a siren moving away). By 1920, 36 of 41 galaxies for which data had been obtained, were seen to be moving away from our galaxy, the Milky Way. Current data puts the fraction much higher.

It was suspected long ago that our galaxy, the "milky way", is shaped like a "grindstone" (a disk). This hypothesis was published by one Thomas Wright in 1750. Better telescopes have revealed six distinct spiral arms. We are on the inner edge of the fourth (Orion) spiral arm. See Fig. 4.1. The rotating disk is about 100,000–120,000 light-years in diameter (with a "halo" outside), but the thickness of the disk is only about 6000 light years. Within the galaxy there are anywhere from 100 to 400 billion stars. Outside, there is nothing at all for a long distance. Our galaxy is like a small island of dry land (so to speak) in a large ocean of nothing, at least nothing visible. The next nearest galaxy (Andromeda) is 700,000 light years away. Our isolation is not unique. Galaxies, in general, are separated by even greater distances (of the order of 10 million light years).

© Springer International Publishing Switzerland 2016
R. Ayres, *Energy, Complexity and Wealth Maximization*, The Frontiers Collection,
DOI 10.1007/978-3-319-30545-5_4

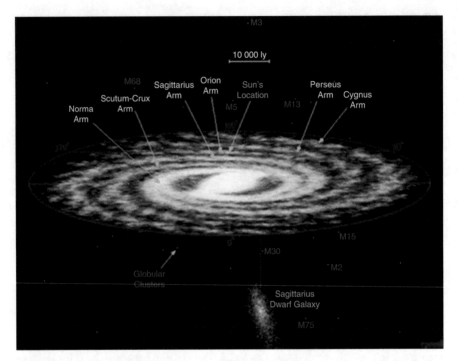

Fig. 4.1 The Milky Way Galaxy. *Source*: http://www.atlasoftheuniverse.com/galaxy.html

Starting in 1920 **Harlow Shapley** determined their actual velocities, using the variable Cepheid star periodicity data as a reference to determine their red shifts.[1] In 1924 **Edwin Hubble** (using Cepheid data from Henrietta Levitt) measured the distance to nine galaxies and found them to be very far away, compared to earlier notions of astronomical distance. Distances began to be expressed in "light-years", meaning the distance light travels in one of our Earthly years. By comparison, the distance between the Earth and the Sun is only a few "light minutes".

The Slipher-Shapley-Hubble observations, and others that followed, found that the "red shift" of starlight increases with distance. This means that virtually all galaxies in the sky are moving away from our galaxy, and—as it turns out—from each other.[2] As telescopes have improved since the 1920s and astronomers have

---

[1] Cepheid stars have time-varying brightness that is closely related to their pulse periodicity. The regular pulsation is now understood to be due to a dynamic contest between temperature and light pressure vs. gravity. A "halo" of ionized helium becomes doubly ionized (and opaque to light) when the star is compressed and hottest, but only singly ionized (and transparent) when it expands and cools. As the star cools energy (as light) escapes and the star cools further, until gravity takes over and recompresses it.

[2] There is one exception: it was discovered in 1977 that our galaxy, the "milky way" is actually moving toward the Andromeda nebula at the relative speed of 600 km/s (Smoot and Davidson 1993). This appears to be due to gravitational attraction. It is evidence that the large-scale structures of the universe are more "organized" than previously thought.

been able to "see" further and further into the past, they have also observed that the red shift Z in light spectra from the more distant galaxies is consistently greater than the red shift from the nearer galaxies. This "Doppler effect" implies that the universe is expanding. This *recession* velocity, as it is called, is consistent with Einstein's original (1915) version of general relativity. More recent data on quasi-stellar objects (quasars) from the distant past suggests that the recession velocity is actually increasing. However this data is not without challenges.

The first scientist to postulate a single point in time for the beginning of the universe was **Georges Lemaitre**, a Belgian astronomer and priest (Lemaitre 1950). Extrapolating the observed motions back in time led to the conclusion that the expansion must have started as a giant explosion. Lemaitre called it the "primeval egg". It has since been relabeled "the big bang" or BB by George Gamow (Gamow 1947). He also derived a formula for the age of the universe, based on the rate of expansion, and calculated the so-called Hubble constant (1927) 2 years before Hubble. At first the BB was thought to be between 25 and 20 billion years ago. (The current estimate, based on the recent WMAT survey in 2010, is 13.77 billion years.) In 1929 Hubble arrived at the same conclusions as Lemaitre and published in a more prestigious journal.

In 1931 **Fritz Zwicky** noted an astronomical anomaly when he was studying the Coma cluster of galaxies, 300 million light years away. He was able to calculate the mass of the stars by methods based on Eddington's fundamental work on star luminosity (Eddington 1926). But Zwicky saw that the galaxies were rotating around each other so fast that they should have been gravitationally unstable. He calculated that the mass needed to keep the cluster from flying apart was 20 times greater than the mass he could "see" based on the mass-brightness relationship.

Zwicky was thus the first to postulate "dark matter", but his work was ignored at the time. Other astronomers later arrived at similar conclusions, but were also ignored. What turned the tide was work in the 1970s by Vera Rubin with Kent Ford. They analyzed data from over 200 galaxies and found the same results that Zwicky had noticed (Rubin 1997). Other experiments have confirmed the overall conclusion that luminous matter does not have enough gravitational mass to keep the galaxies from flying apart.

Occasional extremely bright stellar outbursts have been observed for centuries, but in 1934 Fritz Zwicky postulated that giant stars would die as *supernovas* that would subsequently collapse into neutron stars, emitting "cosmic rays". A super-nova emits as much radiation in a few days as a whole galaxy. They are not common. The last one in our galaxy was observed by Kepler (as well as Chinese and Korean astronomers) in 1604. But there are billions of galaxies. Zwicky began searching for them (and coined the term). Starting in 1935 he eventually found 120 examples before he died in 1974.

Zwicky and Walter Baade also predicted the existence of *neutron stars* from former supernovae. (They have since been found.) The supernovae exhibited a common luminosity curve, with a peak very nearly at the same frequency. This suggested a common nuclear process in all of them. Baade therefore proposed the use of supernovas as standard "candles" for purposes of astronomical distance measurements, since distance is simply proportional to the absolute intensity (or brightness) of the "candles". That astronomical tool, in turn, has enabled astronomers to calculate distances much more accurately than previously.

Intense radio frequency radiation was observed in many parts of the sky during the 1950s. In 1965, in the course of research related to satellite telecommunications, Arno Penzias and Robert Wilson discovered microwave background radiation, from all directions in the sky (Penzias and Wilson 1965). This microwave radiation corresponded to a black-body at a temperature between 2.5 and 4.5 K (since revised to 2.7 K). This discovery caused a sensation in the cosmological community (and a Nobel Prize for the discoverers) because it seemed to confirm the reality of the modern Big Bang theory, suggested by George Gamow in 1946 and expressed mathematically by Gamow and his student Ralph Alpher who did most of the work, while Bethe, listed as co-author as a joke, did none: (Alpher et al. 1948). Previously influential theories of "continuous creation" were already in doubt because they could not explain the Lemaitre-Hubble inference of expansion from a common source, or the evidence of isotope abundances (Bondi 1960; Hoyle 1950).

Astronomical radio sources were gradually associated with light sources, in the mid-1960s. As telescopes improved, some distant sources were found to have very large redshifts, implying that they are extremely far away. To be so bright, while so far away, it was necessary to assume some extraordinary physical process, probably dating from the early universe. The favorite hypothesis among astronomers is that some very large "black hole" is gravitationally attracting and accreting ("swallowing") vast quantities of mass. The radiation was not from inside the black hole, of course, but from the accretion disk around it. The energy of the radiation is presumed to be from the gravitational acceleration of debris within the surrounding disk.

These objects, discovered by Maarten Schmidt in 1963, were originally called "quasi-stellar objects" (QSO) or radio sources. The word "quasi" meant that the spectral lines in the light from these sources are much broader than "standard" stellar spectral lines. They are now known as "quasars", and they are, by far, the brightest objects in the sky. About 200,000 of them have been identified by the Sloan digital sky survey (SDSS) as compared to billions of galaxies and trillions of stars http://cas.sdss.org/dr5/en/proj/advanced/quasars/query.asp. The quasar in the constellation Virgo (3C273) has a luminosity (brightness) no less than 4 trillion times our sun.

Quasars were instrumental to the discovery of "gravitational lenses", first observed in 1979. The image of a quasar behind the gravitational mass of an intervening galaxy can be distorted. A survey of all known gravitational lenses, the Cosmic Lens All Sky Survey (CLASS) has surveyed 10,000 radio sources. http://www.aoc.nrao.edu/~smyers/class.html. It has found evidence of "lensing" (gravitational magnification, like a lens) for 1 out of each 700 quasars, or around 280 examples of lensing. It is this data, combined with statistics about "ordinary" galaxies (of which there are many billions), that seems to imply that around two thirds of the energy in the universe is "dark", contributing to the expansion of space, but otherwise not interacting with ordinary mass.

A crucial feature of quasars, already mentioned, is that they tend to have very large redshifts. The redshift is usually denoted Z where Z is the ratio of observed ("redshifted") wavelengths (based on recognizable spectral lines) to "normal" wavelengths minus 1. Without going into unnecessary detail, large Z is assumed to imply that the emitter is moving away from us at a very high velocity. A new survey, the Baryon Oscillation Spectroscopy Survey (BOSS)) is specifically looking for quasars with high Z values. http://mnras.oxfordjournals.org/content/456/2/1595.abstract.

Because the universe is expanding, the Hubble constant and the velocity then tell us how far away the emitter is located. These data are the basis for the standard Hubble-based estimates of the quasar's distance from the Milky Way. Of the first 50,000 quasars identified by the Sloan digital sky survey over 5700 (11.5 %) had Z-values greater than 2. Now that the Sloan Digital Sky Survey (SDSS), in its 15th year, has found over 200,000 quasars, mostly farther away, it seems likely that the fraction with Z-values greater than 2 is even higher. N.B. as will be noted later, this fact is essentially inconsistent with Einstein's general theory of relativity (EGR), which does not allow high values of Z.

In 1967 Franco Pacini predicted that a rotating neutron star with a magnetic field would emit a beam of "synchrotron radiation" along its magnetic axis (which need not be the same as its axis of rotation) (Pacini 1967). A few months later Jocelyn Bell and Antony Hewish found a pulsating star or *pulsar*, which produced a gamma ray flash regularly every 1.33 s. Hewish was awarded a Nobel Prize in 1974 (with Martin Ryle) for this discovery, but the actual discoverer, Jocelyn Bell, was not.

Pulsars seem to be explained by the Pacini model. To date 1800 of these objects have been found in the sky, and the theory is fairly well advanced. http://www.universetoday.com/25376/pulsars/. The radiation may be visible light, X-rays or even gamma rays and the periodicity can be seconds or milliseconds. The intensity of pulsar radiation decays in times of the order of decades.

Also in 1974 Joseph H. Taylor and Russell Hulse discovered a binary neutron star. This phenomenon led to the first positive, albeit indirect, evidence detection of gravitational waves, for which they received the Nobel Prize in 1993. However in January 2016 a more convincing discovery was announced, based on results from a collaboration of several institutions known as the Laser Interferometer Gravitational wave Observatory (LIGO). The current paper claims detection of a gravitational disturbance due to the collision and merger of two large black holes, totaling 65 solar masses, during which three solar masses disappeared in the form of gravitons. The paper was published in Physical Review Letters and signed by 1000 authors. It reflects a project that was conceived back in 1975, and led since then primarily by theoreticians Kip Thorne of Cal Tech and Rainer Weiss of MIT, plus experimentalist Ronald Drever, at a total cost of more than $1.1 billion from the National Science Foundation. It will undoubtedly be accepted.

In 1979 bursts of gamma rays were traced to a neutron star with a very strong magnetic field. Now 21 of these so-called "magnetars" have been found in the Milky Way, with 5 more awaiting confirmation. http://earthsky.org/space/magnetars-most-powerful-magnets-in-the-universe. Magnetars can exhibit field strengths of the order of 10^{15} G, as compared to the Earth's mere 0.5 G. A theory of these objects was offered by Robert Duncan and Christopher Thomson in 1992. They have rather short active lives, in cosmological terms, around 10 million years. The bursts of gamma rays seem to result from crustal "quakes" (analogous to earthquakes) as the magnetar shrinks. But it is very difficult to account for such fields if we assume that they are generated by electric currents in the neutron star based on Biot-Savart relationships. Electric currents in a neutron star? Would that be a version of superconductivity? But superconductors expel magnetic fields.

In 1986 Valerie de l'Apparent, Margaret Geller and John Huchra published results of a survey of 1100 galaxies that appeared to be on the surfaces of large empty "bubbles". They found extreme inhomogeneities that could not be reconciled with existing theories (de Lapparent et al. 1986). More such unsettling discoveries came rapidly. These include stars, galaxies, clusters, super-clusters, ribbons, shells, and so forth. These objects are quite concentrated in some regions of space, whereas other regions are enormous voids. For instance, the work of Tully and Fischer has showed that, of 5000 galaxies nearer than 120 million light years from the solar system, almost all are strung along "filaments", like Christmas lights (Tully and Fischer 1987). The filaments are less than 7 light years across, which is tiny by inter-galactic standards. The rest of space is essentially empty of visible or detectable matter.

When the galactic mapping project was extended by Tully et al. to a distance of 1.5 billion light years, enclosing around two million galaxies, the filament pattern emerged again, as "ribbons" consisting of dozens of "super-cluster" filaments. Tully has identified five "super-cluster complexes", each containing millions of trillions of individual stars. The density of mass in one of the super-cluster complexes is 25 times greater than the density outside.

In 1989 a larger structure called the "Great Wall" was discovered by Geller and Huchra (1989). This sheet of galaxies stretches 700 million light years in length, 200 million light-years across and only 20 million light years thick. The "great Wall", consisting of over 5000 individual galaxies, corresponds to just one of the five super-cluster complexes found by Tully and Fischer.

Even larger structures, like concentric "shells" (or Russian dolls) have been found since then by T. J. Broadhurst et al. (1990). They have looked even farther into the past, out to 7 billion light years. The structures are organized in the form of filaments, thin bands, and walls, surrounding large voids. The matter density in the voids is less than 10 % of the density of the visible universe as a whole, while the filaments and bands are much denser.

Recent work, based on evidence from one of the few "untouched" meteorites ever found, indicates that the protoplasmic disk of dust and hydrogen "collapsed" into the sun-and-planets structure much faster than was previously thought. This seems to have been due to the presence of very strong magnetic fields, from 5 to 54 µT (up to 100,000 times stronger than the field in interstellar space). What was the source of this powerful magnetic field? Nobody knows.

Recent (since the 1980s) observations of the red-shift in starlight and quasars, interpreted according to Einstein's general relativity (EGR) theory, have yielded an unexpected result: it seems the Hubble expansion of the universe is accelerating.

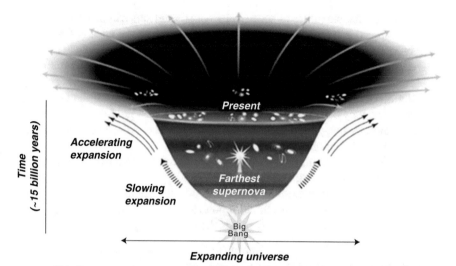

Expanding universe

This diagram reveals changes in the rate of expansion since the universe's birth 15 billion years ago. The more shallow the curve, the faster the rate of expansion. The curve changes noticeably about 7.5 billion years ago. when objects in the universe began flying apart at a faster rate. Astronomers theorize that the faster expansion rate is due to a mysterious, dark force that is pushing galaxies apart.

Fig. 4.2 The acceleration of expansion: Dark energy. *Source*: https://commons.wikimedia.org/wiki/File:Dark_Energy.jpg#/media/File:Dark_Energy.jpg

That acceleration can (apparently) only be explained by postulating a sort of "dark energy" that permeates all of space and acts like negative pressure i.e. a propulsive force like anti-gravity. An artist's view of current dominant theory is shown in Fig. 4.2.

The Cosmic Background Explorer (COBE) was a NASA satellite experiment initiated, in part, to verify (or disprove) early evidence of an unexpected "cosmic quadrupole", based on evidence from instruments on high altitude balloons. The COBE satellite was launched in 1989 and remained in orbit 3 years. It measured small temperature differences between different parts of the sky. The major result was that significant differences ("wrinkles in space time") were observed (Smoot and Davidson 1993).

The follow-up on COBE is the more detailed cosmic temperature map produced by NASA's Wilkinson Microwave Anisotropy Probe (WMAT), led by Charles Bennett and a huge team of astronomers and other technicians. The satellite was "parked" from 2001 to 2010 in orbit at the second Lagrange Point, 1.2 million miles from the Earth's "thermal noise" in the direction away from the sun. The thermal sensitivity of the radiometers was increased (to 10^{-6} K) with a much narrower angular resolution than COBE. The major results of this experiment are listed below:

- A fine resolution (0.2°) microwave temperature map of the cosmos.
- Measured the age of the universe accurately (13.77 billion years) to within 0.5 %.

- Determined the curvature of space to be flat (Euclidean) within 0.4 %.
- Determined that particles of ordinary matter (baryons) make up only 4.6 % of the universe.
- Determined that "dark matter" (not made of atoms) constitutes 24 % of the universe.
- Determined that "dark energy", in the form of a cosmological constant—perhaps more accurately called "anti-gravity"—constitutes 71.4 % of the universe.
- Discovered by mapping polarization that the universe was "re-ionized" earlier than previously thought.
- Discovered that the amplitude of density variations of the universe is greater at large scales than at small scales.
- Discovered that the distribution of amplitude variations across the sky is a symmetric bell curve, consistent with predictions of the "inflation" hypothesis. Hence the WMAP experiment supports the inflation hypothesis.

The WMAT experiment also determined the composition of the universe in its infancy (when it was 380,000 years old) as follows:

- atoms 12 %
- neutrinos 10 %
- photons 15 %
- "dark matter" 63 %
- "dark energy" 0 %

See Fig. 4.3. This comparison is curious, not least because it allows no share in the present universe for photons or neutrinos, while "dark energy" was somehow "created" long after the BB. This result raises more questions than it answers.

The observed density of the universe (based on visible light) turns out to be only 0.02 of the theoretical density that would have been required to account for "flatness" of space (i.e. the fact that it neither collapsed nor exploded during the first micro-second after the BB.) This discrepancy led to the current theory of "dark matter": i.e. gravitational mass that does not emit photons or otherwise interact with most kinds of "conventional" matter (protons, neutrons, electrons). The current situation (2016) is that only a third of the necessary mass (to account for flat-ness) seems to be accounted for.

One point of interest is the theoretical possibility that the dark matter consists of black dwarves or neutron stars, i.e. the "ashes" of dead stars. This seems to be ruled out by the fact that the small stars that do not explode—hence end as black dwarves—tend to burn slowly so most of them should still be visible as white dwarves. The only way a large majority of the stars created soon after the BB could be "dead" would be if they were all in a fairly narrow size-range, i.e. several solar masses, but not enough to become supernovae. The mystery is still unsolved, although there is a lot of speculation about weakly interacting massive particles (WIMPS) and other interesting objects that need not be listed here, since none of them have been seen.

Fig. 4.3 Estimated distribution of matter and energy in the universe today and when the CMB was released. *Source*: https://en.wikipedia.org/wiki/Dark_matter

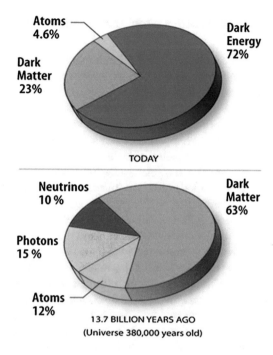

The mystery of "dark matter" (or anti-gravity) and its partner, "dark energy" remains unresolved, as yet. No plausible candidates for the missing mass have been observed up to now, despite the discovery that neutrinos do have mass (2015 Nobel Prize to Takaaki Kajita and Arthur McDonald). N.B. this is contrary to the "Standard Theory" of high energy physics (quantum chromodynamics, or QCD). As regards "dark matter", many possibilities have been suggested, and most of them to date have been dismissed for different reasons, with the possible exception of those neutrinos. The current estimate is that neutrinos might contain as much mass as stars. That would only account for a small fraction of the dark matter. But it is now known from the Kajita and McDonald's experiments that neutrons actually "flip" from "flavor to flavor" during flight. This behavior is not accounted for by the Standard Model. Perhaps, when the SM is suitably revised, it will turn out that neutrinos were more massive than assumed in the past, or—being massive—that most of them are attracted to neutron stars.

Otherwise the theory of dark matter is at an impasse. That leaves hypothetical "weakly interacting massive particles" (WIMPs). However, in a paper not yet published, Nassim Haramein and Val Baker suggest that the "zero-point" energy of the vacuum might offer an answer (Haramein 2016). Could there be a role for tachyons? There are certainly speculations along that line. Or could there be a problem with Einstein's general relativity (EGR) theory, which is the source of the flat universe problem? I think that may be the case. Read on.

4.2   Quantum Mechanics and Relativity

In the year 1905 Albert Einstein published three important papers. All three were revolutionary. One described Brownian motion, the apparently random motion of very small particles being bombarded (as we now know) by other small particles—atoms and molecules. Since atomic theory was in its infancy and many important physicists (such as Mach, Planck and Ostwald) did not believe in atoms, it was a key step in the development of statistical mechanics and atomic theory.

The second paper took the step that Max Planck was too hesitant to make in 1901, arguing that electromagnetic waves of frequency ν can also be seen as particles, called *photons*, with energies $h\nu$, where h is the so-called Planck constant. This quantization of light, as photons, correctly explained the photo-electric effect, thus earning Einstein his (only) Nobel Prize. (He deserved several more).

In the same year, Einstein also published a third paper to explain the apparent inconsistency between Newtonian mechanics and Maxwell's theory of electromagnetism (Einstein 1905). It postulated that the laws of physics must be the same in all "inertial" (non-accelerating) "frames of reference"—in a flat Euclidean space—and that *the speed of light must be a constant in every such frame.* The term "frame of reference" can be confusing, but you can image an x-y chart that is attached to a trolley moving on a track in Zurich in reference to one that is standing still. His point was that no matter how fast the trolley moves, even near the velocity of light, the laws of physics must look the same on both trolleys.

This basic assumption was (amazingly) enough to derive the *Lorentz transformation*, which described how measurements of distance and clock time vary between observers traveling at different velocities (viz. the *length contraction* and *time dilation*) without assuming the *luminiferous aether* that Lorentz and Poincaré (and others) had tried unsuccessfully to save (Lorentz 1904). Einstein's "special theory of relativity" had some other remarkable implications, including relativistic mass and the astounding relationship between energy E and mass ($\mathbf{E} = \mathbf{mc^2}$) where c is the velocity of light. (Note the similarity to the Newton-Leibnitz formula for kinetic energy $\mathbf{E} = \mathbf{mv^2}$, where v stood for momentum or *vis viva*.) Note also that $\mathbf{E} = -\mathbf{mc^2}$ is also consistent with the theory. Paul Dirac made use of in 1928 when his quantized relativistic equations of motion predicted positrons (Dirac 1928).

The equation $\mathbf{E} = \mathbf{mc^2}$ showed that matter and energy are truly inter-convertible; this means that *matter and energy are different aspects of the same essence.* This formula also led to the remarkable idea that the substance of atomic particles, electrons, protons (and neutrons, which had not yet been discovered) are also composed of pure energy in a very condensed form. (This, in turn, means that natural wealth consists of structures of condensed—"frozen"—energy).

The law of Conservation of Energy, which is the First Law of Thermodynamics, is really the law of conservation of <u>mass-energy</u>. In quantum theory as applied to the world of "strong" and "weak" nuclear forces, the interchangeability between mass and energy is crucial: particles can be annihilated by anti-particles, nucleons can be transmuted into other nucleons, emitting or absorbing radiation (nuclear fission and fusion), and so on.

Now back to quantum mechanics. The first major accomplishment of quantum theory, in the next few years after publication of Einstein's special theory, was the explanation of the Balmer-Ritz formula for the spectral emissions of the hydrogen atom. This was determined for the first few spectral lines by **Niels Bohr** (1913). Bohr postulated that the emissions were light quanta (photons) associated with *electrons jumping between discrete orbital states*. Bohr also formulated the "principle of complementarity" meaning that entities can display apparently contradictory aspects (e.g. particle vs. wave). He was later a very influential participant (and umpire) in most of the discussions and arguments about quantum mechanics, in the mid-1920s.

The next breakthrough was **Louis de Broglie's** recognition (1923) that Bohr's quantization of atomic "orbits" can be explained by the particle-wave duality: de Broglie realized that *a stationary orbit can also be regarded as a spherical standing wave, with a particular wavelength and a corresponding frequency*. This insight explains why only certain "orbits" are allowed. It is because they correspond to discrete numbers of standing waves. In effect, Planck's idea for explaining blackbody radiation (1901) could then be adapted to the excited states of individual atoms. But de Broglie's theory of orbitals as spherical "standing waves" was still incomplete. It did not explain the valence shells of heavier atoms.

That was left for **Wolfgang Pauli's** "exclusion principle", one of the greatest insights of modern physics. The exclusion principal says that no two fermions can occupy the same quantum state. *Fermions* are particles that occupy space, and exclude others, in contrast to *leptons*, which are super-posable on each other. This means that just two electrons with opposite "spin" directions constitute the innermost "shell" around the nucleus. Each "spin" direction is a quantum number, but it shouldn't be thought of like the spinning of a top.[3] Pauli's exclusion principle finally explained the spectrum of the hydrogen atom, including excited states. But the exclusion principle is generalizable to other situations, including quantum chemistry. It forced physicists to accept that particles and waves are the same in essence, just as mass and energy are the same in essence. The Pauli principle plays a crucial role in cosmology.

There was a veritable explosion of important theoretical developments in quantum theory during the years 1925–1927. The key names (in addition to Bohr and Einstein) were **Werner Heisenberg, Max Born, Wolfgang Pauli, Erwin Schrodinger, Paul Dirac** and **Enrico Fermi**. David Hilbert, Hermann Minkowski, **John von Neumann** and **Hermann Weyl** were the mathematicians who provided

---

[3] By "quantum numbers" he meant electric charge and spin-direction, because each of them must be conserved in reactions. The concept of quantum numbers has since been extended to include other "either/or" variables, such as parity and "iso-spin". Iso-spin is related to a conservation law applicable to strong nuclear forces and quarks.

the new analytical tools. Weyl, in particular, invented the idea of "gauge theory", which is fundamental to high energy physics today.

Heisenberg is particularly associated with the so-called "uncertainty principle"

$$\Delta x \cdot \Delta p = h/2\pi,$$
$$\Delta E \cdot \Delta t = h/2\pi$$

and Δ is the symbol that means 'an increment' (of uncertainty) of the variable. Schrödinger introduced the non-relativistic quantum-mechanical wave equation, borrowed from the Hamiltonian expression in classical mechanics, viz.

$$H \Psi = E\Psi,$$

where H is the "Hamiltonian" from classical dynamics, expressed as an operator, Ψ is a probability function or "wave function" and E is the energy. Max Born introduced the probabilistic interpretation of wave-functions, which made them much more comprehensible to ordinary folks.

Paul Dirac adapted the Schrodinger equation to relativistic electrons by using a different expression (from special relativity) for the Hamiltonian. It was Dirac's analysis that predicted the existence of *positrons* (by reformulating the equation to allow for $E = -mc^2$) while Fermi predicted the *neutrino* to explain beta decay and nuclear fission, based on Pauli's suggestion that a very small electrically neutral particle must be involved.

Nuclear physicists (e.g. Hans Bethe and Carl Friedrich von Weizsäcker in 1938) soon included nuclear fusion of light elements as the energy source in stars and uranium fusion as a possible source of energy (and explosive power) on Earth (e.g. Otto Hahn and others in 1939). Both nuclear weapons and nuclear powered electric generating plants are applications of this knowledge.

During the war years (1939–1945) several of these scientists were involved in the development of nuclear weapons. In any case, pure research ceased. In the postwar era numerous fundamental theoretical contributions have been made, first in quantum electrodynamics, or QED (Schwinger, Feynman, Dyson), and subsequently in "electro-weak" theory of Yang, Lee, Wu, so-called grand-unification theories (GUTs) and the current "standard model" (SM). Most high energy theorists claim that this theory now fully accounts for three of the four fundamental forces of nature (electro-magnetism, the "electro-weak" force and the "strong" force). However there are doubts and doubters. The fourth force, gravity, is still not fully integrated into the theory. I will have more to say about that in a later section.

Before moving on, I need to revert to the "zero-point" energy (*Nullpunktenergie*) or "vacuum energy", mentioned above in connection with "dark energy". The idea originated with Planck, who built on his own work on quantized black body radiation, starting in 1901. By 1913 he had calculated that the average energy of a quantized oscillator at very low temperatures could not be zero, nor could it be exactly $h\nu$ (where ν is the frequency of the oscillator) and h is Planck's constant. It would be non-zero and would differ from **hν** by a term similar to Planck's original correction to the classical Rayleigh-Jean equation for black body radiation.

A few years later, Walter Nernst suggested that Planck's hypothetical quantum oscillators might populate all of space. In 1927 Heisenberg proved that Planck's formula was a consequence of the uncertainty principle, i.e. that an oscillator could never be found in any particular state without violating the principle. It follows that if space consists of quantum oscillators, they must each have non-zero ground state energies.

Obviously if there are oscillators filling the vacuum, the zero-point energy of the vacuum must be very large. Not surprisingly, a lot of people are investigating the potential for "tapping" this energy (if it exists). If it does exist, it might account for the "dark energy"—or even the "dark matter"—that cosmologists are currently searching for (Haramein 2016).

The pattern of star brightness vs surface temperature, known as the Hertzsprung-Russell diagram, was compiled around 1910 by the efforts of many astronomers. The so-called "main sequence" reflects the fact that small stars "burn" more slowly, at lower surface temperatures, and last a lot longer than big ones. Observations revealed that a star with $60 \times$ solar mass would only last about 3 million years, while a star with the mass of 10 of our sun's mass would last ten times as long (32 million years). Cutting the mass by a factor of 3 (to three solar masses) brings the lifetime up by more than another factor of 10 (370 million years). Cutting the mass in half again (to 1.5 solar masses) lengthens the lifetime to 3 billion years, and cutting by a third, to the size of our sun, triples its life expectancy to 10 billion years. Our sun is about halfway through its expected life of 10 billion years.

Back to the thesis of this book: about 13.77 billion years after the Big Bang, the universe, as a whole, is still expanding and getting cooler. This cooling process means that, in some sense, it is less and less capable of "doing work" (whatever that means) on the galactic scale. The *entropy* of the expanding, cooling universe is steadily increasing. As Sir Arthur Eddington once noted, entropy is "time's arrow" meaning that time and entropy increase together. According to Ludwig Boltzmann's "order principle", entropy is a measure of disorderliness. Low entropy substances (like diamonds, quartz crystals or DNA) are orderly. High entropy stuff (like industrial wastes) is disorderly.

But from another perspective the interesting question is: How and why does the universe become increasingly orderly while continuing on its slow way along the "arrow of time" to its ultimate "heat death"? The existence of very large-scale structures in the cosmos—too large to be explained by "wrinkles in time" was noted above. The stars are not distributed randomly in galaxies. (See Fig. 4.1). On the contrary, the stars of our galaxy are largely confined to a planar disk, where they exhibit a well-known spiral structure, as though the stars were strung out along a chain or filament that is being gradually "wound up" around an ultra-dense core (where some astronomers believe there may be an ultra-massive "black hole"). As already mentioned, the galaxies themselves appear to be strung out along similar chains on a much larger scale, and the pattern is repeated on still larger scales. The existence of order and self-organization in the solar system and on the Earth is evident.

4.3   The Black Hole in Physics

Here I warn you readers, in advance, that what follows in the next few pages depends largely on references to books and articles that may be hard to find or hard to read (and controversial in some cases). I cannot possibly do justice to the technical questions at issue in a few pages. All I can do is to note the questions and point to accessible sources. You can skip this chapter altogether, if you wish, without losing much of the overall message of this book.

Forgive the pun in the title of this section, but it is apposite. A major controversy brews at the intersection of cosmology and theoretical physics that doesn't get the attention it deserves. Apart from having some personal interest in the topics, I can't defend the main thesis of this book without touching on some of these questions, however arcane they may be. This part of the chapter is about that controversy, because some of the "standard" assumptions of cosmology—and their implications—may turn out to be wrong. It is partly about "Big Bang" (BB) vs "continuous creation", partly about the "standard model" of nuclear forces, and partly about the BB itself.

For historical reasons, the BB theory is now closely interwoven with the so-called "standard model" (SM) of nuclear forces. The SM, known as "quantum chromo-dynamics" (QCD), is based on "quark" models invented independently in 1964 by Murray Gell-Mann and George Zweig to explain protons (Zweig called them "aces"). The underlying "gauge theory" was provided in 1972 (Fritzsch and Gell-Mann 1972). The definition of a "gauge theory" is too difficult to explain in a book like this. It depends, in turn, on an arcane branch of mathematics called "group theory", which can be used to describe symmetries of various kinds. I have included some further explanation as Appendix B. In any case, the math is so difficult that its extreme difficulty is sometimes taken as an explanation of why QCD doesn't seem to work in some cases. If you still care, you will need to do some reading. I would recommend a recent paperback book called *The Particle at the End of the Universe* (Carroll 2012). I would also recommend *Fearful Symmetry* by Stewart and Golubitsky (Stewart and Golubitscky 1992) and *Symmetry* by Herman Weyl (Weyl 1952).

A few more words of background may be appropriate here, before getting back to cosmology. The universe consists of space and particles that occupy space and have half-integer "spin" (called "*fermions*", named for Enrico Fermi). Other particles with zero or integer "spin" don't occupy space but transmit forces. Those are called "*bosons*" (named for Satyendra Nath Bose). There are two kinds: the Higgs boson, which is non-zero in empty space, and "gauge bosons", which are force carriers. The gluons are the carriers of the strong nuclear force. The W bosons which are charged and the Z boson, which is not, carry the electro-weak force, the photon carries electromagnetic forces and the graviton carries the gravity force (if it is a force).

The fermions are subdivided into two groups, *quarks* and *leptons*. There are six leptons: three with charge -1 (the electron, the muon and the tau particle) and three neutrinos with no charge and no mass (except possibly the tau neutrino, which might turn out to have mass after all). A recent article suggests that the electron neutrino might be a tachyon (i.e. that it moves faster than light speed, having an imaginary mass). Don't ask me what that means. Here is the reference, if you are interested (Ehrlich 2015).

Bosons are not territorial and can be superposed on each other. The photon and graviton are massless and uncharged. Electro-magnetism, and its particle-wave duality, is now quite well understood. The quantized version, developed (independently) by Richard Feynman, Freeman Dyson and Julian Schwinger is called quantum electrodynamics (QED). It has only one force carrier, the photon. QED makes extremely accurate predictions and is justifiably regarded as a successful theory.

Not only do fermions "occupy" space; but space itself is quantized into "Planck volumes". Hence two fermions cannot have the same quantum number or occupy the same Planck volume of space. On the other hand, force-carrying bosons can be piled on top of one another (superimposed) without limit (Carroll 2012, p. 28). As already mentioned, the rule of non-superposition—or single occupancy—is known as the *Pauli exclusion principle* in quantum mechanics. *It is what makes solids solid.* All solid, liquid or gaseous objects, from atoms to galaxies, are composed of fermions excluding other fermions from "their" personal space.[4]

Atoms are not elementary particles, although when atoms were first postulated by Democritus et al. and again in the nineteenth century, it was assumed that they were indivisible. We now know that they are merely the smallest particles we can weigh and "see" (with the appropriate equipment). They consist of combinations of nucleons (protons and neutrons) and electrons, which are all fermions. Atoms have mass, but they are electrically neutral because the number of electrons must match the number of protons. Atoms can be "ionized" by knocking off an electron, but ions do not remain that way for long, except at extremely high temperatures, in which case they constitute a *plasma*.

Almost all of the mass of atoms belongs to the *nucleons*. The nucleons (protons and neutrons) are each composed of elementary particles, called *quarks*. Nucleons are combinations of quarks, in sets of three, such that the net charges always add up to 1 (protons) or 0 (neutrons). This is a conservation law corresponding to a symmetry. Mesons are also composed of pairs of quarks. Quarks come in six "flavors" (masses) called u ("up"), d ("down"), s ("strange"), c ("charm"), t ("top"), b ("bottom"). Only the two lightest quarks, u and d, are needed to form nucleons. Quarks also come in 3 "colors" (red, green, blue) which are analogous to

---

[4] Confession: I have a personal history with fermions. My PhD thesis (1958) was entitled "Fermion correlation: A variational approach to quantum statistical mechanics" (Ayres 1958).

electric charge. The force carriers, *gluons*, are massless color carriers that can transfer both color and charge between quarks. There are nine possible color combinations, but only eight allowed types, since an electric charge of −e cannot be carried between two quarks.

Particles that interact with electromagnetic forces must have electric charge, which can be positive or negative. Quarks can have electrical charges +2/3 or −1/3 in units of the electron charge, e. Anti-quarks have the opposite charges. Only certain combinations of quarks with charges adding up to 0 (for neutrons) or +1 (for protons) are allowed. That is another conservation law with a corresponding symmetry. Mesons are bound pairs of quarks and anti-quarks (of different "colors"). Every particle has an anti-particle, but the anti-particles of charged particles have charges of the opposite sign.

We now have a collection of newly discovered conservation laws to explain the inter-convertibility relationships of six *quarks* (in three "colors"), six *leptons*,[5] and the eight force carriers or "messenger particles". They include photons for the electromagnetic force, the heavy muons (W^+, W^-, Z^0) for the electro-weak force, and "gluons" to carry the strong force. These force carriers are collectively known as *gauge bosons* because "gauge theories" require symmetries.[6] All of them have integer spin. They also have corresponding anti-particles, which are rarely seen except in the cloud chamber tracks of a high-energy collision.

In the summer of 2013, physicists apparently plugged a major chink in the standard model of nuclear forces. In brief, the super-collider experiments at CERN, near Geneva, produced seemingly definitive results (there are arguments about this): It was announced that the elusive Higgs boson (known as the "God particle"), had been detected with better than 99.99 % probability (Fig. 4.4). It fits the "standard model (SM)" that now supposedly explains the unification of the three fundamental forces (electromagnetic, "weak" and "strong" nuclear forces).

The standard model (SM) of nuclear forces (QCD) explains a large number of experimental results, including the masses of all the known heavy particles, including the Higgs boson. However it is important to point out that QCD does *not* explain all the experimental results, contrary to the impression one gets from the supporters of that theory. The discoveries by Kajita et al. and McDonald et al., that neutrinos do have mass, is one example. The SM also fails to explain around 30 other experimental results, of which 20 are actually inconsistent with QCD.

For instance, a number of experiments suggest that there must be a massive object in addition to the three quarks within a proton or neutron (Comay 2014). QCD does not correctly predict the mass of the proton; the quarks are not heavy enough. Physicists tried to get around the difficulty by assuming that the missing mass was carried by *gluons*, even though QCD says that the force carriers should be massless, and no gluon (or quark) has yet been seen or measured. Even the evidence

---

[5] The six leptons are electron e, electron neutrino (νe), muon (μ), muon neutrino ($\nu\mu$) and tau (ι) and tau neutrino ($\nu\mu$).

[6] The word "gauge", here, refers to "gauge symmetry". Symmetry-breaking changes the theory.

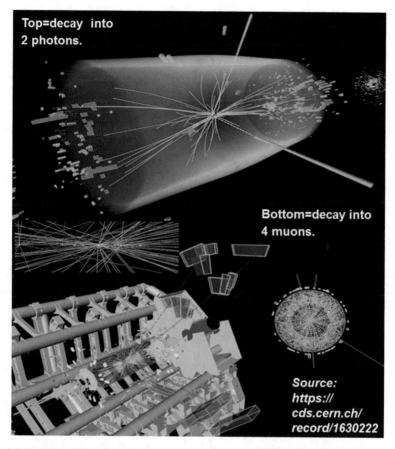

Fig. 4.4 Candidate Higgs Boson decay events at the super-collider. *Source*: https://cds.cern.ch/
record/1630222

for the Higgs boson (126 ± 5 GeV) indicates a range of uncertainty three orders of
magnitude wider than was expected according to QCD (Comay 2014).

An important feature of QCD, as applied to the early history of the universe is
the phenomenon known to physicists as "*symmetry-breaking*". The symmetries in
question correspond to conservation laws for quantum numbers in "*gauge theories*". If there is one important idea that needs to be summarized from the standard
theory, it is that "*symmetry breaking*" *differentiates the world into compartments
and makes it more complex*. A simple example is the phenomenon of condensation.
The hot vapor is homogeneous and isotropic until it condenses into droplets.
Suddenly there is a boundary between the gas and liquid phases, and the boundary
is a sharp gradient. On one side of the boundary the motions are still random and
disorderly, as of a gas. But on the other side there is suddenly some short-range
order. When the liquid cools further there is another phase change as the liquid

solidifies or crystallizes. The orderliness increases again, as crystals form. The structure of a crystal exemplifies both short and long-range order.

I agree (before you say it) that the symmetry-breaking of a single unified force (that is really a kind of average) into two or three subsidiary forces with very different strengths and ranges is not really like a phase change in a substance. There is, however, a major branch of mathematics, called *group theory*, that deals with mathematical symmetries of various kinds. This branch of mathematics has become very important for theoretical physics in recent years. As I said, the so-called "gauge transformations" are based on symmetries. The "standard theory" of nuclear force—see Appendix B—is based on symmetries. "String theory" and "super-string theory" are also entirely based on this branch of mathematics (Kaku and Thompson 1997). However, those theories are not important for this book. On the other hand, symmetry-breaking is important, not only in nuclear physics, but for all of evolution.

A reasonably familiar example of symmetry-breaking is the phenomenon of "bifurcation" that occurs in some non-linear differential equations. A bifurcation in mathematics can be described as a sharp transition from a region where the solution is unique to a region where there are several alternative solutions to choose from. In physics there is a bifurcation from a simple symmetric state (like a gas) to an asymmetric state (like a liquid), or from a liquid to a solid. Other examples of bifurcations include Heisenberg's theory of ferromagnetism (1928), Bogoliubov's theory of super-fluidity (1947), **Bardeen**, Cooper and Schriefer's theory of super-conductivity (1957), and Yang, Lee and Wu's work on non-conservation of parity (1965). Boldest of all was Philip W. Anderson's (1962) theory of what is now called the "Higgs mechanism" for mass-generation from the vacuum (Anderson 1962).

Philip Anderson did not work out a relativistic model for such transitions. But models for so-called "gauge bosons" were proposed almost simultaneously by six men in three different groups in 1964 (F. Englert, R. Brout, **P. Higgs**, G. S. Guralnik, C. R. Hagen, T. W. B. Kibble. These six men were joint winners of the Sakurai Prize in 2010). Francois Englert and Peter Higgs shared the physics Nobel Prize in 2013.

Theoretical applications of the so-called "Higgs mechanism" to other groups of particles followed. It is now thought by most particle physicists that all the particles in the universe were originally (at time zero) massless and there were no broken asymmetries. As the universe cooled and expanded a symmetry-breaking mechanism came into operation some time before 10^{-10} s. This symmetry break created an entity known as the "Higgs field", known only for its incredible consequences. The theory, as it stands, seems to say that the kinetic and thermal energy released by the original BB was somehow released from "the vacuum" by the Higgs field as the universe cooled and expanded. In other words, the Higgs field produced something from nothing. Experts will probably object to this simple wording. But a more precise statement would lead into too many complicated byways.

There is a wonderful cartoon that can be found in virtually every physics lab (and plenty of other places) in the world. I used to see it nearly every day at IIASA. It shows two elderly scientists (you can tell by the beards, bellies and specs) in front of a blackboard covered with mathematical symbols. One of the scientists is explaining it to the other. In the middle of the blackboard there is a gap with the words "A miracle occurs". The other scientist remarks, in all seriousness "I think you need to be more explicit at this point". I think they were talking about mass creation. The miracle that happened was that all particles, starting with the quarks, acquired their mass by interacting with the Higgs field.

The likelihood of exploiting this possibility on a laboratory or industrial scale (or in a future space-ship) seems very remote, but that doesn't stop people from speculating about the possibility of extracting useful energy from the vacuum. After all, the same thing was said of the potential for exploiting nuclear energy back in the 1930s. I agree that this analogy is faulty, but . . .

Now, back to the BB theory. Remember, it started with George **Gamow** and Ralph Alpher in the 1940s, inspired by the earlier ideas of **Lemaitre** in the 1920s and the evidence of expansion from Hubble's work (Lemaitre 1950). Gamow incorporated newer astronomical evidence from **Zwicky** and others, together with new knowledge of nuclear physics and nuclear processes in stars.

But in the 1940s and 1950s there was a competition among "continuous creation" ideas (e.g. Bondi and Gold 1948; Hoyle 1950). The Big Bang theory seems to have accelerated rapidly after the discovery of the microwave "background" radiation, by Penzias and Wilson, in 1965. Elementary particle experiments, and theory, both evolved rapidly in those years. The BB theory was set forth most persuasively by Steven Weinberg in his lecture at Harvard in 1973. That was followed by his best-seller *"The First Three Minutes"* (Weinberg 1977). In that book, Weinberg explained in detail what must have been happening during that time, according to then-current understanding.

There were problems, of course. One of them was the mysterious "smoothness" of the observed background radiation. Another was the fact that the age of the universe, based on Hubble's work on the rate of expansion, was too short to explain the supposed origin at a singularity. In 1980, MIT physicist **Alan Guth** postulated that the very early universe might have expanded faster than the speed of light (the word is "superluminal") after the first 10^{-43} s, doubling in size 60 times, by a factor of 10^{12} or so, from a size smaller than the volume of a proton to the size of a golf ball. The inflation stopped when the universe reached the age of 10^{-33} s (Guth 1981). Alan Guth called it "inflation".[7] Maybe this was what those two cartoon physicists were talking about.

This expansion is now thought to have taken place by an interaction with the "Higgs Field" suffused throughout space. While the inflation of space was many times faster than the speed of light, Einstein's equations don't put any limit on the rate at which space itself can expand. But the expansion, being faster than light, could travel inside the space-bubble, and this had an interesting implication: *the*

---

[7] In fairness, it seems that Alexei Starobinsky had the same idea earlier in Moscow and some anomalies in Guth's version were later "fixed" by Andrei Linde and also, separately, by Andreas Albrecht and Paul Steinhard.

outer parts of the bubble were causally disconnected from the starting point. For instance, the very first particle-anti particle pair could have been separated before they were able to annihilate each other.

If temperature had meaning, at time zero, and if it could have been measured on a cosmic thermometer, it would have been higher than 10^{33} on the Kelvin scale. When the universe was just 10^{-41} s old the unified nuclear force—possibly transmitted by ultra-heavy bosons not yet observed or even predicted—split into an electro-weak force and a strong nuclear force. This is called the "GUT transition" (GUT means "grand unification theory") because the force unification—a kind of symmetry—broke when the two forces became distinguishable in terms of range.

The rapid expansion of the universe after the BB was accompanied by equally rapid cooling. When the universe was 10^{-10} s old, and the temperature had dropped to $10^{15}°C$, the Higgs field generated mass. A "soup" of quarks (and anti-quarks?) "condensed" out of the primordial Higgs field. For some reason, the local symmetry between baryons and anti-baryons was broken. (One part in a billion would have sufficed). More specifically the "universe" now contains baryons (which includes all quark-based particles) but no anti-baryons. But there are symmetry-breaking reactions in the Standard Model that could account for "baryo-genesis" -the antithesis of matter-genesis- as first proposed in 1967 by Andrei Sakhorov. After all the anti-baryons disappeared (from our part of the universe), the remaining baryons could hang around a little longer. The short-range "strong" nuclear forces (part of GUT) induced them to combine—with the help of force carriers called "*gluons*"— into the more stable particles: *baryons* (consisting of three quarks) and *mesons* (consisting of quarks combined with anti-quarks) (Gell-Mann 1964; Zweig 1964). See also Fritzsch (1984).

About that time, the electro-weak force also split again into the medium-range "weak" force as we know it, with its massive force-carriers W^+, W^- and Z, and the familiar long-range electro-magnetic force—transmitted by *photons*—that gives us electric power, electric light, electric motors and computers. Electrons and positrons must have also "condensed" about this time. Most of the pairs annihilated each other and disappeared but a few electrons seem to have been left over, also thanks to the "causal disconnection" during the inflation.

Miraculously, the number of surplus negatively electrons (after the positrons disappeared) almost exactly matched the number of positively charged hadrons (protons), resulting in net zero charge of matter, at least in this part of the universe. This curious coincidence would be explained if the "condensation" of matter had produced only neutrons at first, and if protons and electrons had both been produced by neutron decay (catalyzed by the "weak" force) when the "soup" was still very dense. That sequence would have produced matter with net charge of zero. But did it? If the net charge is not zero, would we know? I'm sure this bit of speculation on my part will offend some expert.

The "inflation" idea solved several problems, notably the "flatness" problem (how the universe managed to exist without either collapsing or exploding) and the anisotropy of background radiation ("wrinkles"), mentioned above. Cosmologists now mostly accept the Guth inflation theory. But nobody that I know of has a plausible explanation for it.

The QCD theory is the hottest topic in theoretical and high-energy physics today, and it is normally regarded by physicists (and presented by the media) as a virtually complete "theory of everything" and an explanation of the Big Bang. That is a large overstatement. It is understandable that high energy physicists are under pressure to justify the expensive research facilities, and experiments, such as the highly publicized search for the "Higgs boson", using the proton "super-collider", at CERN, and other particle accelerators around the world. No doubt, fundamental research creates new knowledge that may pay off in an unguessable way in a distant day. And it is highly probable that this search throws light on the early history of the universe. But there are serious inconsistencies and gaps.

The main message of the next few pages is that, while most physicists and astronomers assume that the BB was a fact, its explanation is still only a theory and that there is strong evidence that the theory is seriously incomplete. Moreover, although Albert Einstein is generally treated by younger scientists as the ultimate authority (as Newton once was), there are also very serious doubts about the Einstein version of general relativity (EGR), which—in turn—is a pillar of the BB theory.

It is helpful to start by listing the successes and failures of BB theory. The successes are (1) an explanation of the Lemaitre-Hubble expansion, (2) explanation of the 2.7 K background radiation discovered by Penzias and Wilson and (3) explanation of the (approximate) abundance of light elements. But the gaps—I hesitate to say failures—include the following:

- The "flatness" problem: According to Einstein's theory, a dense universe would be gravitationally closed, causing an eventual collapse, while a less dense universe would expand forever. The in-between case (a "flat" universe) seemed to be the only one that could account for the fact that the universe had neither collapsed nor exploded within the first microsecond. This problem was "solved" by the Guth "inflationary" phase in the expansion (10^{-36} to 10^{-33} s) (Guth 1981; Guth and Steinhard 1984). The WMAP experiment has now determined that the universe is indeed flat (Euclidian) within 0.4 %.
- The "Horizon problem": before the inflation every part of the universe was in contact with every other part, hence the near-uniformity of the background radiation. The Guth "inflation" broke the contact, resulting in "wrinkles." Wrinkles were detected by the COBE satellite, and its successor WMAT, but they are very tiny. It is hard to see how they could explain the large-scale structures that exist in the cosmos.
- The "Creation problem": When the "inflation" ended (10^{-33} s) the "vacuum energy was transformed instantaneously" (thanks to the Higgs field) into conventional mass-energy. Quantum fluctuations may have created the "seeds" of later stars and galaxies.

Note that the first three problems listed above are all "solved" by Guth's inflation theory, although that theory itself poses very difficult questions. In fact the inflation itself is a mystery unless it can be explained somehow in terms of the Higgs Field.

But there are several more problems:

- Structure formation problem: The structures and voids in the cosmos are too big and too slowly moving to have evolved during the life of the universe (13.77 billion years). If the universe is a lot older—or not expanding at all—other conceptual problems arise.
- The abundance problem: The predicted abundances of ^{4}He, ^{3}He, D and ^{7}Li are inconsistent with observed abundances (Lerner 1991).
- The 20 or so experimental results, several of them concerning the mass of the proton, that are inconsistent with the "standard" model of nuclear forces (Comay 2014).
- The "black hole" problem: These objects (if they exist) imply the existence of singularities and other implications that are completely inconsistent with "normal" physics (as Einstein himself said several times). Nothing in the standard theory of nuclear forces can account for neutron stars, yet they almost certainly exist. In fact pulsars are neutron stars.

I should add that according to the standard model of nuclear forces, gravity is regarded as a force in the same sense that electro-magnetism and short-range nuclear forces are. The SM says that "gravitons" must exist as "carriers" of that force, just as photons are carriers of electromagnetic forces and heavy particles (W, Z and so-called "gluons") are carriers of stronger shorter range strong forces. There is some speculation that gravitons would be emitted during supernova events or when neutron stars or large "black holes" collide. In fact, the signals detected a few months ago by the two new LIGO antennas are being interpreted as the results of a collision between large black holes. But if those signals are due to gravitons, it would seem to mean that gravity is a force in the same sense as the others. That, in turn, would seem to imply that gravity is *not* just curvature of space after all. Stay tuned.

At this point I need to bring Einstein back into the story. During the 10 years after 1905, Einstein tried to generalize his 1905 paper on special relativity, which explained that the Lorentz transformation was derivable simply from the assumption that all physical laws must look the same to riders on all trolleys, regardless of speed. But for 10 years, he had been wondering: what if one of the trolleys was accelerating? He replaced moving trolleys by accelerating elevators. I am guessing that Einstein knew about the Coriolis "force" (explained back in 1835). It explains the trade winds and the rotational direction of water running out of a bathtub or of hurricanes and typhoons: anti-clock-wise in the northern hemisphere, clockwise in the southern hemisphere. Einstein would have realized that the "Coriolis force" is an illusion, not a force, and that a bubble in that water (or a balloon in the jet stream) would just be following a geodesic in a curved space.

He must have thought: could gravity be a similar illusion? Could the gravitational "force" in a falling object simply be the effect of a curvature in space? What might have caused the curvature? Einstein postulated that mass (or mass-energy) might be the cause of the curvature. In 1916 he published his paper on the "general theory of relativity" (Einstein 1916). To rid the "special theory" of that restriction (of "non-accelerating" frames), he asserted that the laws of physics should be the same in an accelerating elevator and a gravitational field. Why? Einstein reasoned that if the elevator accelerates up, the acceleration feels like stronger gravity (and *vice versa* going down).

Einstein postulated an equation between a metric tensor (describing the curvature of space-time) and a tensor describing the distribution of mass-energy in space. His metric tensor was a reduced version of the Riemannian tensor, called the Ricci tensor (Ricci and Levi-Civita 1901). When he worked out the key implications, his new theory seemed to imply that gravity is actually not a force, but an illusion due to the curvature of space. One problem for the theory is that it is not clear how "dark matter" appears in the mass-energy tensor. Would it be uniformly distributed throughout space? If not, where is it?

Einstein's general theory of relativity (EGR) had three immediate predictions: the first was that light from a distant star passing close to the sun should be deflected by 1.8 arc-sec. The second was that solar spectra should be *red-shifted* (towards longer wavelengths) by 1.29 arc-sec, of which 1.19 are explainable by the special theory (Newton's laws) and the other 0.1 arc-sec due to the sun's gravitational field. Together that amounts to 2.1 parts per million, a small but measurable shift in wavelengths.

The astronomer Arthur Eddington led an expedition to West Africa in 1919 to test the first prediction on this list, during an eclipse of the sun. Einstein's prediction that light would bend around gravitational masses was confirmed—within 10 %— by Eddington (and later by two other groups). The second prediction about gravitational red-shift (in weak gravitational fields) was also verified.

The third test of EGR has to do with the orbit of the planet Mercury, which is highly elliptical and is *precessing* in the direction of rotation. In other words, the perigee and apogee of the orbit are also gradually rotating. The measured rate of precession of the orbit is 1.39 arc-sec per orbit or 575 arc-sec per century. The Newtonian theory of gravity (due to gravitational interaction with the outer planets) accounts for only 1.29 arc-sec per orbit, or 532 arc-sec per century. The difference of 43 arc-sec per century should be explained as a relativistic (GR) effect.

But, it turns out that EGR correctly accounted for the 43 arc-sec per century, *but not the other 532 arc-sec per century*. Why is that? The problem is that EGR is inherently a 1-body theory; it cannot be applied to a 2-body problem, still less a many-body problem. The most famous implication of EGR, by Karl Schwarzschild (also in 1916), rocked the physics community (Schwarzschild 1916).[8] It was a rigorous solution of Einstein's equations for a very special 1-body case. Schwartzschild showed that for a static, dense, uncharged, non-rotating, spherically symmetric star with mass m and radius **r**, there are two solutions: An "exterior solution" (outside the star) and an "interior solution", inside the star. The speed of light outside the star is

---

[8] He was a German astrophysicist who volunteered to join the army on the outbreak of war and who died of battle-related auto-immune disease in 1916, aged 43, a few months before his results were published.

$$[1 - 2(m/r)]$$

where $\mathbf{m}$ is a "normalized" mass defined as $\mathbf{m} = \mathbf{MG}/\mathbf{c}^2$. In this formula $\mathbf{M}$ is the mass of the star, $\mathbf{G}$ is the gravitational constant and $\mathbf{c}$ is the velocity of light, all in metric units. This effect is called the *gravitational red-shift*. It follows that when $\mathbf{m}/\mathbf{r} = \frac{1}{2}$ and (in normalized units) the speed of light goes to zero. This is usually interpreted as meaning that the gravitational curvature of space would trap the light inside the star.

The point $\mathbf{r_s} = \mathbf{2\ m}$ is called the "Schwartzschild radius", hence the concept of a "black hole" (so-named later by John Wheeler). The surface defined by $\mathbf{r_s}$ is called the "event horizon" in black hole lingo. Evidently physical laws may be very different inside an event horizon, as suggested e.g. by John Taylor's book (Taylor 1973).

For the case $\mathbf{m} > \frac{1}{2}\mathbf{r}$ or $(\mathbf{r} < \mathbf{2\ m})$ the "interior solution" becomes imaginary. There is another anomaly buried in the "interior solution" (inside the star) that was not discovered immediately. When $\mathbf{4/9} < \mathbf{m/r} < \frac{1}{2}$, there is an "event horizon" *inside* the star, at the radius $\mathbf{r} < \mathbf{r_s}$, where $\mathbf{r_s}$ is the Schwarzschild radius. In this case

$$\mathbf{r}/\mathbf{r_s} = [\mathbf{9} - \mathbf{4r_s}/\mathbf{m}]^{1/2}$$

According to the theory, this internal "event horizon" starts at the origin ($\mathbf{r} = \mathbf{0}$) and moves out as $\mathbf{m/r}$ varies from 4/9 to ½ until it becomes the event horizon for the whole star when $\mathbf{r} = \mathbf{2\ m}$ (Wald 1984, p. 138).

But a star cannot function with an interior event horizon. Why? Simple: because that would prevent electromagnetic forces from working inside the star. (Remember that electromagnetic forces are carried by photons.) Physicists have concluded that the star would be unstable in such a case, whence the collapse into a black hole singularity would actually occur when $\mathbf{m/r} = \mathbf{4/9}$.

The $\mathbf{m/r}$ ratio (in normalized units) at the surface of our sun is 2.12×10^{-6}. However there are a lot of extremely compact stars ("white dwarves" and neutron stars) "out there" that have much higher densities. For instance, the white dwarf that accompanies the bright star Sirius has a density 300,000 times that of the Earth. The density of our sun is 1.4 g/cm^3 or 1.4 tonnes per cubic meter. This is 1.4 times the density of water. A "standard" white dwarf has a density of 1.8×10^6 g/cm^3 or roughly 2 million tons per cubic meter. This is because white dwarves have already gravitationally collapsed as a result of having "burned" all their hydrogen. A neutron star is a lot denser than a white dwarf: 3×10^{14} g/cm^3 or 300 trillion metric tons per cubic meter. That density is close to the Schwarzschild condition ($\mathbf{m/r} = \mathbf{4/9}$).

Einstein argued (more than once) that his equations should not be assumed to apply in such extreme density regions (e.g. Einstein 1936). Yet physicists found the idea fascinating. In 1939 J. R. Oppenheimer (who later directed the Manhattan Project that developed the atomic bomb), wrote a paper with a graduate student, H. Snyder, arguing that—in the case of time-dependent equations—*the mass inside a black hole would continue to shrink all the way into a point-singularity* (Oppen-heimer and Snyder 1939). This paper was the true origin of the "black hole" idea.

Einstein responded to this publication, once again, by arguing that the equations should not be applied to such extreme cases (Einstein 1939). His paper concluded with the words:

> The essential result of this investigation is a clear understanding of why the "Schwarzschild singularities" do not exist in physical reality ... due to the fact that matter cannot be concentrated arbitrarily. And this is due to the fact that otherwise the constituting particles would reach the velocity of light.

He could have added that shrinkage is also limited by the Pauli exclusion principle, as I mentioned earlier.

After that, nobody challenged Einstein on this issue during his lifetime. But, since then, things have changed. Despite Einstein's skepticism, most astronomers and most theoretical physicists today, believe firmly in the existence of black holes, even though they cannot be seen directly (by definition). Subrahmanyan Chandra-sekhar was one of the first astronomers to assert that supermassive black holes actually exist in galaxies (Chandrasekhar 1964). Chandrasekhar in the US and Lev Landau in Russia independently concluded that a collapsing star larger than 1.4 solar masses would necessarily become a neutron star with a core density equal to the density of atomic nuclei (see Fig. 4.14 in Sect. 4.3).

There is another mystery concerning neutron stars. The substance making them up is sometimes referred to as "neutronium", a compact mass in which protons and electrons have been forced to combine into neutrons. But neutrons have a long (by elementary particle standards) but finite mean lifetime of about 14 min. Then they spontaneously decay into protons and electrons, with the emission of an electron, *plus an anti-neutrino and a photon*. The so-called "electro-weak" nuclear force accounts for this decay process. The reverse process has not been demon-strated, although it must be possible, since all nuclear processes are reversible in principle. But, in that case, it would seem that *anti-neutrinos must be absorbed*. Then there is a question as to where they might come from? Lacking a source of anti-neutrinos, a neutron star would have to shrink into a mixture of protons and electrons, i.e. an ultra-dense plasma. Or the protons would have to decompose into their component quarks. The electrodynamics of such a substance are unknown.

Accepted EGR theory says that a neutron star larger than a few solar masses (9.4 according to one estimate) will overcome the resistance due to the Pauli exclusion principle (the *neutron degeneracy pressure* or NDP) and collapse all the way into a black hole.[9] NDP is the same principle that makes atoms seem to be solid. Actually, it has been calculated that NDP is only sufficient to prevent gravitational collapse for stars with masses less than 0.7 that of our sun. (This is known as the Tolman-

---

[9] The Pauli exclusion principle in quantum mechanics says that no two fermions (like neutrons) can occupy the same quantum state. Hence, all but two of the neutrons in a neutron star need to be in highly excited states, whence they must move continuously at speeds closely approaching the speed of light.

Oppenheimer-Volkoff limit.) All known white dwarves and neutron stars are heavier than this limit and should therefore become black holes, according to EGR, according to Schwartzschild.

Every galaxy is now assumed (by most astronomers) to have a black hole at its center. According to indirect evidence (gravitational effects and luminosity) there is a supermassive black hole (Sagittarius A*) in the center of our galaxy. Sagittarius A* is said to have the mass of 4.1 million suns, or roughly a million times bigger than the Chandrasekhar limit, at which all the electrons and protons in the atoms presumably recombine to become neutrons. (But where did those pesky anti-neutrinos come from?) There are apparently still bigger "black holes" in other galaxies, and—maybe—vastly bigger ones in some quasars. An alternative theory is mentioned below.

At this point, I need to mention one other implication of the Schwarzschild (EGR) solution for gravitational redshift. As already mentioned, the speed of light given by the exterior solution is $1 - 2\mathbf{m}/\mathbf{r}$ where (again) $\mathbf{m}$ is the "normalized mass" of the star (defined as $\mathbf{m} = \mathbf{MG}/\mathbf{c^2}$) and $\mathbf{r}$ is the distance from the center of the star. Then the ratio of local to normal wavelengths of that light, inside the star, is simple:

$$\mathbf{Z} = [1 - 2\mathbf{m}/\mathbf{r}]^{-1/2}.$$

As $\mathbf{m}/\mathbf{r}$ approaches its limiting value of 4/9 at the interior "event horizon", the ratio of the wave-length dilation approaches 3 and *the gravitational red-shift* $\mathbf{Z}$ *approaches its upper limit, which is* $\mathbf{Z} = 2$ *(for EGR)*, even as the wavelength ratio $\mathbf{r}/\mathbf{r_s}$ becomes infinite. For a full and clear explanation see Bjornson (2000, Table 8.5 et seq). Given the large number of quasars with redshifts $\mathbf{Z} > 2$ as found by the SDSS and BOSS quasar surveys, and some with $\mathbf{Z}$ as high as 5, this fact by itself seems like rather conclusive evidence against EGR, or at least against black holes.

Halton Arp is an astronomer whose "sin" (which got him expelled from the inner circle) was to collect evidence suggesting that many (most) quasars were a lot closer to us than the current BB doctrine (as sketched in the previous paragraphs) asserts. For more on this see his books (Arp 1987, 1998). For the mainstream astronomers, large observed redshift values are the basis for assuming that quasars are extremely distant (up to 12 billion light years). Arp has collected evidence that many high-Z quasars are associated with low-Z galaxies. In other words, quasars with very large red shifts, moving away from us, are located in galaxies with much lower red shifts.

This implies that the quasars are intra-galactic phenomena, possibly involving a "split" between two enormous radiating masses that are now flying apart in opposite directions at high speed. (If that were true, there should be other radiating masses moving in our direction, which have not been observed, perhaps because nobody has been looking.) Arp's theory to explain this is a version of quasi-continuous creation. It is far too controversial and complicated, to explain here. I only mention it to give you a headache, if you haven't got one already.

There is yet another possible explanation of the high-$\mathbf{Z}$ quasar observations. I will explain it later in connection with Yilmaz general relativity (YGR). Before doing so, it is helpful to summarize, quickly, the known weaknesses in EGR. One weakness, mentioned already, is that it is inherently a 1-body theory that cannot be used for 2-body or many-body calculations. This limitation showed itself in connection with the precession of the orbit of Mercury. Iterative many-body solutions in EGR would require extremely complex mathematics involving millions of terms and requiring powerful computers that were not available until after Einstein's death in 1955.

The computer programs start by assuming approximations of the elements of the 4×4 Ricci metric tensor, from which the elements of the 4×4 Einstein tensor are computed and compared with the desired elements of the mass-energy tensor. The tensors are presumably symmetric, so there are ten independent simultaneous non-linear equations to be solved. The differences result in adjustments to the original guesses as to the metric tensor elements and the procedure continues iteratively until there is a match.

A second problem is that EGR is "over-constrained". In principle, the so-called "covariant derivative" (an extension of ordinary scalar derivatives to vectors and tensors) of the energy-momentum tensor should be zero (the so-called Bianchi identity). This requires that the covariant derivative of the energy-momentum tensor must also be zero. An independent requirement, to satisfy the conservation of mass-energy, is that the derivative of the tensor density of the energy-momentum tensor should also be zero. Many theorists apparently assume that the two requirements are equivalent. However, Landau and Lifshitz have shown that they are not equivalent, *meaning that those two independent constraints can be inconsistent* (Landau and Lifshitz 1973, p. 280). Imposing an extra condition can yield multiple answers depending which constraint is ignored (Bjornson 2000, Appendix J). Because of the non-linearity of the equations, there is no guarantee that the iterative solution procedure mentioned above will converge to a unique answer.

So it is an over-statement to say that the equations of EGR can be solved, even with the help of a super-computer, except in special cases. However, one special case where the solution is simple (but wrong) is the case of two parallel plates of infinite dimensions. The so-called Cavendish experiment performed in 1798 proved that the two plates attract each other. This historic experiment established the gravitational constant G with an error of only 1.4 %. But the Einstein theory has no gravitational mass-energy tensor for the two plates. So the EGR "solution", in that (admittedly non-physical) case is that the gravitational attraction between the plates is zero! The lack of a gravitational tensor is the reason for the inapplicability of EGR to the 2-body problem, or the many-body problem. This was proved 20 years ago by Carroll Alley (Alley 1994).

A third problem is that EGR does not always conserve matter plus energy[10] whereas YGR does so automatically. For an explanation of why this is so, see Bjornson (2000, Appendix E).

---

[10] Such is the god-like authority of Einstein, some people have concluded that perhaps energy is not conserved after all e.g. Mirowski (1989, p. 82).

A final weakness of EGR is that it is not quantizable. Why? Because it is not a *gauge theory with an associated symmetry and conservation law*. This is related to the question of whether gravity is really a force, like the other three forces, or whether gravity is an illusion due to the curvature of space.

Evidently EGR has enormous implications for cosmology. Einstein's original 1916 paper predicted that the universe is dynamic, not static i.e. that it could be expanding or contracting. Einstein couldn't believe in this possibility, so he fiddled the equations by adding a repulsive "cosmological constant"—effectively the same as "dark energy"—to make the universe static (1917). In 1922, Alexander Friedmann, a Russian mathematician, revisited Einstein's original dynamic model. He found a new set of dynamic solutions, by assuming that the universe is *isotropic*—looks the same in all directions at all times—with uniform distributions of mass and energy (Friedmann 1922). In this case, the universe could expand. Expansion would then be governed by the density of mass in the universe.

Alexander Friedmann's results showed that there is a critical density for which the universe can be flat, but that if the density is greater (positive curvature) it will collapse; if it is smaller (negative curvature), the universe can expand forever. (The recent WMAT survey, successor to COBE, has reported that the universe is indeed flat to an accuracy of 0.4 %). In fact, Friedmann also showed that there is a solution of Einstein's equations such that an expanding universe could have been born in a singularity and that the expansion would accelerate. Einstein had introduced his "cosmological constant" to prevent expansion. Later (after seeing Friedmann's solution and Hubble's results) he took it out and apologized for his "greatest error". In fact, red shift data from quasars seems to indicate that the expansion is actually accelerating (hence the "dark energy" theory). However, Halton Arp's work casts doubt on this.

As I mentioned a few paragraphs back, there is an alternative theory of general relativity, espoused by Huseyin Yilmaz. I call it YGR. The static version of YGR was published in 1958 (Yilmaz 1958). Fifteen years later a dynamic version was published (Yilmaz 1973). YGR is a many-body theory, not a 1-body theory. Another difference is that YGR is a gauge theory, so it is quantizable, in principle, unlike EGR (Yilmaz 1986, 1995).

Recall that EGR is an equation between the so-called Einstein-Ricci curvature tensor (on the left hand side) and the stress-energy tensor (a multi-dimensional object that encompasses all forms of energy) on the right hand side There is no contribution to the stress-energy tensor due to gravity. The main difference in the two formulations is that YGR includes a gravitational stress-energy term on the r.h. s of the equation. This enables the YGR to respect the principle of local conservation of energy-momentum, which EGR does not do (even though all other fundamental force fields in nature must.) YGR also makes it possible to solve 2-body and many-body systems.

This modification allows a mathematical simplification. Comparing YGR with EGR, for the static, spherically symmetric (star) case—the same one solved by Schwarzschild—the tensors are diagonal and YGR yields the following:

$$\text{Speed of light}: \quad \mathbf{exp(-2m/r)}$$
$$\text{Clock rate}: \quad \mathbf{exp(-m/r)}$$
$$\text{Wavelength}: \quad \mathbf{exp(+m/r)}$$

The most interesting feature of YGR cosmology is that there is no limiting value of **m/r**, which means that *the gravitational red-shift **Z** can be much larger than is the case for EGR.* There is no positive value of *r* where the speed of light goes to zero. As a consequence, YGR is consistent with much larger gravitational red-shifts than EGR. Hence the supposed evidence of accelerating expansion of the universe—and the need for "dark energy"—is no longer needed. This may resolve the "quasar crisis" that was mentioned earlier.

Moreover YGR—unlike EGR—does not allow singularities, "event horizons" or black holes. In fact, Einstein himself questioned the validity of EGR in regions of high density: He wrote:

> One may not therefore assume the validity of the equations for very high density of field and of matter, and one may not conclude that the 'beginning of the expansion' [of the universe] must mean a singularity in the mathematical sense. All we have to realize is that the equations may not be continued over such regions. (Einstein 1956, p. 129)

According to the standard theory of nuclear forces, in ultra-high gravitational fields the atomic nuclei must lose their structural coherence, condensing into an ultra-dense solid probably consisting of quarks. But the BB theorists have not tackled this question seriously, probably because most of them have accepted the "black hole" (as a singularity) theory as fact.

Where EGR predicts a super-massive black hole (e.g. in the center of our galaxy) YGR predicts a "gray hole" where light "slows down" to near (but not absolute) zero velocity. *So massive objects "falling" into that "gray hole" are actually moving at infinitesimal velocities.* This "gray" home has no definite surface. It is a cloud of bodies falling slowly toward the gravitational center of the cloud. The difference is undetectable from a distance.

Another interesting feature is that YGR cosmology requires a mass density for the flat universe equivalent to nine hydrogen atoms per cubic meter of space, assuming the observed Hubble expansion rate of 20 km/s per million light years. More important, the mass density required by YGR just happens to be the same as the "critical" mass density ρ_{crit} for a "flat" universe, in the Big Bang theory, as predicted back in 1922 (Friedmann 1922), viz.

$$\rho_{crit} = 3H_0{}^2/8\pi G$$

I don't think that coincidence is accidental. The density needed to produce the above Hubble expansion rate is about 9.6 hydrogen atoms per cubic meter, which is reasonably consistent with recent measurements. This density is much greater than the mass implied by stellar luminosity i.e. the mass associated with visible stars that are shining. Thus, like EGR, YGR requires a large amount of "dark matter", to account for the gravitational attraction of rotating galaxies.

There have been several other versions of GR theory, notably by Robert Dicke and his student Carl Brans (in order to accommodate a modified version of the Mach principle[11]). But all of them have been rejected as a result of laser ranging experiments (suggested by Dicke himself) between the Earth and the moon (using reflectors placed on the moon by astronauts in 1969 and 1971). However YGR remains in contention to those tests.

Michael Ibison has conducted a "cosmological test" of YGR, concluding that YGR in its pure form has serious weaknesses that might (and only might) be resolved by a theory that incorporates a matter creation process comparable to the steady state models of Gold, Bondi, Hoyle and Arp (Ibison 2006). The key point is that the process of matter creation cannot be swept aside, when it comes to the general theory of relativity, and hence, of gravity.

So, to summarize the contest (excuse the word) between EGR and YGR from a cosmological perspective:

- EGR implies that gravity *is* curvature; this makes it hard to fit into the QCD theory of forces. YGR merely suggests that gravity *causes* curvature of space.
- The two theories agree that "dark matter" must exist to explain the gravitational stability of galaxies and other cosmological objects; neither offers any clue as to what it might be. As noted in Sect. 4.1, BB theory is unable to account for this dark matter so far, except in terms of hypothetical "weakly interacting massive particles" (WIMPS).
- EGR also requires "dark energy" to explain the apparent acceleration of the expansion of the universe. YGR does not. YGR allows larger gravitational red shifts, and is reasonably consistent with Halep Arp's data.
- EGR implies that "black holes" exist, which implies that point singularities are legitimate in physics; the BB was a point singularity. YGR does not require point singularities (hence no "black holes"). Note that astronomical observations cannot distinguish black holes from gray holes. Can gray holes collide or produce gravitational waves in some other way? I don't know.
- EGR is consistent with (and essential for) the BB theory. YGR is not consistent with BB. It requires something like "continuous creation" of mass throughout space-time (which is why it does not require dark energy).

It should also be said that recent astronomical evidence is increasingly suggestive that the universe is actually much older than the BB theory implies (de Lapparent et al. 1986; Geller and Huchra 1989; Broadhurst et al. 1990). The very large structures—clusters of galaxies, bands, voids, etc.—are moving much more slowly than the speed of light (about 500 km/s). At that rate, it would have

---

[11] The Mach principle (attributed to Ernst Mach, but first stated in the eighteenth century by George Berkeley) says—more or less—that local physical laws are determined by the large scale structure of the universe. Einstein used this idea, without attribution, in a more precise form.

taken 150 billion light years for these structures to evolve. They could not have evolved, based on momentum and gravity, during the 13.77 billion years since the BB. Skeptics argue that this fact, by itself, falsifies the BB theory (Lerner 1991).

Can continuous matter creation be reconciled with the BB? Not easily, for sure. There would have to be an explanation for the observed microwave background radiation temperature of 2.7 K and for the Hubble red shift observations. There would have to be a new theory to explain continuous matter creation in interstellar space. There would have to be a new theory to explain the recent gravitational wave observations. But, I can't help thinking that YGR is part of the answer.

Moreover, it does appear that the so-called "plasma cosmology" of Hannes Alfven and his followers can explain a lot about these large structures; essentially that they are shaped by electromagnetic forces as much or more than by gravity or the BB (Alfven and Falthammar 1963; Alfven 1966, 1981). Basically, Alfven's theories depend on two facts that were originally disputed, or ignored, but that have subsequently been proven to be correct. The first is that space is not "empty" but is permeated by an electrically conducting plasma. This has been confirmed by a series of NASA probes since 1970.

The second fact is that interstellar space is also permeated by magnetic fields. Physical motion of a conductor (the plasma) through a magnetic field generates electric fields, which—in turn—generate magnetically contained filamentary plasma currents (Peratt and Green 1983). (See Fig. 4.5). These plasma filaments,

Fig. 4.5 Attraction between filamentary currents in plasma. *Source*: http://www.plasma-universe.com/images/thumb/7/71/Birkeland-pair-twisted.png/200px-Birkeland-pair-twisted.png

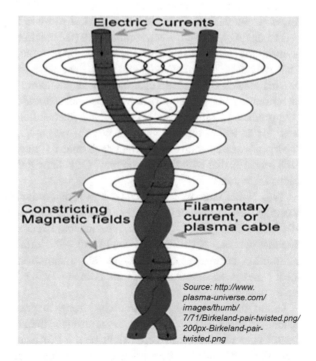

Electric Currents

Constricting
Magnetic fields

Filamentary
current, or
plasma cable

Source: http://www.
plasma-universe.com/
images/thumb/
7/71/Birkeland-pair-twisted.png/
200px-Birkeland-pair-
twisted.png

as they might be called, constitute a kind of cosmological "object" that can be analyzed in its own right—so to speak—like planets, stars, galaxies, quasars and so on. If it should turn out that massive magnetic monopoles—analogous to electrons—actually exist, those magnetic fields in space would be easier to explain.

Alfvén used these ideas originally to explain cosmic rays and the aurora borealis (a consequence of the solar wind interacting with the Earth's magnetic field). His theory of cosmic rays essentially posits an intergalactic "cyclotron", where the acceleration of charged particles is driven by an electric field contained by a magnetic field (Alfven 1936). That terminology, if not the idea, was certainly inspired by the success of the famous research device of that name, invented by Ernest O. Lawrence in 1929. Later, Alfvén's ideas explained why galaxies form in the first place (thanks to a galactic version of the so-called "pinch effect") and why the galaxies take their characteristic spiral shape (Peratt 1986).

Another word on weakly interacting massive particles (WIMPs), mentioned earlier in connection with "dark matter": Just over a year ago (as I write) a new particle, predicted back in 1931, was "discovered" (and photographed) in a "synthetic" magnetic field. It seems to be the Dirac magnetic monopole (Ray et al. 2014). James Clerk Maxwell asserted that magnetic monopoles do not exist, and most physicists today still insist that these particles do not exist. Perhaps the work in the paper cited above is not conclusive. A "synthetic magnetic field" is not a real magnetic field. However, is it conceivable that this particle, previously unseen (like the Emperor's new clothes), might account for one or more of the experimental results that do not "fit" into the so-called standard model of nuclear forces (Comay 2014). I would bet (a modest sum) on that.

The analogy with the electric monopole (the electron) is obvious. The symmetry is intriguing, but puzzling: Are there two opposite magnetic "charges" (N and S)? That would suggest that magnetic field lines have directionality. We know that moving electric charges along a line produces a "wraparound" magnetic field, with directionality but no need for a magnetic pole. Should a moving magnetic monopole produce a comparable "wraparound" electric field, with directionality but no need for an actual electric charge? What if two magnetic monopoles of opposite "charges" meet? Would they attract each other? If they met, would they annihilate each other? Is that why they are so rare? Or is there a Pauli exclusion principle that keeps them apart?

Two Nobel laureates, Paul Dirac and Julian Schwinger, tried hard to develop quantized electrodynamic theories consistent with magnetic monopoles (Dirac 1948; Schwinger 1969). However, it seems that Dirac's theory included an assumption that the magnetic monopole would "interact" with an electric field. Maybe that assumption is unnecessary. Without that assumption, it seems that the revised theory makes reasonable predictions (Comay 1984). However, nobody seems to have noticed.

Given the evidence for extremely strong magnetic fields ("magnetars") in space—in some cases trillions of times stronger than the Earth's magnetic field—I can't help wondering if there might be a lot of heavy magnetic monopoles out there that don't interact directly with electric fields, hence do not absorb or emit photons. Could the magnetic fields in space be explained by distributions of mono-poles? Could magnetic forces account for the strong interstellar forces keeping galaxies and clusters from flying apart? This question belongs in the domain of plasma physics. (I know the experts will object to this unwarranted speculation on my part, but it is irresistible.)

Plasma cosmology differs from BB cosmology in other ways, especially in regard to nucleosynthesis. This not the place for even an over-simplified explana-tion. What I can say is (1) that plasma (consisting of electrons, protons and other nuclei) constitutes 99 % of all ordinary mass, (2) that the plasma cosmology envisions another category of large scale "objects" in the universe, along with stars and galaxies. This category consists of filaments consisting of stable, self-contained bundles of magnetic field lines enclosed by revolving electrons. And (3) filaments tend to attract each other and (4) filaments in plasma can act as agents to synthesize elements. According to Alfvén, this is what happened during the formation of the galaxies (Alfven 1966).

Already mentioned is the fact that the large galactic clusters are not moving nearly fast enough to have originated only 13.77 billion years ago. Given that different regions of the universe were already too far apart to have influenced each other after 10^{-33} s or so, one would expect to see significant anisotropies in the background microwave radiation. But until 1991 there were no visible anisotropies. The COBE satellite experiment did find some, however, and its follow-up from WMAT (Fig. 4.6) shows significant anisotropies (in the temperature range of ± 200 μK).

This "cosmic map" seems, at first glance, to have taken care of one of the chief remaining objections to the standard Big Bang (BB) theory. However, the anisot-ropies are also consistent with the "foam-like" nature of the universe, where enormous galactic structures like clusters, "ribbons", "shells" and the "Great

From blue=cold to red=hot

Fig. 4.6 The cosmic map. *Source*: NASA/WMAP Science Team—http://map.gsfc.nasa.gov/media/121238/ilc_9yr_moll4096.png

Wall", are interspersed by magnetic fields and plasmas, much of which cannot be explained by the standard BB theory. In particular there is a "cold spot"—possibly a colossal void—that is very hard to explain in terms of the standard BB theory.

Reverting from the galactic to the microscopic scale, it is important to bear in mind that the BB "model" assumes a single "symmetry breaking" event at a point-singularity, at which all the energy of the entire universe was "released" (from the vacuum) by the Higgs Field. That total includes both the so-called "dark matter" and "dark energy" which—according to BB—account for 95 % of the total mass/energy leaving only 5 % for ordinary mass and energy. The argument for "dark matter" is that visible mass/energy is not enough to account for the gravitational stability—the apparent strength of the gravitational fields—around most galaxies. The argument for "dark energy" is that the expansion of the universe is apparently accelerating (based on red-shift analysis of quasars), which seems to require a propulsive force to compensate for gravity.

As the universe expanded, tiny density inhomogeneities, left over from the "inflation" era, evolved into macro-inhomogeneities of unimaginable scale. Galaxies consisting of large numbers of stars formed. Galaxies, in turn, belong to clusters, and the clusters are linked to super-clusters, filaments, bands and so on. These galactic structures are hard to explain, as will be seen a little later.

Earlier theories of "continuous creation" of the universe (e.g. Hoyle) were blown away by the Penzias-Wilson discovery of background radiation. That discovery essentially confirms that something like a Big Bang did take place. It did not, however, explain it. Since 1965 BB theory it has rarely been questioned among cosmologists. Yet Hannes Alfven and Ilya Prigogine (both Nobel laureates), as well as Yilmaz and others, have expressed doubts about the BB (Lerner 1991). The subject is not quite closed. On the other hand, continuous creation does not account for that background radiation or for the backward extrapolation of the observed expansion.

Where do I stand? When I began to write this book I didn't doubt that there must have been a BB. It explains a lot, and the alternatives seemed to me to be unconvincing. Now I am not quite so sure. In fact, I now think that Einstein did make a mistake and that too many physicists have been afraid to challenge EGR. The same holds for QCD. It gives the right numbers for many reactions but I can't help thinking something is missing there, also. Magnetic monopoles? No, probably not. I am just keeping an open mind.

What I am sure about is that, BB or no, the aging of a star—our sun—has been doing work and creating complexity and order. It has also been maximizing entropy. There is much more to say on that later.

4.4 Nucleosynthesis: Love Among the Nucleons

The early history of the universe after the Big Bang (BB), according to current understanding, is roughly as follows: At zero time nothing existed: no mass, not even photons. At time zero-plus, radiation and matter were in thermal equilibrium and there were no significant gradients or asymmetries. In fact at the singularity (the BB) it is thought that particles were massless, and that they only acquired mass, charge, spin and quantum numbers due to "symmetry-breaking" and interaction with something called the Higgs field.

Total energy should have been zero, because matter and anti-matter cancel. But since the universe now has matter, but no antimatter,—at least as far as we can tell—the cancellation may have been imperfect. In that case, it would seem that the law of conservation of energy was violated—or didn't exist. Or else, the anti-matter went in one direction ("south"?) while the matter went "north". Nobody knows, or ever will know.

Or, what if the positive energy of the radiation field, along with the matter that was created at time zero, was compensated by the "negative energy" of the gravitational field of that mass. That would satisfy the overall conservation of energy law. But Newton's laws treat gravitational attraction as positive "potential energy", convertible by any falling apple into kinetic energy. It is a puzzle for theorists that I won't attempt to resolve here.

Moving on, when the universe reached the age of 10^{-4} s, the "soup" consisted of photons, electrons, positrons, neutrinos and quarks. At the end of the first second the quarks had combined to form baryons (protons, neutrons) and mesons. When the cosmic clock reached 100 s the temperature of the "soup" was down only a few billion degrees. (About 23 of those 32 zeros in the power of ten had already vanished). And then, under the influence of electro-magnetic forces, the protons and electrons combined further and formed hydrogen atoms, which were immediately destroyed by collisions with hot photons. As the expansion continued, the temperature of the universe declined more gradually. For the next 700,000 years, free electrons were either annihilated by positrons, or they combined with protons to make electrically neutral hydrogen atoms releasing neutrinos.

In 1948 Ralph Alpher, George Gamow and Robert Herman used the nucleosynthesis theory (see Sect. 4.3) to predict a background (residual) radiation at a temperature of 5 K. That work was forgotten. Later, Robert Dicke independently postulated the existence of a background radiation in the micro-wave spectrum "left over" (so to speak) not from the BB. The first estimate of the temperature of that radiation was calculated by his research associate, P. J. E. Peebles, who estimated (in a preprint) that it should be 10 K. Almost simultaneously came the discovery by two radio-astronomers, Arno Penzias and Robert Wilson, of a microwave background radiation corresponding to a thermal temperature of between 2.5 and 4.5 K (Penzias and Wilson 1965).

We now realize that this radiation originated from a time about 700,000 years after the BB, when free electrons had disappeared (having all been annihilated by positrons or combined with protons in hydrogen atoms). By that time the temperature of the universe had cooled to 3000 K, becoming "transparent" to light for the first time (Weinberg 1977). This was the transition between a radiation-dominated regime to a matter-dominated regime. In the radiation-dominated regime, most of the energy in the universe was in the form of photons, and the photons were hot enough to disrupt atomic nuclei as soon as they were formed. In the matter-dominated regime—where we live now—the photons had cooled enough to allow protons and neutrons to combine, and soon after, to allow hydrogen and helium atoms to exist. After that time, most of the energy in the universe was, and is, either in the form of matter or some "dark" form that does not interact with matter.

According to BB most of the hydrogen (and helium) atoms in the early universe were created around that time. Mass was 75 % hydrogen and 25 % helium. It could not have been perfectly uniform or homogeneous. Quantum fluctuations would have created microscopic density variations. The denser regions would have been gravitationally unstable, meaning that they tended to attract more gravitational mass, exaggerating the differences (and creating gradients). During the next 100 million years or so, the dense regions became still denser, thanks to gravitational and electromagnetic forces, (and hotter, due to collisions) until they became dense and hot enough to allow nuclear fusion reactions to take place, emitting radiation.

The stars began to shine. Figure 4.7 traces these events. Plotting luminosity (the shine) against star surface temperature leads to the famous Hertzsprung-Russell diagram (Fig. 4.8). Note that the larger the star, the lower the surface temperature

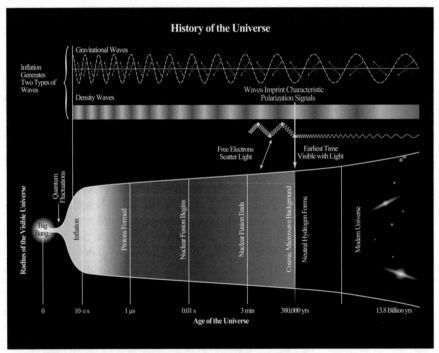

Source: By Yinweichen - Own work, CC BY-SA 3.0,
https://commons.wikimedia.org/w/index.php?curid=31825049

Fig. 4.7 History of the universe. *Source*: By Yinweichen—Own work, CC BY-SA 3.0, https://commons.wikimedia.org/w/index.php?curid=31825049

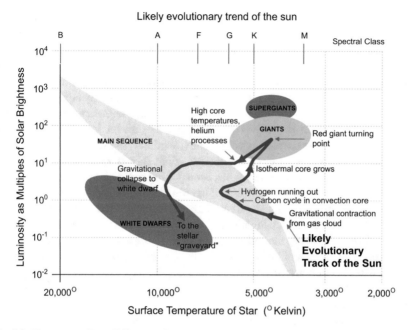

Fig. 4.8 Hertzsprung-Russell diagram of the stars. *Source*: Adapted from Hoyle and Wickramasinghe 78, Figs. 11 and 12

and the redder the light. Our own sun will eventually expand as it cools, becoming a red giant, swallowing up the whole inner solar system, before it finally collapses and becomes a white dwarf. The blue line describes the expected life history of a star like our sun.

When a large star that is not efficiently "mixed" runs out of fuel, the original balance between radiation pressure and gravitational attraction is broken. The gravity starts to dominate. Then the star starts to shrink. The shrinkage increases the internal pressure and internal temperature, accelerating the fusion of remaining hydrogen and other light elements, up to iron. This process can maintain the radiation pressure enough to balance the gravitational attraction, for a while. But eventually gravity "wins" and the star collapses. If the star is a giant, the collapse can be very violent, generating a compression shock-wave and lot of kinetic energy. In fact 99 % of the energy released is in the form of neutrinos. The result is an extreme compression in the core, resulting in very high temperatures and pressures lasting only a very short time in cosmic terms.

Under those conditions fusion processes are accelerated, using up the remaining hydrogen "fuel" rapidly, causing an increase in brightness. During that time endothermic nuclear reactions take place, producing heavier elements. But that endothermic process of heavy element production also uses energy, which accelerates the end-game. When a very massive star reaches the end of its life, that sequence of events happens very fast—in fractions of a second. The kinetic energy of the collapse—together with the last burst of fusion energy—results in a colossal explosion that can dissipate as much energy as a whole galaxy. We know those explosions occur, because astronomers see them, from time to time. Chinese historians recorded a big one in the year 1054 CE. It was bright enough to read by for several months. We know they produce heavy elements because the heavy elements exist, and there is no other way they could have been produced.

The heaviest elements are inherently unstable, which is why some of them are radioactive. *But none of the known radioactive decay sequences ends with iron.* Hence all the iron in the universe—and there is quite a lot—must have been produced by fusion processes in young stars, while all the heavier elements were created by fusion in the cores of aging super-giant stars during the explosion.

The hypothesis of "dark energy" has been put forward since the 1990s to explain astronomical measurements of events (supernovae) that occurred billions of years ago and far away. Those observations indicate that the "red shift", which determines expansion velocity, changes non-linearly with distance. (If the expansion rate were uniform over time, the red shift would increase linearly with distance). But observations of "standard" supernovae[12] shows—or seems to show—that the rate of expansion is not constant, but is accelerating (as indicated by events from very far away and long ago). Data from gravitational lenses, since 2006, have confirmed this acceleration.

So how is one to explain the apparent acceleration? The answer accepted by cosmologists is "dark energy". The effect of "dark energy" is supposed to be a kind of negative pressure, i.e. repulsion that uniformly suffuses all space. This is counter-intuitive, but is actually just a trick to make the equations of Einstein's General Relativity (EGR) fit the data.

Astronomical observations from type 1-A supernovae by the High Z supernova search team (1998) and from the Supernova cosmology project (1999) set the fox (so to speak) among the chickens. The results have been subject to a lot of verification (and interpretation), resulting in the award of three Nobel Prizes to the project leaders (Saul Perlmutter, Brian Schmidt and Adam Riess) in 2011. It seems that acceleration of Hubble expansion is now deemed to be a fact. But, again, there are doubters. Could it be that the observed red shift is an artifact of Einstein's theory (EGR)? Might YGR give a different result? The answer to that is 'yes'. The standard model combined with Einstein's general theory of relativity can be used to calculate the amount of "propulsive energy" (or negative pressure) needed to accelerate the ordinary mass of the universe (as much as has been observed). It is of some interest that the dark energy appears to be very similar in behavior (within 10 %) to Einstein's "cosmological constant". The current estimate is that "dark energy" constitutes 68.3 % of the total energy in the observable universe. The remainder consists of 26.8 % attributed to the energy equivalent of dark matter possibly neutrinos and 4.9 % to ordinary matter (using $E = mc^2$).

---

[12] Type 1-A supernovas are known as "standard candles" in the trade, because their luminosity is both extreme and consistent.

As I mentioned in Sect. 4.1, Sir Arthur Eddington was the first to suggest that stars might be powered by hydrogen fusion into helium (Eddington 1920). He had no idea how fusion could take place, in detail, of course. That was explained later by Hans Bethe. He was also the first to relate the brightness of stars to mass (1924) and to atomic composition (1926). (He guessed wrong, probably based on geological information about the Earth, that the stars would consist mostly of iron. See Fig. 4.9.) In 1928 **George Gamow** calculated the probabilities of nuclear fusion reactions as a function of temperature (the Gamow factor). In the early 1930s Atkinson and Houtermans showed that nuclear transformation processes do take place in stellar interiors.

The next step was the two papers by Hans Bethe in 1939. Bethe worked out the details of both the proton-proton chain reaction and the catalytic carbon-nitrogen cycle (Bethe 1939). The latter process, which is more important for the fusion of heavier elements, was first suggested by Carl Friedrich von Weizsäcker (1938). The six step carbon cycle is as follows: $C^{12} + H = N^{13} = C^{13} + \nu$; $C^{13} + H = N^{14} + H = O^{15} = N^{15} + \nu = C^{12} + He^4$. (H is a hydrogen ion proton). The symbol ν refers to a gamma ray emission. Since the carbon C^{12} (and the other isotopes) are recreated at the end of the cycle, they are effectively just catalysts. The two gamma rays are the net energy output of the cycle. See Fig. 4.10.

It was the idea of the primordial "hot" BB, dominated by radiation rather than matter, that led George Gamow and his student, Ralph Alpher, to postulate a generic process of "nucleo-synthesis" to explain the creation of the elements (Alpher et al. 1948). The paper was from Alpher's doctoral dissertation. Gamow

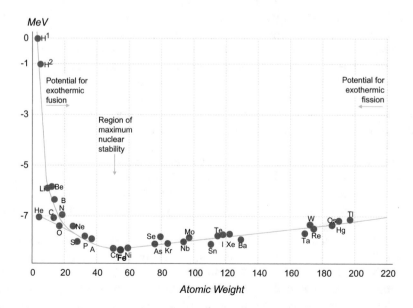

Fig. 4.9 Nuclear binding energy (potential) per nucleon. *Source*: Adapted from Aston 35, 36; Bainbridge 32, 33

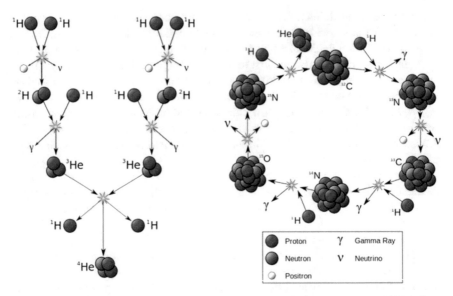

Fig. 4.10 The proton-proton chain and the carbon-nitrogen cycle. *Source*: https://i.kinja-img. com/gawker-media/image/upload/s–NMQ2c75w–/c_scale,fl_progressive,q_80,w_800/hkqovttxw s3qmkwajihs.jpg

suggested the topic. Bethe's name was added (with his permission) as a pun on the first three letters of the Greek alphabet. It was good PR. Everybody read that paper.

Their scheme works more or less as follows: the centrifugal radiation pressure from fusion processes in young and middle-aged stars that started from hydrogen and helium balances the centripetal magnetic and gravitational forces for millions (or billions) of years. But, as the star gets older, the hydrogen-helium fraction decreases, while the fraction of heavier nuclei like carbon increases. Meanwhile the radiation energy output of the fusion process as a whole gradually decreases because the fusion of heavier nuclei—the ones lighter than iron—releases less and less energy. Finally the star consists mostly of iron. Then the centrifugal radiation pressure gradually begins to decrease and the star starts to shrink, becoming smaller and denser.

Gamow and Alpher thought their theory could explain the creation of all the elements, although it turned out later that it only accounted for the first few light elements, through beryllium. However they also predicted the existence of microwave background radiation, later discovered by accident in 1965. That was an important prediction.

Fred Hoyle extended the theory of nucleo-synthesis to the heavier elements (Hoyle 1947, 1954). It is now known that carbon fuses with carbon to produce silicon; then silicon fuses with silicon to form iron, nickel and cobalt. In fact the temperature of a newly collapsed star will be of the order of 100 million K, hot enough to synthesize heavy elements, as already noted. It is estimated that 99.9 % of all stars end their lives as "white dwarfs" (because helium fusion still continues and they are still white hot). When the helium fusion and the carbon and silicon fusion are finished, eventually, they cool off and become "brown dwarfs", which are essentially the first stage of a process of gravitational compression leading to neutron stars. Some of those—still spinning thanks to leftover angular momentum—can be seen as "pulsars". Only one in a thousand gravitationally collapsing stars becomes a supernova and spreads its matter around the galaxy.

Physically, if the star is massive enough, the gravitational collapse compresses the atoms, forcing the outer electrons into contact with the inner protons, producing neutrons and neutrinos. Neutrinos normally move at nearly the speed of light, but in a strong gravitational field they, too, slow down thanks to the *gravitational redshift*. This process begins in the core of a collapsing star. Normally neutrinos barely interact with other particles, but in a gravitational collapse of sufficient magnitude, there is a kind of "tsunami" of neutrinos moving outward. This causes a chain reaction that speeds up the electron-proton reactions, not dissimilar to the chain reaction in a nuclear explosion but on a much greater scale. A fraction of those neutrinos is absorbed by the dense mass of the outer shell of the star, raising the temperature and pressure very suddenly. That neutrino wave is presumed to be the direct cause of a supernova explosion.

The current understanding of the composition of a red giant star just before becoming a supernova is shown in Fig. 4.11. It resembles an onion, with distinct

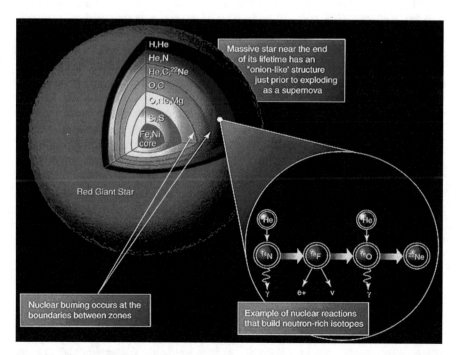

Fig. 4.11 The structure of an "old" red giant before exploding as a supernova. *Source*: https://en. wikipedia.org/wiki/Stellar_nucleosynthesis

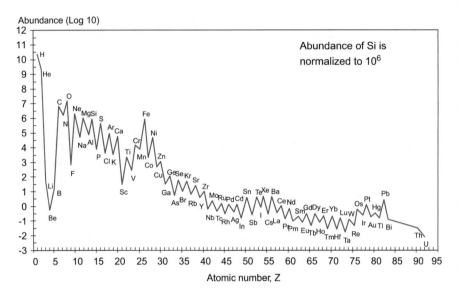

Fig. 4.12 Solar system abundances. *Source*: https://commons.wikimedia.org/wiki/File:
SolarSystemAbundances.png#/media/File:SolarSystemAbundances.png

layers. Note that there is virtually no hydrogen left, except in the thin outer layer, and most of the mass consists of heavier elements including carbon, nitrogen, oxygen, and so on. The fusion reactions occur along the boundaries of the layers. These elements will be widely scattered by the supernova explosion. It seems that lithium, beryllium and boron were not actually produced in stars, but probably resulted from fissions of heavier elements in clouds, induced by energetic cosmic rays (also from supernovae).

The final challenge was to explain the relative abundance of the elements in the universe, based on the theory of the nucleosynthesis. This effort was completed in a famous survey paper, known as BBHF, for the initials of the four authors, published in 1957 (Burbidge et al. 1957). It is important to note that the abundances are logarithmic, which means that the peaks and troughs of the abundance curve are very far apart. The peak at iron is directly correlated to the minimum point in Fig. 4.9, which corresponds to maximum stability. The saw-tooth pattern in Fig. 4.12 reflects the greater stability of atoms with even atomic numbers *vis a vis* their neighbors. That, in turn, is due to the Pauli exclusion principle which allows pairs of electrons with opposite spins to "cohabit" an orbit.

Figure 4.13 shows the origins of the elements. Only hydrogen and helium were produced in the BB. The next three (comparatively rare) elements (lithium, beryllium, boron) seem to have been produced by cosmic rays interacting with heavier elements. When the core of the pre-supernova star consists mostly of iron and other elements heavier than helium, the outward radiation pressure plus the centrifugal force from rotation that keeps gravity at bay, gradually disappears and the star undergoes a gravitational collapse. If the collapse is rapid enough the remaining

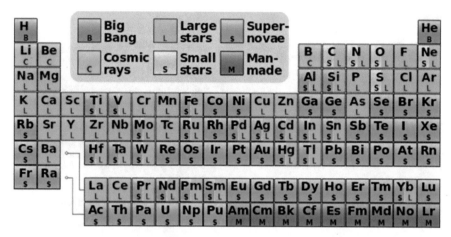

Fig. 4.13 Periodic table showing origin of elements. *Source*: https://en.wikipedia.org/wiki/Stellar_nucleosynthesis

light elements undergo fusion resulting in an explosive release of energy, i.e. a supernova. That, in turn, blows all the lighter materials in the outer shells away, while leaving the dense iron-rich core, spinning to preserve the angular momentum of the original star.

Neutron stars may also account for *pulsars*. It was suggested long ago that rotating neutron stars with very strong magnetic fields can emit light, X-ray or gamma radiation along the magnetic axis. This angular momentum is presumably left over from the original pre-supernova, before it blew off its outer layers. Compressing the angular momentum of a large star into a much smaller one tends to increase the speed of rotation. It is known, by the way, that neutron stars can accrete new mass. This can happen, for instance, if the original star was one of a binary. The additional mass can hide or bury the magnetic field thus weakening it (and extending the lifetime) of the neutron star, or pulsar.

The final challenge was to explain the relative abundance of the elements in the universe, based on reaction probabilities and the stability of reaction products. But (according to theory) in the core itself the protons and electrons under intense pressure may recombine to form neutrons, so the remnant becomes a neutron star.[13] See Fig. 4.14. Apparently the neutron stars tend to have masses equivalent to two of our sun's mass, radii about 10 km (1/30,000 that of the sun), densities at the surface like the density of atomic nuclei and temperatures—when young—around a million K. Why such uniformity? The answer is probably that stars with smaller masses

---

[13] Beta-decay converts neutrons into protons and electrons, plus neutrinos. It occurs (very rarely) inside the nuclei of atoms, involves the emission of neutrinos or anti-neutrinos and kinetic energy. I wonder if at high enough temperatures and pressures, inside a massive star the process may be inverted, converting protons and electrons back into neutrons. The process known as "electron capture" is a clue. The source of neutrinos and anti-neutrinos is still mysterious.

Fig. 4.14 Hypothetical
structure of a neutron star.
Source: http://www.
physics.montana.edu/
people/faculty/link-bennett.
html

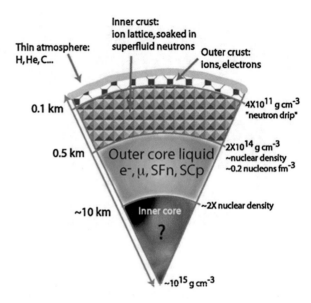

don't blow up at all, while Einstein's General Relativity theory (EGR) says that if the mass is greater than 1.4 solar masses, it will collapse all the way into black holes. YGR says that the black holes don't exist, which implies that the collapse produces an ultra-dense neutron star, or possibly a quark star. Yilmaz did not elaborate.

To summarize: as the universe expanded and as the temperature dropped, some of that "original" energy "condensed" into mass. And, as time has gone on, the simple mass (hydrogen) has been transformed into other elements. Not only that, the (relatively) homogeneous primordial gas has also evolved into stars, galaxies, clusters, bands and so forth. The stars, in turn, may have planets and the planets may have moons.

In short, the system evolved and became more, and more complex. This evolution reflects symmetry-breaking on every scale from the atomic to the galactic. Symmetry-breaking has become rather central to both modern high energy physics and cosmology. (See, for example Barrow 1991). It seems that symmetry-breaking is also very relevant to subsequent evolutionary complexity-generating "phase changes", especially in the realm of chemistry and biology, as described in later chapters of this book.

The ultimate in complexity, of course, is the human brain (Sagan 1977). But brains cannot exist without bodies, and bodies require sustenance, which has to be obtained by various means from other complex organisms, and/or by manipulating and utilizing inorganic materials. The accumulation of useful animals and useful material objects was the first form of human wealth.

4.5   The Sun and Solar System

As indicated earlier, stars are created by a process of condensation of a gas cloud. At the beginning of time (after the BB) the gas cloud was a combination of hydrogen atoms with some helium. By the time our sun was created, about 4.57 billion years ago, a great many stars (in many galaxies) had been created and died. Some of the smaller ones were still shining, but the most massive among that first generation had exploded as supernovae. Those explosions distributed mass, in the form of dust. For some reason a lot of that dust was (and is) in the form of silica.

The dust clouds resulting from those explosions are widely distributed throughout space, and they can be seen by modern telescopes. Figure 4.15 is an example of such a cloud. (Many pictures of interesting clouds, from orbiting NASA telescopes can be found on the Internet.) New stars are formed by gravitational condensation of such clouds. Our sun is one of them.

Our Sun is an average "main sequence" star; not big enough to explode as a supernova when the time comes, but not too small. It has a long life expectancy (10 billion years or so). It is located toward the outer fringe of a galactic spiral arm that we see as the Milky Way. The Sun is nearly white in color, which means it has a surface temperature of roughly 5800 K. While it varies slightly from time to time and from place to place, the surface of the Sun is hotter than any continuous process

Fig. 4.15 A gas cloud giving birth to new stars. *Source*: A gas cloud in Casieopeia, 7000 light years from Earth. The *red color* indicates that organic compounds are present. http://photojournal.jpl.nasa.gov/catalog/PIA03096

on Earth. The internal temperature of the sun, due to fusion processes in the core, is much higher. There are no solids or liquids at those temperatures, nor even gases consisting of neutral molecules. The material substance of the sun, and every star, consists of a *plasma*: a mixture of free electrons, free protons, neutrons, deuterons, tritium nuclei, helium nuclei and the nuclei of some heavier elements.

This solar plasma is not homogeneous or uniform. There are enormous temperature, density and pressure gradients. Somehow the gradients resist change. To overcome the gradients the plasma "boils"; it is a turbulent fluid, both mechanically and electromagnetically. Plasma is a nearly perfect electrical conductor in certain directions (along "frozen–in" magnetic field lines) and highly resistant to current flow across the field lines. When electrical currents flow through a plasma, they generate powerful magnetic fields capable of "pinching"—and cutting off—the current-carrying filaments. The magnetic fields help to contain the superheated plasma in the solar core. To dissipate the heat produced in the core (which has to go somewhere) a star continuously emits radiation (including visible light), from its core to its "photosphere" and then from the photosphere into surrounding space. But from time to time it also hydro-magnetically propels jets of charged particles, like flares.

Your child may ask: What makes the Sun shine? Even though we can't go there, we think we know the answer from indirect evidence. This is called the Standard Solar Model, or **SSM**. In the very center of the star nests a "fusion chamber" with a diameter of nearly 300,000 km. This is about two fifths of the diameter of the sun, which is about 700,000 km altogether. This central fusion chamber is contained not by metal walls but by a combination of gravitational attraction and magnetic fields. In this fusion chamber, 600 million tons of hydrogen nuclei are fused into helium nuclei every second.[14] Some of the helium is also fused into heavier elements, such as boron, beryllium, lithium, neon and carbon. The density of mass in the fusion chamber is 150 times the density of water. The density of particles is such that photons cannot escape quickly. It has been estimated that a photon produced in the core takes a million years to get out of the Sun, although it only takes 500 s to travel from the surface of the Sun to light up the Earth.

The mass difference between the hydrogen nuclei and the fusion products (helium nuclei and others) amounts to about 4 million tons, per second. (About 1 million tons per second boils off the photosphere of the sun in the form of "solar wind".) The Sun is losing weight. In a reversal of the primordial formation of matter from energy in the Big Bang, this mass difference **m** is converted back into energy **E** in the form of electromagnetic radiation. The energy contained in that radiation is

---

[14] The fusion process is fairly complex, starting from hydrogen nuclei (protons), then to neutrons, then Helium 3 nuclei and finally Helium 4 nuclei. The "recipe" for a Helium 4 nucleus is: 4 protons (hydrogen nuclei) minus 2 positrons, 2 neutrinos and 2 photons. The two positrons emitted along the way eventually combine with two electrons, producing two more photons. Two other electrons can combine with the helium nucleus to form a helium atom. The star as a whole remains electrically neutral, but internally the positive and negative charges can be somewhat separated, resulting in electric voltages, currents and magnetic fields.

given by Einstein's energy-mass equation $\mathbf{E} = \mathbf{mc}^2$, where $\mathbf{c}$ is the constant velocity of light. The power generated this way is defined as the *solar luminosity* $\mathbf{L}$ which has the value 3.845×10^{26} W or 3.845×10^{14} TW. By comparison, all the power stations on Earth combined produce less than one terawatt of power.

The Sun radiates this energy away at a surface temperature of 5777 Kelvin (K). It spreads outward into space as a continuous spherical flow of electromagnetic radiation, which consists of high-energy photons plus a "solar wind". The energy flux from the Sun is continuous. But imagine it as an expanding wave-front. When this wave-front hits the top of Earth's atmosphere, after having traveled the (average) distance $\mathbf{D}$ of 150 million km (93 million miles), it has been diluted considerably by distance. But sunlight still carries power to the Earth at the rate of 1.367 kW per square meter. This number, the *solar constant* $\mathbf{S}$, is the luminosity $\mathbf{L}$ defined above divided by the area of the expanding wave at distance $\mathbf{D}$ or ($\mathbf{S} = \mathbf{L}/\mathbf{4D}^2$).

The total amount of power the Earth receives continuously from the sun is about 10^{17} W, or 10^5 TW. That is a very tiny fraction of the sun's total output. But it is still about 10,000 times the total amount of power generated by the burning of fossil fuels on Earth throughout the period from the discovery of coal to the end of the twentieth century. There is evidently no shortage of potential solar power reaching the Earth. The problem is to capture and utilize it to sustain life.

I should note that the SSM is still only a theory, not a fact. There is an alternative theory (not taken seriously by cosmologists, but hard to dismiss out of hand): it is the Pulsar Centered Sun (PCS). This seems to explain some of the problems of the SSM, such as the "faint young Sun" paradox, and the surprisingly large amount of iron, nickel and even heavier elements both in the sun and the inner planets of the solar system. This heterodox idea, very briefly, is that our sun may be a second generation star that was created when a neutron star—left over from an earlier supernova—blew itself apart. The pieces of the dead star accreted a lot of hydrogen and other light elements, thus recycling itself. The powerful magnetic field of the interior pulsar would have been diluted or "buried", becoming effectively invisible.

The "faint young Sun" paradox is that, according to the SSM, our Sun should have been 30 % cooler 5 billion years ago. This is taken for a fact by most scholars. But there are several lines of evidence that are difficult to reconcile with a cooler Sun. For example, there are some indications that the surface temperature of the earth 4.5 billion years ago may have been as high as 80 °C, which is not consistent with the cooler Sun.

Another strand of evidence favoring the PCS—and one I find intriguing—is that several of the organic pigments associated with early organisms, such as the purple bacteria, are most sensitive to light emitted at a somewhat higher—not lower—temperature than 5800 K (Michaelian and Manuel 2011). That suggests a cooling Sun, not a warming one. The PCS theory postulates a 5 % cooling.

One more unsolved problem with the SSM is the **chirality** problem. Chirality is Greek for "handed-ness". It has been mentioned earlier in connection with the "weak" nuclear force. It is the property of breaking mirror-symmetry: i.e. not being super-posable on a mirror image. Your right hand and your left hand are obvious examples of chirality. Whereas all known chemical synthesis routes lead to equal probability of L and R chirality, it is a fact that amino acids in living organisms are all "left handed" (L) whereas the nucleotide bases, RNA and DNA are always "right-handed" (R). One of the few ways to explain this would be if the light under which early life evolved was circularly polarized. That happens to be exactly the kind of light a rotating neutron star would have emitted.

Obviously the PCS theory is not much more than a wild idea at present. But wild ideas occasionally turn out to be right. This one can't be dismissed, for now.

4.6   The Elements Needed to Sustain Life (and Technology)

Most people are aware that the most important element in living organisms is carbon. This is because carbon has a unique ability to form chains by combining with itself. All the crucial molecules in living organisms are built around carbon chains.[15] Carbon dioxide (CO_2) is the original source of all carbon in plant and animal life. To be sure, hydrogen, oxygen and nitrogen are also pervasive in organic molecules, while phosphorus plays a central role in genetic materials, cell walls and energy transport. All living organisms seem to have started in water (H_2O) and all cells contain water. Hydrogen, derived from water by photosynthesis and respiration is also a major component of almost all the important molecules of living organisms. In fact 80 % of our body mass is water.

But there are other elements essential to life. Nitrogen, phosphorus, sodium, potassium, calcium, iodine, magnesium, chlorine, sulfur, and iron are among them. Nitrogen is an essential ingredient of all amino acids, from which all proteins are constructed, and sulfur is also an ingredient of several of the amino acids. Hence all proteins contain nitrogen and sulfur as well as carbon, hydrogen and oxygen. The genetic molecules, DNA and RNA, are held together by "backbones" of sugar molecules linked by phosphates. Cell walls consist of lipids linked to phosphates. The energy metabolism in all known organisms, from the most primitive single-cell organisms to ourselves, requires a pair of phosphorus compounds: adenosine diphosphate and adenosine tri-phosphate, better known as ADP and ATP. I will say more about them later.

Calcium and magnesium are needed for shells, bones and teeth as well as other functions. Iron is an essential constituent of hemoglobin, the protein in animal blood. Iodine is required by the thyroid gland, which regulates the metabolic rates of animals. Hydrochloric acid is made in animal's stomachs for digesting food. Sodium and potassium are essential blood pressure regulators. Moreover, every health-food addict knows that there are other important trace elements, such as boron, cobalt, fluorine, manganese, copper, zinc, selenium and chromium that are

---

[15] Silicon also has this capability, although no known living organisms are silicon-based.

needed by some organisms (including humans) at least in tiny amounts. Some are heavier than iron. So where (your child might ask) did they come from?

The answer for elements heavier than helium is *nucleosynthesis* in stars, as discussed in Sect. 4.3. But there is a different question: though quite a few of the known elements are utilized by living organisms, the majority, especially the elements heavier than iron (atomic weight 56), are not. The spontaneous (*exothermic*) nuclear fusion process—from hydrogen to helium—that powers our relatively young Sun continues from helium through boron, lithium, fluorine, carbon, nitrogen, oxygen, silicon and so on, producing heavier and heavier nuclei. The exothermic fusion process ends with iron nuclei. Of the 92 elements found naturally in the Earth's crust, 66 elements are heavier than iron. They include cobalt, nickel, copper, zinc, molybdenum, tin, silver, gold, lead, and uranium, the heaviest natural element in the periodic system. (Plutonium and other transuranics are "artificial" in the sense that they were produced (on Earth) only by human activity.)

But these heavy elements, as explained earlier, can only be produced by *endothermic* fusion processes at temperatures above 10^8–10^9 K, much higher than those in the core of our sun today. How does this happen? When the balance between radiation pressure and gravity changes slowly, the star may shrink gradually into a "white dwarf". But if the star is large enough, the contraction is faster: It can endure a sudden, violent collapse. When that happens, the temperatures and pressures at the center of the dying star rise and accelerate the fusion process. The remaining exothermic fusion energy is released very quickly. The result is a violent explosion: a nova or supernova. During the short period of super-compression, super-high temperatures occur and some endothermic fusion also takes place. This produces the nuclei of elements heavier than iron. Probably the heavy elements in the Earth's crust are all products of a supernova that occurred 4.55 or 4.6 billion years ago somewhere in our galactic "neighborhood".

In recent years astronomers have discovered that there are several generations of stars, and that most of the ones now shining in the sky are not from the first generation. It seems the first stars, which began from clouds of hydrogen and helium, died relatively early, leaving clouds of solar "ashes" that included some heavier elements. Then new stars were formed in which some heavier nuclei were present from the beginning. The greater the fraction of heavier nuclei (lighter than iron), the weaker the fusion process is overall; the cooler a star is, the more long-lived it will be. Our own Sun is one of the most recent stars to be formed, being much younger than the universe as a whole.

Meanwhile, all the elements heavier than iron, both those needed for life and those not needed (or actively toxic), were created by fusion processes in a now-extinct star, *but not in our Sun*. In the Sun's atmosphere, to be sure, there are some traces of heavy elements that were not made in the Sun itself, but would not be present if the Sun had been formed from hydrogen and helium alone.

That prehistoric supernova also left other atomic debris, of course. It is not known how long it took for the "ashes" of the supernova to agglomerate into clouds and finally into planetary form, collected in the gravitational field of our youthful Sun. The best guess is that the planet Earth was "born" about 4.5 billion years ago.

It is known from isotopic analysis (mainly of carbon isotopes 12, 13 and 14) that living matter did not appear on Earth for another 600 million to a billion years thereafter.[16] But we can say for sure that the Earth—and everything on Earth itself, including the atoms in our bodies—has already been processed through the inside of a supernova. To review, the material of the Earth and the other planets, and the material in all living organisms, including our bodies, was the product of an extremely energy-intensive process in an explosively dying star. We can never hope to duplicate that process in a laboratory or factory on Earth.

4.7   The Terra-Forming of Earth

Five and a half billion years ago, an observer (if such a being can be imagined) would have seen a young Sun surrounded by a dense cloud of interstellar dust probably consisting partly of hydrogen and partly of the debris from supernova explosions. Recall the photo of such a cloud (Fig. 4.15) from Sect. 4.4. The Sun may have been formed by the gravitational collapse of the same interstellar dust cloud that formed the planets. The fact that the heavy elements are much more concentrated in the Earth's crust (and presumably in other planets) than in the Sun itself suggests the possibility of the latter, but that would introduce other difficulties. The cloud must have gradually evolved into a set of discrete concentric "shells", or a disk configuration in a single preferred plane, somewhat like the pattern of our galaxy, the "Milky Way". As mentioned earlier, the disk-shape is explained by electromagnetic forces in plasmas (Alfven and Falthammar 1963, Alfven 1981)

As millions of years passed, there would have been some condensation of mass in the dusty disk, as denser parts of the cloud began to form "gravity wells" and collect their own "halos" at distances from the sun corresponding to the "shells". After several hundred million years, those clouds must have condensed into solid "proto-planets", with smaller "moons" rotating around them within their gravity wells. The Earth was born 4.54 billion years ago, according to current theory.

In the case of the Earth's single moon, we now understand that it was created soon after the Earth itself, as a result of some cataclysmic convulsion—probably a collision with a fairly large mass, roughly the size of Mars. (The "impactor" must have been that big to account for the angular momentum of the system.) This collision is now thought to have occurred about 4.51 billion years ago. A large chunk of low-density matter was carved out of the outer layer of the original Earth. This mass of fragments then became a ring (like the rings of Saturn) before finally

---

[16] For reasons that are not well understood, the photosynthetic carbon fixation process mediated by the enzyme ribulose-1,5-bisphosphate carboxylase discriminates against the heavier isotopes of carbon. Consequently 13C is depleted by about 25 % in organic carbon, as compared to inorganic carbon. Thus the ratio of the isotopic composition of carbon becomes a definite indicator of how long ago the photosynthesis occurred.

aggregating into a satellite. This theory accounts for why the Moon has no dense iron-nickel core like the Earth.

The distance between Earth and Moon was then only about a tenth of the present distance, and the Earth rotated much faster than it does today, producing enormous gravitational tides. When the Moon was 1.5 billion years old, about 3 billion years ago, the tides were still 300 m high and the lunar day was about 8 h long. In fact, the tides have effectively locked the Moon into its present configuration such that one side faces the Earth permanently while the other side of the Moon is permanently invisible to us. The Earth now rotates once every 24 h and the Moon now rotates around the Earth every 27+ days, from a distance of 384,400 km (average). The Moon's rotations around the Earth account for the tides, as well as eclipses.

In short, by 4.5 billion years ago our existing planetary configuration was set: a concentric set of planets revolving independently of each other, but within in the planar disk. The outer planets condensed in the cold, and kept most of their hydrogen, helium, and ammonia, while the inner plants are denser with more heavy elements. All of this sorting-out activity created order (a pattern) from disorder by "self-organization". The material condensation process that created the planets left the rest of the space around the sun relatively empty and clean.[17]

The word "clean" is relative, because the space outside the disk was—and still is—full of the objects known as meteorites and comets, many of which still remain. In fact, for the first 700 million years after the formation of the Earth (from 4.5 billion years ago to 3.8 billion years ago) there were at least 20 million major asteroidal impacts—one every 35 years on average—which left imprints all over the surface of the Earth and the Moon. This period is called the late heavy bombardment (LHB). Could life have been created on Earth under that bombardment? The short answer is: difficult but not absolutely impossible. The earliest possible date seems to be 4.4 billion years ago, while the *likely* latest date is 3.85 billion years ago after the end of the LHB, while the latest *possible* date consistent with fossil evidence is 3.4 billion years ago: quite a range (Lenton and Watson 2011, Chap. 1).

The proto-planets were seismically active at first because, as they accumulated mass from the outside, the insides heated up due to pressure from the weight of the surface layers above. Most of the interior heat comes from radioactive decay due to spontaneous (exothermic) fission of the radioactive elements, notably uranium isotopes (233 and 235), left over from some earlier supernova. As a matter of interest, it was the discovery of radioactivity (c. 1895) which explained the molten core of the Earth and revised earlier assumptions about its age: without that heat from radioactive decay the Earth would have cooled down to its present temperature in a mere 10 million years or so, rather than the 4.5 billion we now believe.

---

[17] The asteroids came later. They were once thought to be fragments of a planet that once existed between Earth and Mars, but which blew up, long ago, for unknown reasons. Now they are thought to be fragments that never "coagulated" into planets, probably due to gravitational disturbances from the giant planets Jupiter and Saturn. The Earth's moon may have resulted from that collision, according to some authorities.

Very early in the accretion process, the heaviest elements sank and accumulated in the center of the planets, while the lighter elements like silicates and carbonates of magnesium, sodium and calcium floated to the surface. The magma is not homogeneous in composition. Some parts are hotter or denser than others, for a variety of reasons. There is a constant "churning", even today, as the magma finds weak spots in the solid crust above, and intrudes or erupts. Vulcanism was much more intense when the Earth was young because the internal heat source that drives vulcanism (radioactive decay) was three times more intense than it is now. The insides of the planets were (and still are) hot enough to liquefy, but the outer shells solidified and very gradually got thicker, despite the LHB. The interior liquid (magma) is an active chemical reactor that plays an important role in several of the abiotic material cycles. There is even a theory, strongly advocated by Russian and Ukrainian petroleum geologists, but dismissed by most Westerners, that petroleum and natural gas are formed by non-biogenic reduction processes in the magma (Kudryavtsev 1959; Gold 1993, 1999; Glasby 2005; Lollar et al. 2006). The dispute is partly ideological, which makes it a risky topic for an outsider such as myself. However I need to take the abiotic theory somewhat seriously, for reasons that will be clear shortly.

Some of those volcanic eruptions in the earlier history of the Earth were colossal by present-day standards. They also carried dissolved gases to the surface, including water vapor, carbon dioxide, methane, ammonia, and hydrogen sulfide. The existence of all mineable deposits of metal ores, from iron and aluminum to copper, zinc, lead and the platinum group of metals (PGMs) is due to various geochemical and hydrothermal processes taking place at the interface of the magma and the crust (Rubey 1955; Ringwood 1969).

For example, we know that porphyry copper deposits, consisting of crystals of chalcopyrite (copper sulfide) and other minerals—the ones mainly being mined today—were created by cooling and crystallization from hydrothermal water expelled from magmatic intrusions associated with volcanoes. This is why most of the mineable copper deposits in the world are located around the "ring of fire", especially along the tectonic margins of the Pacific Rim (Kesler 1994). Those mineable deposits are a major part of the natural wealth of the planet. I will have more to say about preserving that wealth in the last chapter of this book.

The two planets nearest the sun were (and are) very hot. The one nearest the sun, Mercury, has almost no atmosphere at all. That is because the planet is very small, and the gravity well is not very deep. Hence hydrogen and helium molecules just boil off into space. The next planet is Venus, about the same size as Earth. It has a dense and very hot atmosphere full of carbon dioxide, with traces of ammonia, nitrogen or nitrogen oxides, sulfur oxide, water vapor and other gases. In the atmosphere of Venus, water exists only as vapor (steam) at a temperature of 477 °C. There is no pure oxygen in the atmosphere and no liquid water on the surface of Venus. Venus could not support carbon-based life as we know it.

On the other hand, Mars is too cold, although the possibility of some liquid water in the distant past has not been ruled out. In the atmosphere of Mars there is scant water vapor. The vast majority of water on Mars is in the form of ice or is trapped underground.

Earth is the only one of the ten planets that is suitable to retain liquid water on the surface.

Water is of critical importance, not only for life, but for geology. The hydrological cycle (see Sect. 6.2) is a powerful driver of terraformation. Erosion is responsible for much of the surface topography—the Grand Canyon for one example—but also for the dissolved mineral salts in the ocean, from sodium chloride to magnesium, bromine and lithium. In biology, water is the source of hydrogen in the formation of carbohydrates $(HCO_2)_n$ by photosynthesis and by-product oxygen for respiration. Water is also the carrier of hydrogen for abiotic methane production in the magma that accounts for some of the methane in volcanic gases and the methane clathrates under the ocean sediments. More on that later.

The Sun was a little fainter and 15 % cooler 4.5 billion years ago than it is now. The Earth was wrapped in a uniform gray layer of clouds. The atmosphere then probably consisted mainly of nitrogen, methane, water vapor, and carbon dioxide. It is not known for sure how much free hydrogen was in the atmosphere. The carbon dioxide was up to 1000 times more concentrated than it is in the atmosphere we breathe today. There was also some hydrogen sulfide and ammonia present. There was no free oxygen in the Earth's atmosphere when life first appeared. Consequently there was no ozone in the stratosphere to block the intense UV radiation. In fact the appearance of oxygen, and the ozone layer that followed, were, arguably, the most significant of the several "great revolutions that made the Earth" (Lenton and Watson 2011). That event occurred somewhat before 2.7 billion years ago.

If it were not for the carbon dioxide and water vapor in the atmosphere, the Earth's temperature today would now be below the freezing point of water, $-18\,°C$. (That is the temperature at which incoming solar insolation and outgoing infra-red (IR) radiation would balance.) But in reality the Earth was quite hot by today's standards. In fact, the early crustal temperature was about 85 °C, not far below the boiling point of water. It was that hot partly because of the more intense "greenhouse effect" at that time. The Greenhouse effect arises from the fact that carbon dioxide, methane and water vapor all absorb infra-red (IR) radiation—heat—from the Earth's surface and then re-radiate it back to the Earth. These gases behave like a thermal blanket.[18]

Gradually the water vapor in the Earth's original atmosphere condensed and precipitated as rain to form oceans, with additional water arriving from interplanetary space in the form of comets (Frank et al. 1990; Frank 1990). There is indirect but strong evidence (from NASA) that small comets of around 100 tons, mostly ice, still enter the Earth's atmosphere about 20 times a minute. Cometary

---

[18] As a matter of interest, Jean-Baptiste Joseph Fourier (1824) was the first to realize that the Earth would be much colder than it is if it were not somehow being warmed by its atmosphere. John Tyndall was the first to discover that nitrogen and oxygen cannot be responsible, whereas carbon dioxide and water vapor both absorb infra-red radiation very effectively (1859).

water may add up to one inch to the oceans every 10,000 years, an undetectable amount. But that amounts to 100 in. (2.5 m) of new water every million years and in 4 billion years—you can do the arithmetic. Clearly, there has been enough, or more than enough, water from comets to fill the oceans, even if the rate of influx was considerably less than Frank and his colleagues estimated. On the other hand, recent evidence casts doubt on this theory because the water on the comet 67P/Churyumov–Gerasimenko (studied by ESA's Rosetta spacecraft) has three times the deuterium content of ordinary water on Earth. If water did not come from comets, perhaps it came from asteroids, probably during that LHB period. The amount of water we have today (covering 70 % of the Earth's surface) seems to be an accident.

And so, somehow, Earth became the blue planet it is today. Oceans surrounded land masses formed by "wrinkles" in the Earth's crust as it gradually shrank. Clouds of water vapor joined the nitrogen, methane and carbon dioxide in the atmosphere. Hot basaltic rock from volcanic eruptions, especially in the ocean, came into contact with water from time to time. Whenever that happened, incompletely oxidized (or reduced) compounds like NH_3, H_3P, FeO, or SiO grabbed oxygen from the water molecules (H_2O) and released molecular hydrogen (H_2). These reactions may still be happening in the hot Earth's mantle (for example $3Fe_2SiO_4$, or fayalite, $+2H_2O \rightarrow 2Fe_3O_4 + 3SiO_2$, or quartz, $+2H_2$). Higher hydrocarbons may be synthesized from H_2 by an inverse of the coal liquefaction process that the Germans used during WW II, at high temperatures and pressures. A similar reaction, at temperatures above 500 °C, starts from olivine (Mg_2SiO_4) and fayalite (Fe_2SiO_4) with water and carbonic acid yielding serpentine ($Mg_3Si_2O_5(OH)_4$), magnetite (Fe_3O_4) and methane (CH_4). Serpentine and magnetite are commonly found in basalt, suggesting that such methanation reactions may have occurred.

N.B. I am not asserting that these reactions do account for all, or even a large fraction, of the fluid hydrocarbons in the Earth's crust, although other credible authors have done so (e.g. Gold 1999). However, I think there is a fairly plausible case that such reactions may account for the methane clathrates. In the early nineteenth century Sir Humphrey Davy discovered by accident that some gases mixed with water crystallize under pressure into solid "clathrates". Methane is one of these gases. The chemical formula of methane clathrate is $(4CH_4) \cdot (23H_2O)$. The solidification process is similar to freezing, but the methane molecules are trapped in cages of water molecules. Since the clathrate is less dense than water, it will float. Hence clathrates are generally found only in, or under, frozen ocean sediments ("perma-frost"), although they can form outcrops. Methane clathrates occur in polar water where the surface temperature is below freezing or in sediments at depths greater than 300 m and less than 2000 m. in the oceans where the bottom temperature is around 2 °C (or possibly in deep freshwater lakes like Baikal). The temperature-pressure conditions under which methane clathrates exist are shown in Fig. 4.16.

The origin of methane clathrates is thought to be from natural methane "leaks" from undersea vents or faults (Max 2003; Hoffmann 2006). Estimates of the quantity of carbon trapped in methane clathrates under the permafrost and in the

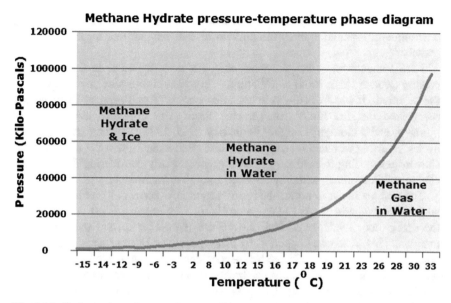

Fig. 4.16 Clathrate phase diagram. *Source*: Wikimedia. https://upload.wikimedia.org/wikipedia/commons/thumb/b/b4/Methane_Hydrate_phase_diagram.jpg/. 260px-Methane_Hydrate_phase_diagram.jpg

oceans range from 1000 gigatons (Gt) to 22,000 Gt with a probable median figure of 10,000 Gt. That is an enormous number, considering that 50 Gt would be enough to increase the methane in the current atmosphere by a factor of 12. That would have a climate warming potential equal to all of the CO_2 that is currently present in the atmosphere.

Hydrogen, the lightest element, quickly rose to the top of the early atmosphere. Under the intense ultraviolet (UV) radiation up there (and with no protection from the ozone layer, which didn't exist yet) molecular hydrogen H_2 or methane CH_4 were easily broken up. The atomic hydrogen was then ionized and becomes a plasma of free protons and free electrons. (These small particles escape easily from a small planet's gravitational attraction.) This process of hydrogen-loss continued as long as free hydrogen was being created by reactions such as the one described above ($FeO + H_2O \rightarrow FeO_2 + H_2$) or the breakup of hydrogen-containing gases such as CH_4 (methane), NH_3 (ammonia), H_2O (water) and H_2S (hydrogen sulfide) under the intense UV irradiation in the upper atmosphere.

In chemical terms, the loss of hydrogen from the atmosphere is primarily responsible for the slow but continuous shift in the so-called reduction/oxidation ("redox") balance from "reducing" to "oxidizing". A "reducing environment" splits the oxygen from metal oxides and recombines it with hydrogen (yielding water) or combines hydrogen with carbon dioxide yielding water and methane. An oxidizing atmosphere reverses the process. Generally speaking the hot magma below Earth's crust is a reducing environment that splits the H_2O molecule, producing molecular hydrogen and methane. By contrast, the surface of the Earth today is an oxidizing environment, especially since the buildup of free oxygen due to photosynthetic activity.[19]

Here I should recognize the work of C-J Koene in France in the nineteenth century (Koene 1856, 2004) and Vladimir Vernadsky, in Russia in the early part of the twentieth. It was Vernadsky, in his famous book "The Biosphere", who clearly recognized that the Earth's oxygen-rich atmosphere of today, as well as the nitrogen, had a biological origin (Vernadsky 1945, 1986). The so-called "nutrient cycles", starting with carbon-oxygen, are discussed in Sects. 6.6–6.9. However, the shift from reducing to oxidizing conditions—especially considering the methane-clathrates—has almost certainly not been smooth or monotonic.

The first geologist to notice the phenomenon now known as "continental drift" was Alfred Wegener, in 1912 (Wegener 1966). Wegener originally suggested that the ocean floor tends to expand, away from the center toward the continental margins. This idea was supported by Arthur Holmes but not accepted by most other geologists until the 1950s when new evidence came along to support it. Now it is known that the Earth's crust has broken into six large tectonic "plates", plus a number of smaller ones (Fig. 4.17). As the crust cools, the plates push and slide along each other, and in some cases, a plate slides under another (subduction), or rams against another (continental collision) causing mountain uplift. The plates do not have uniform thickness or composition: the crust under the continents is thicker, and the rock is primarily granite, whereas the crust under the oceans is thinner, and the rock is predominantly iron-rich basalt. There it is forced to "dive" into the subduction zones along those margins. This explains why there is so little sediment in the oceans and why oceanic basaltic rocks are much younger than continental granite rocks. The stresses generated by that process is what causes earthquakes and vulcanism concentrated along the so-called "ring of fire".

Apart from those geological facts, which are now quite well documented, there are some unexplained mysteries. One of them is the cause of the several periods of glaciation, including extreme cases known as "snowball earth" when geological evidence suggests that all or almost all of the Earth's surface was covered by ice, with temperatures comparable to Antarctica today (−50 °C) and ice covering everything, except (perhaps) a band of "slush" around the equator. The first glaciation that we know of (Pongola) occurred about 2.9 billion years ago. We know very little about it, except that it happened.

There is a theory, based on the global energy balance, proposed by Mikhail Budyko, that explains fairly well how a glaciation becomes global thanks to the interaction between temperature and albedo, once it passes a critical limit (Budyko 1969, 1988). According to Budyko's model, if solar radiation input declines by only 1.6 %, the ice sheet will extend south to 50° latitude. When that happens, the

---

[19] There is some dispute as to the rate of hydrogen loss and whether the early Earth's atmosphere was very rich in hydrogen (strongly reducing) or neutral on the acid-base (pH) scale. At present the balance of opinion favors the reducing atmosphere.

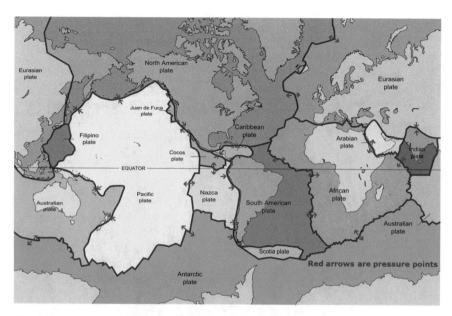

Fig. 4.17 Major tectonic plates c. 1980. *Source*: "Plates tect2 en" by USGS—http://pubs.usgs. gov/publications/text/slabs.html. Licensed under Public Domain via Commons—https://com mons.wikimedia.org/wiki/File:Plates_tect2_en.svg#/media/File:Plates_tect2_en.svg

feedback loop becomes unstoppable and the glacier will continue its advance all the way to the equator.

Another series of several glaciations coincided with the second rise in atmospheric oxygen levels, between 775 million years ago and 730 million years ago (Kaigas glaciation). This was followed by the Sturtian glaciation from 715 to 698 million years ago, and the Marinoan glaciation 665 to 635 million years ago. The range here does not mean the glaciers lasted that long, but that the exact dates—based mainly on the ages of rocks—are hard to pin down more precisely. It is likely that only single-cell organisms could survive those "snowball" episodes, although it is possible that there were "safe havens" e.g. near the equator or near submarine vents such as "black smokers" (Kirschvink et al. 2000). The advent of multi-cell organisms and the "Cambrian explosion" followed after the end of the Marinoan glaciation.

The Makganyeke case mentioned above (and others) also provides evidence that the Earth's crust has not always been wrapped around its axis of rotation exactly the way it is today. In fact there is a lot of evidence that the Earth's crust and mantle are loosely connected with the underlying mantle and the solid iron-nickel core. What that means is that the North and South poles appear to "wander" from time to time, and that those shifts coincide roughly with magnetic polarity reversals (Evans et al. 1998). There is other evidence of relatively sudden shifts in the location of the poles *vis a vis* the Earth's surface. The history of the Earth can be summarized in a pie chart (Fig. 4.18).

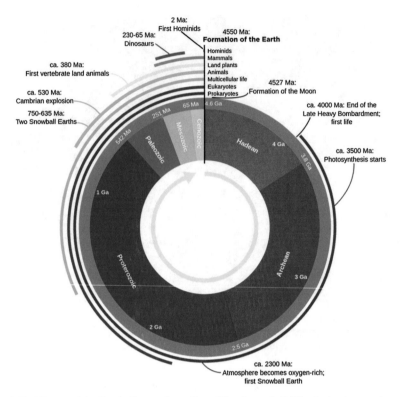

Fig. 4.18 History of the Earth. *Source*: https://en.wikipedia.org/wiki/Geologic_time_scale

The history of the Earth's atmosphere has been of interest since the nineteenth century. In fact, a recently resurrected book in French (1856), now translated into English, traces the long decline in the carbon dioxide concentration in the atmosphere (Koene 1856, 2004). Harold Urey was perhaps the first in recent times to explore this area (Urey 1952).

Life on Earth for most of its history consisted of single-celled organisms clinging to bare rock or populating the oceans. It was only within the last 10 % of the planet's history that multi-celled organisms appeared (during the "Cambrian explosion" after the last "snowball Earth"). Nobody knows how it happened, although there have been some interesting ideas; (e.g. Evans et al. 1998; Kirschvink and Raub 2003). If you were to plot biological complexity over time, there would be periods of gradual overall increase, "punctuated" by a series of setbacks and a few catastrophic "extinctions" when large fractions of all living species disappeared suddenly. Complexity increased after each of these extinctions. From a distance, it looks as if the extinctions were followed by periods of extraordinary biological variability. Whether there is a causal link between the extinction and what followed is hard to say.

The climatic history since the last of the snowball Earth episodes is sketched in Fig. 4.19. I found it on the Internet, in a blog by a climate skeptic, Paul Macrae, but no source was given. I suspect it was based in part on Prothero's book on the eocene-oligocene transition, which Macrae cites several times (Prothero 1993). I include it here to make the point that factors other than carbon dioxide concentration in the atmosphere (quite possibly including methane) have been responsible for global climate change in the past. Clearly "something" happened 450 million years ago, in the middle of the Paleozoic and again in the Mesozoic between 320 and 260 million years ago and finally around 150 million years ago. Whatever it was may be happening again now.

I wonder if the warm periods preceding glaciations could have been driven by periodic methane "burps" from the methane clathrates. It is not clear what could have caused such a "burp", but the most obvious possibility is an undersea volcanic eruption. The sudden (in geologic terms) release of a lot of methane (and carbon dioxide from the volcano itself) would be followed by a dramatic increase in "greenhouse" warming, since methane is much more potent in that regard than carbon dioxide. The warmth would result in an explosion of biomass. The biomass produces oxygen as a by-product. Gradually, over time, the methane in the atmosphere would be oxidized to carbon dioxide, which would subsequently be absorbed by the oceans. The declining atmospheric carbon dioxide level would result in cooling. That increases the solubility in water of carbon dioxide and thus accelerates the cooling. After a point the cooling could trigger Budyko's ice-albedo mechanism.

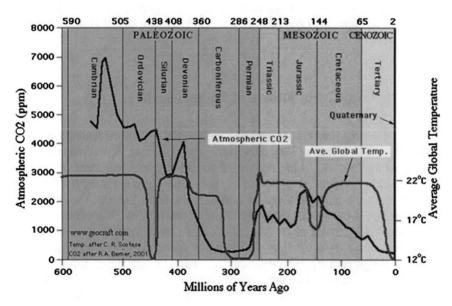

Fig. 4.19 Long historic relationship between carbon dioxide and temperature. *Source*: http://www.paulmacrae.com

The gradual shift from reducing to oxidizing conditions has been accompanied by gradually increasing acidification of the oceans, as more and more carbon dioxide is dissolved. Most of the carbon dioxide originally in the atmosphere must have been first dissolved in the oceans. Subsequently, most of that was deposited on the ocean floors as insoluble calcium carbonate (limestone). That material is gradually removed from the ocean floor by subduction into the magma where the CO_2 is released again and re-enters the atmosphere via volcanic eruptions. The oceans contain about 5000 times as much carbon dioxide as the atmosphere, which means that the atmospheric level depends, above all, on the solubility of CO_2 in the oceans. That, in turn depends on the surface temperature of the water.

The warmer the water, the less additional CO_2 it can absorb. As warming occurs, less of the "new" CO_2 emitted each year (by combustion of fossil fuels) is removed, leaving more as a greenhouse gas (GHG) in the atmosphere and accelerating the warming process. But, warmer water temperatures would also increase surface evaporation, rainfall and erosion rates. Clay minerals washed into the sea would then react with the dissolved CO_2 causing it to precipitate as $CaCO_3$ (limestone) and thus tending to re-stabilize the system (Holland 1978).

There are now four molecules of carbon dioxide in the form of insoluble carbonates for every free molecule in organic detritus. The early oceans must have had a pH of around 11, which is quite alkaline. Currently the pH of the ocean is closer to 8.3 due to the large amount of dissolved carbon dioxide (carbonic acid). As it happens, rainfall averages 5.5 pH, which is quite acid; the neutral point is pH 7. In other words, the Earth's surface "redox" chemistry has shifted gradually from alkaline to acid, and the shift continues. (The shift has potentially catastrophic future consequences for all calcium-fixing marine organisms, from coral reefs to shellfish. This constitutes one of the ten or so "planetary limits" discussed in the final chapter.)

4.8   The Long-Term Future of Planet Earth

For several reasons it seems worthwhile to point out that the current situation of our planet is by no means a long-term steady-state. In fact, it has already been mentioned that our sun, being one of the so-called "main sequence" of stars, is now about 5 billion years old, and that it is now about halfway through its probable lifetime (based on mass) of about 10 billion years. Since Earth was newborn, 4.5 billion years ago, our sun has increased in brightness by 30 %. This is due to the fact that a significant amount of its original hydrogen has already been converted by thermonuclear fusion into helium, which is four times denser than hydrogen.

Surface brightness depends on surface temperature, which depends on the rate at which the solar mass is "burning", which depends on pressure, which depends on density. The details do not matter. The point is that (according to standard theory) the Sun is getting hotter, and therefore the amount of solar energy illuminating the top of the Earth's atmosphere is slowly increasing. If this continues, the Earth will eventually get too hot to retain its water in liquid form. It will become uninhabitable in 0.5–1.5 billion years or so (Lenton and Watson 2011, p. 90).

The Sun is also losing mass through its corona at a rate of about a million tons per second. Even so, the total mass loss of the Sun is today less than 1 % of the original mass when the Sun was new. This coronal evaporation process produces the "solar wind", which blows hot particles on everything in the solar system. (It could theoretically provide impetus for space-ships using "solar sails"). The solar wind consists of electrons and protons that "boil off" the Sun's super-hot corona. The average velocity of the "wind" particles is 400 km/s. but particles ejected from "coronal holes" are much faster. The coronal holes seem to be a sort of magnetic cannon that shoots protons into space.

When the solar wind arrives at Earth orbit, most of the charged particles are deflected by the Earth's magnetic field. (Not every planet has a magnetic field. Jupiter and Saturn do, but Venus and Mars do not.[20]) Because of the Earth's magnetic field, the charged particles (and cosmic rays) that manage to penetrate from the solar wind are trapped, creating the two Van Allen Belts, discovered in 1958 (Fig. 4.20). The outer belt is electron dominated, the inner one is proton dominated. The charged particles travel along magnetic field lines, but as the field lines come together at the Earth's magnetic poles, the field becomes stronger and the particles are reflected. They "bounce" back and forth doing no harm.

But there is another problem. The interior of the Earth is gradually cooling, as the radioactive elements (uranium, thorium, etc.) continue to decay. One long-term consequence of that cooling is that the Earth's crust grows thicker, volcanism decreases, and tectonic activity gradually declines. As the continental plates move more slowly, the rate of subduction of sediments from the oceanic basins into the hot interior decreases. This slows down the rate at which carbonates in the sediments can be reduced ("cooked") to release their CO_2 in the reducing environment of the magma. In a few words, the rate at which carbon dioxide is recycled to the atmosphere by volcanism is slowing down. It is already clear that atmospheric carbon dioxide levels are far lower than they were at earlier times in the Earth's geological history. More and more of the Earth's carbon stock is being sequestered in biologically unavailable forms.

Another consequence of the cooling in the core may be that the "geo-dynamo" that produces our protective magnetic field is "running out of steam". Apparently, the magnetic field that enables migrating birds and turtles—and ancient mariners— to locate themselves is created by electric currents in the liquid mantle that surrounds the Earth's solid nickel-iron core. These electric currents result from convective mass flows in the liquid metal, also mostly iron and nickel, surrounding the solid core. This system of convective flows and currents is called a "geo-dynamo". The radioactive decay of trans-uranic elements keeps the mantle hot

---

[20] Without a magnetic field to deflect it, the solar wind interacts directly with the top of Venus' atmosphere, sometimes causing huge planet-sized storms called "hot flow anomalies" that would also kill any living organisms.

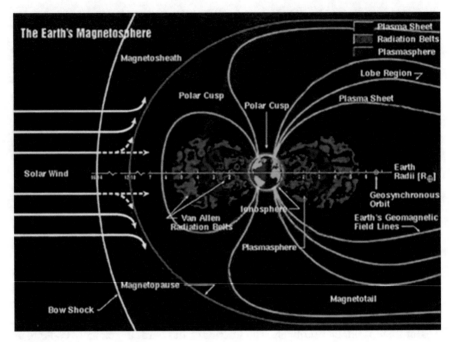

Fig. 4.20 The Earth's magnetosphere. *Source*: http://gbailey.staff.shef.ac.uk/researchoverview. html

and drives the convection currents. These, in turn, provide the energy for those electric currents, and thus for the magnetic field. As the cooling due to radioactive decay continues, the magnetic field strength can be expected to decline, bringing the solar wind ever nearer.

Actually, some geological evidence suggests that the Earth's magnetic field has reversed polarity many times in the past. Magnetic reversals seem to have occurred about once in 5 million years, but the system is gradually becoming less stable. The Earth's magnetic field has become measurably weaker just in the past century. Indeed, the magnetic poles—which are not at the rotational poles—are wandering around at a rate of several kilometers per year. The current expectation is for a magnetic polarity reversal (when "south" becomes "north" and "north" becomes "south") every 200,000 years or so. See Fig. 4.21.

Moreover, there is evidence of "magnetic striping" of the oceanic basalt. It turns out that the magnetic fields embedded in neighboring "strips" have opposite polarity, meaning that each strip represents a field reversal. Nobody knows what causes the reversals. But I can't help wondering if the field reversals are analogous to the swings of a pendulum, except that the magnetic field reversal is more closely analogous to tightening a spring and then releasing it. This has some plausibility given that the Earth's magnetic field is created by electric currents generated by the

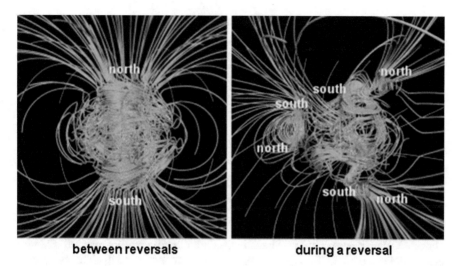

between reversals   **during a reversal**

Fig. 4.21 Magnetic field lines before and during reversals. *Source*: https://upload.wikimedia.org/wikipedia/commons/e/e5/NASA_54559main_comparison1_strip.gif

twisting and untwisting motion of the crust and mantle with respect to the nickel-iron core.

In fact, it appears that a polarity "flip" can be expected with the next few thousand years. We don't know how rapidly this might happen. But, whenever it happens, the magnetic field strength presumably decreases to zero, before reappearing in the opposite direction. This reversal would be accompanied by increased radiation levels at the top of the atmosphere from the solar wind. Fortunately, the ozone layer should be unaffected, or even enhanced. There is no evidence that major extinctions have been caused by magnetic field reversals in the past. Then again, there is no evidence to the contrary, either.

However, the cooling of the interior of the Earth is relatively slow, and during the next few thousand years it seems that the human propensity to burn fossil fuels will keep the planet warm enough. Maybe too warm. Here I should make reference to three alternative theories of what may happen in the future. One is the *Gaia* theory, which says that life tends to co-evolve with the planet, so as to maximize its stability and resilience against perturbations (Lovelock 1972, 1979; Lovelock and Margulis 1974). The opposite theory, which has been called the *Medea* theory[21] argues that life is inherently suicidal, in that it tends to exploit all available resources until they are exhausted (Ward 2009). One might call it the "Easter Island" theory. There is no conclusive evidence on either side. Neither of those theories provides any useful information about the specific industrial processes that could contribute to (or prevent) that end point.

---

[21] Medea was the legendary Greek wife of Jason (seeker of the golden fleece), who killed her own children.

The third theory, *Thanatia*, is not a theory of what will happen or would happen without human intervention. It focuses on the specific process that would lead to a "dead" Earth, in which all available exergy resources have been used up by human economic activity, *vis a vis* the present state (Valero Capilla and Valero Delgado 2015). It could be called the Business-As-Usual (BAU) future. The "circular economy", discussed at the end of Part III is, in effect, the "anti-Thanatia" scenario.

4.9   Summary of Pre-biotic Evolution

To the best of our understanding, in the beginning there was nothing but "the primordial vacuum". Suddenly there was an indescribable "event": the all-encompassing matter-anti-matter symmetry was broken. The Higgs field appeared. Positive (radiant) and negative (gravitational) energy appeared. Photons and particles separated. Strong and weak nuclear forces became distinct, along with their carrier-particles. Fermions and bosons became distinct. Positive and negative charges became distinct. Space "inflated" and cooled at tachyonic speeds, resulting in the disruption of causality between particle-anti-particle pairs. Quarks and nucleons were created, collided, annihilated. Finally the longest lived (fittest?) survivors started to get together. Electrons and neutrinos combined with protons, first to make neutrons, then hydrogen atoms, then deuterons and helium atoms. The universe became transparent to radiation.

After about 700,000 years the universe was a very hot, dense, plasma consisting of photons, neutrinos, electrons, protons, neutrons, hydrogen nuclei, deuterium nuclei, and helium nuclei. The net charge may have been zero. There were no anti-particles left. Electromagnetic (EM) forces dominated. Small density and charge fluctuations resulted in electric currents and magnetic fields. Gravity enhanced these density differences. After hundreds of millions of years, these density fluctuations became proto-stars. Gravitational forces created the distinguishable stars (and galaxies) while nuclear forces initiated the exothermic fusion process that created heavier elements (up to iron) and made the stars shine.

The smaller stars lived the longest, cooled gradually and died a gentle death as "white dwarfs", rich in iron, or neutron stars. Bigger stars suffered a more violent gravitational collapse when their "fuel" of light elements was nearly exhausted. As the core temperatures and pressures rose, elements heavier than iron were also created endothermically at that time. Then the final explosions (supernovae) scattered the residual mass, resulting in clouds of "dust" containing light and heavy elements, mixed with hydrogen and helium not yet incorporated into stars. The collapsed neutron cores of heavy stars may have become "black holes", if Einstein's equations are correct (still debatable). Otherwise, they may be "gray holes" if Yilmaz' theory is correct.

The star creation process continued. Some second generation stars resulted from accretion of new hydrogen around neutron stars or white dwarves. Others resulted from gravitational forces on "virgin" hydrogen-helium clouds. EM forces (much stronger than gravity) certainly played a role in that condensation process. Some young stars collect planets and asteroids as they grow. Our Sun is one of those.

The Earth is unusual, having an iron core with its own protective magnetic field, a considerable endowment of both light and heavy elements from a previous supernova, an internal heat source powered by radioactive decay. That heat drives internal geochemical reactions that created mineable concentrations ("ores") of those elements, and a large amount of water on its surface. It is just far enough from the Sun so that the water remains liquid. That, in turn, enables a powerful hydraulic terraforming engine. This combination is what made life possible. It is rare, but almost certainly not unique in the universe.

Chapter 5
The Origin of Life

5.1 Exogenesis?

The late heavy bombardment (LHB) of Earth, between 4.51 and 3.84 billion years ago, was mentioned in Sect. 4.6. The LHB is called the "Hadean" period, during which the Earth was still molten and had no solid crust. The bombardment seems to have ceased fairly abruptly, as if the whole solar system had moved through a "cloud" of planetoids and asteroids and finally emerged on the other side. It was succeeded by the calmer and cooler "Archean period", when the crust solidified and life got established on Earth. The possibilities range from 4.4 billion years ago to 3.2 billion years ago, with the likely date 3.8 billion years ago (Lenton and Watson 2011). The likely end to the LHB coincides with when the first banded iron formations (BIF) appear, along with graphite particles in which the characteristic ratio of carbon $_{12}C$ and $_{13}C$ isotopes—the basis of "carbon dating"—suggests that photosynthesis was going on. The later estimate is when cyanobacteria were definitely present in the oceans.

However, a still unresolved problem is whether all of the necessary molecular precursors for living organisms were actually created on Earth, during the 200 million years after the end of the late heavy bombardment (LHB). Or did they come from space? One expert has commented:

> ... if we assume that all the heavy bombardment was delivering millions of death-dealing blows to the Earth similar to the one which took care of the dinosaurs 65 million years ago, there was not a great window of time for life to have emerged [on Earth]. The question is whether the comets were the source of life as well as death? (Greenberg 2002)

The answer to that question is important. It could be "yes", if the primitive self-reproducing organisms and supplies of organic "food" molecules were produced in space and delivered by comets or asteroids that subsequently made a "soft" landing on the surface of early Earth. The idea was put forward in 1961 by **John Oro** (Oro 1961). A similar notion seems to have been put forward by **Francis Crick**; The idea

© Springer International Publishing Switzerland 2016
R. Ayres, *Energy, Complexity and Wealth Maximization*, The Frontiers Collection,
DOI 10.1007/978-3-319-30545-5_5

was worked out later in some detail in the "life cloud" theory of Fred Hoyle and N. C. Wickramasinghe (Hoyle and Wickramasinghe 1978).

A more radical theory, "pan-spermia" first proposed a century ago, was that life itself may have originated in space or on comets and subsequently got distributed around the universe (Arrhenius 1903, 1908). This theory has few adherents today. However, for present purposes the relevant part is whether and how organic molecules such as sugars and nucleotides could have been produced in dust clouds during the formation of the solar system. (The possibility that it happened elsewhere in inter-stellar space is a lot less plausible.)

The "life cloud" theory cannot be ruled out, although most theorists currently dismiss it. The reason is simple: Organic materials are found in about 86 % of meteorites (known as chondrites), of which there are over 9000 specimens in collections around the world. Over 600 organic compounds, including amino acids, have been identified in chondrites. The so-called carbonaceous chondrites, in particular, can be quite rich in organics. Some meteorites are 5 % organic. The famous Murchison meteorite that landed in Australia in 1969 contained sugars, notably dihydroxyacetone, amino acids and various sugar acids and alcohols. Spectroscopic analysis of Halley's Comet suggests that its mass is almost one third organic. Interplanetary dust—collected in space—is typically around 10 % organic. But when the dust enters the atmosphere, almost all of that organic component is burned up. So how did the organics get down to Earth? Or did they? There are possibilities, but nothing plausible has turned up as yet.

Still, a new subfield of chemistry (ion-molecule chemistry) at zero pressure and near zero temperature, led by William Klemperer and his students at Harvard, has evolved to explain the existence of organic molecules in space (Solomon and William Klemperer 1972; Herbst and Klemperer 1973; Dalgamo and Black 1976; Watson 1978). At the N. N. Semenov Institute for Chemical Physics in Moscow, V. I. Goldanskii and others also contributed significantly to the field of low temperature chemistry (Goldanskii et al. 1973; Goldanskii 1977, 1979; Goldanskii and Vitalii 1997; Morozov and Goldanskii 1984).

Quite dense interstellar clouds are known to exist out in space, between the stars. The dust makes up something like 2 % of the mass of visible interstellar clouds (Fig. 5.1). It is responsible for the "dimming" of visible light (and UV radiation) as it passes through a cloud. Some of the clouds are dense enough to block the light of stars behind them entirely. There are many astronomical photographs showing gaps or blots in the sky that can only be caused by dust clouds. Figure 4.15 in Sect. 4.4 showed one of them. In fact, the stars themselves—like our Sun—were formed when clouds of "dust" became dense enough to collapse gravitationally.

During their dense pre-collapse period (which lasts hundreds of millions of years), interstellar clouds get gradually denser. The small solid particles in the dust become "attractants" on which the hydrogen atoms in the cloud can (and do) attach themselves. Once on a surface—like ice-skaters on a pond—they have fewer degrees of freedom than in three dimensional space, and correspondingly lower entropy. This facilitates interactions and catalytic transformations. It is now known that most of the hydrogen molecules (as opposed to the bare atoms) that exist in

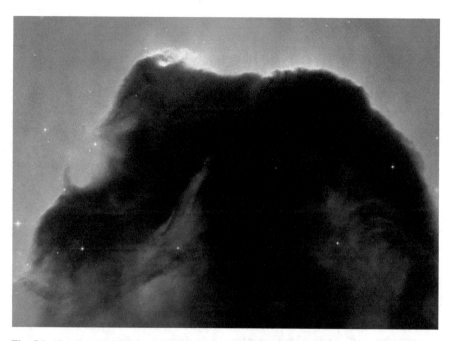

Fig. 5.1 The Horsehead Nebula as seen by the Hubble Telescope. *Source*: https://en.wikipedia.org/wiki/Cosmic_dust

space are produced on the surface of these dust particles. Heavier molecules are known to be created the same way.

It is tempting to look for a chemical version of the nucleosynthesis story, i.e. a graph showing the binding energy of various molecular bonds, such as H-H, H-C H-N, C-C, C-O, etc. It is certainly true that those bonds are quite strong, but not nearly as strong as nuclear binding forces. Larger molecules (like heavy elements) can exist because the strength of the binding energy at low temperatures exceeds the centrifugal forces that become dominant at higher temperatures.

However, there is no simple rank-ordering of molecules in terms of chemical binding energies, as in Fig. 4.9, because of three complications. One is that temperature (kinetic energy) makes a much greater difference between chemical combinations. The binding energy at near-zero temperature is not the same as the binding energy at room temperature. The second complication is that ions and free radicals are much less stable (more reactive) than neutral molecules. And the third complication is that a lot of reactions between molecules occur in aqueous solution, because water molecules have a tendency to ionize other molecules—again, as a function of temperature. All of these complications (and others) explain why chemical synthesis is much more complicated than nucleosynthesis.

Having said that, microwave spectroscopy since 1968 has revealed that dust clouds contain a wide variety of organic molecules (some of which are also found in comets and meteorites). From 1968 to 1980 50 different compounds had already

Illustrations of ice-covered silica grains in dust clouds where
chemical reactions can take place, driven by UV radiation

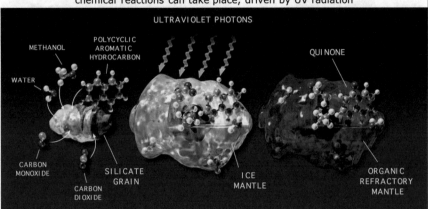

Fig. 5.2 Dust particles as chemical factories. *Source*: http://www.astrochem.org/sci_img/
icegrain.jpg

been identified in space by their microwave spectrograms (Green 1981). The most
common inorganic (no carbon) two element compounds in that group are hydroxyl
(OH) (not surprising) and four combinations of much heavier elements, sulfur and
silicon (SiO, SiS, NS and SO). Of the organic (carbon-containing) pairs, CH
appears both in neutral and ionic (CH^+) form, followed by cyanogen (CN), carbon
monoxide (CO) and carbon mono-sulfide (CS). There is a theory that a lot of these
compounds were created on the surfaces of silica dust particles, under a coating of
ice. See Fig. 5.2.

Molecules involving three atoms that have been observed include water (H_2O),
hydrogen sulfide (H_2S), sulfur dioxide (SO_2) and a curious nitrogen-hydride ion
(N_2H^+). Other 3-atom compounds include ethynal (CCH) with two carbons,
hydrogen cyanide (HCN) and hydrogen cyanate (HNC), formyl (HCO) and its
ion (HCO^+) plus carbonyl sulfide (OCS). In the case of 4-atom molecules, all
except ammonia (NH_3) involve carbon. All the rest, but one, include hydrogen.
All but one of the 5-atom and more complex compounds in that group involve both
hydrogen and carbon, plus either oxygen or nitrogen or both. The exception is
(SiH_4). The other compounds with 5 or more atoms are organic and involve the
elements needed for life (H, C, O, N and S).

The current number of confirmed molecular species found in space is
206 (according to Wikipedia, as of 2014). Of them, 17 are deuterated (with
deuterium instead of normal hydrogen). Of the 189 non-deuterated species,
79 have 5 or more atoms and 60 have 6 or more atoms. Another eight species
have been reported in the scientific literature but not yet confirmed. Larger molecules may also be present in space but they are not yet readily identifiable by current
methods in microwave spectroscopy (Laane 2009).

According to one theory, polymers were gradually created on comets or asteroids by a sequence of pairwise combinations somewhat analogous to nuclear fusions (unlike "normal" chemical reactions where two chemicals meet and recombine to form two other chemicals). As in the nuclear fusion case, the starting point is atomic hydrogen, followed by molecular hydrogen, followed by heavier compounds involving C, O, N and S.

The most important intermediate molecule is methanol (CH_3OH) which has been characterized as the major chemical pathway to complex organic molecules in interstellar space. It is formed on surfaces where a carbon monoxide molecule can combine with two hydrogen molecules.

Accordingly, astronomers at several institutions have started looking for concentrations of methanol in the neighborhoods of stars. It turns out that methanol concentrations are highest near a few newly formed stars—stars similar to our Sun 4.5 billion years ago—but much lower in other parts of space (Whittet et al. 2011). The methanol "sweet spots" that have been found recently seem to be related to the rate of carbon-monoxide accumulation *vis a vis* hydrogen. In short, the creation of organic "pie in the sky" as organic food for the first living cells on Earth is increasingly plausible. It certainly cannot be ruled out today.

Here I need to mention a shocking new discovery that occurred while I was writing this section (July 2015). In brief, the Rosetta space probe, which physically explored a comet known as 67P, found molecular oxygen in the cloud of outgas (coma) constituting its tail (Bieler 2015). The comas of most comets consist mostly (95 %) of three gases, CO, CO_2 and H_2O. Until now it has been assumed that any molecular oxygen in the dust cloud from which the solar system was formed, would have reacted with hydrogen and disappeared. The fact that molecular oxygen still exists inside that comet mass means that it must have been there when the comet was formed 4.61 billion years ago. This fact is inconsistent with all prior theories of the formation of the solar system.

One possible implication is that this comet was created from the debris of a planet where there was photosynthetic life (because oxygen is a by-product of photosynthesis). Is it possible that such a planet was swallowed up by the supernova that created all the heavy elements with which the Earth is now endowed? I'm sure there will be speculation along these lines.

5.2 The Origin of Organic Monomers

In Earth's early oceans, carbon dioxide was captured and converted into carbohydrate $(CH_2O)_n$ by several different photosynthetic reactions, summarized by the equation

$$CO_2 + 2H_2A + \text{photon} \rightarrow (CH_2O)_n + H_2O + 2A$$

where A is one of several possible ions, involving hydrogen, sulfur, iron or oxygen. The process of producing carbohydrate from by reducing hydrogen sulfide or ferrous iron by photosynthesis was an early source of carbohydrate molecules later usable as food by the first living organisms. The sulfur deposits and ferric iron oxide deposits now being exploited are leftovers. The process of splitting water molecules to combine H with carbon dioxide, releasing free oxygen is the most energy intensive of these reactions, so it was the last to evolve.

Photosynthesis in plants, via chlorophyll, yields oxygen as a secondary product, which is how it accumulated in the atmosphere. At the time, oxygen was a waste product, and it was actually toxic to the early photosynthesizers. Luckily the oxygen was quickly used up, at first, either by recombining with free hydrogen or by oxidizing combining ferrous iron in the oceans to form ferric iron. But when the atmospheric oxygen began to build up, all life was threatened. Fortunately, a new kind of single cell organism appeared. This new organism consumed carbohydrates and used oxygen for its energy metabolism, by a new metabolic process called *respiration*. So photosynthesizers and respirators now live in a rough balance, keeping the oxygen and carbon-dioxide levels in the atmosphere more or less constant. This process is called the carbon-oxygen cycle.

But first things first: life did not appear in a single step. The carbon-based molecular building blocks of living systems, such as amino-acids, peptides and sugars, could not have been created by any spontaneous exothermic (self-energizing) process starting from simple molecules like CO, CO_2, H_2O, CH_4, CN or NH_3. An exergy flux was certainly needed. Apart from the question of "where" such molecules could have been generated, we need to consider the sequence.

J. D. **Bernal** coined the term "*biopoiesis*" to give the process a name. He broke it into three stages (Bernal 1951). The first stage was the generation of simple organic compounds (monomers) from inorganic molecules. The second stage was spontaneous synthesis of large, self-replicating molecules (polymers) from the monomers. The third stage was the enclosure of a self-replicating collection of polymers inside a protective cell wall with the property of allowing useful molecules to enter while expelling toxic "waste" molecules e.g. Eigen (Eigen 1971; Eigen and Schuster 1979).

Some experts think that Bernal's second and third stages may have occurred in the reverse order, or (somehow) simultaneously (e.g. Morowitz 1992, 2002). The question can be characterized as "metabolism vs. replication". The point of the argument is that the self-replicating polymers—"naked genes"—in Bernal's third stage are unlikely to survive to reproduce themselves outside an impermeable cell wall that also encloses an exergy transformation system. On the other hand, cells can't reproduce without information carriers: genes (or an equivalent precursor). This leaves the question: what came first—the chicken or the egg?

However, nobody argues about the priority of Bernal's stage 1. There are five main categories of organic *monomers* essential to life that must somehow be accounted for. They are as follows:

Sugars are the components of all carbohydrates. There are three categories: *Mono-saccharides* (glucose, galactose, ribose and fructose) contain one carbon ring. *Disaccharides* (sucrose, maltose, lactose) are molecules with two carbon rings. Polysaccharides (glycogen, starch, cellulose) are molecules containing three or more carbon rings. Sugars are the sources of energy for many metabolic processes. *Carbohydrates* are simple chains of sugars, cellulose being a familiar example. Sugars also constitute the structural backbone of many important molecules, including the nucleic acids (RNA, DNA) which carry genetic information.

Amino acids, are biologically important organic compounds composed of amine (–NH_2) and carboxylic acid (–$COOH$) functional groups, along with a side-chain that varies from acid to acid. **Amines** are derived from ammonia (NH_3), which was certainly present in the early atmosphere, as well as oceans (as ammonium NH_4^+ ions). They are the building blocks from which **peptides** and eventually **proteins** are constructed. Proteins are simply long chains of peptides, folded over in complex ways. Proteins play many roles in higher organisms but in a simple primitive cell, proteins are needed primarily as **enzymes** (organic catalysts). Enzymes facilitate every chemical reaction within the cell, including photosynthesis and nitrogen fixation. There are only 20 amino acids in proteins, although the laws of chemistry permit a large number. Over 500 are known. Thus, the limited number of amino acids found in nature is another one of the crucial facts yet to be explained.

Nucleotide bases are the molecules constructed from amino acids and phosphate groups, from which information-carrying polymers, especially **nucleic acids** are constructed. The five nucleotide bases are **purines** (adenine (A) and guanine (G)) and **pyrimidines** (cytosine (C), thymine (T) and uracil (U)). **Nucleotides** are constructed from these five bases, plus ribose (sugar) and phosphates. The nucleotide chains consist of 5-carbon ribose, connected by phosphate (PO_4) groups. Most phosphates are insoluble except by strong acids, and the origin of soluble phosphates that could be mobilized by prokaryotes and early eukaryotes, in the early oceans is an open question.

Nucleic acids (RNA, DNA) are the information storage and reproduction systems of the cell. Among other things, the nucleic acids contain the "codes" that control the synthesis of proteins (enzymes) which, in turn, govern all metabolic and reproductive activities. Other important molecules are also constructed from nucleotide bases.

Energy-carriers, adenosine triphosphate (ATP) and nicotinamide adenine dinucleotide (NADH) are both constructed from the base adenine, the sugar ribose and three phosphate groups.

Note that the last three categories involve the element phosphorus, even though it is not, by any means, one of the commonest elements.[1] Let me start with the

---

[1] Phosphorus has atomic number 14, between silicon and sulfur (both of which are considerably more common).

energy carriers. This mechanism—which is found in all known living organisms—also involves the element phosphorus, in the form of phosphate (PO_4^+) groups (radicals). Specifically, the mechanism involves the molecules adenosine diphosphate (ADP) and adenosine triphosphate (ATP). A single phosphate (PO_4^+) radical can accept an electron from an excited source, such as a chlorophyll molecule. It then attaches itself weakly to the other phosphates in an ADP molecule, converting it to ATP. The ATP carries that PO_4 radical to a site where an increment of energy is needed. At that site the "extra" phosphate (PO_4) group loses its electron and is detached (converting the ATP back to ADP).

The reason this "trick" works is (1) that the PO_4 combination (with or without an extra electron) is extremely stable and very hard to break up. It is also quite insoluble. Moreover, (2) a single PO_4 does not readily combine with other organic groups, except the lipids (discussed later in connection with polymers). But (3) PO_4 groups do combine tightly, *in pairs*, with the nucleotide base, *adenine* (which consists of five linked HCN molecules), forming ATP. Finally (4), a single PO_4 will combine (weakly) with ADP in preference to a lipid. As to why the phosphates have these interesting properties, only quantum chemistry can explain. The essential point is that ADP/ATP play no role except as a carrier of electrons, via a detachable PO_4, from one site to another.

Now, back to the question of origins. Where did all this phosphorus come from? Recent discoveries suggest that phosphorus may have come to earth in meteorites during the late heavy bombardment (LHB). It came in the reduced form—*schreibersite* $(FeNi)_3P$—which is found in a wide variety of meteorites. There was no free oxygen on Earth during the late heavy bombardment, but this mineral happens to be soluble. In water, the nickel and iron atoms are detached from the phosphorus atoms. Moreover, atomic P is so reactive that it steals oxygen from oceanic H_2O, forming *phosphite* (HPO_3^{2-}).

Ortho-phosphates (minerals containing oxidized PO_4 groups) existed on Earth in the early crust and mantle, but there are no known chemical pathways leading to organic compounds, such as phospho-lipids or ATP. On the other hand, such pathways have been found starting from phosphite (Pasek 2008). Phosphite is around 1000 times more soluble than the ortho-phosphates, which also favors this possibility. Finally, if this idea is confirmed, it adds support to the notion that organic molecules were also synthesized in space and that they were delivered to the earth via meteorites during that same late heavy bombardment.

As mentioned in Sect. 4.5, the most generally accepted theory of evolution—at least until recently—held that the pre-biotic atmosphere of Earth probably consisted mostly of carbon dioxide with some hydrogen, methane (CH_4), ammonia (NH_3), and traces of hydrogen sulfide (H_2S). The presence of hydrogen, methane, ammonia and hydrogen sulfide made it a "reducing" atmosphere because those hydrogen-rich gases would steal oxygen away from other oxides, except incompletely oxidized metals, carbon dioxide (CO_2) and water (H_2O).

The Russian biochemist **Alexander Oparin** first suggested (in 1924) that primitive H-C-N-O compounds, such as amino acids, were probably produced by natural processes, such as ultra-violet radiation or electric currents, in a reducing atmosphere (Oparin 1938). A few years later the British biologist **J. B. S. Haldane** also noted that the first organic molecules were most likely produced in an oxygen-free reducing environment (Haldane 1985). Haldane and Oparin both suggested that ultraviolet solar radiation was the source of the activation exergy, and that the organic synthesis probably took place primarily in the oceans. It is called the "hot dilute soup" hypothesis of life creation.

Harold Urey at the University of Chicago was one of the first chemists to think about the early atmosphere of the Earth, and he argued that free hydrogen was probably also present (Urey 1952). His graduate student **Stanley Miller** then decided to simulate the reducing atmosphere and test the Haldane-Oparin "hot soup" hypothesis. See Fig. 5.3. Miller built an apparatus and ran an electric current through it. Sure enough, some amino acids (glycine and alanine) were found, demonstrating the possibility of amino acid synthesis reactions in a reducing atmosphere (Miller 1953; Miller and Urey 1959). Actually when the products of the Miller-Urey experiment were re-examined years later with more sensitive equipment, no less than 23 amino acids were discovered. Many later experiments with different amounts of the gases and different temperatures and pressures have shown that most, perhaps all, of the 20 amino acids found in living organisms could have been created in electrical storms, or in the "hot soup" itself, in early Earth (reducing) conditions, especially in the absence of free oxygen (Abelson 1966).

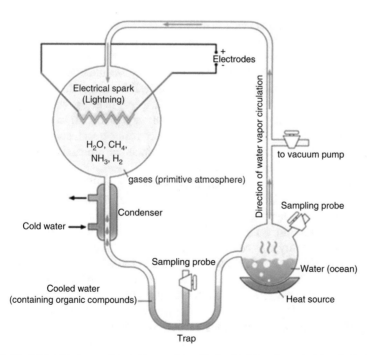

Fig. 5.3 The Miller-Urey experiment. *Source*: https://upload.wikimedia.org/wikipedia/commons/thumb/5/54/Miller-Urey_experiment-en.svg/2000px-Miller-Urey_experiment-en.svg.png

As it happens, quite a few other amino acids were also produced at the same time in those experiments. It is not clear why only those 20 are incorporated in living systems. That is another unresolved question. So much for the amino-acid synthesis mechanism.

Unfortunately, "activated" molecules (with electrons in excited states, or missing altogether) tend to de-activate very quickly (in fractions of a second) in water, due to hydrolysis. A problem for both the "hot soup" theory and the hydrothermal "black smoker" variant is that they assume chemistry in the liquid phase. The odds that an activated molecule in a liquid will have time to randomly encounter and interact with another activated molecule is much too low to account for an accumulation of complex organics such as biochemists believe must have existed. In particular, they do not explain spontaneous polymerization—such as the creation of peptide chains—that normally require a protein catalyst of some sort.

This is where catalysis must be invoked. Catalysis is a process whereby a substance acts as an intermediary for a reaction between two other substances, without being changed itself. A host of substances have catalytic properties for specific reactions. Metal ions are catalysts for many reactions. Metallic sulfides have been suggested as a possible substrate. There is a recent theory (the "pizza theory") that the polymerization process could have taken place on the surface of a mineral with especially honeycomb-like structures with a very high surface to volume ratio. (Volcanic pumice would be an example.)

Clay was first suggested by A.G. Cairns-Smith (Cairns-Smith 1984), but has since been ruled out. A good candidate is iron pyrite (FeS_2 known as "fool's gold") because it has a positively charged surface (unlike clay), that binds anionic particles but allows them to move about (Wachterhauser 1988, 1992). There are several factors favoring surface chemistry. A molecular combination on a two-dimensional surface drastically reduces the number of possible microstates compared to a combination that occurs in a three-dimensional space. In effect, the entropy change is less on a surface (Maynard-Smith and Szathmary 1995). To say it another way, the probability of synthesis reactions actually occurring are a lot higher on surfaces. Reaction kinetics are more favorable, and the inherent probability of three molecules meeting is better than in a solution.

The point is that the first living organisms, about 3.6 billion years ago, needed a source of energy ("food") for metabolism already available. It now seems quite likely that primitive synthesizers utilized energy from reactions between oxygen and hydrogen or sulfur compounds from volcanic vents (hot springs) in the deep ocean. This possibility was confirmed by direct observation of over 3 km deep in the Galapagos Rift, west of Ecuador (Corliss and Ballard 1978), and since then in many locations. This evidence has led Freeman Dyson and Thomas Gold (separately) to hypothesize that life on earth originated in the deep ocean, rather than in the warm surface waters (Dyson 1999 (revised ed.); Gold 1999). Later organisms used partially oxidized compounds (especially ferrous oxide) in the water to synthesize carbohydrates and other organics. There specialized organisms, called chemo-litho-autotrophs, still live near hot springs. Moreover, there are bacteria living in rock.

The CH_2O radicals can be linked together to form carbohydrates and sugars, as well as cellulosic structural materials. Note that this reaction can take place in the presence of radiation with the right energy content (photon frequency) without a catalyst or enzyme. Of course—as every chemist knows—what the flux of radiant energy can do, it can also undo. So, to build up a stock of CH_2O radicals in the oceans (or elsewhere) there needs to be a storage and protection mechanism of some sort. Deep water is one possibility. A dust cloud in space is another.

5.3 From Monomers to Polymers

The next stage in Bernal's sequence, polymerization, almost certainly did take place on the Earth itself (Bernal 1951). Large organic polymers are constructed by linking the monomeric "bricks" together by a kind of "glue", consisting of carbon atoms or methyl (CH_2) groups (and, in some cases, phosphate (PH_4) groups). Carbon atoms, and CH_2 groups, are uniquely inclined to link to each other in chains. Cellulose is the most obvious example of chains formed of sugar monomers. The reaction processes are all endothermic, requiring an exogenous source of exergy (photons).

A brief digression: The reaction processes, at this stage, were mostly catalyzed by crystals or metals that were either unchanged or regenerated in the process. The regeneration process for reactants x and y, producing z with a catalyst C is as follows:

$$x + C \rightarrow xC$$
$$y + xC \rightarrow xyC \rightarrow Cz + z$$

Many reactions are uniquely catalyzed by one metal, and many of the metals with catalytic properties are among the 38 so-called "transition metals" in groups 3–12 of the periodic table (including all 17 of the so-called "rare earths"). What this means is the practicality of combining the light chemical elements into stable complex compounds, like chains, seems to be enabled, or at least enhanced, by the presence of many of the heavier elements in the periodic table, even if those elements are not necessarily embodied in the organic compounds in living organisms.

It is important that these carbon-based chains, once formed, are extremely stable. Stability is the key to building chemical structures that endure constant bombardment by high-energy UV photons.

It happens that there are many more papers in the literature about genetic self-replication than about cell formation, the spontaneous creation of two-dimensional phospho-lipid arrays that might then form themselves into cell walls. This is probably attributable to the intense scientific interest created by the discovery of DNA by James Watson and Francis Crick (Watson and Crick 1953).

Be that as it may, a crucial precursor of three dimensional cells was two dimensional membranes, consisting of phospho-lipid polymers. They are the constituents of all cell walls. The "building blocks" ("bricks") of these arrays consist of a lipid (fatty acid) unit attached to a phosphate (PO_4) unit. The inter-molecular forces between these "bricks" makes them bind to each other with the phosphates at one end and the lipids at the other end, rather like the blades of grass in a lawn. Moreover, these phospho-lipid arrays display the tendency to bind together in that way doesn't stop with pairs. Triplets with three phosphates at one end and three lipids at the other are more strongly bound than pairs, foursomes are more tightly bound than triplets, and so on.

The phosphate group at one end of each monomer has a positive charge (an electron is missing) and it attracts water molecules. However the other end of the chain has a negative charge that repels water molecules. Hence the phospholipids form a film on water. Surface tension causes the film to fold up and form a cell enclosing a droplet. The closed surface is stable because it has no unattached "edges": every monomer is linked to others of the same kind. The surface is likely to be *spherical*, rather than some other shape, thanks to *surface tension* (e.g. Rashevsky 1948). This is the same mechanism by which rain-water condenses into spherical droplets.

It seems reasonable to assume that the main requirement for a 2-dimensional membrane of phospho-lipid "bricks" must be the availability of the component molecules. The exact mechanism by which such molecules were formed in the pre-biotic "soup" is unclear, although the peculiar chemistry of phosphorus must have been a crucial factor. So, skipping over millions of years and a myriad failed "micro-experiments", I suppose that there came a time when non-living "cells", consisting of mixtures of organic polymers contained within protective polymeric "skins" came to exist in large numbers, and that some of these contained just the right mix of components for the next stage. How did this occur? We don't know (yet).

Bernal's third stage of chemical evolution is, perhaps, the most interesting. It is the enclosure of mixtures of large organic molecules (e.g. sugars, amino acids and nucleotide bases) within a protective membrane (also a polymer). It happens that all living cells are protected by one kind of membrane, a bipolar phospho-lipid polymer.[2] See Fig. 5.4.

Given the finite thickness of the "sheet" there is obviously a minimum to the volume enclosed by the "container". That minimum presumably accords with the smallest observable cells. (It can presumably be calculated from the dimensions of the molecules, but I am not aware of the details of any such calculation.) More important, there is also a maximum cell size, beyond which it tends to divide into two smaller cells (meiosis). This phenomenon has also been explained, in terms of surface tension, by Rashevsky (op cit.).

---

[2] The only organisms that do not require a protective membrane are viruses. However, viruses are only active ("alive") after penetrating a cell.

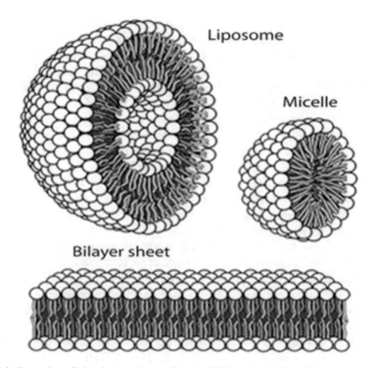

Fig. 5.4 Formation of bi-polar membranes. *Source*: Wikimedia, Mariana Ruiz

It appears that those cell walls may have had a function in catalyzing the synthesis of larger molecules. The idea is that organic molecules in solution were adsorbed by the phospho-lipid membrane and synthesis reactions were catalyzed on the surface of iron sulfide particles (Morowitz 1992). This would reverse Manfred Eigen's supposed order of priority (Eigen and Schuster 1979).

Seen in retrospect, the essential function of cellular wall membranes is to maintain a controlled interior (*autopoietic*) environment where complex information carriers can survive and replicate. Such an environment is necessary to contain and protect the large highly organized macro-molecules which carry out metabolic functions from the chemical chaos outside. The wall permits nutrients—small molecules—to enter (osmotically), and to let unwanted (toxic) metabolic wastes escape to the "outside".

Without this ability, life-processes would quickly self-destruct. British chemist Peter Mitchell first suggested that these membranes also play an essential role in the synthesis of adenosine tri-phosphate (ATP), the universal energy carrier molecule that is found in every known living organism (Mitchell 1961; Hinkle and McCarty 1978). He was awarded the 1978 Nobel Prize in chemistry for his "chemi-osmotic theory", which is now generally accepted.

As regards the primary source of exergy, the chains may have been dependent on prebiotic photosynthetic reactions not as yet identified, or (more likely) on sulfur-iron chemistry by primitive chemo-litho-autotrophs in the neighborhood of deep hydrothermal vents. As I have noted previously, the basic chemistry of life requires either a continuing flux of free energy (exergy) inputs from somewhere "outside" or a stock of free energy (such as fossil fuels) from previous accumulations. The original flux of free energy must have been solar radiation, but at first the flux may have been interrupted by an inorganic accumulation process.

Whether the polymerization process occurred in the deep ocean or near the surface, it is the unique ability of carbon to combine with itself in stable chains or rings that accounts for many of the more complex organics, including the nucleotides. For instance, adenine, one of the nucleotide bases, is just a combination of five copies of the inorganic hydrogen cyanide molecule HCN, linked to itself as $(HCN)^5$. Similarly, adenosine diphosphate (ADP), which is essential for energy metabolism in all species, consists of a 5-carbon sugar molecule (ribose) linked to adenine and three phosphate groups.

Nucleotides are the essential building blocks of nucleic acids. Admittedly, interesting synthesis experiments have been done in laboratories. For instance, Leslie Orgel has demonstrated the spontaneous formation of a sequence of 50 nucleotides in a "DNA like" molecule, starting from simple carbon compounds and lead salts. However, what can be done in a test tube under laboratory conditions would not necessarily happen under natural conditions. The difficulty with the "hot soup" hypothesis (already mentioned) is that unprotected nucleotides, and chains, decompose very rapidly in water due to hydrolysis. This difficulty is one of the most persuasive arguments in favor of an extra-terrestrial process of organic synthesis, along the lines discussed above.

A. G. Cairns-Smith (cited earlier) offered a theory of the origin of nucleotide replication. He postulated that life started not with carbon-based organic materials but with self-replicating silicon-based crystals (Cairns-Smith 1984, 1985). The idea is that it is much easier to imagine how a dissipative chemical structure could persist in the presence of a solid "template" than merely floating in a liquid solution. The complex surfaces of certain natural minerals offer ideal templates for the formation of complex carbon-based compounds. According to the "template theory", replication of the crystals led to replication of the associated organics. Gradually, by evolutionary selection, the carbon-based compounds "learned" to replicate without the help of the silicon crystals. This scheme has been dubbed "the genetic takeover".

Polymers, such as polysaccharides, polypeptides and polynucleotides, are an essential component of all known organic life. Fatty acids, cellulose, proteins and DNA/RNA are all examples of polymers. At some stage in chemical evolution simple organic compounds (monomers) must have been concentrated and combined into chains or sheets to form polymers.[3] It seems that one might account for

---

[3] Of course, this required a reasonably high concentration of organic molecules. It was assumed before the deep ocean "hot spring" theory that the concentration process occurred mainly around the edges of ponds, by evaporation and freezing.

polysaccharides in a warm organic "soup" by a mechanism, such as that postulated by Oparin, given the availability of protein catalysts (enzymes).

The preferred theory of polymerization today seems to depend on some sort of autocatalysis. The concentration of polymers in the primitive ocean was subsequently built up by some as yet unknown mechanism. One possibility has been explored by Sydney Fox, who has shown that when high concentrations of aminoacids are heated under pressure, a spontaneous polymerization occurs, yielding "proteinoids" (Fox 1980, 1988). This is also consistent with the possibility that (instead of life originating in shallow ponds) organic life may have originated deep in the ocean in superheated regions near volcanic vents.

The synthesis of large molecules in large numbers had to depend on autocatalytic chemical cycles that reproduce important molecules (and provide a surplus to compensate for replication errors). The importance of catalysis has been mentioned already. It is possible that a reaction product may also be a catalyst for that reaction. This is called autocatalysis. It is very important for replication of living organisms. There is no way to know what the earliest autocatalytic cycles looked like, because they have long been superseded. But it is possible that the main source of "food" for the first cells was one of the sugars, produced by an autocatalytic reaction. A plausible example is the "formose" reaction, first described in 1861 by Butlerow (1861). The system is catalyzed by a divalent metal ion (probably Ca^{2+}), as shown in Fig. 5.5.

A version of the formose reaction that accounts for specific products was discovered in 1959 (Breslow 1959). It starts from a combination of two formaldehyde (HCHO) molecules, forming glycol aldehyde (plus two hydrogens). This

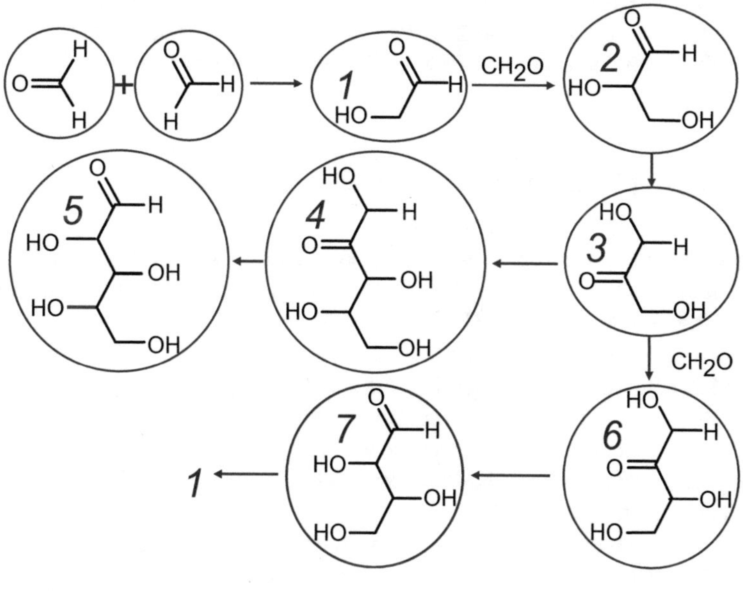

Fig. 5.5 Formose reaction

reacts with another HCHO to become glyceraldehyde (plus another H^+). Then the glyceraldehyde combines with the glycol aldehyde to form ribulose, which becomes ribose after an isomerization. The overall reaction leading to ribose is as follows:

$$5(HCOH) \rightarrow (CHO)_5 + 5H$$

The reaction is catalyzed by its own end product (glycolaldehyde in this case). So once it gets started it accelerates and continues as long as the feedstock (formaldehyde) lasts. It is crucial that the process be truly autocatalytic, i.e. that it produces extra copies of the original catalyst molecule. If not, errors of replication—side reactions—would gradually consume the catalysts and the process would stop (Maynard-Smith and Szathmary 1995).

There are many remaining puzzles, notably how ribose—among the 40 or more sugars—could be produced in large enough quantities and why the decay of unstable sugars should leave formaldehyde as an end product. However, where there is one simple autocatalytic reaction, there must certainly be others. Modern organisms depend upon more complex versions of such autocatalytic cycles. Two of them will be described later.

It is unclear whether the autocatalytic formose reaction (or some cousin) could have taken place in the open water, even on a friendly mineral surface like iron sulfide. The alternative possibility is that it took place inside a primitive "cell" protected by a polymeric membrane. One suspects that the phospho-lipid monomers of that membrane must have been produced by a self-catalyzed reaction comparable to the formose reaction discussed above. However, no such self-catalyzed reaction to produce phospho-lipids, inside or outside of cells, has yet been suggested, unless the work of Pasek, mentioned earlier, fits the bill (Pasek 2008). This would seem to be a key missing piece in the puzzle.

5.4   Self-replication of Macromolecules (Genes)

I should begin with a very new theory, published by Jeremy England (England 2013). It is based on a recently discovered theorem in non-equilibrium thermodynamics, the Jarzynski equality (Jarzynski 1997) and the Crooks fluctuation theorem (Crooks 1999). The Crooks theorem states a rigorous equality between a ratio of forward and backward process probabilities and the expression $\exp[(W - \Delta F)/kT]$ where W is work done, ΔF is the change in Helmholtz free energy, k is the Boltzmann constant and T is temperature. The exponent is proportional to the entropy S. It will be zero in equilibrium, when the forward and backward probabilities are equal.

England's theorem uses these theorems to explore the case of atoms of different species, free to migrate (in a solution or a gas), given an external source of source of exergy (and a heat dump). It concludes that these atoms (or monomers) will *automatically* seek and tend to find stable molecular structures that maximize stability (binding energy). Meanwhile the search process itself simultaneously maximizes entropy production (England 2013). Not only that, it appears that *the system has a high probability of becoming autocatalytic* and capable of replicating itself. Comparable results have been obtained experimentally, using coated microspheres (Zoravcic and Brenner 2015). It is too early to say how general this new result is, or how accurately it can predict biochemical process outcomes. However, it is already clear that under certain conditions self-replication is achievable by fairly simple chemicals in non-living systems.

Having noted the possibility of a major breakthrough, let me now focus on "standard" evolutionary theory, as understood in 1995. The authors were an elderly Englishman (John Maynard-Smith) and a young Hungarian (Eörs Szathmáry) (Maynard-Smith and Szathmary 1995). They identified eight major transitions in evolution, known as the "SMS transitions" after the initials of the authors. Each of these transitions depended on the prior one. Their list started with the transition from self-replicating molecules (Bernal's second stage) to populations of such molecules in compartments (i.e. Bernal's third stage).

The crucial step, probably the most difficult of all, was cellular replication. Whereas some of the later SMS transitions, such as multi-cellularity, have occurred several, or even many, times in biological history; this one probably happened only once. SMS describes it as the transition from unlinked replicators to chromosomes (linked genes). But a few introductory words are needed.

Erwin Schrödinger, one of the founders of quantum mechanics, pointed out in his Dublin lectures 1943 that the first self-reproducing entities (macro-molecules or cells), had to be relatively unaffected by ultraviolet (UV) radiation, and with strong chemical bonds making them relatively unreactive with oxygen. Yet they needed error-correcting, self-repairing and replicating capability (Schrödinger 1945). Those hypothetical macro-molecules or cells were also storehouses of chemical energy (exergy) in the form of sugar molecules, plus catalytic "digestive" apparatus (to move energetic phosphate ions from ADP to ATP and back) and thus enable reproductive activity.

As von Neumann pointed out, a storehouse of information about the structure of the macro-molecule or cell itself was needed. That information, in all cells of living organisms today, is contained—indeed embodied—in the genetic substance called **deoxyribonucleic acid (DNA)**. But there was an auto-catalytic molecular predecessor, ribonucleic acid (RNA), that had the capability of catalytically replicating itself. RNA is a single—not double—helix of the nucleotide bases (adenosine, guanine, cytosine and uracil). The double helix, DNA, which is less unstable, came later.

A possible mechanism for direct protein replication without the DNA machinery has been postulated by Robert Shapiro (Shapiro 1986). It requires that there be an external "support" for the protein molecule that is to be reproduced, and an enzyme to recognize and match amino acids at both ends. There is little direct evidence for this hypothesis, but it does suggest a reason why such a small number of amino acids (or the next larger building block, **exons**) are found in proteins.

The minimum elements of a self-reproducing system (of any kind) were first identified in 1948 by the mathematician John von Neumann, viz.

- A **blueprint** (i.e. stored information)
- A **factory** to fabricate the components and assemble them
- A **controller** to supervise the operations of the factory
- A **duplicating machine** to copy the blueprint.

Von Neumann neglected to mention an energy (exergy) source. This takes us to the topic of photosynthesis, which is complicated enough tó deserve a separate section, Sect. 5.7. Of course, von Neumann—a Hungarian mathematician—characterized all of these items in fairly abstract mathematical language (von Neumann 1961–1963).

Perhaps the first attempt to construct a theoretical model with the elements suggested by von Neumann was a simple autocatalytic synthesis model called the "chemoton" proposed by Tibor Ganti in the 1970s (Ganti 1979, 2003). For a comprehensive discussion, see Maynard-Smith and Szathmary (1995). Ganti's idea was to create an abstract model of a cell with the minimum characteristics of life. These are: (1) a chemical synthesis system capable of producing all the materials needed by the cell, including the other main components (2) a bi-layer "skin" or membrane (see the end of Sect. 5.3) separating the cell from its surroundings, but capable of allowing certain substances ("nutrients") to enter and others ("waste") to leave, and (3) an information carrier allowing the entire system to replicate itself. The schema is shown in Fig. 5.6. He did not distinguish the "blueprint" from the "duplicating machine", although von Neumann did so.

The chemoton scheme for self replication is quite general, but as applied to the single cell organisms that dominated Earth for 2 billion years, the key "nutrients" were carbon dioxide (CO_2), water (H_2O) and one of the reducing agents like hydrogen sulfide (H_2S), methane (CH_4), ammonia (NH_3), or ferrous iron oxide (FeO). The wastes were S, N_2, Fe_2O_3, or (finally) O_2. Deposits of elemental sulfur and iron ore, as well as atmospheric oxygen are possible evidence of the existence of such a cell. (As a matter of some interest, it appears that most atmospheric nitrogen probably originated in space; the contribution from ammonia metabolism would have been very small.)

At least one version of the chemoton has been synthesized in a computer. It performs all the functions described, including growth and replication and it retains its coherence in a variety of solutions (Csendes 1984). N.B. The synthetic chemoton-in-a-computer did not require any enzymes. Hence it seems to fill the gap between autocatalytic chemical synthesis and "RNA world", as discussed later.

The biological counterparts of von Neumann's four elements are easily recognized. As expressed in biological organisms today, the "blueprint" is the DNA, the cytoplasm and its enzymes are the "factory", the specialized *replicase* enzymes play the role of "controller" and the DNA replication process is the "duplicating machine". The mechanism for information replication need not be described in detail here, though the unravelling of the universal *genetic code* has been one of the greatest milestones of twentieth century biology.

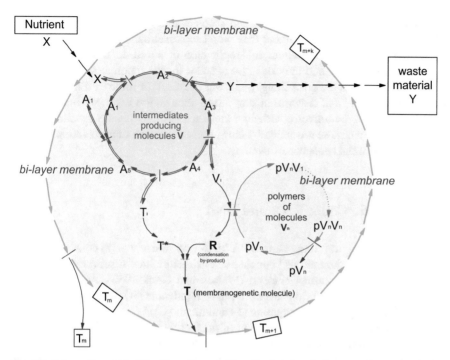

Fig. 5.6 Schematic model of the Chemoton: a self-replicating chemical system. *Source*: Adapted from Smith and Szathmary (1995, Fig. 2.2, p. 21)

I should not pass on without mentioning one of the still unresolved mysteries of biological evolution. The catalysts of living organisms today are enzymes. Enzymes are proteins. Yet the process of manufacturing proteins, in cells, requires enzymes. How could proteins have been produced in early organisms without enzymes? We don't know. It is a puzzle.

The choice between continuous and discontinuous theories remains unresolved. However, a significant "straw in the wind" may be the recent discovery that the prevalence of *chiral purity* (the conspicuous violation of mirror symmetry with regard to the "right-handedness" or "left-handedness" of organic molecules, mentioned above) might be explained by the possibility that the core of the sun is a rotating neutron star, originally one of a binary pair, emitting circularly polarized light. Astronomers are not sympathetic with this idea. But otherwise chiral purity can only be explained by a catastrophic—uniquely single event—type of model (Morozov and Goldanskii 1984).

A remaining difficulty in accounting for life is whether replication was already "invented" in space (e.g. as viruses), or whether that last step occurred on Earth. If the former is true, there must be many other planets in the universe with carbon-based life. However the odds that life on other planets would share the same amino acids, the same nucleotide bases, the same genetic code, the same energy transfer mechanism (using ADP and ATP) and so on would seem to be extremely low. There may be aliens out there but they won't look like us.

If the "invention" of cellular self-replication occurred on Earth, the question remains why it happened only once (to account for the uniqueness of the genetic code, chirality, and the other things). If it was the result of some extremely unlikely circumstance, and that circumstance occurred once on the Earth, we may conceivably be alone in the universe, although there are so many planets around so many stars that uniqueness seems unlikely. However, this avenue of speculation does not help us unravel the sequence of events.

5.5 Genetic Code: RNA and DNA

The basic information storage unit **DNA** (**deoxyribonucleic acid**) consists of a kind of twisted ladder; the famous "double helix" structure, first discovered in 1953 by **Francis Crick** and **James Watson** (Watson and Crick 1953). As mentioned in Sect. 5.1, the five nucleotide bases are **purines** (adenine (A) and guanine (G)) and **pyrimidines** (cytosine (C), thymine (T) and uracil (U)).

The "rungs" of the ladder are pairs of nucleotide bases (A $\leftrightarrow$ T, G $\leftrightarrow$ C). Each base is anchored to a ribose (sugar) molecule on the helical frame. The sugar molecules are linked longitudinally by phosphates. Information is stored in the longitudinal sequence of bases. Each set of 3 bases corresponds to a 3-letter sequence of the four letters A, T, G, C. There are $4^3 = 64$ different possible 3-letter sequences. These 3-letter sequences are called **codons**. They are the basis of the genetic code. Each codon corresponds to one of the 20 amino acids from which all proteins are constructed.

Note that one of the five possible bases, uracil (U), appears only in the single-strand nucleic acid RNA, as a direct substitute for thymine (T). It never appears in double-stranded DNA. The equivalent pairings between corresponding strands of RNA and DNA are therefore

$$(U \leftrightarrow A, \; A \leftrightarrow T; \; G \leftrightarrow C).$$

The replication procedure is simple in principle: a special enzyme (protein molecule) simply "unwinds" the double helix and separates the two strands. These are chains of nucleotides. Because of the strict pairing rules (A $\leftrightarrow$ T; G $\leftrightarrow$ C) each strand can then be used as a template to construct an exact duplicate of the original DNA, given a supply of unattached nucleotide bases. This reconstruction is done by other special enzymes.

The translation of DNA information into protein structures is slightly different. It begins with the synthesis of single strand of nucleic acid (called *messenger RNA* or *mRNA*) by the appropriate base-pairing rules (except that U replaces T in each codon). This process is called "transcription". The mRNA contains the same coded information as the corresponding strand of DNA. The mRNA is then "decoded", codon by codon, by a protein molecule called a **ribosome**, which is roughly the analogy of the assembly worker in a factory. The materials (amino acids) for constructing new proteins have to be brought to the assembler by carrier molecules called "transfer RNA" or tRNA.

Suppose, for instance, the ribosome is decoding the 3-letter sequence AUC in the mRNA. It happens that this particular codon corresponds to the amino acid **isoleucine**. The ribosome needs a tRNA molecule that exactly fits the AUC codon, but in reverse order. This is an **anti-codon**. By the pairing rules, one of the anti-codons for isoleucine is GAU. When the ribosome encounters GAU (or one of the others that match) it detaches the amino acid from the tRNA and reattaches it to the protein "work-piece". When the ribosome encounters a "stop" instruction on the mRNA chain it releases the complete protein. The standard protein synthesis hypothesis of molecular biology is therefore

$$DNA \rightarrow RNA \rightarrow tRNA \rightarrow protein$$

It is important to observe that the standard hypothesis does *not* admit any flow of information from proteins back to DNA. Such a reversal would, in principle, permit *learned* information to be retained permanently in the genes. This process was once postulated by the nineteenth century biologist Lamarck, who didn't know about genes. It was later revived by the discredited Soviet biologist Lysenko. But there is no experimental evidence for such a learning process and geneticists today rule it out.

The foregoing paragraphs do not cover all of the key synthesis processes by any means, but they do summarize what is known of the most important one. It is an inescapable fact that nucleic acids and proteins are interdependent from an evolutionary perspective. Therefore, it is necessary to explain not only how both proteins and DNA were generated from non-living matter, but also how such a **process** came into existence. Hence there is a classic "chicken and egg" problem to be resolved. How could DNA be generated in the first place without protein replicase enzymes? How, on the other hand, could enzymes (proteins) be produced without DNA, RNA and ribosomes?

In the context of biological evolution, the nucleic acids (building blocks of polynucleotides) were probably a precondition for the first appearance of self-replication. This is because of their known ability to catalyze reproduction through template action. However, a number of template-catalyzed reactions would have to be coupled together via chains of autocatalytic processes. One scheme to get around the difficulty is the "*hyper-cycle*", postulated by Manfred Eigen and Peter Schuster is (Eigen and Schuster 1979; Eigen 1971). Like Oparin, they assumed a primordial "soup", already containing peptides or small proteins and fatty acids, as well as nucleotides. (In other words, they assume away the difficulties noted previously). They also assume a prototype RNA molecule formed by accident, possibly assisted by the proteins in the soup.

Eigen and Schuster then postulate that the genetic message for a complex protein can be modularized and that each module (an exon, presumably) is autocatalytically replicated in the soup and evolves independently. But they suppose that the modules (e.g. exons corresponding to genes) are also inter-related in a larger cycle; that each exon catalyzes the formation of the *next* exon in the sequence. Such a system is stable if (and only if) the sequence is closed on itself, like a **chemoton**. This is the hyper-cycle, and its material manifestation is the RNA molecule. The system can also evolve and grow in complexity.[4] Thus, in effect, Eigen and Schuster postulate Darwinian evolution at the molecular level.

But the question remains: How can an RNA molecule "learn" to synthesize complex proteins? The problem is to explain replication without replicase enzymes that perform the error detection and correction function. Under such conditions one would expect a high natural error rate. The high copying-error rate limits the maximum number of nucleotide bases that can be copied successfully to the order of 100 (Casti 1989). That is about right for a single amino acid. But it is nowhere near enough to encode a protein complex enough to be an enzyme. In other words, without having enzymes in the first place, the replication error rate is too high to reproduce enzymes.

The high error rate of RNA reproduction corresponds to rapid mutation. In fact the HIV virus, with only nine genes is based on RNA. It is the error-prone-ness of RNA reproduction that accounts for the high mutation rate of HIV. The frequency of copying errors in HIV virus reproduction is 100,000 times greater than it would be for a modern cell. Note that a bacterium typically has around 500 genes, and some of them are devoted to producing error-correcting enzymes that enable the bacterium to reproduce itself with reasonable accuracy. The next stage would be a eukaryote call, that has 10 to a 100 times more genes than a bacterium and up to a 1000 times as much DNA material. The problem of minimizing copying errors (i.e. error correction) in complex organisms (such as ourselves) becomes far more difficult. In fact, it is the core of the difference between prokaryotes and eukaryotes.

Walter Gilbert (1986) has suggested a scheme for RNA replication in the absence of replicase enzymes. He postulates auto-catalysis by the RNA molecule itself, leading to proliferation and evolutionary improvements; followed by protein synthesis, and finally to DNA synthesis. There is some experimental evidence of autocatalysis by RNA; (e.g. Orgel 1983). But for both Eigen and Gilbert the original synthesis of RNA itself remains unexplained. However, biological evolution, which coincided, to a large extent with chemical evolution, began with the appearance of DNA (along with other associated chemical species) or, more properly, with the

---

[4] Unfortunately, the hypercycle is not stable against all possible perturbations (Niessert 1987). Computer simulations have uncovered three types of possible discontinuities that can result from unfortunate mutations (copying errors): (1) One RNA molecule learns how to replicate faster than others but "forgets" to catalyze the production of the next one. (2) One RNA molecule changes its role and skips one or more steps in the chain, leading to a "short-circuit". (3) One population dies out due to statistical fluctuation. It turns out that the first two types of catastrophe increase in probability with population size, while the third type has a high probability if the population is low. This implies a finite lifetime for a hypercycle, unless, of course, the hypercycle succeeds in evolving a protective mechanism such as a phosphor-lipid membrane.

emergence of the DNA replication cycle in single-cell prokaryotes roughly 3.6 billion years ago.

The dominant current theory of the origin of life today is called "RNA-World". It holds that ribonucleic acid (RNA) was actually the halfway house or precursor to DNA. This is because the RNA molecule is able not only to replicate itself but also to catalyze amino-acid polymerization and peptide ligation, which are essential for protein manufacturing in the cell. Moreover, several of the co-factors, such as adenosine triphosphate (ATP), are essentially constructed from nucleotides. It seems likely, therefore, that RNA came first and DNA came later.

The main difference between DNA and RNA, in this context, is that RNA is more reactive and replicates faster, but is less durable and less reliable as an information carrier. In effect, there is a tradeoff between reproduction rate and error-rate (cost). Part of the explanation seems to be that the substitution of thymine for uracil reduces the error probability. (Thymine is identical to uracil except that it has an extra methyl group attached. That methyl group effectively blocks a weak spot where other free radicals might attach themselves, with unpredictable consequences.) At first, the faster replication rate "won" the competition for food, but later the greater durability of DNA seems to have tilted the competition, leaving RNA as an evolutionary remnant (like the human appendix or the tailbone).

Whatever the detailed mechanism for self-reproduction at the chemical level turns out to be, it is increasingly clear that the processes can best be described in terms of information storage and transfer. In all likelihood, the final theory will be expressed in the language of information theory. Manfred Eigen emphasizes that the theory of selection is ultimately derivable from classical (or quantum) statistical mechanics by adding a parameter representing *selective value* for each information state. In fact, he characterizes his own approach as "*a general theory which includes the origin or self-organization of ('valuable') information,*[5] thereby uniting Darwin's evolution principle with classical information theory" (Eigen 1971, p. 516).

In fact, these words are generally applicable to modern molecular biology. It seems almost superfluous to add that Eigen's notion of "selectively valuable" information is essentially identical to what I have previously called D-information (Ayres 1994). The notion of "value" takes on more precise meaning as we move later in this book into the realm of social science and economics.

5.6 Information Transfer: The Genetic Code

A brief recapitulation is appropriate here. As already explained, the genetic DNA molecule is a double strand polymer, constructed of a pair of helical "backbones", which are actually more like the hand rails of a spiral staircase (Watson and Crick

---

[5] Eigen's italics.

1953). These helical backbones—or hand-rails—are also cross-connected by "base pairs" of nucleotides linked by hydrogen bonds. The base pairs are like the steps of the spiral staircase. There are just four possible pairings: adenine-thymine (A-T and T-A), and guanine-cytosine (G-C and C-G)). These four pairs (A-T, T-A, C-G, G-C), connecting the two backbones, are like the stairs of the spiral staircase. The pairs constitute the four "letters" of the genetic alphabet. Sets of three letters are like "syllables"; they are called "codons". There are 27 possible three letter combinations (codons) for each set of three letters but since there are four letters, there are $4 \times 27 = 108$ possible codons.

Each of the 20 amino acids from which all proteins are constructed corresponds (maps) to at least one of the codons, while some of the amino acids are mapped to as many as six codons. It is not known why this is so. But as far as is currently known, only 61 of the possible codons map to any amino acid, while 1 codon corresponds to the instruction "start" and 3 to the instruction "stop". This leaves 43 codons with no known function. It is thought that they may have had other "meanings", in the past, but now they appear to be "junk" with no function except to punctuate the meaningful sequences. Needless to say, what now appears to be "junk" may later turn out to have important but unknown functions.

Sequences of codons are analogous to sequences of syllables in the words in a written language; the words are called "*axons*". The axons, ranging from 20 to 120 codons in length, correspond to groups of amino acids. An axon of length 40 could be a code-word for a group of between 5 and 10 amino acids. There appear to be somewhere between 1000 and 7000 different axons (words) in genetic material. Each gene consists of 15–20 axons, or somewhere in the range of 60–120 amino acids. (There are also long sequences of amino acids, located between some axons, called "introns", that appear to have no meaning today, but that might be evolutionary leftovers.) Thus, each gene may be analogous to a sentence in written language. Several genes could correspond to a paragraph. However, the sentences in the paragraphs may contain meaningless letters or syllables (the introns) unlike paragraphs in written languages.

The set of mappings noted above (details can be found in textbooks) constitute the "standard" genetic code, which applies to virtually every form of life on Earth. There are a few minor variations of the code, but even the variants are quite similar to the primary one. (There are also a few proteins, found in extremely rare creatures, that have incorporated an extra amino acid in their proteins.) However, it is a great puzzle as to why there is only one code, when in theory there could have been a huge number—around 1.5×10^{84}—of possible genetic codes (mappings). (Don't ask how this number was calculated. The point is that the number of possible mappings is very large, but only one of them is in use by virtually all living organisms. For that matter, there is another puzzle: why do (almost) all proteins contain just 20 amino-acids when there could easily have been 30, 40, 60 (or more) different amino acids? Homo-chirality is yet another puzzle. Chirality is like right-handedness (dextro, or D-) or left-handedness (levo, or L-) as applied to amino acids and sugars, respectively. In theory the pre-biotic synthesis of amino acids both L and D types are equally probable. But in living cells all amino acids are left-handed

(D) while all sugars are right-handed (R). How could this have happened? There are several hypotheses to explain these mysteries, but none has yet been generally accepted.

Speaking of proteins, it is worth noting that a protein molecule might contain as few as 20 to as many as several thousand amino acids. (The muscle protein titin contains nearly 27,000 amino acids.) The well-studied bacterium $E.$ $coli$ contains 2400 different proteins, with an average length of 320 amino acids. The number of possible combinations of that length is then 20 (the number of different amino acids) to the power 320 which is a hyper-astronomical number (10^{416}) which is far, far greater than the number of atoms in the universe (around 10^{78}).

Humans have more different proteins than $E.$ $coli$, more than 10,000. But most organisms are much simpler than humans. It is also clear that most species that have existed in the past 3.5 billion years are now extinct. Suppose that there were 100 million species, each having 10,000 different proteins and lasting only 100 years before being replaced by another species with 10,000 different proteins, the number of proteins that ever existed on Earth would be 3.5×10^{19}. In reality, the number of proteins that ever existed was much, much less, because the mutation that creates a new species rarely changes more than one or at most a few proteins, while the average lifetime of a species is more like a million years, not just a hundred. Still, based on this upper limit, it has been calculated that the number of proteins that has ever existed since the origin of life on Earth, as compared to the number of possible proteins (based on 20 amino acids), is much, much smaller than the fraction 10^{-397}. In other words, effectively zero. Why? Nobody knows why so few of the possible proteins are needed.

There is another difficulty, already mentioned, namely the problem of copying errors. In theory, the replication produces an exact copy, but in reality errors can (and do) occur. Think of it as a communication system. Every replication is like sending and receiving a message. Every message contains a tiny bit of static. If it were that simple, the message would be lost in static after a few dozen replications. But this doesn't happen. There must be an error- correction system at work. Such systems are possible. They have been implemented in large computer systems. But how it works in biology we don't know yet.

The laws of thermodynamics, especially the second law, come into play again at this point. It seems that the entropy produced by a dissipative process is proportional to the ratio of probabilities of the reaction vis $à$ vis the probability of its reverse (Jarzynski 1997; Crooks 1999). In terms of the "hill" metaphor, a reaction in which the electrons from the donor go "downhill" to the acceptor is exothermic and highly probable. The reverse reaction is "uphill" and very improbable.

These probabilities are determined (in part) by the stability of a configuration. For example, when two ions in a solution, each having many possible microstates, come together to form an insoluble compound, the number of microstates of the compound (its entropy) and the probability of reversal are near zero. But the ratio of probabilities (the entropy dissipation) of the reaction is very high. So the formation of very stable molecules such as DNA, is an illustration of dissipation maximization and replication error minimization.

Evidently stability is a measure of evolutionary fitness. The more stable a configuration the longer it will survive. What we don't yet understand adequately is the chain of improbable reactions that resulted in the creation of an RNA or DNA macro-molecule.

5.7   Oxygen Photosynthesis

The first living cells, the "last universal common ancestors" (LUCAs), appeared between 3.6 billion years ago and 4 billion years ago. The first cells without a nucleus (prokaryotes), were heterotrophs, meaning that they depended for metabolic energy (exergy) on organic molecules extracted from the environment. A distinct exothermic "digestion" process, glycolysis (which means "sugar splitting") must have also evolved during that period. Glycolysis breaks up carbohydrate molecules and stores the chemical exergy in the form of adenosine triphosphate (ATP).

One of the most remarkable features of biology on Earth is that the energy carriers ATP and ADP, as well as chemical cousins NADPH and NADP, are all based on the nucleotide molecule adenine. This specific energy transfer mechanism is common to all known living species. Whenever work has to be performed in a certain place within a cell, ATP is transported there, where it and converted into its cousin adenosine diphosphate (ADP) plus an inorganic phosphate (PO_4^+) and a jolt of exergy that powers some other downstream reaction. The phosphate ion is then recycled by a photon into the next molecule of ATP.

Some of those organic "food molecules" in the early environment were produced by abiotic photosynthesis processes, either in space or on the Earth, driven by solar photons in the ultraviolet (UV) range (Sect. 5.1). But at some point the photosynthesis processes that converted carbon dioxide (CO_2) to carbohydrates (CH_2O) were internalized. We don't know how that happened.

We do know that they also needed initial stockpiles of some "reducing agent" (electron donor). The possibilities include hydrogen gas, methane, ammonia, atomic sulfur, hydrogen sulfide, atomic iron or ferrous iron. Several of these were actually utilized, by different species of primitive bacteria.

A crucial problem was that none of those reducing agents could be completely recycled. Hence, the original stockpiles were gradually depleted. For instance, recycling of hydrogen could occur by methano-genesis of organic detritus in the oceans, but only half of the hydrogen contained in the feed is recovered this way. (And there was further loss as organic materials were buried under sediments.) All of the other "legacy" reducing agents were probably also recycled to some degree in the oceans, but the recycling process is always imperfect, resulting in insoluble chemical combinations like sulfates and sulfides and burial in sediments.

The largest supply of reducing agent in the early oceans was ferrous iron (FeO). As it happens, the soluble ferrous iron was not yet exhausted at the time oxygenic photosynthesis was "invented" about 2.7 billion years ago. But the remainder in the oceans was oxidized and deposited as ferric iron—now iron ore—during the next

300 million years after oxygen buildup began. The oxygen in the atmosphere eventually killed the organisms that utilized ferrous iron as a reducing agent, namely the **stromatolites**.

For readers with some knowledge of chemistry, the key point is that converting carbon dioxide molecules to carbohydrate molecules involves moving electrons "uphill" against an energy (voltage) gradient. This occurs by taking electrons from a "donor" and moving them along a pathway (that may involve several intermediate chemical reactions) to an eventual "electron acceptor". The electron "donor" is the reducing agent and the "acceptor" is the end product, usually expressed as a generalized carbohydrate (CH_2O). The adverse voltage gradient has to be overcome by reactions with a molecule in an excited state that has absorbed a high energy photon from sunlight.

The scheme can be expressed by the general formula:

$$CO_2 + 2H_2A + photon \rightarrow CH_2O + H_2O + 2A$$

where A (an electron donor) can be any one of a variety of elements or molecular groups—ions or radicals—that are recycled in the overall reaction sequence. The most likely early candidates for A were elemental sulfur or iron, or incompletely oxidized compounds thereof. The A molecules must have been absorbed through the phospho-lipid cell wall (Fig. 5.4). Other chemical complexes inside the cell extracted carbon from dissolved carbon dioxide and taking hydrogen from H_2A (a hydride such as hydrogen sulfide) to make a sugar (ribose).

Some early proto-bacteria split hydrogen sulfide to synthesize carbohydrates while releasing elemental sulfur. The underground sulfur deposits of Louisiana and Texas are probably leftovers from prebiotic bacterial photosynthesis based on H_2S from undersea vents.

Unfortunately, excited electronic states resulting from absorption of a high-energy photon don't last long (only micro-seconds). So the "uphill" reactions have to happen quickly, and there must be a pre-existing pathway for the electrons to move from the original "electron donor" to the final "electron acceptor". Needless to say, the greater the voltage difference, the more difficult it is to find chemical reaction sequences that make it possible.

Having said this, we come to arguably the greatest and most unlikely evolutionary invention, namely **oxygen photosynthesis**. In this process, the electron donor A in the above equation is the oxygen in water (H_2O). The extremely stable water molecule is split, separating the oxygen from the hydrogen, recombining the hydrogen with carbon dioxide (CO_2), yielding oxygen as a waste. These new organisms are known as **cyanobacteria** or "blue-green algae".

What makes this photosynthesis process so much more complicated than its predecessors is that it cannot be done in a single step. Two completely different reaction sequences are involved.[6] These two sequences, which have been labeled

---

[6] For details see Chaps. 7 and 8 of Lenton and Watson (2011).

a The Calvin cycle has three phases.   **b** The reaction occurs in a cycle.

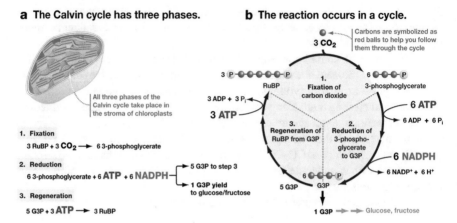

Fig. 5.7   Calvin Cycle for photosynthesis. *Source*: https://roboplant.files.wordpress.com/2013/06/calvincycle.jpg

Photosystem I (P I) and **Photosystem II (P II)**, probably evolved quite independently in different environments.[7] But the water-splitting oxygen photosynthesis system now utilized by all plants has somehow managed to combine the photosystem used by the green bacteria with the photosystem used by the purple bacteria. Two different photon excitations are required to kick the electrons up the steep gradient from the donor (water) level to the final acceptor (carbohydrate) level.

The photosynthesis mechanism in plants is known today as the Calvin cycle for its discoverer, **Melvin Calvin** (Calvin 1956, 1969). See Fig. 5.7. If you are not a chemist, skip the next two paragraphs. The cycle starts with a molecule of carbon dioxide and one of water. The carbon dioxide molecule then combines with a 5-carbon sugar (ribulose-5-biphosphate, or RuBP for short). They combine to form a 6-carbon intermediate that is unstable and immediately splits into two 3-carbon molecules of 3-phosphoglyceric acid (3PGA). The enzyme catalyst that makes it happen is called *RuBisCO*. Its chemical name is ribulose-1,5-bisphosphate carboxylase/oxygenase (better known as chlorophyll). RuBisCO is the most common protein on the planet Earth.

Each of these 3-carbon molecules then reacts with adenosine triphosphate (ATP), which loses a phosphate and is converted to the diphosphate (ADP). The molecule of 3PGA then becomes a molecule of diphospho-glycerate (DPGA). This molecule then undergoes a similar reaction with a co-enzyme (nicotinamide adenine dinucleotide phosphate), better known as NADPH. The result is 2 3-carbon sugars, called glyceralaldehyde-3-phosphate, known as G3P. This is the stuff that is later converted to fructose diphosphate and still later, to glucose, sucrose and other

---

[7] Each of these photosystems has several variants. PS I is found in green sulfur bacteria (such as chlorobiacians, using H_2S as a donor) and also in heliobacteria (nitrogen fixers found in wet paddy fields). P II is found in purple bacteria (proteobacteria) around hot springs and vents. Some of the proteobacteria use sulfur as a donor, and some not).

things. However, only one molecule of G3P out of every six is available to make sugars; the other five molecules are needed to reconstruct the original 5 carbon sugar, RuBP. Overall the Calvin cycle formula is:

$$6CO_2 + 6H_2O + \text{photons} \rightarrow \text{Glucose} + 6O_2$$

The energy of the solar photons is thus transformed into the chemical energy of glucose. Glucose is very stable and can be stored in cells for future metabolic use. It is interesting to note that the basic oxygen-photosynthesis process, "invented" 3 billion years ago—although not in the modern Calvin cycle form—has provided almost all of the useful energy and organic material produced on the Earth after that time. Current carbon dioxide fixation by photosynthesis in the biosphere amounts to roughly 1 GT/y, about half in the oceans and half on land. This rate of fixation consumes 100 W/kg or 100 TW of solar power.

Though the blue-green algae (cyanobacteria) produced oxygen, the oxygen was rapidly recombined with hydrogen-rich compounds like hydrogen sulfide, methane and ammonia, or ferrous iron, in the reducing environment. There was only one source of free oxygen in the atmosphere for a long time, even after anaerobic photosynthesis came along: Intense UV radiation at the top of the atmosphere caused photolysis of water molecules yielding atomic oxygen and atomic hydrogen. As mentioned earlier, atomic hydrogen was escaping from the primitive atmosphere, leaving atomic oxygen behind. Some of it was carried into the oceans by rain.

To recapitulate, oxygen photosynthesis (to give it its full name) uses energy from photons to split water molecules (H–O–H), discarding the oxygen as a waste but keeping the hydrogen atoms and combining them with carbon dioxide to make glucose, the main source of food for all plants and animals. We humans can, in principle, burn the biomass for fuel, thus "closing" the carbon cycle.

But wouldn't it be interesting if—instead of making glucose—the catalytic properties of chlorophyll could be harnessed just to split the water molecule, yielding hydrogen and oxygen as co-products, i.e. to do very efficiently what electrolysis does less efficiently today? The result would be a hydrogen-based energy system, powered by the sun, generating no wastes at all, and in particular, carbon dioxide as a combustion product. Could this be a future target for genetic engineering?

5.8 The "Great Oxidation" and the "Invention" of Respiration

For a billion years or more after the first living cells, carbon dioxide was gradually being removed from the atmosphere and the oceans, partly by photosynthesis and partly by abiotic processes. Some of the carbon dioxide was dissolved in the oceans.

In water, it ionized and became carbonic acid $H_2 CO_3$. Carbonic acid combined with alkali metal ions, like CaO and MgO, in the ocean, resulting in the creation of mineral carbonates like limestone, magnesite, and dolomite.

After anaerobic photosynthesis (Sect. 5.7) got established, the biomass of self-sufficient photo-synthesizers started to increase. This engendered a slow but steady "rain" of organic detritus drifting to the bottom of the oceans. The detritus was covered by sediment, buried and sequestered.

There was also (and still is) a purely geophysical process that also removes carbon dioxide from the atmospheric. It is the "weathering" of newly exposed basaltic rock, mainly silicates (minerals with SiO_3 groups in them), from the Earth's crust. This weathering process—which incidentally produces the silt that sequesters detritus under the sea, occurs in places like the Grand Canyon, as the river erodes its channel ever deeper. But, primarily, it occurs in the most recent mountain ranges (such as the Himalayas) where fresh basaltic rock is being exposed as a consequence of the collision of two massive tectonic plates.

Each year some CO_2 is permanently removed from the atmosphere and buried in the form of carbonates by this geophysical mechanism. The precise amount being sequestered by weathering each year is unknown but it seems to be close to 1 gigaton (GT) of CO_2 per year, which is about the same as the annual rate of CO_2 fixation by photosynthesis. This is small compared to the 30 GT/year now being added to the atmosphere by the combustion of fossil fuels, but over geological time it is large. And a large fraction (perhaps half) of each annual contribution from weathering is subsequently dissolved in the oceans.

This is why mineral carbonates resulting from that deposition contain 60 million GT of CO_2 in carbonates, compared to just 38,400 GT of CO_2 in the oceans and 720 GT in the atmosphere. While the weathering process is now roughly balanced by the biological processes of photosynthesis and respiration, it has operated for at least 3 billion years, whereas the photosynthetic fixation rate was much smaller in the past. Hence the bulk of the CO_2 in mineral carbonates is attributable to weathering, not photosynthesis. The amount of carbon in mineral carbonates in the Earth's crust is estimated to be about four times the amount of organic carbon in sedimentary rocks (shale), and perhaps 20,000 times the world's total coal reserves.

Nevertheless for the first billion years or so after anaerobic photosynthesis was "invented" (3.5 billion years ago) the Earth's atmosphere remained nearly free of oxygen because every molecule produced by photolysis and hydrogen loss, or by photosynthetic activity, was immediately recombined with some other element as fast as it was produced. But a little before 2.5 billion years ago, the oxygen "sinks", such as the hydrogen-rich gases in the "reducing" atmosphere (methane, ammonia, hydrogen disulfide) and the ferrous iron in the oceans, began to run out. Molecular oxygen began to build up in the atmosphere 2.4 billion years ago, albeit very slowly.

During this oxygen buildup, the biosphere, as a whole became increasingly unsustainable. Oxygen (in several forms) is reactive and toxic to all anaerobic organisms, even at low levels. On top of that, both carbon dioxide and methane, the two most effective greenhouse gases (GHGs) in the atmosphere, apart from water vapor, were significantly depleted. *Consequently the global temperature began to drop.* That cooling had predictable consequences for organisms adapted to relatively high temperatures. It was the first, and arguably the greatest, of the "great extinctions" that have wiped out a large fraction of all the species alive at the time. This, and the later extinctions, constitute some of the evidence for the *"Medea"* theory (that life inevitably destroys itself) (Ward 2009).

There must have been a trigger to shift the oxygen level suddenly (geologically speaking) from parts per million during the first 300 million years after the invention of photosynthesis, to parts per thousand just before the "great oxidation" occurred. The trigger could have been the creation of an ozone layer in the stratosphere. This would have inhibited UV radiation and enabled a sharp increase in biomass that—in turn—sucked out more CO_2 and further accelerated oxygen production. That resulted in more cooling. Or something else happened that sharply reduced the methane content of the atmosphere.

The "Great oxidation" was not exactly an "event" as we understand the word, although we regard it as the end of the Archaean age and the beginning of the Proterozoic age. It consisted of a slow (50 million year) rise in the oxygen content of the atmosphere, along with ozone, followed by a peak and a slow decline. The oxygen (and ozone) peak occurred between 300 and 500 million years after the advent of the photo-cyanobacteria, or about 2.4 billion years ago. The so-called Huronian ice age, consisting of three advances and retreats, probably began about the same time.

A few years ago it was generally assumed that the warming at the end of the Huronian glaciation was due to carbon dioxide accumulation, possibly from volcanoes, but now it appears likely that volcanic emissions of methane may have been partly responsible. Methane happens to be a much more potent greenhouse gas (GHG) than carbon dioxide. Nobody knows what caused "snowball Earth" to melt. However, after the "great melt" there was a burst of new biological innovations.

After the ice ages, carbon dioxide must have accumulated in the atmosphere. Moreover, oxygen began to accumulate in the atmosphere once again, probably because the oxygen photo-synthesizers increasingly out-competed other organisms. But molecular oxygen is toxic, while atomic oxygen and related free radicals (resulting from ionizing radiation) are even more toxic. A new biological "use" for the oxygen was badly needed. Without such a mutation, all living organisms would eventually have been poisoned by their own waste products.

I do not suggest that the "invention" that saved the day was intentional in any sense, although that is consistent with Lovelock's "Gaia" theory of co-evolution (Lovelock 1979). I think it just happened. It was a lucky accident. In the early cells, glucose was split (glycolysis) into lactic acid and ethanol before regenerating the original carbon dioxide. This process is better known as *fermentation*, still used by yeasts to produce bread dough, not to mention the organisms that produce beer and wine.[8] What happened was a mutation that enabled organisms to oxidize glucose

---

[8] Ethanol is an intermediate product of fermentation. (Nowadays the ethanol is utilized, for example, in beer, while the carbon dioxide is what makes the little holes in bread.)

directly. This "invention" is what we call *respiration*. It is very hard to pinpoint exactly when it took place, but based on indirect evidence it appears some prokaryotic bacteria had developed the chemistry of respiration by 2 billion years ago. However, this invention didn't become critical until the atmospheric oxygen level had risen significantly. At that point the respiration system was probably adopted by the eukaryotes.

The chemistry of respiration may have been another case of two processes combining fortuitously. The combination enables oxidation of carbohydrates, fats and proteins, with only CO_2 as a waste product. It is known as the *citric acid cycle* (or the Krebs cycle for its discoverer **Hans Krebs**, awarded the Nobel Prize in 1937). The cycle begins—like fermentation—with **glycolysis** (the splitting of glucose into two molecules) by an enzyme. But then, the intermediate products, lactic acid and ethyl alcohol, are further oxidized by a sequence of 10 stages. The waste product is CO_2.

The organic catalyst for the process is citric acid (tri-carboxylic acid) which is consumed and later reconstituted. Along the way the energy carrier adenosine diphosphate (ADP) is upgraded to adenosine triphosphate (ATP). The energy carrier adenosine triphosphate (ATP) is recycled as the corresponding diphosphate

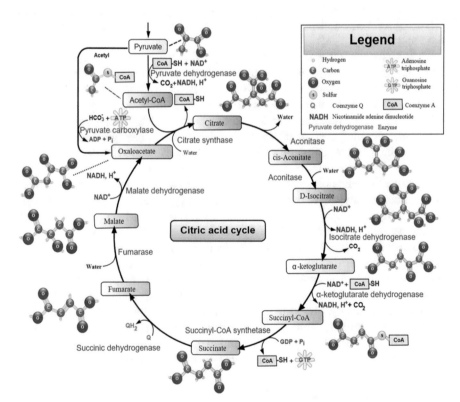

Fig. 5.8 The Krebs Cycle and respiration. *Source*: http://www.tutorvista.com/content/biology/biology-ii/respiration/cellular-respiration.php

(ADP) at the point where the energy is needed by some other process, such as flexing a muscle. The details are far too complex for this book (it took two Nobel laureates, and plenty of others, a number of years to work it all out.) See Fig. 5.8.

The metabolic advantage of respiration over fermentation is enormous, and it translates directly into better performance at the cellular level. The respiration process generates 18 times more free energy (exergy) from glucose than anaerobic fermentation, because it oxidizes the lactic acid and ethanol that fermentation leaves behind. For each molecule of glucose "food" it produces 36 molecules of the energy carrier ATP, as compared to only 2 for the first stage (glycolysis) alone. This huge energy advantage became a competitive advantage between species. It led to an eventual displacement of anaerobic organisms from all habitats where oxygen was available. Sometime between 1.5 billion years ago and 1 billion years ago, the basic energy metabolism of the biosphere, and subsequently the atmosphere, stabilized as an early version of the *carbon cycle*. I discuss that cycle in more detail later (Sect. 6.6).

Yet all of these phenomena are consequences of very small (in relative terms) imbalances between solar influx at the top of the atmosphere and IR radiation from the Earth into outer space. The details are still being worked out.

5.9   Evolution Before the Cambrian Explosion

Evolution did occur during the Archean age, of course. About 3.5 billion years ago the last universal common ancestor (LUCA) died out, leaving two biological "empires", consisting of bacteria and archaea, both of which are single cell *prokaryotes* (cells without a nucleus). For reasons not yet understood, the bacteria survived but the archaea died out. However, the archaea somehow spawned another empire, the *eukaryota* (cells with nuclei), from which all multi-celled creatures ultimately appeared. (See the "Tree of Life", Fig. 5.9). We do know that all prokaryotes had a globular shape, contained within a sort of membrane. Cell division was a physical consequence of growth, which is why most cells are microscopic: More specifically "a spherical cell which produces a substance (such as a protein) becomes unstable upon exceeding a critical size" (Rashevsky 1948, p. 127 et seq). We know very little about the archaea.

The development of *eukaryotes* (cells with true nuclei) was the key evolutionary development of the Proterozoic period. In fact, it was revolutionary. This happened after the "great melt" that ended the Huronian "snowball earth" episode, about 1.85 billion years ago. This evolutionary step seems to have resulted from evolutionary "mergers" between different, but symbiotic, species. Probably the cellular nuclei, and other interior "organs" like mitochondria, organelles and chloroplasts, were all originally independent organisms (bacteria or viruses) that were either engulfed by prokaryotes, or invaded them, and evolved a symbiotic relationship enabling their own survival (Margulis 1970). That was also when the DNA established itself as the information storage, control mechanism and heredity system.

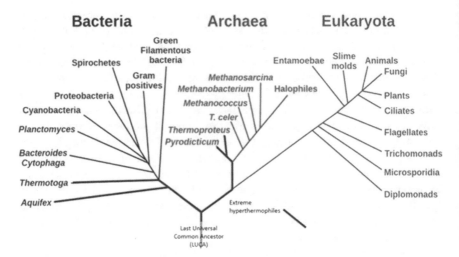

Fig. 5.9   Phylogenetic "Tree of Life" (J. D. Crofts). *Source*: https://en.wikipedia.org/wiki/Tree_of_life

The evolutionary "invention" of a specialized gene-carrying nucleus opened the way to meiosis, sexual reproduction and complexity (biodiversity and multi-cellularity). But it took another 600 million years before multicellular organisms appeared in the fossil record. The first fossil evidence of sexual reproduction was also about 1.2 billion years ago. But life was still strictly confined to the oceans, and most of it was microscopic, in scale, except for some colonies. It was a further 600 million years before the animal and vegetable kingdoms separated and another 50 million years before both animals and plants began to live on land. That only happened 550 million years ago, 3 billion years after the first living cells.

For the first 3 billion years, or so, after life began on Earth, populations and diversity grew slowly, if at all. There were setbacks, as when "ice ages" occurred. But gradually the collection of living organisms that constituted "life" was capturing "free energy" (exergy) in the form of sunlight or geothermal heat, and converting it into a stockpile of organic substances, mainly carbohydrates and lipids. Some of that stockpile was in the form of biomass, and some was in the form of metabolic wastes accumulating in sediments. What the living organisms were doing with that captured exergy was mostly body growth and maintenance (metabolism) but a component was for cellular reproduction.

Cellular replication can be thought of as information processing, i.e. implanting genetic information into copies of itself. Suffice it to say that evolution has resulted in enormous biodiversity, with a corresponding increase in both embodied genetic information and structural information (including information stored in the brains of higher animals). This is why some modern biologists have chosen to use information theory to explain evolution (Brooks and Wiley 1986–1988; Hirata and Ulanowicz 1984; Ulanowicz 1986; Kay 1984; Schneider and Kay 1994; Schneider and Sagan 2005).

This is not the place to try to explain and compare the various theories, even if I were competent to do so. Their common feature is the argument that diversification is an information maximization or entropy minimization (treating information entropy and physical entropy as equivalent). But these concepts must be used with care. In any case, it is clear that the entropy of the universe is always increasing, although the rate of increase may vary. The interesting fact is that disorder (entropy) in a local subsystem may decrease at the same time, thanks to self-organization created by the dissipation of exogenous fluxes of exergy (Prigogine 1976). The theoretical question remaining is whether this can be regarded as the consequence of an explicit extremum principle, or not.

5.10   The "Cambrian Explosion"

Much of the early history of evolution is still speculation. The so-called "pre-Cambrian" era ending 550 million years ago is not well documented, partly because there is not much of a fossil record to look at dating from that time. However the following (Cambrian) epoch, with its explosion of multi-cellular life forms, began about 530 million years ago. There are fossils of trilobites, the first fish and the first insects. About 434 million years ago, plants and fungi began the conquest of the land. Amphibians soon followed.

I suggested earlier that the proliferation of environmental gradients following the "great melt" may have contributed to the acceleration of speciation. In fact, the conquest of the land may have been a major cause of the explosion, given that it was accomplished by organisms that accelerated the rock-weathering process. The resulting increase in phosphorus mobilization could have allowed protective innovations based on calcium phosphate (like teeth, shells and bones) to develop.

For reasons not well understood, the carbon-dioxide content of the atmosphere rose again rather sharply about 350 million years ago. Could it have been the result of a series of huge volcanic explosions? We don't know. But at that time the climate was much warmer and more humid than today and the carbon dioxide level in the atmosphere was much higher than it is today. This led to the so-called "carboniferous" era. From about 363 million years ago to 251 million years ago, a remarkable 112-million year period unfolded during which plant growth proliferated on land to an extent that is hard to imagine today. Everything that grew seems to have been huge. There were millipedes a meter in length and cockroaches as big as rats. Huge forests of giant ferns and shave-grass proliferated in warm, swampy freshwater regions.

When the trees in these forests died, they fell onto the swampy ground and were buried by the debris of the following years. Decay was probably slower than it would be today, because efficient decay specialists had not yet evolved. Many generations of plants formed layers of dead vegetation, which, in turn, were overlaid by sediments of inorganic material washed down into the low lying swamps from surrounding higher ground. Much of the dead biomass (humus), sealed off from the oxygen of the air, did not rot away. A good part of the energy (exergy) stored in those carbohydrates was conserved by being compressed, heated and transformed into peat. As more layers of sediment piled up on the organic deposits they sank further down, coming under increased heat and pressure. They were further transformed, first into lignite (brown coal), then bituminous (hard) coal and finally to anthracite.

But 251 million years ago there was another drastic change. It is known as the Permian-Triassic extinction. Something caused the catastrophic disappearance of 90–95 % of all marine species existing at the time. There was also a slightly less severe—but still radical—terrestrial extinction. Nobody knows what caused this event. There is no evidence of a meteorite, so the most likely cause of this extinction is now thought to be the colossal series of volcanic eruptions that occurred over 700,000 years, at the right time, and left an area of 2 million square kilometers covered by basalt: the so-called "Siberian traps". However it is not at all clear how massive volcanism in the far north could have caused the global extinction, especially of the marine species. There are several competing theories but none is totally convincing.

Other lesser, but still important, mass extinctions took place every few million years thereafter. Nobody knows why. Glaciation, volcanism, asteroid collisions may or may not have been involved. Yet speciation accelerated and biological diversity recovered (eventually) after each extinction, though the mix changed. One consequence in the oceans was a shift in the "balance of power" between sessile and mobile species, resulting in more mobile predators (sharks) and more effective protection (shells) for immobile species.

The first dinosaurs appeared around 225 million years ago, followed by an explosion of other vegetarian saurians. Armored stegosaurs turned up 175 million years ago. Flowering plants (angiosperms) followed 130 million years ago, with all the new mechanics of pollination by insects. The first mammals came 115 million years ago. The forests bloomed. Bees appeared 100 million years ago. Ants arrived 80 million years ago. The dinosaurs "peaked" about 68 million years ago when the fearsome meat-eating *Tyrannosaurus Rex* showed up in the fossil record.

A question evolutionary biologists must ask themselves (again) is: why so many species? If Darwinian evolution is a competition—which it is—then why do we need so many different types of organisms? Why not only a few, e.g. the most efficient single-cell algae, the most efficient zooplankton, the most efficient consumer of zoo-plankton (a whale?) and the most efficient killer whale at the top of the food chain. The answer is already implicit in the question. Each of these species has evolved specialized capabilities that enable it to capture and digest food from the one beneath it. But every advantage has a cost. Specialization also involves simplification.

Consider the pattern of succession after a forest fire. The first plants to sprout will be very fast growing species that produce flowers and seeds in a few days. You see them in your window box. Why don't those plants cover the Earth? The answer is that other seeds will also sprout and generate plants with higher stalks and deeper roots (and longer lives). As soon as they grow they crowd out the smaller ones. Then why don't these plants cover the earth? The answer is the same: still later come the larger, longer-lived plants with even deeper roots. They will collect more of the available soil moisture and still taller stems (thanks to longer lives) to intercept more of the rays of the sun. As they grow, the forest giants shade out the smaller species. But as the system evolves, and as the giants grow old, it also becomes less resilient and increasingly vulnerable to a fire or disease epidemic. See Holling's ecosystem cycle Fig. 5.10 (Holling 1986).

But survival isn't only about maximizing photosynthetic efficiency, or water capture and storage efficiency at a specific location and moment in time. Some organisms will specialize to do better at high temperatures, others will specialize to do better at low temperatures. Some leaves and stems will resist wind, or drought, or fire, better than others. Some will grow from seeds, others from roots. Some seeds will sprout only after fires; other will need to go through the guts of a bird or insect in order to sprout. Some of these specializations are only needed in rare

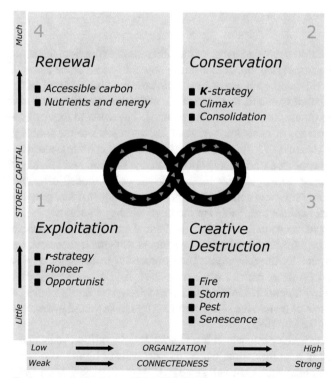

The four ecosystem functions and their relationship to the amount of stored capital and the degree of connectedness. The arrowheads show an ecosystem cycle. The interval between arrowheads indicates speed; i.e. a short interval means slow change, and long interval rapid change.

Fig. 5.10 Hollin's ecosystem cycle. *Source*: Holling (1986, Fig. 10.5)

circumstances. Some organisms live by "hitch-hiking" on others. Some live off the wastes or dead bodies of others. Among animals, there are specialized grazers with extra stomachs to digest cellulose. Some, like silkworms or monarch butterflies, can only digest particular plants. Some predators (such as raptors) also specialize in their prey. Some specialize in building complex structures to live in to protect their eggs or their young. Others specialize in finding and penetrating those safe-havens. Every specialty requires specialized capabilities with corresponding disabilities and costs.

Ecosystems have room for all kinds of specialized organisms because the conditions for "fitness" vary from deep sea to shallow sea, from sea to land, from mountain to plain, from one latitude (or altitude) to another. They also vary by time of day, by season of the year, from sunshine to rain and from one glaciation to another. This is exactly why gradients in space, temperature, humidity and time generate bio-diversity. Biodiversity is a form of natural wealth. It is bio-diversity that has constituted part of the capital stock underlying human wealth.

5.11   Since the Asteroid

Sixty-six million years ago there was another mass extinction, the so-called Creta-ceous-Paleogene extinction. Scientists are now fairly sure of the cause of that one: a large asteroid landed somewhere in what is now the Gulf of Mexico. The resulting explosion caused tsunamis that circled the globe and flooded many of the swampy areas where dinosaurs thrived and reproduced. The asteroid impact lifted hundreds or even thousands of cubic kilometers of soil and rock into the stratosphere, where much of it remained as dust for several years. The consequence was a dramatic (temporary) cooling as those stratospheric clouds blocked and reflected the sunlight back into space. The cold must have made the cold-blooded dinosaurs (and their eggs) easy prey for more active warm-blooded animals that had previously evolved in peripheral habitats unsuitable for the huge, clumsy saurians.

Thus began the so-called Tertiary geological era. The losers included all of the saurians, except for the early birds. The winners were the insects and warm-blooded animals, especially the mammals. They proceeded to occupy every "niche" then existing and (in time) created some new ones. And bio-diversification continues. There are now around 1.7 million different species on Earth, notwithstanding the worrisome rate of (called the sixth extinction by Leakey and Lewin) now under way (Leakey and Lewin 1995).

During the tertiary era, which lasted until about a million years ago, there was a second peak of coal formation, when large deposits of lignite were formed. During the Carboniferous and Permian geological eras, a large fraction of the carbon dioxide in the atmosphere had been removed and sequestered, while the oxygen level of the atmosphere increased dramatically. Coal, petroleum and natural gas were also formed from the remains of marine plants and animals, especially plankton. These organic remains were laid down mainly in coastal regions near or under salt water. They were eventually covered by sediments from the rivers and sealed off in much the same way as dead swamp vegetation was covered by sediments from erosion. These sediments eventually accumulated formed new layers of sedimentary rock. Over millions of years, these animal remains also underwent chemical changes similar to those that produced coal and became the liquid (and gaseous) hydrocarbon reserves we now exploit.

Over the millennia, the burial of carbon in organic refuse and the resulting decrease of carbon dioxide in the atmosphere gradually reduced the greenhouse effect that accompanied the carboniferous era and allowed the cold-blooded dinosaurs to thrive. Gradually the average temperature of the Earth's surface has declined to a level that is relatively comfortable for us hairless humans. The global temperature fluctuates around $+15\ °C$ in the temperate zones (lower during glacial epochs and higher during the interglacial periods). The average surface temperature of the Earth varies by around $5\ °C$, from highest (interglacial) to lowest (glacial).

A cooling climate has been favorable for warm-blooded terrestrial mammals, despite their need for more food than cold-blooded creatures in the same niche. Apparently the temperature regulatory capability enables warm-blooded creatures to operate at optimum efficiency all the time, whereas cold-blooded competitors cannot. Cold-blooded reptiles and their cousins the alligators and crocodiles, tend to survive now only in the warmest regions. The air-breathing, warm-blooded mammals have taken over the terrestrial Earth and occupied almost every possible ecological "niche" from seed-eating rodents to grass-eating ungulates to fierce meat-eating predators such as the saber-toothed tiger. They have even re-conquered the oceans. Only the birds (warm-blooded but egg laying) remain without competition in the air.

A few words about ecological evolution, *per se*, need to be added here. It has been evident since the work of Darwin that "fitness" at the species level pertains—in part—to the utilization of exergy resources from the environment. Green plants all obtain metabolic energy from the same process (oxygen photosynthesis), but they specialize by adapting to circumstances in a variety of ways. Primary producers like the grasses and the forest giants try to monopolize the available sunlight (if moisture is available) or, like cacti, they store water when it is available, and hibernate at other times. Leaf shapes vary, partly to accommodate heat or cold or to resist wind. Fruits and flowers vary, to attract animals—from squirrels to insects and birds—that will spread their seeds. Vines use tree trunks for support, where there are trees. Some species, like Sequoias and orchids, obtain moisture from the air. Plants that grow in cold climates survive by specializing in "anti-freeze" leaf designs. But every specialization comes at a cost, which makes it less fit in other circumstances. This is why there is no single "best" plant species for every environment.

Plants do not live forever. (Long life also has a cost.) When plants die naturally or are consumed by grazers, they become food for other species that recycle the carbohydrates, fats and proteins in plant biomass. It is much more efficient, in energy (exergy) terms, to construct new proteins by decomposing the proteins from dead biomass into their component amino acids, than to start from photosynthesized glucose as the primary producers must do. This, of course, is the survival strategy of the grazers, the parasites and the saprophytes. It is also the strategy of higher level predators in the food chain. But all of them compete by means of some specialization.

All of the species in the biosphere are "exergy maximizers"—and consequently "entropy maximizers"—at the species level, albeit under a wide variety of different constraints. Yet, paradoxically, the ecosystem *as a whole* does not necessarily maximize primary exergy consumption. On the contrary, it seems to maximize the life-cycles of organic molecules (such as amino acids) by minimizing entropy production per unit of exergy consumed.

5.12   Down from the Trees

One group of mammals (hominids) learned to live in trees, developing binocular vision, prehensile fingers and opposable thumbs for grasping branches, picking fruits and climbing. During the mid-Pliocene (from 2.5 to 3.6 million years BCE) the Earth was 2–3 °C warmer than it is now, the Arctic Sea was 12 °C warmer, the atmospheric CO_2 level was between 380 and 450 ppm and the sea-level was between 16 and 26 m higher than it is now. There was no Greenland ice sheet as yet.

Not only that, the North Pole was not where it is now. A recent expedition to a lake called El'gygytgyn (in the language of the local people)—or simply Lake E— in northeast Siberia near Chukotka has revealed some unexpected facts. First of all, that lake, about 100 km north of the Arctic circle, was formed by a meteorite strike 3.6 m years ago. Today, the lake is 12 km in diameter, it is 170 m deep, with 350–400 m of sediments under it, all deposited since the lake's formation. *It was never covered by a glacier.* That is why the sediments under the lake were untouched, making it possible to obtain meaningful core samples.

Such cores were obtained by an expedition led by Prof. Julie Brigham-Grette in 2009. The cores go back to the beginning of the Pliocene. They have been analyzed carefully. The seeds and plant remains found (e.g. of five different pine trees) reveal that the area was heavily forested. It was never ice covered year-round.[9]

However, it seems that 5 million years ago (at the beginning of the Pliocene), thanks to drifting of the Earth's crust with respect to the rotational axis of the earth, the south pole settled in Antarctica and that continent froze solid. The ice-albedo mechanism is quite straightforward: once the snow on the ground remains in place over the summer, the white stuff reflects the sunlight and cooling accelerates (Budyko 1988). Huge masses of ice spread from the north and south poles over

---

[9] See http://works.bepress.com/julie_brigham_grette/ http://www.gazettenet.com/home/13512822 -95/groundbreaking-research-in-arctic-leads-umass-professor-julie-brigham-grette-to-head-polar-research.

the southern ocean and northern parts of the continents (but not over northeastern Siberia) and glaciers crept from high mountains onto plains.

The glacial period, driven by Milankovitch orbital oscillations (Milankovitch 1941 [1998]) began about 2.5 million years ago. There have been 22 glacial episodes, roughly 100,000 years apart. At its maximum during the last glacial era the sea level was nearly 200 m lower than it is today, because so much water was trapped as ice. The Black Sea and the Mediterranean were inland lakes, cut off from the oceans. The Red Sea was dry. Siberia was connected by land to Alaska (permitting overland migration). The British Isles were connected by land to Western Europe. Animal migration between continents became possible at times.

There is a controversial theory—Earth Crust Displacement Theory—that the freezing of Antarctica occurred much more recently (in fact 9600 years BCE), and that it occurred as a result of slippage of the Earth's solid crust over the underlying plastic asthenosphere from a point several hundred kilometers away from the current pole (Hapgood 1958). If it happened, it would probably have been partly due to the asymmetry of the North-American glaciers, resulting in a shift of weight that could (in theory) trigger a change in the Earth's rotational axis. Albert Einstein wrote a foreword to Hapgood's book, although he questioned whether the weight of ice over North America was sufficient for this slippage.

Evidence favoring Hapgood's theory includes evidence of sediments from a river in Antarctica, only 4000 years ago, and fossil evidence of an ancient beech forest 2 million years old, that was found recently at a location only 400 km from the present location of the south pole at an altitude of 1800 m (Webb and Harwood 1991). Moreover, the supposed time of this displacement event (9600 BCE) corresponds with a wave of megafauna extinctions around the world, including North American elephants, Siberian mammoths, cave lions, saber-toothed tigers and others. At about that same time, agriculture emerged in several parts of the world, possibly because land suitable for hunting and gathering was disappearing.

The Hapgood theory does not actually contradict the current geological theory of tectonic plates. It postulates that the plates shifted together, albeit with some slippage along the boundaries. Hapgood has been discredited by mainstream geologists, largely for two reasons. In the first place, evidence cited in its favor (mainly by "Atlantis" enthusiasts) has depended on some ancient maps (e.g. the Piri-Reis map of 1513) that supposedly identified a continent near the south pole, where Antarctica now is located. Recent scholarship largely dismisses these maps as valid evidence. In the second place, and more conclusive (to some), is the fact that some Antarctic ice cores are at least 2 million years old. In fact, paleo-geologists insist that the poles have not migrated more than about 5° over the past 130 million years, although greater shifts seems to have occurred earlier.

This seems at first glance to preclude the possibility of a rapid shift from temperate to polar conditions. However, on examination of Hapgood's maps, it can be seen that some parts of what is now Antarctica might have been located in a significantly warmer climate, suitable for forests, while other parts were then, and still are, ice-covered (Flem-Ath 1993). This would account for both the possibility of a flowing (not yet frozen) river 4000 years ago and very old ice cores.

In the northern hemisphere, the Hapgood maps suggest that the pole might have migrated from a point at the north end of Hudson's Bay to its present location in the middle of the Arctic ocean, much closer to the location of Lake E. I do not think that Hapgood's theory can be dismissed out of hand, given its potential importance as regards human evolution from Africa. Neither do I endorse it.

The Africa story is that, as a consequence of the cold north-flowing current from frozen Antarctica, the rainfall patterns and the climate of East Africa changed from tropical forest in the Pliocene to mixed forest and savannah (Price-Williams 2015). At that time the hominids split into distinct groups. Most of them stayed in the shrinking forests and retreated with them, evolving into the baboons, chimpanzees, orangutans and gorillas we see today. All of those species are actually quadrupeds, even though they can climb and hang from branches.

But one group came down out of the trees and learned to walk (and run) on two legs, leaving the other two limbs—replete with fingers and opposable thumbs—free for other purposes. Those purposes included making artificial "teeth and claws" (i.e. stone tools). The first bipeds out of the forest (genus *Australopithecus*), evolved around 7 million years ago, according to the time series in Fig. 5.11. They had brains about 500 cm^3 in volume. That was only a little larger than the brains of chimps or gorillas (~450 cm^3). The brains of those bipedal hominids evolved also. But larger brains had survival value. As time went on, the hands evolved further and the brains increased in size.

Genus *Homo* first appeared around 2.5 million years ago. The skulls of *Homo habilis, H. erectus, H. afarensis* and a variety of cousins, have been found in Africa, the earliest being 1.8 million years ago. At that time they had brains of 600 cm^3 volume, and they made crude stone chopping tools, probably used to get at the

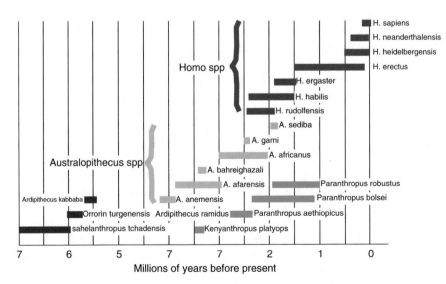

Fig. 5.11 Hominid species distributed through time. *Source*: https://en.wikipedia.org/wiki/Punctuated_equilibrium

marrow in large animal bones. There is a first class mystery about recent discoveries in the country of Georgia. Four complete and one partial skull of *H. erectus* (or one of the cousins) have been discovered in the garden of a monastery at Mdasini, south of Tbilisi (Price-Williams 2015). Those skulls would have accommodated brains of 600 cm^3 and have been dated as 1.8 million years of age by several methods, along with the remains of other flora and fauna found at the site. Yet all the other discoveries of that age were found in southern Africa, and there is no indication (or theory) to explain how *H. erectus* could have got from southern Africa to Georgia.

By 1 million years ago, genus *Homo* had spread from Africa to the Middle East and China. He had a larger brain (close to 1000 cm^3) and was making much more sophisticated stone tools (such as arrow heads). Whether these are all different species or not is disputed. The authors of Fig. 5.11 accept that they were different enough to be separate species but the differences are all within a very narrow range. *Homo habilis, or erectus* (or other species') skulls from that era have been found in France, Germany, China (Peking Man) and even Indonesia. In China, these human-oids not only made high quality stone tools; they also made fires, judging from the evidence of charcoal near the bones.

Homo sapiens, with bigger brains, appeared around 500,000 years ago. The sub-species *H. sapiens neanderthalensis* emerged 400,000 years ago. Our sub-species of *Homo sapiens sapiens* is much more recent. Our direct ancestors seem to have appeared about 100,000 years ago (and the last of Neanderthals disappeared about 35,000 years ago, although it is now known that all Europeans share Neanderthal genes.)

Surprisingly, humans still share 98.4 % of our genes with the two species of chimpanzee which differ by only 0.7 % from each other. This accounts for the title of Jared Diamond's books "The Third Chimpanzee" (Diamond 1991). Our species diverged from the "chimp" line about 10 million years ago, according to the "DNA clock" developed by Sibley and Ahlquist in the 1980s (based on bird data) (Sibley and Ahlquist 1984, 1987; Sibley et al. 1990). But the 1.6 % of our genes that differ from chimps account for several skeletal differences (for walking in two legs), canine teeth, face shape and the much larger human brain.

It has important to emphasize that brain-to-body-weight is actually a more important measure than absolute brain size, when comparing different species in terms of intelligence. Chimps are comparable, or a little smaller than humans in body size, whereas gorillas are considerably larger. This suggests that chimps are actually a little closer to us, in intelligence, than gorillas.

Human (*H. sapiens*) brains now average 1350 cm^3 in volume, almost three times larger than the brains of chimps and gorillas. Neanderthal brains were even larger, up to 1900 cm^3. But the larger cerebral cortex also required larger skulls, which resulted in (comparatively) abbreviated parturition and a need for many years of extended *post-partum* nurturing care. This was the starting point for the development of family-based tribal societies based on long-term family relationships. Long lifespans facilitate learning from experience, and learning from teaching. Learning allows for specialization of functions, which has survival value for families and tribes living together. The extended life-span of humans may have been enabled by the extraordinary immune and repair systems with which human bodies are endowed.[10]

It is possible that the Neanderthals were killed by some disease that humans survived. It is also possible that their skulls were actually too large for easy childbirth, resulting in a reproduction rate that was too low for survival in competition with a faster breeding sub-species. There are other scenarios.

For reasons not entirely clear, the climate of Earth has cooled significantly in the last 5 million years or so. The "icing" of Antarctica has kept the sea level significantly lower than it was previously. Each of the "ice ages" lasted 10,000 years or so, and there were warmer periods (interglacials) in between. During warm periods, when the ice was melting, huge rivers of melt-water carved deep valleys and filled lakes—both above and under the ground. During the warmer periods the temperature rose by five or so degrees, glacier ice receded, the sea level rose, and the land was covered by lush forests at first. Later, the rivers shrank and the underground waters receded, so some of the land was left dry to become grasslands or deserts. But the glaciers returned after some time—nobody knows for sure why[11]—and this sequence has repeated itself several times.

Since end of the last glaciation, about ten thousand years ago, the "greenhouse" has kept the average surface temperature of the Earth near this level, albeit with considerable variability from season to season and from latitude to latitude. This "comfort zone" may not last much longer, however. Today we humans are consuming fossil fuels that took millions of years to accumulate in a matter of decades. Details of this process are given in Chap. 6. In short, we humans are rapidly reversing the slow process of carbon sequestration that occurred when the Earth was warmer so many millennia ago. The consequences for the atmosphere, and the global climate, may be very dangerous, because they are occurring so rapidly in geological terms. In fact, a century from now the Earth is likely to be much less habitable, for humans, than it was a century ago and still is, today.

---

[10] A science writer, possibly Isaac Asimov, once observed that the longest human lives, measured in heart-beats, are significantly longer than the longest recorded lives of other animals. This implies that the aging process (per heart-beat) is slower, which suggests that resistance to microbial attack, and metabolic repair mechanisms, are more efficient.

[11] The most plausible theory cites complex cyclic variations in the tilt of the Earth's rotational axis *vis a vis* its orbit around the sun. These are called Milankovitch cycles, after their discoverer (Milankovitch 1941 [1998]).

Chapter 6
Energy, Water, Climate and Cycles

6.1 The Earth's Energy (Exergy) Balance[1]

The sun, which continuously energizes our solar system, radiates about 4×10^{14} terawatts (TW) of power, in all directions. Since a terawatt is a thousand Gigawatts (GW) and a million Megawatts (MW), that is an extremely large amount of power. The unit of power, terawatt (TW), is convenient because it is about the right size for discussions of energy use in the global economy. The Earth intercepts only a tiny fraction of that enormous solar output, viz. 174,260 TW, or 340.2 ± 0.1 W per square meter of the Earth's silhouette or intercept surface.

It should be mentioned again that there is also a source of heat within the Earth's crust. It amounts to only 0.013 % of solar heat input. But geothermal heat drives endothermic chemical reactions in the mantle and tectonic activity, including earthquakes, volcanic eruptions and subduction. The combination of subduction and eruption enables some chemical recycling. It probably accounts for the recycling of CO_2. Thus carbonates, sulfates, nitrates and possibly phosphates deposited in sediments are "reduced" in the magma to carbon dioxide or carbon monoxide, sulfur dioxide (or sulfides), nitrites (or ammonia) and phosphides or possibly phosgene. These drive exergetic biological (metabolic) processes at low temperatures in the oceans and the atmosphere. Terrestrial volcanism today averages only 0.3 GW of power, while submarine volcanism provides 11 GW and conduction adds 21 GW, or 32.3 GW in all. These numbers are insignificant exergetically in comparison to the 174,260 TW that arrives continuously from the sun.

Geothermal heat has had important consequences in the past, both to recycle carbon dioxide via subduction of carbonates and vulcanism and (perhaps) to provide a suitable underwater environment for the first living cells. Again, later,

---

[1] This topic was pioneered by the Russian geoclimatologist, Mikhail Budyko (Budyko 1956, 1969, 1974, 1977, 1988).

© Springer International Publishing Switzerland 2016

R. Ayres, *Energy, Complexity and Wealth Maximization*, The Frontiers Collection,

DOI 10.1007/978-3-319-30545-5_6

undersea vents may have enabled single cell organisms to survive the "snowball Earth" episodes. The solar influx (1.353 J per square meter, on average) consists of radiation ranging from relatively long waves (infra-red), including the visible part of the spectrum, to ultraviolet (UV) radiation. The sun can be regarded as a radiating "black body" at a temperature of about 5800 K. (The word "black" is not a mistake. A "black body" is black in the sense that it absorbs and emits radiation but does not reflect any). However, the Earth is by no means a black body in thermal equilibrium. That point needs to be emphasized.

Because the earth is round, sunlight is much more intense when it arrives from directly overhead, than at an acute angle, such as near the poles. The result is unequal heating between the poles and the equator. Because heat tends to flow from warmer to cooler places (thanks to the second law of thermodynamics) warm air and warm water from equatorial regions tends to flow towards the poles, while cold air and cold water flow towards the equator. Because the Earth is also rotating, these currents are diverted away from simple North-south directions into East-west directions, to the right in the North and to the left in the South. These currents are especially apparent in the form of "trade winds", "westerlies" and the weather-pattern controlling "jet streams" in the stratosphere. (See Fig. 6.1).

The E-W diversion is caused by the so-called "Coriolis force" (explained mathematically by its discoverer, Gaspard-Gustave Coriolis in 1835). It explains the direction of the rotation of cyclones, hurricanes and typhoons. In fact, the "force" is an illusion: air and water are merely following *geodesics* (the shortest distance between two points) in a rotating frame of reference. If you read Sect. 4.3 you recall that Einstein's General theory of relativity says that gravity itself is the shortest distance between two points in a curved space.

Though the earth receives radiation from a source at a temperature of 5880 K, the Earth (as a whole) radiates heat back into space at an equivalent black-body temperature of 255 °K or −18 °C. The temperature difference between incoming and outgoing radiation represents available Gibbs free energy capable of doing work. The kinds of work that are done include moving air (about 350 TW) and moving water (about 45 TW). Figure 6.2 shows the major radiation fluxes, according to the latest data (Stephens et al. 2012).[2]

On average, 165 ± 6 W/m^2 (48.5 %) of the 340 W/m^2 solar influx to the top of the atmosphere reaches the Earth's surface where it is temporarily absorbed by air, land or oceans. Of the reflective component (100 W/m^2 or 29.4 %), 27.2 ± 4.6 W/m^2 is reflected by atmospheric dust; (5 ± 5) W/m^2 is reflected by clouds; and 23 ± 3 % W/m^2 is reflected by land or ocean surface. The remaining 75 ± 10 W/m^2 (22 %) is absorbed by the atmosphere. Of that part, 16–18 % is absorbed by water vapor, ozone or dust and 4–6 % is absorbed by condensed water vapor in the form of clouds.

---

[2] The results shown in Fig. 6.2 are significantly different from earlier and more widely published results compiled by Trenberth et al. (2007), especially with respect to evaporation and rainfall. The new results, based on more sophisticated sensors, are more trustworthy.

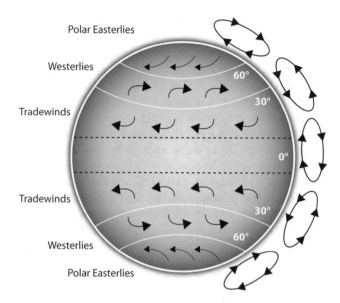

Fig. 6.1 Coriolis "forces" in relation to latitude. *Source*: http://images.flatworldknowledge.com/wrench/wrench-fig15_001.jpg

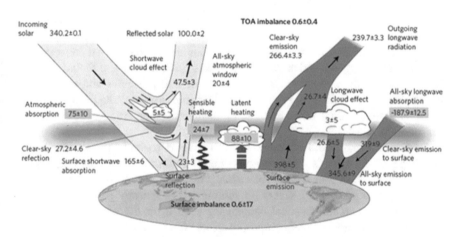

Fig. 6.2 Solar and infrared radiation fluxes. *Source*: https://curryja.files.wordpress.com/2012/11/stephens2.gif

Altogether, about 70.6 % of the incoming radiation is absorbed, at least temporarily. The 29.4 % that is reflected is what would be seen from the moon or an alien spaceship. The technical term for that reflected component is *albedo*. Most of the 239.7 ± 3.3 W/m² re-radiation is in the infra-red (long waves), at a much lower temperature than the influx. The largest part of that low temperature radiation (38 %) is re-radiated from clouds, while 17 % is from water vapor and other trace gases in the atmosphere, and 15 % is infra-red (IR) radiation directly from the Earth's surface.

Exergy balance requires that all of the influx of solar energy (the 70.6 % that is initially absorbed) must be re-radiated (at lower frequencies). If there is a persistent imbalance, it follows that the Earth must heat up. It is unquestionably heating up (we know this from measurements of ocean temperatures), but there is an argument about where the excess incoming heat is going.

The difference between high temperature short-wave solar exergy inflows and low temperature infra-red (LW) exergy outflows is "free energy" that is energy available to do mechanical or chemical work. The work it does is to drive cyclic terrestrial atmospheric, hydrological, geological and biological processes. Most of the work done is converted to waste heat by internal friction or other loss mechanisms. But those processes can (and do) also create and maintain increasingly complex structures and patterns that constitute a kind of stable global order.

The global biosphere is part of a self-organizing process that occurs far from thermodynamic equilibrium, and is maintained by solar exergy flux (Nicolis and Prigogine 1977). Some of the "structures" created far from equilibrium by this process (waves, wind-driven gyres, typhoons) tend to maximize local entropy production by increasing the rate of "destruction" of thermal gradients. However, the exergy captured by photosynthesis and sedimentation simultaneously reduces the *local* rate of entropy-increase.

Among these natural processes, only the biosphere extracts exergy directly from the solar influx and accumulates it as biomass. Some of the biomass is subsequently sequestered in sediments that later become sedimentary rock. That sequestration consists of both organic and mineral by-products (such as bone and shells of marine life). These accumulations are segregated by natural processes and transformed over time into fossil fuels, limestone, shale, hematite (iron ore) and phosphate rock. Some of those accumulations are currently being mined and recovered by humans for industrial purposes.

Some of the photosynthetic work, (as well as some geological work done by the shrinkage of the Earth's crust due to cooling) contribute to the carbon-oxygen cycle, and the "nutrient cycles" discussed later in Sects. 6.6–6.9. All of those material cycles are exergy savers (or minimizers of entropy generation), in the sense that some stable chemical compounds needed for life support—those that would require significant exergy inputs to create "from scratch"—are saved thereby for re-use.

Overall, about 25 % of the energy absorbed at the surface drives evaporation. The hydrological component (40 TW) is discussed later. The photochemical part drives plant photosynthesis, which enables biomass reproduction and growth. Total exergy utilized by plants for photosynthesis is a small fraction (0.023 %) of total solar input to the Earth. It is proportional to evapo-transpiration by leaves (on land) and by algae in the oceans. Evapo-transpiration from the surface of the Earth accounts for 88 ± 10 W/m^2 (as compared to about 90 W/m^2 for other abiotic processes, including evaporation and total absorption). The chemical energy in the form of carbohydrates produced by photosynthesis amounts to 2.9 TW, about six times more than all exergy consumed by humans burning fossil (and nuclear) fuels. However 99.9 % of the solar energy absorbed by the chlorophyll drives evapo-transpiration. Only a tiny fraction drives photosynthesis and produces glucose (and other carbohydrates) from CO_2 and H_2O. Biomass of plants today amounts to 2 billion metric tons (GT) dry weight.

In fact, the winds that result from temperature gradients on the Earth's surface, especially between land and water surfaces, constitute a total power flux of about 2000 TW. This wind, mostly at high altitudes, is not useful. But there is potential usable power output from surface winds of 6000 GW (6 TW) of which only about 300 GW is currently used (as of 2013). Note that the total global wind resource amounts to a little more than 1 % of the solar power incident on the Earth's atmosphere, while the potentially usable fraction of the global wind resource (by - wind-turbines) is estimated to be around 0.3 % of that (Smil 2003).

Ocean waves, driven by the wind, have a power of 2 or 3 TW, of which only a very tiny fraction could possibly be used. Ocean currents, including the "conveyor belt" (next section) driven by temperature gradients and wind, have a power of 1200–2000 GW. Surface currents like the Gulf Stream might amount to 100 GW, of which less than 5 GW might be exploitable. There is also a tidal component, due to the interaction between the Earth's gravity and its rotation. The power of tides has been estimated at 27.3 GW (Valero Capilla and Valero Delgado 2015, Chap. 6).

To summarize a complicated situation (Fig. 6.2), considerable heat flows from the Earth's surface to the atmosphere and from the atmosphere back to the Earth. In energy (W/m^2) terms, the upward IR flow from the surface is the sum of atmospheric absorption (75 ± 10) plus hot air (sensible heat) convected from the surface (24 ± 7) plus the latent heat released by the condensation of water vapor in the clouds (where the heat used for evaporation at the surface is recovered) (88 ± 10). Obviously a lot of the upward heat flux is intercepted by the atmosphere, especially clouds. In addition, there is upward IR flux from the surface, amounting to (398 ± 5), of which 187.9 ± 12.5 is absorbed by the atmosphere.

The IR flow from the atmosphere and the clouds back to the surface—the "Greenhouse effect"—consists of downward IR flux from the clouds plus downward IR flux from the air itself (clear sky). The first component, in W/m^2, is 26.6 ± 5 while the second component is 319 ± 9. As compared to previous estimates (Trenberth et al. 2007), the downward IR flux is larger by 10–17 W/m^2, while the evaporation/condensation rate is also larger by 10 ± 5 or so. This suggests that the role of clouds is somewhat greater than previously thought, although it is unclear what impact, if any, this would have on the climate-warming thesis.

The presence of carbon dioxide, methane and other GHGs in the Earth's atmosphere raises the average surface temperature of Earth by 33°K above the "black-body" temperature of Earth, which—as mentioned already—is 255°K (or −18 °C) to the actual average temperature of 288°K or +15 °C. These numbers refer to an average over the whole Earth. In fact, the heat balance is complicated enormously by differences between the equator and the poles, differences in reflectivity (albedo) and absorption over different surfaces, and differences in cloud cover, among others. The heat balance for the Earth as a whole, therefore,

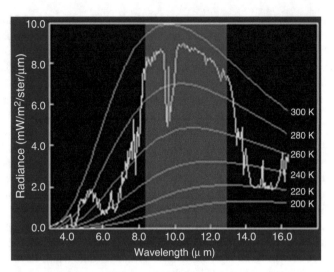

Fig. 6.3 Radiance vs wave length. *Source*: Wasdell 2014. Climate sensitivity and the carbon budget. Apollo-Gaia project

requires significant heat transfers from the sea to the land and from the equator to the poles. Since horizontal thermal conductivity is negligible, these transfers involve physical transfers, notably of water vapor in clouds and ocean currents.

A few technical points: The outward flux of infrared (IR) radiation from the Earth depends on the fourth power of the Earth's effective ("black-body") surface temperature. This is known to physicists as the *Rayleigh-Jeans law*, mentioned in Sect. 3.5. This flux was 7 % lower than the levels at the peak of the last glaciation. Part of this was certainly due to the "greenhouse effect" operating in reverse: lower atmospheric carbon-dioxide (CO_2) concentration (180 ppm) resulting in less heat trapping. Figure 6.3 illustrates the situation. White curves represent black body spectra at temperatures ranging from 200° to 300°K. The yellow curve indicates the Earth's IR radiation spectrum. The area under a curve corresponds to radiative exergy output.

The big dip between wavelengths of 14.5 and 16 μm is the effect of carbon dioxide absorption. Other dips are due to water vapor (5–7 μm) and ozone (9.5 μm). These regions of low emissions—I call them "dips"—are effectively barriers to the IR radiation in those wavelengths that would otherwise be emitted into space. In effect, they act like blankets warming the Earth. From this diagram, it is possible to calculate what would happen if the CO_2 concentration should increase or decrease. It turns out that doubling the CO_2 concentration from 140 to 280 ppm would have the same warming effect as increasing the radiation impinging on the Earth from the sun by 3.7 W/m^2. This doubling, in turn, would raise the temperature on the Earth's surface by approximately 1 °C (Ad hoc Study Group on Carbon Dioxide and Climate 1979).

Note that we have emphasized the importance of energy balance up to this point. There are, however, several important imbalances. The first is small, but cumulative. It is due to biological activity and sedimentation that slowly removes carbon dioxide from the atmosphere and sequesters carbon as hydrocarbons, carbonates and hydrocarbons. This sequestration phenomenon was especially active during the so-called "carboniferous era" that lasted 112 million years from 363 million years ago to 251 million years ago. During that period carbon dioxide levels in the atmosphere were much higher than they are at present. The buried hydrocarbons from that period are the fossil fuels currently being extracted and burned by our industrial society at a rate millions of times faster than they were created.

The second imbalance is de-sequestration. Emissions of carbon dioxide (and other gases) resulting from fossil fuel combustion are now changing the radiation balance, by absorbing IR radiation from the Earth and increasing the downward re-radiation (Gates 1993; Walker and King 2009). The average temperature of Earth is now rising, albeit somewhat discontinuously (for reasons discussed later), and the global climate is changing in consequence.

The third imbalance results from increased heat absorption in the oceans, as discussed next.

6.2 The Hydrological Cycle

The primitive Earth atmosphere may or may not have contained much water vapor, although it is widely assumed that there were oceans of liquid water. However, the Earth receives a continuous influx of small comets, which is mainly ice (Frank et al. 1990). The annual accretion is small enough to be practically unnoticeable. But it has been going on for billions of years. It seems that water from the tails of dying comets probably accounts for most of the water now in our oceans. It is those oceans that make our planet appear to be blue, as seen from space. The white clouds that float above us are another form of water.

Whatever the source, the stock (mass) of water on the Earth's surface today is very large, in relation (for example) to the mass of the atmosphere. In units of a thousand cubic kilometers (10^3 km^3) the oceans contain 1,335,040 units, while glacial ice embodies 26,350, groundwater contains 15,300 units while rivers and lakes account for 178 units and soil moisture accounts for 122 units (plus 22 for permafrost).

By comparison, the total amount of water vapor in the atmosphere is only 12.7 units (Trenberth et al. 2007). Freshwater accounts for only 2.5 % of the total, but that is mostly ice-caps, glaciers and groundwater. Lakes, rivers, soil moisture and permafrost account for 422 units, only 1.2 % of surface freshwater, or 0.052 % of the grand total. The water content of all living things on Earth—mostly trees—is about 1/9 of the water as vapor in the atmosphere, or 1.12 units of 10^3 km^3 or 1120 cubic kilometers (Shiklomanov 1993).

Water is important because it exists in three states at temperatures that occur on the Earth's surface. The solid state is ice. At various times in the history of the planet, ice has covered most, or all, of the surface. Those episodes have been called "snowball Earth" for obvious reasons. The early "snowball" episodes were probably triggered by the rise of oxygen photosynthesis and the corresponding disappearance of carbon dioxide and/or methane from the atmosphere. It is less easy to account for the subsequent "great melts"; massive volcanic eruptions or asteroid impacts may have been involved. However, it is certain that reduced ice cover, for any reason, led to increased atmospheric humidity and that feedback, by itself, would have accelerated the warming trend once it got started. It is also certain that evapo-transpiration by terrestrial vegetation contributed to increased humidity, even while removing CO_2 from the atmosphere.

But another difference between ice-covered "snowball Earth" and the blue planet we live on today is that ice is virtually static (it flows down hills, but only very slowly) while liquid water and water vapor are much more dynamic. The mobility of water accounts for the hydrological cycle. The hydrological cycle is illustrated in Fig. 6.4.

Some flows, such as groundwater flow, are not quantitatively known. But the most important ones above ground are as follows (in 10^3 km^3/year): ocean evaporation (413), ocean precipitation (373), land precipitation (113), land evaporation-transpiration (73) ocean to land water vapor transport (40) and land to ocean surface

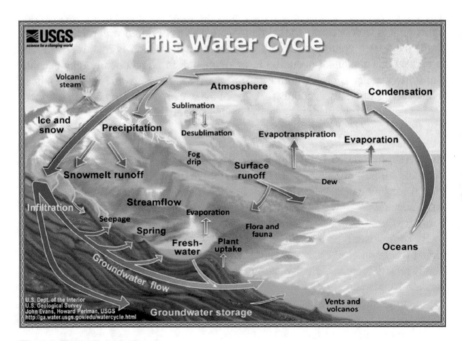

Fig. 6.4 The hydrological cycle

flow (40) units (Trenberth et al. 2007). Of course, the flows vary from region to region and from season to season.

A liter of water weighs 1 kg, so the conversion from volume to mass is easy. In mass terms, annual evaporation amounts to about 505 billion metric tons (Gigatons, GT) per annum. Of this, 430 GT is from the oceans and 75 GT is evapo-transpiration from terrestrial vegetation. On the other hand, precipitation over the oceans amounts to 390 GT, compared to 115 GT on land. Hence there is a net 40 GT transfer of water as vapor from sea to land, matched by 40 GT of surface runoff from land to sea.

Those flows of water between surface and atmosphere, and between land and sea are also flow of heat. The major one is the evaporation/condensation cycle. When water evaporates from land or ocean surface, it carries latent heat into the atmosphere. When the vapor condenses as droplets, that latent heat is transferred to the air. When the droplets coalesce into drops and fall as rain, some heat is returned to the surface.

As we all know, liquid water is mobile. It moves in all three dimensions, from surface to air, from sea to land and from south to north (and *vice versa*). The driving forces, evaporation and condensation, create temperature differences between the land surface and higher altitudes, between the ocean surface and the land surface, and between the ocean surface and the deeper water. Those temperature differences, in turn, create pressure and density differences that must be compensated by airflow or water flow. Wind is the primary cause of ocean wave formation, which is the primary cause of erosion on coastlines. Wind power is estimated to be 1000 TW

Evaporation cools the oceans. Without this cooling, but with the Greenhouse effect operating, it is estimated that the surface temperature of the Earth would rise from its present average level of 15–67 °C. It also increases the salinity of tropical and sub-tropical waters. This plays a role in ocean current dynamics, insofar as heavier (more saline) water may be carrying a lot of otherwise-unaccounted-for heat from the surface down into the deep ocean.

Condensation occurs preferentially over land because of two factors: hills and "hot spots". The majority of hot spots are cities. They are created mainly by local differences in surface reflectivity/absorptivity or local heat production from industrial or domestic combustion processes. Hot spots create vertical convection (i.e. "thermals"). Convective vertical airflows, if they are powerful enough, generate rotational flows or cyclones around the vertical flow. These can become hurricanes, typhoons and water-spouts or tornados). The direction of the airflow, whether clockwise (south of the equator) or counter-clockwise (north of the equator), is determined by the so-called Coriolis effect caused by the Earth's rotation.

Humid air flowing inshore from the sea (usually at night when the land is cooler) is forced to rise when it encounters hills or mountains, or hot spots. When humid air rises it expands and cools. In the presence of so-called "condensation nuclei", the moisture then condenses (releasing latent heat) and forms clouds or fog. Under appropriate conditions, clouds produce rain or snow. Net evaporation cools the oceans, while net condensation warms the land. This influences the land-water temperature gradient, which controls the annual monsoon weather patterns in the tropics, for instance. This part of the system is self-adjusting on a comparatively short time scale: if the land cools too much relative to the ocean, onshore winds bring in moisture, and conversely.

From a physics perspective, the influx of solar radiation at the surface of the Earth (not counting reflection) of 161 W/m^2 does several kinds of work. The work done by photosynthesis has already been mentioned: Assuming this corresponds roughly to net absorption of heat, it amounts to 0.9 W per square meter (W/m^2). The other two kinds of work done by incoming radiation are evaporation of water (which consumes latent heat at the surface and releases it during condensation) amounting to about 80 W/m^2. Thermal updrafts, which cause vertical mixing of air (and storms), account for 17 W/m^2. The remainder (63 W/m^2) produces temperature differences (gradients) that drive horizontal winds and ocean currents, both between land and water and between north and south. These counter-gradient movements correspond to the expansion phase of a giant heat engine that is not connected to any output (piston or turbine wheel) except the erosion of solid land surfaces and coastlines. The currents encounter internal resistance and warm the ocean and the air.

The 80 W/m^2 of work that evaporates water moves 40 billion tons (40 GT) of water per year from the oceans to land surfaces, where it is precipitated as rain or snow. Precipitation at altitudes above sea-level—especially higher elevations—becomes runoff: 40 GT of annual runoff corresponds to a hydraulic power output of 11.7 TW, or 11,700 GW (Smil 2008, Table A-16, p. 397). Of this, 1700 GW is regarded as potentially usable for electricity generation. Of that, 750 GW is the nominal capacity of currently installed hydro-electric generating plants (as of 2005 or so), although actual output is much lower because of seasonal flow variability. By comparison, current (2013) global fossil fuel consumption is also around 11 TW. In other words, the energy supplied by burning fossil fuels could all be supplied by the surface runoff, in principle.

Runoff from glaciers deserves special mention, partly because the ice is mostly at high altitudes (above 2000 m) and partly because the low temperature contains a lot of thermal exergy (the exergy that would otherwise be required to produce all that ice in a giant refrigerator). The thermal gradient between Antarctic or Greenland ices and the surface could be used to do useful work, although nobody knows how to exploit that potential right now. As the ice melts, all that potential is lost.

About 65 % of the rainfall on land is taken up by plants where it participates in the photosynthesis process and the carbon cycle. Some water is embodied in bio-mass, but assuming no net growth of biomass, it is subsequently re-evaporated (transpired). The rest runs off the surface in rivers. That flow encounters resistance from friction (and causes erosion). The "waste" heat from all of these air and water flows accounts for part of the low temperature radiation from the earth (396 W/m^2). Heat conducted from the interior of the Earth constitutes the remainder.

The water in the oceans, and its peculiar physical properties, has important effects on the climate, as well as indirect effects on evolution itself. One of those peculiar physical properties is that frozen water (ice) is about 9 % less dense than water at the same temperature. (Most other materials become denser in the solid state.) Hence ice floats, rather than sinking to the bottom. This permits liquid water, and marine life, to survive under ice; which is one of the reasons why life is possible on Earth. Incidentally, floating ice, especially in the Arctic, reflects sunlight much better than open water. This causes cooling. Without that reflection the climate would be warmer than it is now.

The hydrological cycle and associated winds constitute a gigantic Earth-moving engine that grinds and weathers exposed rock, creates soil, and moves it as dust or river sediment. Weathering not only converts rock to soil, but also gradually removes carbon dioxide from the atmosphere by reacting with silicates. Erosion, of course, carries weathered material (soil and silt) from hills to valleys and finally into the sea, where the material may be buried for millions of years, until some tectonic process lifts the submerged area once again. Hydrological processes are of major importance in several of the "nutrient cycles" discussed in Sect. 6.5 et seq.

There are cycles within the hydrological cycle, notably the ocean currents and the *El Nino/ La Nina* events (next section).

6.3   Ocean Currents and Atmospheric Circulation

A (mostly) hidden part of the terraforming "engine" is the system of ocean currents, of which the most familiar example is the Gulf Stream. In tropical oceans, like the Gulf of Mexico, warm surface water evaporates, leaving the salt behind. The salty warm water flows north (or south) on the surface of the ocean, over a colder, deeper layer. As the saline current loses its warmth closer to the poles, it eventually sinks. This is due to its higher density, as compared to fresh water inflows from rivers (like the St. Lawrence) fed by surface runoff and snowmelt. As it sinks, it also carries heat. Thence it flows back towards the equator in the deep ocean, picking up organic detritus on the way. The deep water, now laden with nutrients (such as phosphates) eventually rises to the surface in places like the West Coast of South America, to replace the warm salty water that has moved away. These surface and subsurface currents, taken together, constitute the *thermo-haline* circulation, better known as the "conveyor belt" which transports heat from the equatorial regions to the poles and transports nutrients (especially phosphorus) from the deep waters of the polar oceans back to the surface waters of the equatorial regions.

The main currents constituting the "conveyor belt" are shown from a south polar perspective in Fig. 6.5. The surface currents (red) are balanced by undersea currents (blue). The warm surface currents dive (becoming the "deep-water formation") in just two places, the North Atlantic between Iceland and Norway and the South Atlantic near the Weddell Sea off the coast of Antarctica. The southern ocean is apparently the "gatekeeper" of the deep sea. It is the most important place where carbon dioxide from the surface waters is carried down to the deep sea, where it remains for a long time. This is partly because the water in the southern ocean is cold, which increases the solubility of CO_2. It is also partly because surface

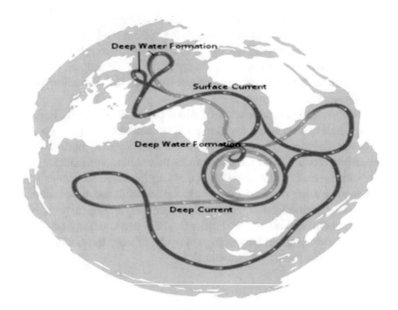

Fig. 6.5 The "conveyor belt" of ocean currents. *Source*: http://www.bitsofscience.org/wordpress-3.0.1/wordpress/wp-content/uploads//greenland-antarctica-seesaw-250x258.jpg

dwelling organisms (plankton) do not utilize as much carbon as they could, probably due to scarcity of other essential nutrients (like iron).[3] In any case, it is known that conditions at the surface of the southern ocean control the rate of sequestration of CO_2 in the oceans.

Meanwhile, the places where warm surface currents such as the Gulf Stream lose their buoyancy and sink can raise the average temperature of nearby land. The Gulf Stream is an important example because of its effect on Western Europe, especially Norway. (Without it, the temperature of the Barents Sea, north of Norway, would be 15 °C lower than it is.)

As the water in the Gulf of Mexico becomes both warmer and more saline, the point where it ceases its northward flow and sinks to the bottom depends (roughly) on which of three different drivers is the dominant one. They are: (1) horizontal thermal convection from the excess heat buildup in the Gulf of Mexico, (2) the excess salinity due to enhanced evaporation in the tropics (which makes the water heavier) and (3) the increased fresh water input from land surface runoff via the St, Lawrence River and the melting of Greenland ice. Which of these effects will be dominant in the future is not yet clear because the Gulf Stream is not contained within sharp boundaries, as a river on land would be. Moreover, its depth, velocity and even its temperature gradient can only be estimated roughly. Also, it varies from year to year.

---

[3] It has been suggested that a deliberate policy of increasing the iron supply to the plankton could help to "manage" the atmospheric level of draw down the CO_2.

The "conveyor belt" does not only warm northwestern Europe. The reverse flow of the deep currents ends with up-welling of cold and nutrient rich water in parts of the tropics, especially off the west coast of South America (mainly Peru and Ecuador). These cold waters rise to the surface at a rate determined by the strength of the surface trade winds known as "easterlies". These winds blow from southeast to northwest along the Peruvian coast and offshore. They normally push the warm surface waters away from the continental coastline, thus pulling the cold deeper waters to the surface in their wake, like a siphon.

When the easterly winds blow at full strength, the cold water rising to the surface carries nutrients that support the great anchovy fishery of Peru (and the fishermen and their families). It also cools the air in the lower atmosphere immediately above the sea. This cool air is too heavy to rise high enough to condense into raindrops. Hence, in most years there is little or no precipitation near the western coast of South America. These areas are consequently extremely arid. The humid surface air moves across the sea surface (warming gradually as it goes) and finally releases its moisture as rain when it encounters mountainous islands in the central or western Pacific (e.g. Polynesia, Melanesia and Indonesia).

At least, that is the normal pattern. However, the complexity of ocean atmosphere relationships is illustrated by reversals, called *El Niño* events.[4] The latter were only "discovered" (first noticed?) in 1950, but this cycle, known officially as the El Nino Southern Oscillation, or ENSO, has become more and more noticeable since then. The ENSO phenomenon is not local to the west coast of South America, although it was first noticed there. The fall and winter of 2015 looks like the strongest such event yet observed. It seems to be driven by a large "pool" of warm seawater that moves from west to east across the Pacific Ocean, and finally accumulates near the west coast of South America. Above this warm water, the atmospheric pressure is below average, and during this period, there can be heavy rainfall and floods over the otherwise arid coastal areas of Chile and Peru.

An *El Niño* event is characterized by abnormally weak surface winds from east to west, resulting in less upwelling. This shift in wind patterns means sharp reductions in the traditional anchovy fish catch in the up-welling zone off Peru and Ecuador. It also means a shift of the rainfall zone from the western to the central or eastern Pacific. That results in drought in Northern Australia and Indonesia. To illustrate the magnitude of the change, "normal" rainfall at Christmas Island in the central Pacific ranges from 0 to 8 in. In *El Niño* years, rainfall on that island has several times exceeded 40 in.

El Niño also affects conditions as far away as the Caribbean and South Africa. For instance, Northern Europe experiences a colder, drier winter, while southern Europe gets a warmer, wetter one. There is no doubt that strong *El Niño* events are very costly in terms of reduced fish catch, floods, storms, droughts and fires in

---

[4] El Niño (the Christ child) was originally the name given to a warm coastal current that appeared off the coast of Peru around Christmas time, each year. Now, however, it refers to the abnormal conditions.

drought impacted areas. Suspicion among climatologists that these events are becoming more frequent, and that they have been triggered by general climate warming, is now very strong. In fact, *El Niño* Southern Oscillation (ENSO) is now regarded as an important part of the global climate system.

There is an atmospheric circulation pattern similar to the ocean's "conveyor belt" but much more dynamic. Its main features are the low altitude tropical "trade winds", the high altitude sub-tropical jet stream and the mid-altitude polar jet streams. The sub-tropical jet stream is located at 30°N (or S) latitude and normally between 10 and 16 km altitude and wind velocities normally exceed 100 km/h. The polar jet stream is lower and faster, with velocities from 110 to 184 km/h. It meanders but the meanders are normally found between 30° and 60°N (or S) latitude. The polar jet stream is said to "follow the sun" meaning that in the northern hemisphere, it tends to be more southerly during the winter, and conversely in the southern hemisphere. The total kinetic energy of all the winds blowing in the atmosphere has been estimated by several authors at about 900 TW (Miller et al. 2011). The possibility for tapping into high altitude winds as a source of renewable energy is occasionally discussed.

It is increasingly clear that weather patterns are determined, to a considerable extent, by the location of the polar jet stream. Extended cold periods or dry periods can sometimes be traced to unusual configurations of the polar jet stream. To mention one example, there are indications that the "dust bowl" in the American mid-west during the 1930s was due to an unusual southern "meander" that stayed over the area for several years.

The jet streams are driven by the pressure generated by the thermal gradient between the equator and the poles. However, instead of being convection currents flowing directly between the equatorial regions and the poles, the air currents are diverted by the Coriolis force (due to the Earth's rotation) into currents from west to east along the boundaries of the air masses (or "cells"). See Fig. 6.6.

6.4 Climate Change

In a steady state equilibrium condition, the first law of thermodynamics (conservation of energy), says that energy inflows from the sun to the Earth in the form of short-wave (SW) light must always be balanced by energy outflows in the form of low temperature (infrared) heat or long-wave (LW) light. Otherwise the Earth would warm up or cool down, as already explained (Sect. 6.1).

However, the Earth is essentially never in thermodynamic equilibrium. There is usually an imbalance attributable to heating or cooling, ice formation or melting, or biomass generation or destruction. This energy imbalance is responsible for climate change. As long ago as 1896, Swedish physicist Svante Arrhenius warned of the possibility (Arrhenius 1896). Normally the imbalances are small and responses to change are slow. But recently the imbalance has increased in a short period of time

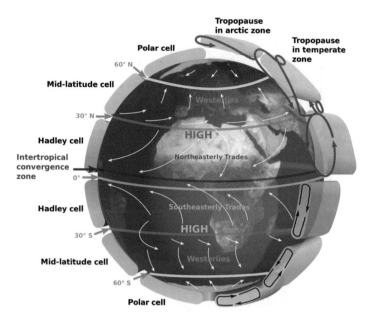

Fig. 6.6 Jet streams and trade winds. *Source*: https://upload.wikimedia.org/wikipedia/commons/thumb/9/9c/Earth_Global_Circulation_-_en.svg/2000px-Earth_Global_Circulation_-_en.svg.png

(geologically speaking) and the climate of the Earth is now changing faster in response.

The increase over the last 50 years in atmospheric CO_2, a major climate determinant, is shown in Fig. 6.7. Bear in mind that, unlike energy, exergy (the useful part of energy capable of doing work) is not conserved. There are exergy gains or losses in natural systems (such as glaciers, forests and deep water) as well as from biological and economic processes. Many of these processes are cyclic, with daily, lunar monthly, annual or even longer periodicity. For instance, the ice ages appear to be periodic, with very long cycles (Milankovitch 1941 [1998]; Hays et al. 1976; Muller and MacDonald 1997). Hence climatic comparisons between now and earlier times in the Earth's history need to be made with extreme care to allow for such cyclic phenomena.

It is important to realize that the oceans are by far the main storage system for heat in the short to medium term, having absorbed 93 % of the increase in global heating between 1971 and 2010 (IPCC 2014). They can absorb or emit heat much faster than solid rock, and can store much more (~1000 times as much) than the atmosphere. Hence the effective heat storage capacity of the top 700 m of the oceans, which exhibits measurable seasonal variation, is much larger than the heat storage capacity of either the atmosphere or the land. This is because thermal conductivity of the ground is very low and vertical convection is almost zero (except during volcanic eruptions), while the mass of the atmosphere is far less than that of the top layer of the oceans. Hence the oceans store much more heat in

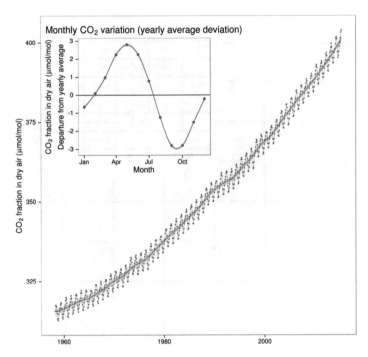

Fig. 6.7 Monthly CO_2 concentrations on Mauna Loa (NOAA). *Source*: https://upload.wikimedia. org/wikipedia/commons/thumb/c/c5/Mauna_Loa_CO2_monthly_mean_concentration.svg/

the summer than the land or atmosphere. This heat is then released during the winter, as the warm currents flow toward the poles.

Until the 1950s it was received wisdom that essentially all of the excess solar heat due to atmospheric greenhouse gas (GHG) buildup would be quickly taken up by the oceans. However, **Roger Revelle**, Director of the Scripps Oceanographic Institute, began to worry about this question. He initiated research on how to measure the rate of thermal uptake by the ocean, both by absorption at the surface, and into the deep oceans via convection currents. The first step was to determine what happens to excess carbon dioxide. The previous scientific consensus was that any excess CO_2 would be quickly absorbed by the oceans, since the oceans have so much more absorption capacity than the land surface or the atmosphere. The research at Scripps changed that picture dramatically (Revelle and Suess 1957). It is now accepted that about 30 % of *excess* carbon dioxide in the atmosphere is absorbed in the oceans, 25 % taken up by photosynthetic organisms (biomass) and the rest remains in the atmosphere. But the equilibrium is not reached for a long time.

The rate of heat uptake of the oceans is not the same as the rate of CO_2 uptake. It has been calculated to be 0.55 ± 15 W/m^2 (IPCC 2014). But the whole issue of oceanic heat uptake remains contentious, since the data up to that time were sporadic and (inevitably) local. Since 2001, the Argo project, part of the Global

Ocean Observing System (GOOS), within the Global Ocean Data Assimilation Experiment (GODAE), has distributed over 4000 expendable bathythermographs (XBTs) around the globe. Each one takes three measurements per month and transmits the data via satellite to the GODAE server in Monterey Cal. Each of the floats covers an ocean area of 10^5 km^2, down to a depth of 2 km. They are distributed from 70°N to 70°S. Obviously the coverage is sparse. However, for the 11 year period 2004–2014, there is a clear warming trend at the rate of 0.23 °C per century.

It seems likely that the rate of heat exchange in the deep oceans (not measured by Argo), depends on the thermo-haline ("conveyor belt") circulation (Sect. 6.3). It is probably much lower than heat exchange on the surface, since mixing induced by storms is less important. The rates of mixing vs. conduction and convection are still not well-known. (The residence time of a water molecule in the oceans is estimated to be 3200 years.) It looks like 90 % of the excess heat warms the oceans, and only 10 % warms the land surface. The heat taken up by the oceans causes thermal expansion of the water, which is one of the two main causes of global mean sea-level (GMSL) rise. This amounts to about 1.7 mm/year since 1901 and about 3.2 mm per year since 1993 (Church and White 2011).

The other two causes of GMSL rise are glacial ice melting and vertical land motion (VLM) resulting from the removal of the weight of glacial ice in certain regions that were once ice-covered, such as Northern Canada and Scandinavia. The VLM adjustment is fairly localized. But in recent years, it appears that about half of the GMSL rise—roughly 1.6 mm/year—is due to the melting of glacier ice (mainly Greenland and Antarctica). The quantitative change in ocean mass—as opposed to temperature—is now being measured by a the Gravity Recovery and Climate Experiment (GRACE), since 2002 (Chambers et al. 2004; Peltier 2004).

A question of considerable scientific importance right now is how sensitive the global climate is to anthropogenic perturbation, in the short and medium term, especially the burning of fossil fuels and the buildup of greenhouse gases (GHGs) in the atmosphere? Or, in other words: How much will the average global temperature of Earth, in equilibrium, rise in response to a doubling of the CO_2 level? And how soon will it happen? The answer is called *climate sensitivity*. In principle this might depend on the starting (reference) CO_2 level or on the starting (reference) temperature. Luckily it appears that the climate sensitivity does not depend on either of those conditions, so it is almost a constant of nature, at least for the Earth. Unfortunately, scientists have not yet been able to determine the Earth's climate sensitivity—constant though it may be—with great accuracy. I will try to explain why it remains uncertain, a little later.

Before I do, I need to introduce another constant: *radiative forcing* is another term for the driver of climate change due to GHG increases above the reference level. The adjective "radiative" conveys the fact that the heating due to a given excess GHG level in the atmosphere causes an equivalent decrease in long-wave (LW) radiation into space from the top of the atmosphere. It is also equivalent to the additional solar heat input needed to raise the equilibrium temperature of Earth by 1 °C. This can be expressed in energy units of watts per square meter (W/m^2) per degree of temperature difference.

The equivalence is expressed by the so-called *radiation damping coefficient*, for the Earth, which is 3.7 W/m^2 per °C. (See Fig. 6.3 in Sect. 6.1, and the explanation there.) That number is also a constant for the Earth as a radiating system, at least within very wide limits. It happens, rather conveniently, that the "radiative forcing" effect of GHGs is proportional to the logarithm of the atmospheric CO_2-equivalent concentration.

This information, plus a little arithmetic, tells us that increasing the GHG (CO_2 equivalent) concentration from 180 ppm back in the glacial maximum period (Previdi et al. 2013) to 280 ppm in the pre-industrial period would be the equivalent of increasing the solar power input at the top of the atmosphere by about 180/280 × 3.7 = 2.37 W/m^2. That, in turn, would have raised the Earth's temperature by about 0.64 °C. But the global average temperature actually increased by around 5 °C. So, by this "top down" argument, there seems to be an amplification factor of 5/0.64 = 7.8.[5]

I say "seems to be" because there are some important uncertainties in the calculation, and it is inconsistent with some "bottom up" estimates in the technical literature. However it deserves serious consideration for a simple reason: *If the climate sensitivity is that large, or anywhere near that large, the next doubling of atmospheric GHG levels (CO_2-equivalent) will result in a larger temperature increase than climate experts at the International Panel on Climate Change (IPCC) are currently expecting.*

The amplification factor, mentioned above, is known to climate modelers as *climate sensitivity*. The next question is: How can we account for such a large sensitivity? Modelers have been trying to estimate climate sensitivity in terms of explicit feedback mechanisms for many years. The first credible attempt (1979) was by a group, chaired by Professor Jules Charney, for the US National Academy of Sciences (NAS) (Ad hoc Study Group on Carbon Dioxide and Climate 1979). They took into account several "fast" feedbacks, such as increased evaporation and humidity, changes in albedo due to surface ice melting (the Budyko ice-albedo model), and lower solubility of CO_2 in the ocean as temperatures rise (Budyko 1969). The Carbon Dioxide Assessment Committee (CDAC) report of the National Academy of Sciences arrived at a climate sensitivity (amplification factor, AF) of 3.1 (CDAC 1983). Most of the general circulation (climate) models (GCMs) still use this "Charney sensitivity". Hansen et al. have carried out a detailed analysis of other feedbacks not taken into account by CDAC, concluding that the Charney sensitivity is too low (Hansen et al. 1984, 1985). Other sensitivity estimates in the literature are in the range of 4–4.5 (Previdi et al. 2013; Lunt et al. 2010).

Two later efforts are worth mentioning. The first is attributed to a group at the Hadley Institute in the UK taking into account changes in the carbon cycle. They estimated a sensitivity of 4.64, about 50 % above the Charney sensitivity (Wasdell 2014).[6] Later James Hansen *et al* made a further correction to take into account

---

[5] I am grateful to David Wasdell for pointing out this relationship, even though I think he relies too much on it.

[6] I was unable to find the original reference, which was not cited by Wasdell.

changes in the continental ice sheets associated with Milankovitch cycles (Hansen et al. 2008, 2013). The revised Hansen sensitivity (according to Wasdell) is 6.2, which is double the Charney estimate.

However, there is still a gap between these feedback corrections and the factor of 7.8 calculated above from historical climate data by Wasdell. The difference between the feedbacks explicitly accounted for by these groups and the factor of 7.8 would have to be explained by various unconsidered Earth System feedbacks not otherwise taken into account, such as methane clathrate emissions and changes in the ocean currents.

Here is what climate scientists think happens when GHG concentration in the atmosphere increases the surface temperature of the Earth. One immediate effect would be increased photosynthesis and more biomass accumulation (partly from CO_2 fertilization), more evapotranspiration, and more water vapor in the atmosphere. Water vapor is the most potent of all the GHGs. This is probably a positive (amplifying) feedback, but a few climate scientists disagree.

A second fairly immediate effect of warming would be to melt the remaining mountain glaciers and further increase the evaporation of water from the warmer oceans. (More amplification?) A third consequence is that as the ocean warms up, it expands—resulting in sea level rise—and (thanks to warming) keeps less carbon dioxide in solution. This puts more CO_2 into the atmosphere. Currently it is estimated that 57 % of CO_2 emissions are dissolved in the oceans (resulting in increased acidity) but a warmer ocean will dissolve progressively less, thus keeping more of the excess CO_2 in the atmosphere. This so-called "carbon-cycle feedback" has different consequences in different GCMs, ranging from 0.1 to 1.5 °C (Previdi et al. 2013).

A fourth possible effect, already noted but not easily quantifiable or predictable, might be to alter the "conveyor belt" of surface and deep ocean currents. For example, this might cause the warm, salty Gulf Stream to abort its northward passage on the surface, and sink prematurely. This would be due to cooling and dilution by melting Greenland ice, plus increased salinity due to salt release from the melting of Arctic sea ice. Increased salinity makes the water denser and causes it to sink sooner than it otherwise would (Wadhams et al. 2003). (This may have happened in the past, during the rapid melting of the Wisconsin glacier.) Were this to happen again, northern Europe could be cooled by several degrees Celsius as a direct consequence of general global warming!

A fifth possible effect of atmospheric warming would be to thaw some of the northern "permafrost" area, both above ground and under the Arctic ocean. Climate warming is happening much faster in the polar regions, than the tropics. (All the climate models show this effect.) The rapid thinning of the ice and likely disappearance (in summer) of the ice in the Arctic Ocean only a few years hence, is confirming evidence. The fact that the poles are getting warmer faster than the tropics means that the north-south temperature gradient is getting smaller and less sharp. That, in turn, is pushing the northern jet stream northward (on average) at the rate of 2 km per year. This would permit both aerobic and anaerobic microorganism activity under the soil surface to accelerate, releasing both carbon dioxide and methane into the atmosphere. The undersea cousin of permafrost on land, methane clathrate, could also start to thaw, releasing methane into the ocean and thence into the atmosphere, resulting in positive feedback.

In 2003 a group of scientists postulated the "clathrate gun" hypothesis, i.e. that a sudden warming might trigger an unstoppable self-sustaining process of methane release that would proceed until all the undersea methane trapped in clathrates is released (Wadhams et al. 2003; Kennett et al. 2003). See Sect. 4.6. There is evidence that methane is now escaping in large quantities from the Arctic East Siberian shelf and the North Atlantic continental shelf (Shakhova et al. 2007; Skarke et al. 2014). Indeed, the methane content of the atmosphere over the Arctic ocean shelf has increased sharply during the past decade (Shakhova et al. 2008, 2010a, b).

If this methane release process should become irreversible, it could theoretically raise the temperature of the Earth ten or more degrees, especially at the poles, melt all the glaciers in the world, raise the sea level by over a 100 m, and probably drive most of the terrestrial species that now exist into extinction. This process may have actually begun already in the Arctic Ocean, but it is not yet taken into account in the general circulation models (GCMs).

Unless compensated for by some other feedback phenomenon, not yet taken into account, some scientists, such as James Hanson, fear that beyond some "tipping point" the system might go into "runaway" mode until some more distant limit is reached. In other words, climate change beyond that point might not be reversible at any temperature friendly to humans (Lenton et al. 2008). Nobody knows for sure where that "tipping point" is (if it exists), or what the trigger mechanism might be The latest report by the Intergovernmental Panel on Climate Change (IPCC) argues that 2 °C is the maximum safe limit (IPCC 2014). This is largely based on ice-core records from 8 glacial cycles from the EPICA project at Dome C in Antarctica, reflecting times in the past when the temperatures were higher (Previdi et al. 2013). Hansen and others now think the safe limit should be lower (Hansen et al. 2013). But that should not be confused with the unknown point at which "runaway" temperature increases might start.

The only obvious negative feedback that could reverse warming over a period of years or decades would be sharply increased high-level cloud cover to increase the Earth's albedo. Clouds already account for 55 % of the sunlight reflected into space from the Earth, and it is clouds that account for the difference between the albedo without clouds (0.15) and the albedo with clouds (0.31). Increased albedo (other factors being the same) could also result from higher micro-particulate (e.g. sulfate) emissions from industry or volcanoes. But sulfate emissions are increasingly being cut at the origin, thanks to pollution regulation, so that is not a likely outcome.

As regards cloud cover, there is one respected climatologist who thinks this is the answer. He has proposed the so-called "iris" theory, that the Earth's climate responds automatically to excess solar radiation input, as the iris of the human eye responds to brighter light (Lindzen and Chou 2001; Lindzen and Choi 2011). The problem is that high level cirrus clouds consist of ice particles. They are nearly transparent to sunlight, but they absorb and re-emit long wave (IR) radiation, thus contributing to the greenhouse effect. On the other hand, cumulus and stratocumulus clouds are good reflectors of sunlight, hence contributing to the albedo. Is there a mechanism to increase the prevalence of these clouds, as a consequence of the atmospheric GHG buildup, thus reducing climate sensitivity? None has been suggested. The recent slowdown in global warming might support Lindzen's "iris" hypothesis, although direct observations have not done so, as yet.

It is important to note that the Earth is definitely heating up (as the Argo data indicates), and that ground temperatures are rising faster than temperatures at high altitudes (based on satellite measurements) and much faster than average ocean temperatures. Hence, in the long run there must be land-sea equilibration mechanisms, some of which are not yet reflected in the GCMs. One obvious possibility is the ENSO phenomenon: it is a mechanism that transfers heat from the ocean surface to the land. The spikes in surface temperatures during El Nino events are consistent with this hypothesis.

A few words about so-called global circulation models (GCMs), are appropriate here. (You can skip this part if models bore you.) These models use physical data on the chemical composition of the atmosphere, ground, sea and air temperatures, wind velocities, rainfall, river flows, cloud reflection, optical spectra, heat capacities and heat transfer (and large computers). The key variable in all these models is the atmospheric concentration of so-called "greenhouse gases" (GHGs), so-called because these gases tend to absorb and re-radiate heat back to the Earth, raising the temperature like an agricultural greenhouse. Of these gases, the most are important of all is water vapor, but it is inherently localized, making it the least controllable and also hardest to obtain good global average data on. It accounts for between a third and two thirds of the GHG effect. The next is carbon dioxide, CO_2, which has been known as a greenhouse gas for more than a century (Arrhenius 1896). That is followed by methane, nitrous oxide and a few industrial chemicals.

Molecule-for-molecule, methane is much more potent than CO_2, but methane molecules do not remain in the atmosphere nearly as long (only about 6 years), compared to a hundred years or more for CO_2. Comparing equal concentrations, averaged over 100 years, the methane is still about 20 times as potent as CO_2 as a GHG. Luckily, there is much less methane in the atmosphere than carbon dioxide, so it is only a third as important, right now. That could change, however, if methane emissions from the methane clathrates in the sediments under the east Siberian shelf start to accelerate.

CO_2 levels during the past 500,000 years have recently been determined with considerable accuracy from ice-core records from Antarctica: See Vostok (Petit et al. 1999) and EPICA (Loulergue et al. 2007). See also (Lenton and Watson 2011, Chap. 18, Fig. 18.3). During each of the last five ice-ages, ice core data indicates that the Earth's average surface temperature dropped about 5 °C below the currently assumed pre-industrial equilibrium level of 15 °C. However, the sea level during interglacial periods was higher by varying amounts, ranging from 2 to 32 m. Thus it appears that feedbacks involving glacial ice have differed significantly between glaciations.

More recently, very accurate records going back 3.3 million years to the late Pliocene have been obtained from boron isotopes in the shells of tiny sea creatures (foraminifera) buried in the sediments (Martinez-Boti et al. 2015). During the Pliocene, average Earth surface temperatures were about 3 °C higher than they are today. However, it appears that the climate sensitivity (relationship between temperature and CO_2 concentration) was roughly the same 3 million years ago when the temperature was higher, as it is today. This fact undermines ideas that positive feedbacks may increase with temperature.

Sea level changes were caused by the fact that the more water was tied up in glacial ice, the lower the ocean level fell. At the glacial maximum of the last ice age, 20,000 years ago, the sea level was 122 m below the current level. During the last interglacial, 125,000 years ago, the sea level was 5.5 m higher than it is today. This might suggest that the last ice age is not quite over. That would imply that some further warming would occur without any human intervention. However, ice core data now indicates clearly that the interglacial peak temperature actually occurred 7000 years ago (5000 BCE) and that the cooling phase was already well under way before human intervention began to change the picture in the past century. See Fig. 6.8.

It must be acknowledged that the so-called general circulation models (GCMs) are still far too crude to be relied on absolutely, especially for local phenomena such as cloud behavior and storms. On the other hand, the existing models do reproduce recent trends over periods of months and longer fairly well. Figure 6.9, which compares the Earth's yearly average heating rate with model simulations, makes this point quite well. The correlation is obviously quite good. This provides limited grounds for confidence.

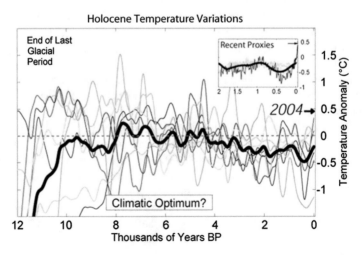

Fig. 6.8 Long-term temperature variations for the Earth. *Source*: http://upload.wikimedia.org/wikipedia/commons/c/ca/Holocene_Temperature_Variations.png

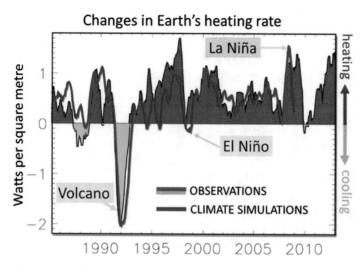

Fig. 6.9 Fluctuations in Earth's heating rate 1985–2010. *Source*: www.ncas.ac.uk/index.php/en/climate-science-highlights/2121-changes-in-earth-s-heating-rate-between-1985-and-2012

Skeptics correctly point out that the GCMs still share several major weaknesses. One is that they assume vertical atmospheric thermal convection without fully taking into account the horizontal component of circulation arising from ocean currents and mixing. Another is that they divide the surface of the globe into grid squares (or icosahedrons) but the minimum size of the grid area (except near the poles) is still much too large. They cannot simulate small scale behavior even within a few kilometers, nowhere near the tens or hundreds of meters one would want to predict local precipitation patterns or storm paths. They cannot accurately predict hurricane probabilities, still less hurricane or tornado tracks. The omission of possible methane emissions from clathrates is also a weakness. The lack of any mechanism to explain the ENSO phenomenon could turn out to be the most important weakness.

Theory says, broadly, that surface warming (and increased evaporation) *must* drive the hydrological cycle harder, making rainstorms, floods, and droughts more frequent or intense. Indeed, this is predicted by all the GCMs. But some new evidence, based on salinity studies, suggest that the exergy flux driving the hydrological cycle is actually increasing about twice as fast as the GCM's predict, i.e. $8 \pm 5\%$ p.a. (Durack et al. 2012). It is not clear what this might mean, although it does not support climate skeptics, unless it could explain increased cloud cover and higher albedo. Recent years have been characterized by increased storminess, very much as the theory predicts.

As a practical matter, the location of any precipitation is important, even vital, for agriculture, tourism and air transportation. If precipitation occurs as snow in the far north, or in high mountains, year round snow cover will increase and more of the solar radiation will be reflected away from the Earth. Thus, increased cloud cover

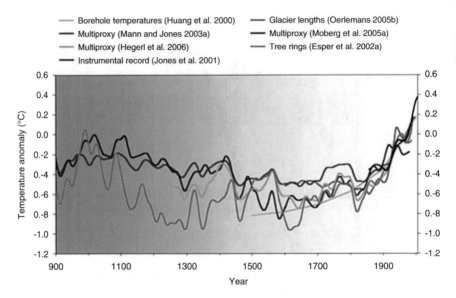

Fig. 6.10 Global temperatures since 900 CE. *Source*: Surface temperature reconstructions for the last 2000 years. Committee on Surface Temperature Reconstructions for the Last 2000 Years; Board on Atmospheric Sciences and Climate; Division on Earth and Life Studies; National Research Council. National Academy Press, ISBN: 978-0-309-10225-4 DOI: 10.17226/11676

and snowfall in the north also tend to increase the albedo and cause additional compensatory cooling.[7] But intense rainfall on mountains results in floods and accelerated erosion (as seen in intense El Nino episodes). In some countries the consequences can be devastating. Unfortunately, GCMs today are not yet capable of predicting geographical details, and may not be so capable for another decade.

Another problem for the GCMs is data. Ideally, a model must reproduce the past, but we do not have data about the past that is comparable in accuracy to what we have today. Historical temperatures have to be reconstructed from other measurements. Temperature estimates over the last 1100 years up to 2000 are shown in Fig. 6.10, but the different estimates vary considerably.

Figure 6.11 shows global surface temperature changes since 1880. The data is fairly reliable (Mann et al. 2016). It shows a rapid and continuous rise in surface temperatures since 1910 except for a slowdown from 1945 to 1975 another slowdown in surface air temperature rise from 1998 through 2013 or so. It is generally accepted that, during the "flat" period (1945–1975) rising CO_2 emissions were approximately balanced by the cooling effect due to sunlight-reflection from aerosols (e.g. smoke) and sulfate (SO_4) ions. Those emissions came from the rapid expansion of unregulated coal-burning power plants, cement mills and steel mills in Europe and the US, during the post WW II recovery and reconstruction period. The

---

[7] The physics is actually quite complicated, and there are still significant uncertainties; research in this area is now very active. See IPCC (1995, p. 201 et seq).

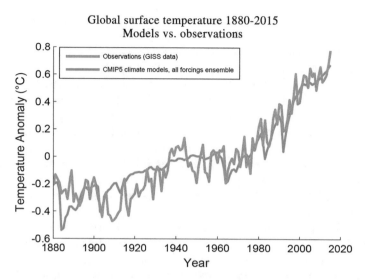

Fig. 6.11 Global mean surface temperatures since 1880 models vs. data. *Source*: http://herdsoft. com/climate/widget/image.php?width=400&height=230&title. Data from Mann et al. (2016)

slowdown after 1995 through 2013 was also probably due to particulate and sulfate emissions due to the ultra-rapid industrialization of China during those years.

The temperature rise, based on surface measurements, continued after that and 2014 was a record which was broken in 2015.

It is true that surface temperature is not necessarily a good indicator of planetary heating. In fact, between 2000 and 2012 the Earth gained 0.62 ± 0.43 W/m^2 (90 % confidence level), or 320 ± 220 TW of heat energy during that period (Allan et al. 2014). Even if the heat added was only 100 TW, it is equivalent to about 20 times the total energy produced (and consumed) by human activity over this period.

Where did all the excess heat energy go? Most of it raised the temperature of the oceans, as confirmed by the ARGO measurements. The accelerated melting of glaciers in Greenland and Antarctica also absorbed a considerable amount of heat, and contributed to sea-level rise. A small fraction may have been absorbed by accelerated forest growth. Any of these mechanisms could have reduced the rate of global (surface) air temperature rise as measured by surface-based instruments.

The question we all ask is: what comes next? Are we in for a few centuries of very slowly rising temperatures, perhaps allowing for oil drilling in the Arctic, wine grapes to grow in Sweden and Siberia and allowing the inhabitants of the overheated tropical countries a safe haven? Unfortunately, that is a pipe dream. There is no way the relatively unpopulated "north"—Canada, Siberia—can (or would) accommodate several billion refugees from the "South", even if they could get there.

Or is there a "tipping point of no return" ahead that could trigger runaway heating, irreversible sea-level rise, droughts, forest fires, floods, probably accompanied by geopolitical chaos (Lenton et al. 2008) ? The Inter-governmental Panel on Climate Control (**IPCC**) was created in the 1980s to consider this question, among others. Since the IPCC started the scientists have generally agreed that likelihood of a "tipping point" increases very rapidly if the global temperature rises by more than 2 °C. If the global average rises by 2° or more, the temperature in the Arctic could increase by 5 °C if not more. The latest IPCC report says that 5° could suffice to melt most or all the Greenland ice. The warm spell 125,000 years ago was characterized by an average global temperature only 1 °C higher than today, but the temperature in the Arctic then was about 5 °C higher than today. That was when the sea-level was 5.5 m higher than it is today. Three million years ago there was an even warmer spell, when the sea level was 25 m higher. All of Greenland and half of Antarctica probably melted in that period (Azar 2008, 2012, pp. 30–31).

The IPCC reports have consistently asserted that the safe limit corresponds to a CO_2-equivalent concentration of 400 parts per million (ppm) in the atmosphere. Since the equivalent CO_2 concentration is already at or above the 400 ppm level (it is 425 as I write this) and the climate warming seems to have slowed down, you may think (like some skeptics) that the argument in the last several paragraphs must be wrong. But that does not follow. The real implication is that the climate system has not yet re-equilibrated. Even the "fast" feedbacks take decades, while others take centuries. If the current 400 ppm concentration level for CO_2 equivalents were to remain unchanged for the next several hundred years, the global temperature would continue to rise until it reached a new equilibrium level. That level would probably be +2.2 ° C above the present level if the current models are correct. (When error bars are taken into account, the temperature range for the new equilibrium would probably lie between 1.5 and 4.5 °C but not above 6 °C, according to the latest IPCC report (IPCC 2013).)

The international scientific consensus, expressed in the reports of the IPCC and many hundreds of other books and articles, is that climate change (warming) is mostly man-made and is happening worldwide, but with much greater effect near the poles [http://www.theguardian.com/environment/climate-consensus-97-percent/2015/jan/22/oceans-warming-so-fast-they-keep-breaking-scientists-charts]. Typhoons that kill thousands of people in the Philippines (for instance) are becoming routine. They are a symptom of climate change that has already begun. There will be more and bigger storms in the coming years. Other changes are occurring:

- The Arctic has been warming twice as fast as the global average. Arctic Ocean ice cover has been declining for years. The winter ice is also getting thinner. Between the end of 1979 and 2010 the September average (annual minimum) was 6.5 million square km. The 2007 IPCC report estimated that the sea ice could disappear entirely within 70 years (IPCC 2007). But the melting rate is accelerating. In 2012 ice coverage was below 4 million square km, the lowest ever by a significant margin. It is possible that much of the Arctic Ocean will be ice-free in the summer by 2025, or even sooner.

- Greenland ice is melting faster than previously projected. The Jakobshavn glacier, the most active in Greenland, has doubled its rate of discharge since 2002. Greenland lost 350 gigatons (GT) of ice in 2012, as compared to an average of 240 GT during 2003–2010. Earlier melt rates were much slower.
- Warming rates will increase as Arctic ice cover declines and more water surface is exposed, because less sunlight will be reflected. (This is the albedo effect.)
- Warming will accelerate as frozen tundra in the far north of Siberia, Canada and Alaska thaws, because anaerobic decay organisms become more active and convert organic material to carbon dioxide and methane. Methane is more than 20 times more potent than carbon dioxide as a greenhouse gas (GHG), so increasing the methane concentration in the atmosphere will speed up the warming.
- The current rate of sea-level rise is about 3 mm per year, or 3 cm per decade. Part of this (about one third) is due to thermal expansion of the water itself as the temperature rises. The rest is from the melting of mountain glaciers, Greenland and Antarctica. The rate of melting of mountain glaciers is quite well documented around the world. The main uncertainties concern Greenland and Antarctica. But melting is virtually certain to increase as global warming accelerates. The projections for the end of the twenty-first century range from 20 cm to 2 m, with a middle range of 0.4–1.0 m depending on the scenario and model. (Admittedly, the models cannot be relied on as one would like.) The sea-level rise, whatever the rate, will not stop in 2100.
- A "mere" one meter sea-level rise means that some island countries like the Maldives will be uninhabitable. Coastal Bangladesh, Indonesia, Viet Nam, the Philippines, Malaysia and China, as well as Florida will be flooded regularly by storms. Many of the most productive agricultural regions (especially for rice cultivation) are river deltas where the land is only slightly above high tide (e.g. the Nile, the Mississippi, the Mekong, the Ganges-Brahmaputra and the Indus). In those areas sea-level rise will cause salt penetration of ground water and salt-buildup in the soil, as well as more frequent flooding.
- Many big cities are on or near the sea coast, and much of the global population and productive capacity is at risk (as exemplified by the recent storm Sandy that hit New York City) in 2012. Dikes are feasible in some cases, and dike-building is a source of employment, but at high cost. Dikes add nothing to economic productivity. Storm recovery efforts add to GDP but not to productive capital assets or wealth.

The "great thaw" that is now in progress will not stop in the year 2100, nor will sea level rise stop at 1 m. The reality is that once those levels are reached, the climate changes will have enormous momentum and will be unstoppable. The time to act is now, not when. In the last interglacial warm period the sea level was 5.5 m higher than it is now. If (when) all the ice on Greenland melts, the sea level would rise 7.1 m. If the West Antarctic Ice sheet also melts, it will add another 5 m or so. But just a 1 m sea level rise would threaten the croplands of 60 million people, according to the German Advisory Council on Global Change (Azar 2008, 2012, pp. 26–27).

Five meters of sea-level rise would be enough to drown New York, London, Tokyo, Shanghai, Mumbai, Lagos, Buenos Aires, Rio de Janeiro, Caracas and scores of other major cities, along with the state of Florida, Long Island, all of the Netherlands and many other low lying places. Huge territories, that currently feed billions of people, would also be underwater, including much of the Nile valley, the Indus delta, the Ganges-Brahmaputra delta, the lower Yangtze valley, the Yellow River delta, the lower Mekong, the Niger delta, the Mississippi delta (probably as far upstream as St. Louis), most of the Netherlands, much of the Rhone and Loire valleys.

Scientists generally agree that the consequences are somewhat unpredictable, but that the worst case scenarios—such as a major release of methane from the melting of permafrost and/or the undersea deposits of methane-clathrates—while very unlikely, would be truly catastrophic. They also agree that, while natural variability (e.g. of sunspots) can conceivably have some effect, by far the greatest part of the drivers of change are anthropogenic, i.e. human-caused. More specifically, the main drivers of climate change are attributable to economic activity—especially the emission of carbon dioxide methane, nitrous oxide and other so-called greenhouse gases (GHGs). While agriculture (rice cultivation and cattle feeding) play a major part, the biggest single driver of climate change is combustion of fossil fuels in factories, homes and cars.[8]

On the other hand, our global civilization (a term some may dispute) faces a potential catastrophe that is *not* immediately apparent to most people, including a large percentage of business and political leaders. One of the barriers for policy-makers is that even those who suffer from typhoons and droughts do not see any direct connection between their problems and the drivers of long-term climate change. The only indications of this potential catastrophe immediately evident to members of the public are a very small increase in average global temperatures that most people scarcely notice (or even welcome), together with news reports of gradually shrinking glaciers, weather reports (with pictures) of increasingly fierce storms, accelerated melting of the summer sea ice in the Arctic Ocean, bleaching of the coral in the Great Barrier Reef and so on. In short, the symptoms of the change are still very remote from the everyday experiences of most people. In fact, every time there is a blizzard or cold spell somewhere, the skeptics leap to the conclusion that climate change has just been disproven, failing to notice that an increase in the frequency of extreme weather events (including blizzards) are actual predictions of the climate models.

Needless to say, but I'll say it once more, all the existing GCMs are still gross simplifications of reality, and there is a constant process of addition, revision and

---

[8] Another important source, that will be much harder to control, is agriculture, especially grass-eating animals—cattle, sheep and goats—that rely on bacteria in their second stomachs to digest cellulose. These bacteria generate methane as a waste product in the process.

correction. This process leads some to doubt that the model conclusions are robust enough to base policy on. However, generic criticism of models as guides for policy neglects the fact that doing nothing ("business as usual") is also a policy choice, and one that is increasingly hard to justify in view of the facts already known for sure. The weight of the evidence can only be understood by developing more quantifiable models to explain the behavior of a very complex system.

Curiously, there is still enough residual uncertainty to allow a few serious scientists to attribute the observable evidence of warming as "natural variation". One of the favorite suggestions involves the frequency of sunspots, which are analogous to tornados but a lot bigger and more powerful. Their rotation enables them to carry excess heat from the interior of the sun to the surface faster than ordinary convection. They can be thought of as large "plasma guns" spraying high energy particles into space, some of which encounter the Earth's atmosphere and provide warming. On the other hand, it is now known that the sun's surface temperature is actually higher when there are no sunspots visible.

Sunspots come and go, and occasionally they disappear altogether for a time, as during the Maunder minimum (1650–1715), as seen in Fig. 6.12. It is argued (by some) that such fluctuations could conceivably be related to periods of exceptional cooling like the "little ice age" in the late seventeenth and early eighteenth centuries. If there is a causal link from sunspots to climate change, it is obscure.

Because the threat of climate change is still seemingly so remote to ordinary citizens, it may be worthwhile to review, briefly, the main arguments that have been put forward by the scientific skeptics and the reasons for dismissing those arguments. (For a list of alternative explanations for climate warming that have been suggested, with brief rebuttals see (Walker and King 2009). I will mention only the most important ones.

Back in the 1970s a number of meteorologists and climatologists expected a new ice age to start soon, rather than an increase in climate warming. Reid Bryson, a meteorologist at the University of Wisconsin and Sir Fred Hoyle—best known as a cosmologist who didn't believe in the "big bang"—were among the most prestigious (Bryson 1974; Hoyle 1981). This theory was partly based on the historical evidence of long glacial "ice ages" separated by shorter inter-glacial periods that happened to be about as long as the time since the last ice age.

Fred Hoyle (and the others) were depending on the reliability of some astronomical clockwork mechanism called "orbital forcing". This theory, known as **Milankovitch cycles**, is based on several long-period changes in the Earth's orbit around the sun, and the tilt of its axis of rotation (Milankovitch 1941 [1998]). However, at least three different factors with different periodicities are involved: (1) orbital shape (eccentricity) with a complicated periodicity averaging 98,000 years, (2) axial tilt (obliquity) with a 41,000 year periodicity and (3) precession ("wobble"), with a 19,000–23,000 year period. There is also an orbital inclination (not considered by Milankovitch) with a periodicity of 70,000 years.

The Milankovitch theory was revived in the 1970s, due mainly to ocean sediment studies that confirmed a long 100,000-year cycle. Despite some problems, the Milankovitch cycles do roughly explain the glacial cycles and could explain long-

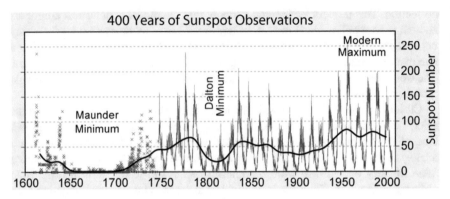

Fig. 6.12 400 year history of sunspot activity. *Source*: https://en.wikipedia.org/wiki/Solar_cycle

term warming and cooling (Hays et al. 1976). There has been additional confirmation of the theory in recent years. Conclusion: *Orbital forcing might explain quite a bit of the variability over the last three million years, but cannot explain any climate changes that occur on a time scale of decades.*

In short, the scientific evidence of anthropogenic climate forcing is even stronger than the latest IPCC Report is able to present, in part due to the lag between research and publication in peer-reviewed journals and in part because of political pressure by governments that are beholden to the fossil fuel industry. The environmental website, *Mother Jones* has pointed out that lobbies for the US coal industry (and the infamous Koch Brothers) and the oil industry, have spent millions of dollars to attack the science itself: their objective was—and still is—to "reposition global warming as theory, not fact" (Azar 2008, 2012, p. 33). The increased media attention to "climate skeptics" is evidence of their success.

The evidence for climate change, in the opinion of most scientists, is really overwhelming. But the vast majority of skeptics don't know all the facts and don't want to know. However, continued denial in the face of clear, irrefutable evidence (such as the bleaching of coral reefs, the retreat of the glaciers and the drastic thinning of Arctic Ocean ice in the summers), is absurd. It is certainly not a legitimate basis for policy-making.

It is by no means clear that the measures now under consideration by governments would be capable of holding the CO_2 concentration in the atmosphere under the 440 ppm limit, or even the 550 ppm limit that some propose. Be that as it may, it is clear that every reasonable effort to reduce CO_2 emissions must be undertaken if climate change is not to go beyond the "point of no return" if it has not so already.

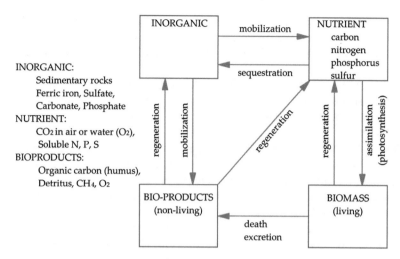

Fig. 6.13 4-Box scheme for bio-geochemical cycles

6.5 Bio-geochemical Cycles

It is convenient for purposes of exposition to deal with the nutrient cycles in functional terms, as illustrated by Fig. 6.13. There are two kinds of cycles, briefly characterized as "short" (biological) and "long" (geological) (Schlesinger 1991). The carbon-oxygen cycle is a combination of the two types.

The cycles differ markedly in terms of the physical and chemical form of the major inorganic reservoir, and the mechanisms for transfer from one reservoir to another. When an element passes from the bio-unavailable inorganic reservoir to the bio-available reservoir it is said to be *mobilized*. In the case of nitrogen, mobilization essentially consists of splitting the di-nitrogen N_2 molecule, which is quite stable.[9] When a nutrient moves in the reverse direction it is *sequestrated*. (In the case of carbon, the term is also used in connection with the accumulation of carbon or carbon dioxide in a reservoir from which CO_2 cannot reach the atmosphere). A nutrient can pass from an organic reservoir (e.g. dead plant material) to an inorganic reservoir by *decomposition* or (in the case of nitrogen) bacterial *denitrification*.

Apart from reservoirs, it is important to identify carriers. For carbon, there are two main gaseous carriers: carbon dioxide (CO_2) and methane (CH_4). The water soluble form of carbon is bicarbonate (HCO_3^-), which is also a carrier. For nitrogen

---

[9] For instance, the oxidation reaction $N_2 + O_2 \rightarrow 2NO$ is highly endothermic. It does not occur spontaneously at normal temperatures and pressures (otherwise the atmosphere could not contain both molecular oxygen and molecular nitrogen). Once NO is formed, however, further oxidation reactions do occur spontaneously— albeit slowly—until the most oxidized form of nitrogen (N_2O_5) is reached. Dissolved in water, this is nitric acid (HNO_3). In thermodynamic equilibrium, oxygen and nitrogen would be combined in this form.

the main gaseous carriers are nitrogen oxides (NO_X) and ammonia (NH_3); most ammonium compounds and nitrates are soluble in water, but the main aqueous form is nitrate (NO_3^-). In the case of sulfur, the inorganic gaseous media are hydrogen sulfide (H_2S), carbonyl sulfide (COS), carbon disulfide (CS_2) and dimethyl-sulfide ($DMS = (CH_3)_2S$) and sulfur dioxide (SO_2); the main aqueous form is sulfate (SO_4^{2-}), but DMS is also quite soluble in water. The only phosphorus carriers are phosphate ions PO_4^{2-}. The aqueous form is phosphoric acid (H_3PO_4).

The second much faster sub-cycle is an exchange between the bio-available nutrient reservoir and living organisms themselves (which constitute a secondary reservoir of the nutrient). The reverse transfer, by decomposition or mineralization, has already been mentioned. In the case of carbon the conversion of CO_2 to its primary biological form (ribose, a kind of sugar) is accomplished by photosynthetic organisms, from algae to trees. Most of the other transfers in this sub-cycle are carried out by specialized bacteria or by enzymes within cells. For instance, nitrogen in living (or dead) organisms is normally in the amine group ($-NH_2$). A few free-living bacteria—notably *Rhizobium*—and some anaerobic cyanobacteria and yeasts have the ability to split the di-nitrogen molecule and *fix* nitrogen in a bio-available form.

It is tempting to compare the preindustrial nutrient "cycles" (C, N, S) with the current fluxes, on the assumption that these three cycles were, or very nearly were, at a steady-state in preindustrial times. Steady-state, in this context, means that each of the major reservoirs remains essentially constant or fluctuates within narrow limits. Inputs and outputs of each species to each reservoir must exactly balance (on average) any chemical transformations from one species to another in a steady-state condition. Even by this straightforward test, as will be seen, none of the nutrient cycles is in steady-state now. However, it seems that the nutrient cycles are seldom in balance for long, if ever, thanks to climate change, great extinctions, and geological processes such as continental drift, uplifts, episodic vulcanism and ice ages that occur over geological time scales. Thus, it can be quite misleading to compare the current state of imbalance with a hypothetical balance condition.

6.6   The Carbon-Oxygen Cycle

The main carbon-oxygen cycle (sometimes called the fast oxygen cycle) reflects the interaction between the biosphere and the atmosphere. Atmospheric CO_2 is biologically transformed by photosynthesis into sugars and cellulose, as pointed out in Chap. 5. Oxygen is produced by photosynthesis, by producing carbohydrate monomers from carbon dioxide and water, via the endothermic reaction:

$$nCO_2 + nH_2O \rightarrow (CH_2O)_n + nO_2$$

where a photon provides the activating exergy. The reverse flow is due to respiration of living organisms and decay of dead organisms. The exothermic reaction equation is

$$(CH_2O)_n + nO_2 \rightarrow nCO_2 + nH_2O$$

with heat given off in the process. As pointed out in Sect. 5.7 the biological purpose of respiration is to provide exergy for other cellular functions by "burning" glucose more efficiently than the primitive cells, depending on fermentation, could do.

The atmosphere is the only reservoir of molecular oxygen. It contains about 1.2×10^6 gigatons (GT) of oxygen (Jacob 2002). A molecule of oxygen produced by photosynthesis "lives" in the atmosphere about 5000 years (on average) before being re-converted to carbon-dioxide.

The "fast" carbon reservoirs are the atmosphere itself, plus the biosphere, the soil and oceans. The atmosphere contains 790 GT of carbon as CO_2 while another 700 GT of organic carbon is embodied in the biosphere and about 3000 GT of organic carbon is embodied in dead plants, soil humus and organic detritus in the oceans. The major reservoirs of carbon at present are estimated to be as shown in Table 6.1.

Aerobic respiration is the reverse of carbon fixation. On a longer time scale, CO_2 fertilization of terrestrial vegetation is a factor (along with temperature) tending to maintain atmospheric CO_2 at a constant level. Rising atmospheric carbon dioxide directly enhances the rate of photosynthesis, other factors being equal.[10] It also causes climate warming, and increased rainfall, both of which further enhance the rate of plant growth, subject to the availability of other nutrients, water, etc.

However, the organic carbon cycle cannot be understood in terms of the biochemistry of photosynthesis alone. Nor does all sedimentary carbon exist in the form of carbonates. There is a significant reservoir of reduced organic carbon (kerogen), buried in sediments, some of which has been aggregated by geological processes and transformed by heat or biological activity to form coal, petroleum and (possibly) natural gas.[11] Of course, it is the geologically concentrated sedimentary hydrocarbons that constitute our fossil fuel resources and which are currently being reconverted to CO_2 by combustion. Methane (CH_4) has its own sub-cycle. In any anaerobic environment[12]—including the guts of cellulose-ingesting animals such as ungulates and termites—organic carbon is broken down

---

[10] In the more general case, the rate of photosynthesis can be expected to depend on the concentrations of all the essential nutrients—especially C, N, S, P—in biologically available form.

[11] The origin of natural gas is currently in doubt. For a long time it was assumed that natural gas was entirely biogenic and associated mainly with petroleum. Now it is known that gas deposits are much more widely distributed than petroleum deposits. It has been suggested by several astronomers that much of the hydrogen in the Earth's crust may have originated from the sun (via the "solar wind" proton bombardment).

[12] To be more precise, an environment lacking nitrates, manganese oxide, iron oxides or sulfates. Recall the earlier discussion of "redox potential" and bacterial sources of oxygen for metabolism.

Table 6.1 Carbon pools in the major reservoirs on Earth

Pool	Quantity GT (gigatons)
Atmosphere	720
Oceans (total)	38,400
Total inorganic	37,400
Total organic	1000
Surface layer	670
Deep layer	36,730
Lithosphere	
Sedimentary carbonates	>60,000,000
Kerogens	15,000,000
Terrestrial biosphere (total)	2000
Living biomass	600–1000
Dead biomass	1200
Aquatic biosphere	1–2
Fossil fuels (total)	4130
Coal	3510
Oil	230
Gas	140
Other (peat)	250

by bacteria. In sediments, these anaerobic bacteria produce "swamp gas" (while the organic nitrogen and sulfur are reduced to ammonia and hydrogen sulfide). In the stomachs and intestines of grazing animals such as cattle and sheep, or in termites, the methane is eliminated by belching or with solid excreta. As noted previously, the existence of free oxygen in the atmosphere is due to the fact that so much organic carbon has been sequestered over eons by burial in silt. Nevertheless, at least half of all buried organic carbon is recycled to the atmosphere by anaerobic methanation. The methane in the atmosphere is gradually oxidized, via many steps, to CO_2. Methane is not recycled biologically, as such. However, at present methane is being emitted to the atmosphere faster than it is being removed.

Researchers dispute the total atmospheric reservoir, since the residence time is difficult to measure precisely. Global atmospheric concentrations are increasing at annual rate of about 1 % per annum. Annual sources currently exceed sinks by around 10 % (Jacob 2002). This is a matter of concern, since methane is a very potent greenhouse gas, much more so than CO_2. Clearly, any mechanism that increases the rate of methane production by anaerobic bacteria will have a pronounced impact on climate, *ceteris paribus*. For instance, the expansion of wet rice cultivation in the orient, together with the spread of cattle and sheep husbandry worldwide, constitute a significant anthropogenic interference in the natural methane cycle—and contribute significantly to the emission of GHGs.

There is a "slow" carbon-oxygen cycle involving mineral (silicate) weathering. The major inorganic reservoir of carbon is sedimentary carbonate rocks, such as limestone or calcite ($CaCO_3$) and dolomite ($CaMg(CO_3)_2$). This reservoir contains more than 10^5 times more carbon than the atmosphere and the biosphere together.

These reservoirs participate in a "slow" (inorganic) cycle, in which, carbon dioxide from the atmosphere is taken up by the weathering of silicate rocks, driven (as mentioned earlier) by the hydrological cycle. This process occurs in a series of reactions that can be summarized by

$$CaSiO_3 + CO_2 + 2H_2O \rightarrow Ca(OH)_2 + SiO_2 + H_2CO_3$$

The calcium, magnesium and bicarbonate ions, as well as the dissolved silica, in the surface waters are carried to the oceans. There the dissolved calcium, silica, carbonate and bicarbonate are either precipitated inorganically or picked up by marine organisms and incorporated into their shells as calcium carbonate and opal.[13] The calcium and carbonate part of the marine system can be summarized as:

$$Ca(OH)_2 + H_2CO_3 \rightarrow CaCO_3 + 2H_2O$$

In due course the organic precipitates and shells drift down to the ocean floor as sediments. Sediments contain around 10^7 Pg of organic C which is gradually compressed eventually being converted by heat and pressure into shale, limestone, chalk and quartz. The sum of the two reactions together is

$$CaSiO_3 + CO_2 \rightarrow CaCO_3 + SiO_2$$

The observed rate of oceanic calcium carbonate deposition would use up all the carbon dioxide in the oceans in about 400,000 years. If that happens, the Earth may experience another "snowball Earth" period, similar to the one that seems to have occurred several times in the past.

However, just as volcanoes eject hot lava and gases onto the surface from under the Earth's crust, there are subduction zones (mostly under the oceans) that carry sediments back into the reducing magmatic layer. Another chemical reaction occurs at the high pressures and temperatures of that region. This reaction reverses the weathering reaction and reconverts sedimentary calcium and/or magnesium carbonate rocks (mixed with quartz) into calcium or magnesium silicate, releasing gaseous CO_2 that is vented through volcanic eruptions or hot springs. Weathering rates are relatively easier to measure (Holland 1978; Berner et al. 1983) compared to outgassing rates (Berner 1990; Gerlach 1991; Kasting and Walker 1992). But insofar as the data is available, the two rates (CO_2 uptake and emission) appear to agree within a factor of two (Kasting and Walker 1992). The fact that agreement is not closer is an indication of the fact that much remains to be learned about the details of these (and most other) bio-geo-chemical processes (e.g. Smil 2001).

In principle, there is a somewhat crude geological mechanism that would tend to keep the silicate weathering rate roughly equal to the volcanic out-gassing rate over very long time periods. A buildup of CO_2 in the atmosphere would lead to

---

[13] Opal is a form of silica used for the shells of diatoms.

greenhouse warming. This increases the rate of evaporation (and precipitation) of water, thus accelerating the weathering process, which removes CO_2 from the atmosphere. The silicate weathering rate is directly dependent on climate conditions, including the precipitation rate and the surface exposure. This would eventually halt the temperature rise. If the atmospheric CO_2 level rises, due to abnormal volcanic activity, there will be an increase in the rate of weathering and CO_2 uptake. Conversely, if the CO_2 level is dropping, so will the temperature and the weathering rate.[14]

In fact, it is asserted by some geologists that the inorganic carbon cycle is sufficient to explain the major features of paleo-climatic history (Holland 1978; Berner et al. 1983). Berner's model failed to explain one major feature, however: the fact that the Earth's climate has actually cooled significantly during the last 50 million years. However, thanks to more recent work there is now a strong presumption that the cause of this cooling is also geological (i.e. tectonic) in origin. In brief, during that period the Indian subcontinent moved north and collided with Asia (actually, about 35 million years ago), forcing the Tibetan plateau to rise.

The rise of this enormous mountain range, in turn, must have sharply increased the rate of CO_2 removal from the atmosphere (Raymo and Ruddiman 1992). In fact, some calculations suggest that the weathering of the Tibetan plateau alone would remove all the CO_2 from the atmosphere in a few hundred thousand years. If that happened, as it seems to have happened 2.5 billion years ago after the biological "invention" of photosynthesis, and before the "invention of respiration, there would be another super-cooled "Snowball Earth".

The geochemical response mechanisms described above are much too slow to account for the strong observed correlation between climate and atmospheric CO_2 levels over much shorter periods. These are part of the "fast" carbon cycle, which is biologically (and now industrially) controlled. The observed seasonal cycle of atmospheric CO_2 is an obvious short-term effect of the biosphere. Photosynthetic activity in the spring and summer in the northern hemisphere normally brings about a measurable reduction in the atmospheric CO_2 concentration.

The carbon cycle as a whole is summarized in Figs. 6.14 and 6.15. The carbon cycle is not now in balance. Whether it was truly balanced in preindustrial times is debatable. However, a preindustrial carbon cycle has been described, where the atmospheric reservoir contained 615 Pg, the terrestrial biosphere contained 730 Pg and the surface ocean biota contained 840 Pg with annual photosynthesis of 60 Pg on land and 60 Pg in the surface of the oceans, balanced by equal reverse flows from respiration and decay (Jacob 2002). In any case, the carbon dioxide level of the atmosphere has been rising sharply for over a century, from a hypothesized

---

[14] If the oceans were to freeze the weathering rate would fall to zero, allowing the atmospheric CO_2 level to rise due to volcanic action. It has been shown that this feedback is sufficient to assure that the oceans would not have been frozen over, even during the Earth's early history when the sun was emitting 30 % less energy than it does today (Walker et al. 1981).

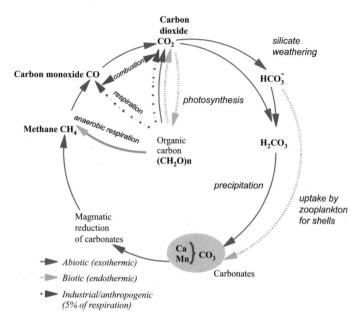

Fig. 6.14 Carbon cycle; chemical transformations

pre-industrial 615 Pg to 790 Pg in the late 1990s. The current level (c. 2015) is several percentage points higher and rising every year.

The Jacob model of the preindustrial carbon "balance" is, by no means, established fact. Smil reports that estimates of the carbon content of late twentieth century terrestrial biomass range between 420 and 840 Pg (as compared to 730 Pg in Jacob's model), whereas a group of Russian biologists have estimated that terrestrial biomass amounted to 1100 Pg 5000 years ago, before human deforestation began (Smil 2001, p. 81). Jacob's implication that the "preindustrial" carbon cycle as of (say) 200 years ago was a steady-state is misleading.

In fact, deforestation and land clearing for agriculture in North America peaked during the nineteenth century. (As it happens, there has been some reforestation in North America during the twentieth century, partly due to agricultural mechanization—no more need to feed horses—and partly due to CO_2 and NOx fertilization—the fact that plants grow faster as the carbon dioxide and nitrogen oxides levels in the atmosphere increases.)

Anthropogenic extraction and combustion of carbonaceous fuels, together with deforestation to clear land for agriculture, have contributed significantly to altering the atmospheric CO_2 balance. About 5 % of total CO_2 emissions from the land to the atmosphere are anthropogenic (Bolin, Doeoes, and Jaeger 1986). The CO_2 concentration is now 400 ppm (2013), 60 % above the preindustrial level (estimated to be 280 ppm). The rate of increase in recent years has been about 0.4 % per year, although it fluctuates. According to ice-core (and other) data, during the quaternary glaciation 2 million years ago, the atmospheric concentration of carbon dioxide was

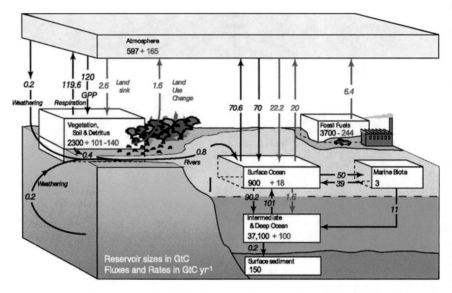

Fig. 6.15 The carbon cycle in the 1990s. *Source*: IPCC, https://www.ipcc.ch/publications_and_data/ar4/wg1/en/fig/figure-7-3-l.png

as low as 180 ppm, while during the Carboniferous era 300 million years ago, it was as high as 7000 ppm.

As of 1990 approximately 5.4 Pg/year of carbon was being converted to CO_2 by combustion processes and transferred from underground reservoirs of reduced carbon to the atmosphere. A further 1.6 Pg/year was attributed to tropical deforestation, for a total of 7 Pg/year (Stern, Young, and Druckman 1992). Roughly half of this excess anthropogenic flux, or 3.5 Pg/year, was known to be accumulating in the atmosphere. At first glance, it would appear that the remainder must be accumulating either in the oceans or in terrestrial biomass. A molecule of oxygen produced by photosynthesis "lives" in the atmosphere about 5000 years (on average) before being re-converted to carbon-dioxide.

Although it was assumed for a while that the ocean must be the ultimate sink; (e.g. Peng et al. 1983), there is increasing evidence that the known atmosphere-ocean transfer mechanisms cannot account for all the carbon disappearance (Tans et al. 1990; Schlesinger 1991; Sundquist 1993). However, simulation models of the carbon cycle, working both "forward" and "backward", have contributed significantly to clarification of the situation.

To summarize, it appears that the most plausible way to balance the fast carbon budget, within historical emissions and parametric uncertainty ranges, is to introduce a terrestrial biospheric "sink" for CO_2—mostly in the northern hemisphere—partially attributable to reforestation thanks to a combination of carbon and nitrogen fertilization (Houghton et al. 1992).

There is physical evidence to support the hypothesis of enhanced plant growth in the northern hemisphere as a sink for atmospheric CO_2. Notwithstanding increased timber and wood-pulp harvesting, the forest biomass of the north temperate zones, North America and northern Europe—and perhaps Russia—is actually *increasing* (Kauppi et al. 1992; Sedjo 1992). This finding was confirmed by the 1994 IPCC *Scientific Assessment* (Schimel et al. 1994). Quantitative estimates now appear to confirm the sufficiency of the N-fertilization hypothesis (Galloway et al. 1995; den Elzen et al. 1995).

A "box model" of the carbon cycle, combining both the chemical and biological branches, has been able to match the history of atmospheric CO_2 concentration since 1850 remarkably well, both before and after 1950. See (Valero Capilla and Valero Delgado 2015, Fig. 10.1; Tomizuka 2009; Valero et al. 2011)

6.7 The Nitrogen Cycle

Nitrogen compounds are extremely soluble In general, vegetation can utilize either soluble nitrates or ammonium compounds, but not elemental nitrogen.[15] Thus, all life depends on nitrogen "fixation". Some of the fixation is attributable to lightning (which induces oxidation) and some by diazotrophic bacteria having a nitro-genase enzyme, that combine atmospheric nitrogen (N_2) with hydrogen to form ammonium ions. Some symbiotic bacteria then convert the ammonium into amino acids used by plants. Other bacteria convert ammonium into other nitrogen compounds such as nitrites and nitrates. Nitrites are toxic to plants but with oxidation they become nitrates, also toxic.

The nitrates are incorporated in a number of important biological chemical and processes, where the oxygen is stripped off and combined with carbon forming carbon dioxide, while the NH_2 ions are embodied in amino acids that combine to form proteins. By the same token, the nitrogen cycle (Figs. 6.16 and 6.17 depends intimately on other denitrifying bacteria that convert the organic nitrogen compounds in dead organisms back into molecular nitrogen.

Because nitrogen is a limiting factor in many agricultural regions, it has been relatively easy to increase output by supplementing natural sources of available nitrogen by the addition of synthetic sources. Smil points out that the amount of reactive nitrogen added to soils now roughly equals the amount fixed by legumes and other nitrogen fixing organisms (Smil 2004). But the chain of transformations resulting in protein for human food is inefficient. One unfortunate consequence is that a lot of soluble nitrogen is lost to groundwater and runoff. For this reason, however, the imbalances in the nitrogen cycle may prove to be extraordinarily difficult to correct—or compensate for—by deliberate human action.

Nitrogen fluxes to the atmosphere (the major reservoir of inorganic nitrogen) are of two kinds. Bacterial de-nitrification from the decay of organic materials, along with de-nitrification of nitrate fertilizers, returns nitrogen to the atmosphere as inert

---

[15] Most bacteria and animals can only utilize organic nitrogen, mainly as amino acids.

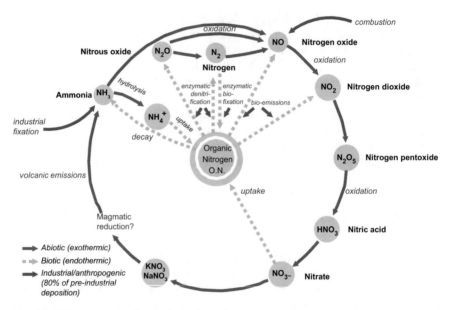

Fig. 6.16 Nitrogen cycle: chemical transformations

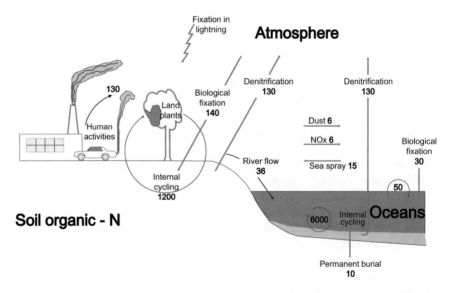

Fig. 6.17 1990s nitrogen cycle; intermediate flux 1012 gN. *Source*: Schlesinger (1991, Fig. 12.1)

N_2 or N_2O. This loss must be compensated by N-fixation. On the other hand, ammonia volatilization from the soil, combustion of organic materials (generating NO_X) and bio-emissions from marine organisms are recycled as "odd" nitrogen. Fluxes from the atmosphere to the land and ocean surface include deposition of nitrate (NO_3) ions (from "acid rain") and ammonium (NH_4) ions. Fluxes to the biosphere include uptake of soluble nitrates and ammonium compounds, recycling of organic detritus (e.g. manure), and bio-fixation by microorganisms. None of these fluxes, except the application of synthetic fertilizers, is well quantified.

Synthetic fertilizers from industrial nitrogen fixation (mainly as ammonia, NH_3) became important only in the twentieth century. Current global annual ammonia-based fertilizer production is 160 million metric tons, or 130 Tg/year of fixed N. Of the 160 Tg of ammonia produced worldwide each year, 16 Tg or so is used to manufacture other industrial chemicals, notably explosives, pesticides, plastics (nylon), etc. The rest is made into fertilizer chemicals, starting with urea. These products eventually become wastes and are disposed of, either in landfills or via waterways. All of this nitrogen eventually finds its way back into the global cycle. China is now the world's biggest producer and consumer, having reached that pinnacle by 1995, from a fairly low level, starting in 1960. China now applies fertilizer at the rate of 190 kg-N/ha, compared to just 60 kg-N/ha for the US. The most intensively fertilized areas receive more than 300 kg-N/ha, enabling a hectare of land to feed 30 people on a vegetarian diet or 20 people on a high protein diet. It is this intensive use that has enabled China to feed itself.

As mentioned earlier, the only global reservoir of nitrogen is the atmosphere, which is 78 % elemental nitrogen gas (N_2). The total stock of molecular nitrogen (di-nitrogen) in the atmosphere is estimated to be 3.8 billion Tg. Di-nitrogen (N_2) and nitrous oxide (N_2O) are not biologically available, because of the tremendous amount of energy needed to break the triple bond holding the two N atoms together (about double the energy required to break the oxygen, or O_2, bond). There are only four sources of biologically available ("odd") nitrogen compounds. These are (1) biological fixation, (2) atmospheric electrical discharges (lightning), (3) high temperature combustion and (4) industrial processes for producing synthetic ammonia.

As already mentioned, only a few very specialized bacteria and actinomycetes (yeasts) can utilize (i.e., "fix") elemental di-nitrogen. There are some 25 genera of free-living or symbiotic bacteria. The most important is *Rhizobium*, which attaches itself to the roots of legumes, such as alfalfa. In addition, there are 60 genera of anaerobic cyanobacteria, such as *anabaena* (leftovers from the early evolutionary history of Earth), and 15 genera of actinomycetes (most prominently, *Frankia*). The range of estimates for biofixation in the literature is from 100 to 170 Tg/year on land and from 10 to 110 Tg/year in the oceans. Overall, biological N-fixation, from all sources, may be as little as 110 Tg/year and as much as 280 Tg/year. The most recent estimates are that 140 Tg is fixed by terrestrial vegetation, plus 30 Tg from marine phytoplankton (Smil 2001).

The rate of natural atmospheric nitrogen fixation as NO_X by electrical discharges is even less accurately known than the bio-fixation rate. Estimates vary from 1.2 to 220 Tg/year, although the latter figure now looks much too high. The most plausible estimate is 20–30 Tg/year (Smil 2001, p. 172). Nitrogen oxides (NO_X) are also produced by high temperature combustion processes. Anthropogenic activities, mostly automobiles and trucks and electric power generating plants, currently generate around 40 Tg/year of NO_X There is a further contribution (estimated to be around 12 Tg/year) from natural forest fires. Finally, of the order 5 Tg/year of ammonia (NH_3) is also discharges by volcanoes and fumaroles, on average, although this flux can vary a lot from year to year.

Evidently, known anthropogenic inputs of fixed nitrogen are already comparable in quantity to (and may even be significantly larger than) estimated natural fixation rates. Assuming the nitrogen cycle to have been balanced in pre-industrial times (an assumption that is probably unjustified) it would follow that inputs and outputs to some reservoirs are now out of balance by a large factor. Of course, it would be very difficult to detect any changes in total atmospheric nitrogen content (i.e. pressure) over any short period of time. However, it is known that nitrous oxide (N_2O) has increased in recent years, from a level at approximately 300 ppb in 1978 (Weiss 1981). If we extrapolate from the graph on p. 135 of Smil's book, it is probably close to 330 ppb today (Smil 2001).

Nitrous oxide is not oxidized in the troposphere. In the stratosphere it is photolyzed yielding N_2 and O, or it is oxidized by ozone to NO_X. In fact, oxidation of nitrous oxide is the major source of stratospheric NO_X. The disappearance rate (of N_2O) by these two mechanisms in combination is estimated to be 10 Tg/year (Weiss 1981; Liu and Cicerone 1984; McElroy and Wofsy 1986). This process has aroused great interest in recent years because of the discovery by Crutzen and others that this set of processes governs the stratospheric ozone level, at least in the absence of chlorine compounds (which also catalytically destroy ozone) (Crutzen 1970, 1974).

N_2O is a co-product (with N_2) of natural de-nitrification by anaerobic bacteria and other microorganisms. Aggregated data on the N_2O/N_2 ratio is scarce. However, such data as does exist (for fertilized land) implies that the ratio of N_2 to N_2O production on land is probably in the range 10–20, with a most likely value of about 16:1 (Council for Agricultural Science and Technology 1976). In other words, N_2O is about 1/16 of the total terrestrial de-nitrification flux. There is no *a priori* reason to assume this ratio should not hold true for marine conditions or pre-industrial times, although it might actually vary quite a bit. It would follow from this assumption that the current apparent N_2O flux of 13 Tg/year should be accompanied by a corresponding N_2 flux of about $16 \times 13 = 208$ Tg/year. (This compares with $2 \times 130 = 260$ Tg/year in Fig. 6.17). However, it must be emphasized that the numbers are subject to considerable uncertainty).

It now appears that de-nitrification of nitrate fertilizers accounts for 0.7 Tg/year of N_2O emissions at present. According to one source, approximately 0.3 % of fertilizer nitrogen is converted to N_2O (Galbally 1985). A calculation by Crutzen sets the figure at 0.4 %, which would correspond to N_2O emissions of 0.35 Tg/year at current fertilizer production levels.[16] An industrial source of N_2O is the

---

[16] Denitrification bacteria reduce nitrates (NO_3) to obtain oxygen for metabolic purposes. They do not metabolize ammonia. Thus the denitrification flux from fertilizers depends somewhat on the chemical form in which it is applied. The N_2O/N_2 proportion depends on local factors, such as

production of adipic acid, an intermediate in nylon manufacture (Thiemens and Trogler 1991). This source could theoretically account for as much as 0.4 Tg/year or 10 % of the annual increase, in the absence of any emissions controls. However, the actual contribution from this source is probably much less. One other possible source of N_2O is explosives. Virtually all explosives are manufactured from nitrogenated compounds (such as nitrocellulose, ammonium nitrate, trinitro-glycerin, and various amines). According to simulation calculations, under conditions of rapid oxidation and decomposition up to 9 % of the nitrogen in explosives may end up as nitrous oxide.

De-nitrification is the complementary process to nitrogen fixation (as utilized by plants). Hence, the terrestrial contribution to de-nitrification must have increased in rough proportion to overall terrestrial and atmospheric nitrogen fixation, taking into account both natural and anthropogenic sources. On this basis, preindustrial natural fixation (approx. 140 Tg/year) has been increased by anthropogenic contributions, taking into account intensive cultivation of legumes, of the same order of magnitude. In other words, human activity has doubled the amount of biologically available (reactive) nitrogen being produced each year.

It is tempting to assume that global de-nitrification should increase proportionally, along with the percentage increase in nitrous oxide (N_2O) emissions, since preindustrial times. This argument is not affected by uncertainties in the $N_2 : N_2O$ ratio. On this basis, it would follow that the overall rate of de-nitrification—including N_2O emissions—could have increased by over 50 % in little more than a century. At first sight, this hypothesis seems plausible. Unquestionably, global agricultural activity has increased sharply over the past two centuries, both in scope and intensity. The nitrate content of riverine runoff from land to oceans has increased sharply. At the same time, the organic (humus) content of most agricultural soils has declined. This would seem to be consistent with the notion of accelerated de-nitrification.

However, there is a problem. The declining organic content of soils is mainly due to plowing and exposure of organic humus to oxygen in the air. Oxidation is mainly responsible for the loss of organic material in soils. But increased exposure to oxygen would, *ceteris paribus*, probably tend to *decrease* the rate of de-nitrification (despite the fact that some N_2O and NO are apparently produced in aerobic soils). It must be remembered that de-nitrification is essentially a process whereby anaerobic bacteria "steal" oxygen from nitrates in the absence of molecular oxygen (air). Thus, the only major agricultural activity that would plausibly result in increased de-nitrification is wet rice cultivation with nitrate fertilizers. On the other hand, the drainage of wetlands would tend to have the opposite effect,

---

carbon content of the soil, acidity and dissolved oxygen. It must be acknowledged that the combined uncertainties are quite large. Thus, for instance, a recent US study sets the N_2O emissions from fertilizer at 1.5 Tg/year, as compared to only 1 Tg/year from fossil fuel combustion. Other estimates in the literature range from 0.01 to 2.2 Tg/year (Watson et al. 1992).

decreasing de-nitrification. In sum, current evidence suggests that while de-nitrification has also increased, it does not keep pace with nitrogen fixation.

The alternative to the hypothesis of compensating global de-nitrification is that reactive nitrogen is now accumulating. For one thing, global nitrogen fertilization (from acid rain and ammonium sulfate deposition) has increased the reservoir of nitrogen in biomes like grasslands and forests that are not cultivated. This explanation would be qualitatively consistent with the observations of increased forest biomass in the northern hemisphere mentioned previously: (e.g. Kauppi et al. 1992; Sedjo 1992). This explanation is now preferred because it also simultaneously provides a satisfactory explanation of the "missing carbon" problem that worried people a few years ago, as mentioned previously.

Anthropogenic nitrogen fixation from all sources—especially fertilizer use and fossil fuel combustion—is certainly increasing quite rapidly. Not all of this excess is immediately de-nitrified to N. It is likely to double again within a few decades. For instance, one group has estimated that the anthropogenic fixation rate will increase from 140 Tg in 1990 to 230 Tg by 2020, with no end in sight (Galloway et al. 1995). The consequences are very hard to predict; certainly they vary from one reservoir to another. One predictable consequence of nitrogen fertilization will be a buildup of toxic and carcinogenic nitrates and nitrites in ground waters. This is already occurring in many agricultural areas. Increased forest and pasture growth rates is another likely consequence already mentioned. But along with the gross fertilization effect, there is a tendency to reduced bio-diversity. Regions where nitrogen availability has been the limiting factor for biological productivity are likely to shrink, or even disappear, to be replaced by regions where other plant nutrients are the limiting factor. This shift could lead to major changes in species composition for both plants and animals. A specific consequence of increased NO_x emissions is predictable: NO_x affects the oxidizing capacity of the atmosphere and, indirectly, increased the tropospheric ozone concentration. This has well known adverse consequences on cereal crop productivity and on human health, especially for people with respiratory problems.

The modern atmosphere reflects evolutionary changes in the *nitrogen cycle*. Early proto-cells probably obtained their nitrogen (an essential component of all amino acids and proteins) from dissolved ammonium (NH_4) ions or an iron-ammonium complex. However, as the soluble iron in the oceans was oxidized and atmospheric oxygen began to accumulate, free ammonia was more and more rapidly oxidized. One group of oxidation reactions yields oxides of nitrogen (NO_x), most of which eventually dissolve in water and form nitrous and nitric acids. These, in turn, can be metabolized by micro-organisms to form nitrites and nitrates.

However other reaction paths for the direct oxidation of ammonia can lead to molecular nitrogen (N_2) and water vapor.[17] Molecular nitrogen is extremely stable and quite unreactive—hence unavailable to plants—except when endothermic oxidation reactions are caused by lightning. Consequently the atmosphere is a

---

[17] The reaction is $4NH_3 + 3O_2 = 2 N_2 + 6H_2O$.

nitrogen "sink". In other words, the ultra-stable molecular nitrogen keeps accumulating. It seems very plausible that some early proto-bacteria were able to obtain metabolic energy directly from the oxidation of ammonia, reducing it to hydrogen gas and molecular nitrogen.

So-called "denitrifying" (more accurately, ammonia metabolizing) bacteria descended from these organisms still abound in the soil (where they are not in contact with oxygen) and play a major role in closing the nitrogen cycle. In fact, the existence of free nitrogen in the Earth's atmosphere may be attributable to these primitive but hypothetical ancestral organisms. As the rate of input of free ammonia to the atmosphere from volcanoes gradually decreased, while the need for reactive nitrogen by living organisms increased, a mechanism was needed to recapture some of the atmospheric molecular nitrogen in soluble and biologically usable form. This ability was acquired, eventually, by specialized micro-organisms blessed with an enzyme called *nitrogenase*. This capability if possessed by nearly 100 genera of cyanobacteria, symbiotic actinomycetes or free-living bacteria, but by for the most important are two genera, *rhizobium* and *bradyrhizobium* that live symbiotically on the root structures of certain plants (the legumes). The details need not concern us here.

However before moving on, we need to point out that natural processes (lightning, decay organisms, and nitrifying bacteria) are not sufficient to provide all the nitrogen needed by the global human population (over 7 billion) that exists today. A few years ago, Vaclav Smil wrote, "Depending on the diet we would be willing to accept, our numbers would have to shrink by 2–3 billion in a world devoid of synthetic nitrogenous fertilizers. This dependence will only increase in the future" (Smil 1997, 2000). When he wrote, human numbers were about 6 billion, which implies that synthetic ammonia already accounts for up to half of all the soil nitrogen needed to support the human population. It happens that nitrogen synthesis is also extremely energy intensive. Synthetic ammonia production consumes, on average, 40 GJ per metric ton. Urea, the most common and transportable form of nitrogen fertilizer, requires an additional 25 GJ/t. Global production of synthetic ammonia, mostly for agriculture, is now over 100 million tonnes per year.

6.8   The Sulfur Cycle

The global sulfur cycle resembles the nitrogen cycle thermodynamically, insofar as reduced forms of sulfur (S, H_2S) are gradually oxidized by atmospheric oxygen, ending in sulfur oxides (SO_2, SO_3) and finally sulfuric acid (H_2SO_4). See Fig. 6.18. Sulfate ions (SO_4^-) are eventually deposited on land or in the oceans in wet or dry form (e.g. as ammonium sulfate). The reverse part of the cycle, which converts sulfur back to reduced states of higher thermodynamic potential, is accomplished either by biological activity or by high temperature magmatic reactions in the Earth's mantle.

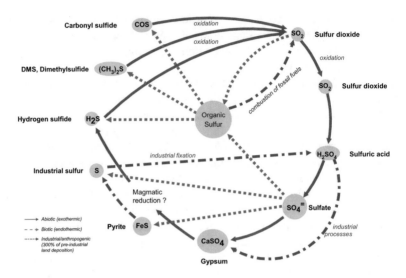

Fig. 6.18 Sulfur cycle; chemical transformations

From another perspective, the S-cycle consists of transformations of insoluble and biologically unavailable forms of sulfur (notably sulfides—pyrites—and calcium and magnesium sulfates) to available forms. These are utilized by organisms and finally returned once again to unavailable forms. From this perspective, the cycle can also be seen as a complex set of transfers between air, land and sea, as shown in Fig. 6.19. Assuming the preindustrial version of the cycle was really balanced (which is open to question) the controlling rate, or "bottleneck" in the system must have been the rate at which insoluble sulfides or sulfates were deposited in oceanic sediments. In the very long run, the average deposition rate must have been equal to the rate at which sulfur was remobilized by preindustrial geochemical processes, with or without biological assistance. It must also equal the preindustrial rate of net deposition of sulfur compounds on the ocean surface, plus the preindustrial runoff from rivers, abrasion of shores, etc.

Various estimates of sulfur fluxes are available in the literature. See Fig. 6.19 for an example. Roughly, the pre-industrial inputs to the land surface must have been about 26 Tg/year, as compared to 84 Tg/year from atmospheric deposition (c. 1980) and a further 28 Tg/year as fertilizer. In short, the sulfur flux to land has more than quadrupled since the beginning of industrialization. It is likely that river runoff has doubled, e.g. from 72 Tg/year preindustrial to more than 213 Tg/year currently. It is clear that the global sulfur cycle is now extremely unbalanced. Inputs to oceans appear to exceed deposition to the ocean bottom by as much as 100 Tg/year (Schlesinger 1991). It must be acknowledged that there is no known or obvious reason to worry unduly about this aspect of the anthropogenic perturbation of the sulfur cycle *per se*, based on current knowledge of downstream effects, except for one problem. Clearly, the oxidation of large amounts of reduced sulfur will continue to acidify the soils and the ocean. Deposition of sulfur oxides—sulfite/

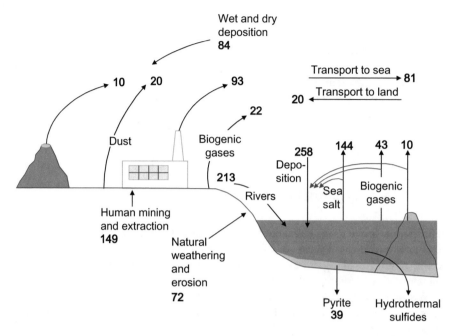

Fig. 6.19   1990s sulfur cycle; intermediate fluxes 1012 gS. *Source*: Schlesinger (1991, Fig. 13.1)

sulfate (SO₃/SO₄)—and nitrate (NO₃) onto the land or water surface does fertilize biomass growth, but it is also "acid rain". This causes measurable changes in the pH of rain-water, and fresh water lakes and streams. Acidification of the oceans is slower, but probably more dangerous in the long run. Ocean pH is currently much higher than it was in the early history of our planet.

Atmospheric haze, consisting mostly of sulfate aerosols in the micron size range, has increased by orders of magnitude over some land areas (China comes to mind). Sulfate haze is due to SO_2 emissions from the combustion of fossil fuels. This was a topic of major concern in the US during the 1980s, due to the association of sulfate particulates with acidification. However, it is not clear whether sulfate haze has increased significantly over the oceans, partly because it is unclear whether oceanic sulfates are attributable to terrestrial industry or marine biology. There is some evidence that marine phytoplankton emit dimethyl-sulfide (DMS), which subsequently oxidizes to sulfate aerosols. The haze has a net cooling effect. The sulfate haze phenomenon is apparently sufficient to measurably increase the Earth's albedo (i.e. it's reflectivity to visible radiation) and, possibly, compensate partially for greenhouse warming in recent decades (Wigley 1989; Taylor and Penner 1994). It is now incorporated in GCMs, where it largely accounts for the fact that climate warming has not occurred as rapidly as the earlier models predicted. On the other hand, sulfate haze is proportional to the current rate of sulfate emissions, whereas the greenhouse phenomenon is a function of the atmospheric concentration, or total buildup, of greenhouse gases. Thus, the greenhouse warming and sulfate cooling cannot continue to cancel each other indefinitely.

Sulfate particulates play a different and much subtler role in climate control, however, since micro-particulates also act as cloud condensation nuclei (CCNs). These nuclei are known to affect cloud properties, and hence to modify their absorption and reflection of radiation of different wavelengths. But the quantitative aspects and even the aggregate magnitude of these effects are not yet well understood.

6.9   The Phosphorus Cycle

Phosphorus is one of the elements, together with water, carbon, nitrogen and sulfur, that is essential to all forms of life. Phosphorus constitutes 0.12 % of the Earth's crust, overall, making it the 11th most common element (Krauss et al. 1984). However, it constitutes only 0.0655 % of the Earth's continental crust, according to the latest compilation (Rudnick and Gao 2004). This discrepancy means that most of the phosphorus (by far) is dissolved in the oceans, or in sediments under the oceans.

It is found in about 200 minerals, but most of them are exceedingly rare. The most common mineral (by far) is phosphate rock $Ca_3(PO_4)_2$ with a crustal concentration of 0.279 %, followed by fluoro-apatite $Ca_5(PO_4)_3(OH)_{0.33}F_{0.33}Cl_{0.33}$ with a concentration of 0.0403 % adding up to 0.319 %. It is relevant to what follows that the apatites are relatively insoluble in alkaline or neutral water, but increasingly soluble in acid water (as the pH declines).

Phosphate rock is sedimentary rock, while the fluoro-apatite is of igneous origin. There is data on five other igneous phosphorus minerals. Francolite $[Ca_5(PO_4)_{2.63}(CO_3)_{0.5}F_{1.11}]$ has a crustal concentration of 0.00435, ten times less than fluoro-apatite. The next one is Amblygonite $[Li_{0.75}Na_{0.25}Al(PO_4)F_{0.75}(OH)_{0.25}]$ with a concentration of 0.00195 %%. The last three (Rhabdofane, Metatorbenite and Vivianite) are much rarer and can be neglected. According to the US Geological Survey, almost all (95 %) of the phosphorus in the Earth's crust is now embodied in the two apatite minerals, of which 87 % is sedimentary and of biological origin (Krauss et al. 1984).

All of that sedimentary phosphorus was originally extracted from igneous rock on land, exposed to weathering (by carbonic acid from rain). The carbon dioxide concentration in the atmosphere was much stronger when the Earth was young, whence the rain must have been much more acid than it is today. The reaction of carbonic acid with the igneous rock yields insoluble oxides (e.g. SiO_2) and carbonates ($CaCO_3$) as well as hydroxides and (barely) soluble phosphorus pentoxide (P_2O_5). Some soluble phosphorus (from humus) still remains in soil where it is now accessible to plants, but most of it was carried to the oceans long ago as phosphoric acid (H_3PO_4) via runoff.

A more recent theory suggests that a reduced phosphorus compound known as *schreibersite* $(Fe,Ni)_3P$ arrived on meteorites during the late heavy bombardment. This may have been the vehicle by which a much more soluble phosphorus compound (phosphite HPO_3^{2-}) got into the oceans and subsequently into organic compounds essential to life (Pasek 2008).

At some stage during the first billion years there was an accumulation of organic molecules in the oceans which reacted with that phosphoric acid. Several important organic molecules, notably the nucleic acids (DNA, RNA) and (ADP/ATP) incorporate phosphorus. Adenosine triphosphate (ATP) consists of a 5-carbon sugar (ribose) linked to adenine and 3 phosphate (PO_4) groups. Moreover, tricalcium phosphate $Ca_3(PO_4)_2$ is the main hard material in vertebrate bones and teeth.

Phosphorus increasingly concentrated in the food chain; the higher an organism is on the ladder, the greater the phosphorus requirement. The dried bodies of protozoans contain about 0.6 % P_2O_5, but the body of a cartilaginous fish (like the skate) contains 0.92 % P_2O_5, the bodies of small crustaceans (with shells) range from 1.4 to 2.4 % and the bodies of bony fish average 1.5 %. We humans average a little over 2 % P_2O_5. About 80 % of that is in the form of bones and teeth (true for all vertebrates); the remainder is embodied in DNA and in the energy carriers ADP and ATP.

According to one calculation, P_2O_5 now constitutes about 0.045 % of the mass of river flows, but less than 0.005 % of mass of the oceans. This is noteworthy because the concentration of dissolved salts (mainly sodium, potassium and magnesium chlorides) in the oceans is much higher than the concentration of those salts in the rivers. That is because those chlorides are accumulating in the oceans and are not utilized extensively in biomass, nor are they transferred back to the land by natural processes. Yet, the concentration of P_2O_5 in the oceans is much *lower* than the concentration in rivers: P_2O_5 accounts for less than 0.005 % of the ocean (0.18 % of the dissolved salts in the ocean).

The most plausible explanation is that most of the phosphorus that moved into the oceans was quickly taken up by living organisms and subsequently deposited in sediments and mineralized as apatite. This would result in an accumulation on the ocean floor of 0.2 mm per year (Blackwelder 1916) cited by Lotka and Alfred (1925). (That estimate was based on data collected prior to the dramatic increase in fertilizer use that has occurred since the war.) But even at the low deposition rate of 0.2 mm/year the accumulation in a billion years would be 200 km, which is clearly out of the question. Hence the true deposition rate must have been much lower, probably because there is more recycling within the ocean than geologists thought.

Since most of the phosphorus in the Earth's crust is in solution or is embodied in that mineral, it follows that it—like the oxygen in the atmosphere—must be of biological origin. If that logic holds water (so to speak) it suggests that *the biomass of the Earth is actually phosphorus limited.*

As I noted earlier, the only significant mechanism for phosphorus recycling to terrestrial life is geological. (A small amount is deposited on land in guano). This is because there is no phosphorus in the atmosphere and no gaseous form of phosphorus that could facilitate transfer from marine to terrestrial reservoirs. Phosphine gas (H_3P) is the only plausible possibility, but it oxidizes much too rapidly in the oxygen atmosphere (not to mention being extremely toxic). Consequently, the terrestrial part of the phosphorus cycle is only "closed" (incompletely) by means of geological processes, plus reverse flows from the oceans to the land in the excreta from seabirds (guano) and anadromous fish (e.g. salmon) that return to their spawning place and are consumed by terrestrial predators like bears.

However there must be significant recycling within the oceans. Calcium phosphate is relatively insoluble, which is why shellfish (for instance) use it for their shells. But animals do have some ability to uptake mineral phosphorus. It is known that solubility is increased by the presence of certain amino acids. (We know that dogs like to chew on bones.) That inherited behavior is unlikely to be accidental. This enables some biological organisms to "digest" calcium phosphates, enabling phosphorus recycling in the oceans. It is a fact that ocean currents bring phosphates from deep water (where the phosphate content is much higher) back to the surface via upwelling. (Upwelling zones, such as the west coast of South America, are famously productive of both fish and guano). About a quarter of the global fish catch comes from five upwelling zones constituting only 5 % of the sea surface.

There is a possibility that organisms in the deep oceans do recycle some phosphorus. The main evidence for this, however, is inference: the upwelling of deep ocean waters (e.g. off the coast of Peru) is known to be a major source of phosphorus for the fish in the upper layers. The upwelling phenomenon is a consequence of surface winds the push the surface along the shoreline, as shown schematically in Fig. 6.20.

In the very long run, of course, there is some leakage from the marine ecosystem to submarine sediments where phosphates do accumulate. Only when formerly submerged areas of the sea floor are raised by tectonic processes are these phosphate accumulations ("phosphate rock") mobilized once again by erosion. It is for this reason that phosphates have been identified as a planetary boundary, with a suggested upper limit of $10\times$ the pre-industrial runoff rate (Rockström et al. 2009). I think the limit for long-term human survival needs to be set much lower.

Needless to say, anthropogenic activity, notably phosphate fertilizers, has enormously increased the rate of terrestrial phosphate mobilization as well as the rate of phosphate loss, via erosion and runoff, to the oceans. While known reserves of mineable phosphate rock appear adequate for the next few centuries (recent studies by the USGS suggest that phosphate resources are much larger than was thought a few years ago) phosphate loss will eventually be a limiting factor for agriculture as we know it. Since there are no mineral substitutes, the only possibility in the long run must be some combination of changing diet, significantly enhanced recycling from wastes and/or fewer humans.

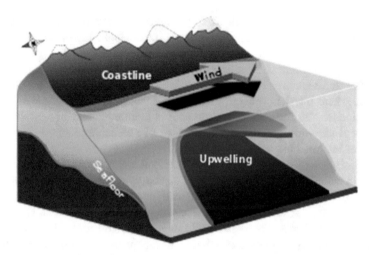

Fig. 6.20 Upwelling brings nutrients (notably P) to the surface. *Source*: https://upload.wikimedia.
org/wikipedia/commons/thumb/7/7f/Upwelling-labels-en.svg/300px-Upwelling-labels-en.svg.
png

6.10   Thermodynamic Dis-equilibrium

One important point needs to be emphasized again with regard to the nutrient
cycles. It is that the atmosphere and lithosphere are emphatically *not* in thermodynamic equilibrium with the oceans. All of the first three cycles discussed above
consist of two distinct branches, a geochemical and a biological branch. In the
geochemical branch reduced forms of carbon, nitrogen and sulfur are gradually
oxidized by a multi-step sequence of processes, approaching the most stable (i.e.,
lowest energy) thermodynamic state. The latter would be one in which buried
carbon and sulfur, and atmospheric nitrogen, combined with oxygen (and water)
to form acids, *viz.*

$$C + O_2 + H_2O \rightarrow H_2CO_3$$

$$H_2O + N_2 + \frac{5}{2}O_2 \rightarrow 2HNO_3$$

$$S + \frac{3}{2}O_2 + H_2O_2 \rightarrow H_2SO_4$$

These acids react with all alkaline species in the environment and eventually the
salts accumulate in the ocean. For instance, if the nitric acid formation reaction,
alone, proceeded to chemical equilibrium it has been calculated that almost all the
oxygen in the atmosphere would be used up and the pH of the ocean would decrease
to 1.5 (Lewis and Randall 1923). More recent equilibrium calculations (below)
have confirmed and refined this worrisome result.

Fortunately, some of these thermodynamically favored oxidation reactions do not occur at significant rates on the Earth's surface—including the atmosphere and the oceans—under current conditions. In the case of carbon and sulfur, this is because biological reduction processes, combined with sedimentation and burial, regenerate and sequester reduced forms (e.g. kerogen and pyrites). In the case of nitrogen, where physical sequestration is not a factor, there are two barriers. The first is kinetic. The formation of nitric acid is thermodynamically favored, to be sure. But it can only proceed by a sequence of reversible reactions involving a number of intermediate oxidation stages.

The first stage ($N_2 + O_2 \rightarrow 2NO$) is quite endothermic. It only occurs in very hot fires (or in the path of a lightning bolt). Thus, the *rate* at which this reaction occurs in nature is very low. Even so, the nitric acid level of the oceans would gradually build up, except for another barrier. There are several enzymatically catalyzed biological denitrification processes that convert soluble nitrates back to reduced forms, including NH_3 and even N_2, thus restoring the non-equilibrium situation. In fact, the nitrogen content of the atmosphere appears to be quite stable.

In the absence of these biological de-nitrification processes, most of the atmospheric oxygen would end up as dissolved nitrates in the ocean. Ahrendts has calculated that an atmosphere-ocean-crustal system (equilibrated to a depth of 10 m) would have an atmosphere of 95 % N_2, with only a trace amount of oxygen (0.3 ppm and a pressure of 0.77 atm (Ahrendts 1980)). The atmospheric oxygen would end up mostly as $NaNO_3$ in the ocean, where it would constitute 0.4 % of the dissolved solids, and the surface layer would consist largely (54.5 %) of silicic acid (H_4SiO_4) (Ahrendts 1980).

If the equilibration in the Earth's crust were taken to a deeper level (e.g. 1000 m), the silicic acid, ferric iron (hematite), and sulfates would be reduced to silica, magnetite, and sulfides respectively. Calcium carbonate and silica would also recombine to produce calcium silicate ($CaSiO_3$) and CO_2 (the reverse of the weathering equation). Essentially all of the sequestered carbon and hydrocarbons in the Earth's crust would be oxidized (releasing CO_2 to the atmosphere). Ammonia and methane would also exist in the atmosphere (Ahrendts 1980). Atmospheric pressure and temperature would then rise. But, in any case, there would be no free oxygen.

In summary, the Earth system is not anywhere near thermodynamic equilibrium. This is lucky for us, since the true equilibrium state—or anything close to it—would be antithetical to the existence of life. It is possible to estimate roughly what Earth would be like if all of the thermodynamically favored chemical reactions went to completion without biological interference. This calculation has been made by Sillèn (1967) and more recently by others. The atmosphere in that scenario would consist almost exclusively (99 %) of carbon dioxide and there would be no molecular nitrogen in the air because all the nitrogen would have reacted with oxygen to become nitrates. In that idealized world, water would constitute only 85 % of ocean mass (compared to 96 % in the actual ocean), sodium nitrate would constitute 1.7 % of the ocean mass (which is many times greater than the atmospheric mass), and sodium chloride would constitute 13 % of ocean mass (as compared to 3.4 % in the actual ocean). Life could not exist in that hypothetical world, if only because of the excess salinity of the oceans.

However in the real world, oxidation of ferrous iron in the Earth's crust is blocked by diffusion resistance in the rock and oxidation of molecular nitrogen in the atmosphere is blocked by the high ignition temperature of intermediate nitric oxides, whence it only occurs in lightning strikes or forest fires. (There is a slow loss, but it is compensated by nitrogen-fixing organisms.) So, disallowing these two oxidation paths, Ahrendts has calculated that there would be a quasi- equilibrium between the atmosphere, hydrosphere and the top meter of crust. This be characterized by an atmosphere consisting of N_2 (74.1 %), O_2 (22.3 %), H_2O (1.9 %), Ar (1.3 %) and CO_2 (0.5 %) and an ocean consisting of 96.57 % water, plus 2.66 % NaCl and 0.77 % a mixture of other chlorides and sulfates (Ahrendts 1980, Fig. 3). These conditions are more like conditions on the actual Earth as we know it.

Allowing the equilibration to include the top 100 m of crust (rather than only 1 m) the oxygen is all used up leaving an atmosphere consisting of 95.24 % nitrogen, the rest being divided between water vapor (2.58 %), Argon (1.66 %) and CO_2 (0.52 %) (Ahrendts, 1980 #37, Fig. 4). In this calculation atmospheric pressure was only 0.77 % of the current atmosphere, largely because of the disappearance of oxygen. But that world is also quite unlike the actual world 4 billion years ago with its high CO_2 level, not to mention hydrogen, methane and ammonia. Evidently that world was also very far from chemical equilibrium.

Very recently a new attack on this problem has been set forth by Antonio and Alicia Valero (Valero Capilla and Valero Delgado 2015). In their work the reference state is not an unreachable chemical equilibrium, but the "dead" state that would occur if all useful mineral ores and natural concentrates (in-homogeneities) on the Earth, including hydrocarbons, were to be used up by another 2000 years of human exploitation (Ahrendts 1980). Their "crepuscular" model postulates burning all the known and probable hydrocarbon resources on the planet (plus other predictable changes attributable to industrial activity). This "dead" state (called *Thanatia*) differs considerably from Ahrendts results. Surprisingly, it results in an atmosphere two millennia hence remarkably similar to the one we now enjoy, with nitrogen constituting 78 8 % and oxygen at 20.92 %. The global average temperature of the Earth would be 17.2 ° C, a little higher than the current average by about 2.7 °C. But, according to the model, the global temperature would rise by a maximum of 3.76 °C during the next few decades before gradually declining.

Meanwhile, the tendency of the biosphere to maintain the Earth system far from thermodynamic equilibrium and favorable to the biosphere is at the core of the "Gaia" hypothesis (Lovelock 1979, 1988). A thermodynamic measure of the Earth's "distance from equilibrium" is the exergy (or availability) content of the atmosphere, ocean and crustal layer. However, until the equilibrium state, or some other reference state, can be specified, it is impossible to determine accurately the exergy of the various chemical substances in the system. We can only say that stored environmental exergy has increased, on average, over geologic time at least until large scale human industrial activity began in earnest, 200 years ago. I will discuss that in Part II.

Since the eighteenth century, there has been a reversal. Environmental exergy capture via photosynthesis has certainly decreased, although probably not yet by a very significant amount. (For example, the amount of free oxygen in the atmosphere is essentially unchanged), and the fraction of sequestered carbon that has been consumed by burning fossil fuels is still infinitesimal compared to the amount stored in shales, not to mention carbonates. Nevertheless, there are potential risks.

Acidification is one of the consequences of environmental changes caused by carbon-dioxide absorption and, most recently, by industrialization. Acids are associated with H^+ ions and bases are associated with OH^- ions. A strong acid such as nitric acid (HNO_3) is one that ionizes easily in water, viz. $HNO_3 \leftrightarrow H^+ + NO_3^-$. Obviously the total number of positive and negative charges (due to overall charge neutrality) remains constant in the universe, the world and the oceans. However, acidification can increase if the number of H^+ ions increases. For this to happen without a corresponding increase in OH^- ions means that there must be a buildup of other negative ions to balance the positive charges.

Acidification currently results largely from the oxidation of other atmospheric gases. In particular, reduced gases such as ammonia (NH_3) and hydrogen sulfide (H_2S) were present in the primordial atmosphere.[18] Ammonia dissolved in water is ammonium hydroxide (NH_4OH), a base. These gases can be oxidized (in several steps) to nitric and sulfuric oxides, respectively. In water, these oxides become strong acids. The oxygen buildup, of course, resulted from the evolutionary "invention" of photosynthesis and the sequestration of carbon (as hydrocarbons) in sediments.

Acid rain is a natural phenomenon, but is closely related to both the sulfur and nitrogen cycles. Hence the anthropogenic acceleration of those two cycles has greatly increased the rate of acidification. The basic mechanism for sulfuric acid generation is that sulfur dioxide (SO_2) from combustion products oxidizes in the atmosphere to sulfur trioxide (SO_3) which subsequently dissolves in water droplets to form sulfuric acid (H_2SO_4). In the case of nitrogen, the sequence also starts with combustion, except that the nitrogen is from the air itself. The two nitrogen oxides, NO and NO_2, are also produced by high temperature combustion with excess air, as in electric generating plants and internal combustion engines operating with lean mixtures. Further reactions with oxygen occur in the atmosphere, by various routes, finally producing N_2O_5 and (with water) nitric acid (HNO_3).

These strong acids ionize to generate nitrate (NO_3^-) and sulfate (SO_4^{2-}) radicals and hydrogen ions (H^+). This process increases soil and water acidity (i.e. reduce the pH). They react immediately with ammonia or any other base. Many metallic ions, including toxic metals, that are currently bound quite firmly to soil particles (especially clay) are likely to be mobilized as the alkaline "buffering" capacity (Ca^{2+} and Mg^{2+} ions) in the soil are used up. Aluminum is one example.

---

[18] To be sure, CO_2 was present in the early atmosphere and carbonic acid (H_2CO_3) is CO_2 dissolved in water. But CO_2 is not very soluble, and the oceans are essentially a saturated solution. Moreover, carbonic acid is a very weak acid as compared to sulfuric and nitric acids.

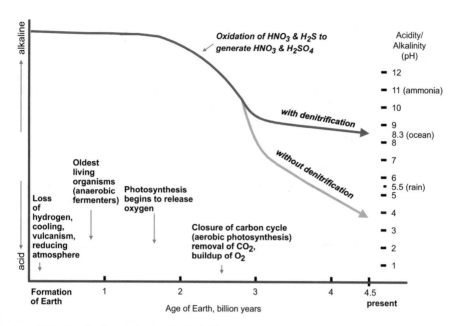

Fig. 6.21 Biospheric stabilization by denitrification

Aluminum poisoning, caused indirectly by acid rain, may be one of the causes of the European *Waldsterben* that has decimated some forests in central Europe. Similar problems may arise in the future as toxic heavy metals like Pb, Cd, As and Hg continue to accumulate in soils and sediments.

Although virtually all of the literature on acid rain pertains to localized effects, it may be important to consider the global implications. The buildup of NO_3^- and SO_4^{2-} ions in the environment is necessarily matched by a corresponding buildup of H^+ ions. This long-term acidification trend has been under way throughout geologic time. On the other hand the carbon dioxide content of the atmosphere, and the corresponding carbonic acid content of rainwater, has been declining, at least until recently. An hypothetical evolutionary acidification trajectory is indicated schematically in Fig. 6.21.

In pre-industrial times, I suspect that the long-term trend toward environmental acidification was mainly driven by two processes. One was volcanic sulfur (H_2S) emissions. These would have been gradually oxidized and resulted in a buildup of soluble sulfates in the oceans until the rate of removal and sequestration by sedimentation (as insoluble calcium sulfate or in organic materials) matched the rate of input. The fact that large amounts of sulfur have been sequestered in this manner is obvious from the existence of enormous deposits of gypsum and, of course, the sulfur content of all natural hydrocarbon deposits, which ranges from 1 to 3 %.

Biological nitrogen fixation must have begun relatively early in Earth history—nobody knows when—since amino acids, building blocks of protein, are constructed from amines (NH_2). That is why anaerobic decay and animal metabolism produces ammonia emissions. The ammonia was—and is—oxidized and partially acidified and converted to nitrates (some of which were left in dry likes), or dissolved in the oceans, most of which are recycled by the biota. But the problem is that all nitrates are extremely soluble in water and thus *cannot* be sequestered in insoluble sediments as carbon and sulfur can be.

Other things being equal, nitrates and nitric acid would have continued to accumulate gradually in the oceans, after oxygen began to accumulate in the atmosphere, thanks to oxidation reactions in the neighborhood of vulcanism and electrical storms. But the problem is that all nitrates are extremely soluble in water and thus *cannot* be sequestered in insoluble sediments (as carbon and sulfur can be). Nitrates and nitric acid would have continued to accumulate in the oceans, other things being equal.

Chapter 7
Summary of Part I: From the "Big Bang" to Nutrient Cycles

The first 13.8 or so billion years of the universe's existence might be thought of as an elaboration of the details of the *Book of Genesis*. God said "Let there be light". And what followed, according to the astrophysical orthodox view, was the "Big Bang" (BB), or perhaps it was the "Big Lightning Bolt" (BLB) since the word "Bang" connotes a loud sound, like what happens when you slam a door, which implies the existence of air to carry the vibrations and an ear to hear, whereas .... OK, that sentence is already out of control, so let's go on to the next one from Genesis (as I remember it): "And there was light". But the guy who trans-scribed Genesis from God's words probably said "light" when he really meant to say "energy", which is what God actually created that day. The identity of mass and energy came later. Unfortunately, the word "energy"—or energeia—hadn't yet been created because Aristotle wasn't born yet, so the author of Genesis said "light".

Anyhow, God created energy, as well as exergy, 13.77 billion years ago, in the blink of a cosmic eye—OK, eyes hadn't been invented yet—but you get the idea. Ever since then, exergy has been destroyed at a rapid rate, and the energy in the universe has been getting cooler (it was very, very hot at first) and also seems to be losing mass. Stars shine because of nuclear fission reactions that convert mass into radiation. Is there another "law" at work? You may wonder why the ultra-hot light from that first BB didn't just radiate away into the void. At first the hot photons were busy creating pairs of particles and anti-particles, which then annihilated each other creating more photons, that created more particles pairs, that ... Eventually things cooled down (to about 3000 °C). At that point the universe seems to have became transparent—allowing some of the radiation to escape. Some of it did just that, of course. Thanks to subsequent expansion, the wavelengths of that radiation got longer and longer. Indeed, Penzias and Wilson detected it as microwave "background" corresponding to a low temperature of 2.7 °K or so.

But some of the little particles, like the quarks, stuck to each other—with short life cycles, to be sure—and made bigger particles, like neutrons, electrons and protons with more mass and longer life cycles. Electrical charges and currents

R. Ayres, *Energy, Complexity and Wealth Maximization*, The Frontiers Collection, DOI 10.1007/978-3-319-30545-5_7

appeared. Particles took on a lot of different sizes and shapes (and colors and flavors) and several different kinds of symmetry that resulted in a variety of different conservation laws that nobody quite understands yet, even after studying at the University of Chicago. The longest lived particles are negatively charged electrons, positively charged protons and neutrons. (The neutrons eventually decay back into electrons and protons).

And, it happens that those long-lived particles *with mass* created what the physicists like to call a *force*, gravity, though Einstein called it the curvature of space. And the presence of mass (hence gravity), in turn, also affects mass-less photons (light quanta) by bending the space through which they move, just as if they had mass. Before mass there was no gravity, but after mass was created there was gravity (or curvature), and that gravity is what kept the mass of the universe as a whole from flying off into the infinite void. Actually, it turns out that the gravity from visible mass (shining stars), even including gravity from "dark matter" that nobody understands, isn't enough to stop the expansion of the universe, because of "dark energy". But that gets us into a discussion of theories of general relativity (EGR vs. YGR) that don't need to be recapitulated again here.

The creation of long-lived electrons, protons and neutrons particles wasn't the end of the "great condensation". The next step was hydrogen gas, a combination of one electron and one proton, kept together by the electromagnetic attraction between opposite electric charges. And the hydrogen atoms were subject to gravity. So, even if they were originally spread uniformly and isotropically through space, quantum fluctuations created in-homogeneities that also grew over time.

Here there is a divergence of opinion. One group (the majority) thinks that gravity alone caused the denser regions of hydrogen to attract each other to make still denser ones and finally the clouds condensed into stars. And, as the aggregations got bigger and denser, the gravitational force became more dominant. Gravity caused pressure and heat, which made the atoms collide with each other (or with neutrons) more frequently and more energetically. You can see where this is going.

The other group (a minority), believes that the gas wasn't just neutral hydrogen, but that it was a *plasma* of charged particles in which electric currents created magnetic fields, and vice versa. They think that it was electromagnetic forces—much stronger than gravity—that led to the creation of plasma in-homogeneities that condensed into matter and stars.

The next thing that happened was thermonuclear fusion. It seems that a proton and a neutron will combine spontaneously to make a deuteron, a photon and an antineutrino, and then two deuterons combine spontaneously to make a helium nucleus plus another photon and another anti-neutrino. Or, a proton combines with a deuteron to make a Helium 3 nucleus, and with another neutron to make a Helium 4 nucleus. Two hydrogens without the extra neutrons don't combine because of the electrical repulsion between the protons in the nucleus, but with the two extra neutrons, the attractive "electro-weak" nuclear forces overcome the repulsion. Not only that, the combination process releases energy in the form of photons. That conversion of mass into radiation is what makes the sun and stars "shine" (emit light). But it also created more elements.

In fact, further atomic fusion combinations—helium fusion to neon, then to carbon and oxygen, then to silicon and finally to iron—give off still more fusion energy. The end of that sequence of exothermic fusions is iron (atomic number 56). Stars (as well as galaxies and other astronomical structures) can be regarded as quasi-steady-state storehouses of mass-energy that—otherwise—would have disappeared into the void long ago as pure radiation. Indeed, the gravitational field of a mass is, itself, a storehouse of potential energy. Or, thinking of it the other way, the mass equivalent of energy is a source of spatial curvature.

But stars have finite lifetimes. They "burn" hydrogen, and when that is all gone they "burn" helium and so on. When all the hydrogen/helium fusion potential is exhausted the star runs out of fuel. Gravity is no longer compensated by radiation pressure, so the star implodes. Several outcomes seem to be possible. Smaller stars may simply cool down and shrink, becoming white dwarves, that cool off to become brown dwarves, without exploding. Other stars—the more massive ones—do explode as a result of processes triggered by the gravitational collapse. They become supernovae. (Notice that this differentiation constitutes a kind of symmetry-break, from a continuum to a dicotomy). The explosions scatter all the outer layers of mass, including any unburned hydrogen and helium (along with the heavier atoms up to and including iron) back into interstellar space, as nebulae. The very heavy core of the supernova star may remain as a neutron star (or possibly a pulsar). If the star is big enough, it may collapse even further into a "black hole" as most astronomers currently believe (but I have doubts, explained in Sect. 4.3).

It also seems possible that some neutron stars are "recycled" as young stars by attracting a new supply of "virgin" hydrogen that was previously too diffuse to become a new star by itself. In that case, the presence of an old white or brown dwarf or neutron star as its center would provide more gravitational attraction, hence higher pressure and density in the inner mantle. That would accelerate the rate of thermonuclear fusion, making the "second-generation" star hotter than a first generation star of the same size. At least, some theorists think so.

The solar system was new 4.5 billion years ago, whether or not the sun was a virgin or "recycled". Since then one of the nine or so planets has become our favorite "blue planet". It must have been formed in a dust cloud that was endowed with quite a lot of heavy metals from an earlier supernova. Earth has amolten iron core—not every planet does—which accounts for its magnetic field. The magnetosphere of the Earth, thanks to the iron core, protects the surface of the planet from the "solar wind" of energetic electrons and protons; another safe-storage mechanism.

Earth also has a lot more water now than it had at first, probably due to comets and asteroids. And it now has an oxygen-nitrogen atmosphere (and an ozone layer) that certainly couldn't have existed 4.5 billion years ago. Most important (from our perspective) is that Earth is now habitable, because its average temperature allows water to be liquid. It is inhabited, by living organisms, whose bodies are largely based on carbohydrates and lipids along with amino acids, phosphates and some other elements. One class of organisms (plants) can split water molecules (using energy from photons) and produce glucose from carbon dioxide. They also produce other bio-polymers (from the glucose) that are autocatalytic and self-replicating.

From the perspective of this book, Earth has also become a significant storehouse for chemical energy, mainly in the form of organic hydrocarbons plus a little (comparatively) in the form of living biomass. The size of that energy storehouse can be estimated (albeit only very crudely) by supposing that every oxygen molecule in the atmosphere were to be recombined with a carbon atom, which is what the oxygen and carbon atoms would "like" to do, if the solar exergy flux—and the second law of thermodynamics—were repealed. But somehow, it doesn't happen. The reason it doesn't happen is part of the story of life.

When life was emergent, it was crucially important for the chemical synthesis, replication, and information processing activities (DNA/RNA)—in short, metabolism—to be protected inside a cell. The cell had to have a relatively impermeable "skin" admitting light and nutrients and permitting wastes to get out. In a rough analogy with the first living cells, the Earth can be thought of as a dissipative structure contained within a boundary layer or "skin". Actually, there are two distinct skins. Each "skin" is a permeable—perhaps immaterial—membrane. The outer "skin" of the Earth is its magnetic field, which deflects the solar "wind" of charged particles (protons) and that would, otherwise, irradiate the surface of the Earth and make organic life impossible.[1]

The inner "skin" of the Earth is the ozone layer. The ozone layer is caused by ultraviolet (UV) radiation interacting with atmospheric oxygen and creating extremely reactive ozone (O_3) molecules. But the ozone, itself, is an effective barrier against that same UV radiation reaching the surface of the Earth. As explained in Sect. 6.1 it was not possible for large organic molecules, and polymers, or complex living organisms to accumulate anywhere except in deep water, before the ozone layer was created by the oxygen buildup in the atmosphere. That must have happened at the beginning of the "Great Oxidation", about 2.3 billion years ago. However, the ozone layer seems to be very well established and stable by now.

The ozone layer can be characterized as a self-organized "dissipative structure", viz. an ordered stable state, far from (thermodynamic) equilibrium and maintained by a flow of useful energy in the form of UV radiation.[2] The "skin" of the habitable Earth is a composite consisting of the Earth's magnetosphere, the two (and

---

[1] Earth is not a permanent magnet, like a lodestone. As noted in Sect. 4.8, the Earth's magnetic field is apparently created by convective currents of molten metal (iron and nickel) around its solid iron core. Cooling of the core may reduce the intensity of the Earth's magnetosphere, and thus permit deeper penetration by the "solar wind". Indeed, magnetic field reversals seem to occur at regular intervals. Such a reversal is expected (by the specialists) in the next few thousand years. It is unclear whether this reversal would have a significant impact on the biosphere.

[2] While the most interesting examples of dissipative structures are living systems, there are a number of simple abiotic examples. The ozone layer is one. Others include the famous Benard cell, which exhibits cooperative behavior that greatly enhances heat transfer across a thermal gradient (Koschmieder 1993). Whirlpools, cyclones and tornados and sunspots are other examples of cooperative physical systems that enhance heat transfer and destroy thermal gradients.

sometimes three) Van Allen radiation belts, and—since oxygen accumulated in the atmosphere—the ozone layer. These are penetrable to solids (e.g. meteorites and comets) but they minimize penetration by high energy cosmic rays, solar wind (protons) and UV radiation.

When the Earth was new, 4.5 billion years ago, like all babies, it was naked. It had no protective clothes. The oceans were just salty water. The rocks were bare. The atmosphere was a mixture of carbon dioxide, methane, and nitrogen gas, because most of the hydrogen would have boiled off. For most of the time since then an observer without a microscope would have seen little change, except for surface water accumulation and several periods of surface ice cover accompanied by very low sea levels. An observer a pH sensor would have noticed that the oceans were growing more acid, while dissolved ferrous (not completely oxidized) iron was disappearing. One with a chemical test kit would have seen a comparable change in the composition of the atmosphere: less methane, less carbon dioxide, but for a long time, no oxygen.

And for the first 500 million years (maybe more, maybe less) an observer with a microscope would only have seen evidence of cellular life near underwater volcanic vents. The carbon-fixing work done by all organisms before photosynthesis might have been less than 50 GW or 0.05 TW. Later, after a hundred million years or so, the microscopist also have seen cellular life here and there in the open oceans, doing photosynthesis with reducing agents such as hydrogen sulfide or ferrous iron and producing sulfur or ferric iron as waste products—without producing oxygen. Yet anoxic photosynthesis probably increased the total carbon-fixation work by a factor of 100. Call it 5 TW.

Then—when exactly we don't know—the great invention of oxygen photosynthesis (OP), by splitting the water molecules, occurred. It gave the organisms with OP an advantage over the others when the oceans ran short of suitable reductants, especially iron ions. Now the carbon fixation rate increased dramatically. Gradually the oxygen by-product of OP accumulated in the atmosphere and killed off most of the anaerobic organisms. Yet OP increased the energy conversion capability of the bio-system as a whole by another factor of 10, to 50 TW.

Some time later, as the toxic oxygen level in the atmosphere reached dangerous levels, another evolutionary invention, *respiration*, came along, perhaps 3 billion years ago. The organisms that could breathe the oxygen and use it to oxidize carbohydrates got a huge metabolic (and mobility) advantage compared to fermenters: much more metabolic energy to spend. It was undoubtedly this new energy availability that enabled micro-organisms to develop mobility, both for predation and avoidance. Only after another billion years would the long-lived observer have seen a diverse population of micro-organisms occupying most of the ocean surface growing bigger, more complex, more mobile and much more diverse.

Then, a mere 550 million years ago, things began to change more rapidly. It was what we now call the "Cambrian explosion". Suddenly (in geological terms) the plant and animal kingdoms separated. Multi-cell organisms with differentiated cells performing different functions appeared. These multi-cellular organisms themselves began to specialize and to occupy different "niches", to exploit different resources. They colonized the land. They grew large. Some of them grew very large. And during the whole of this time, the biosphere learned how to convert more and more of the available solar energy into carbohydrates, lipids and proteins. Colonizing the land probably increased the carbon fixation work output by at least another factor of 2, or so, to the current global level of 130 TW.

For billions of years the biosphere has prospered and grown inside its protective "skin". It now converts a significant percentage of the solar exergy that reaches the surface into carbon-based living organisms, mostly short-lived but some not so short. A much larger percentage of that input drives the weather-climate system, without which the biosphere could not exist. That system is an entropy minimizing cycle. The so-called "nutrient cycles" (including the oxygen-carbon cycle, the nitrogen cycle, the sulfur cycle and the phosphorus cycle) are other examples of entropy minimizing cycles.

Then, a mere million years ago (or less) humans came onto the stage and the "Anthropocene era" began (Crutzen and Stoermer 2000). Our ancestors learned how to be energy predators and energy consumers. In recent generations, we have become very efficient at this lifestyle. Moreover, the protective "skins" of humans, both individually and in social systems, have been radically enhanced. Part 2 looks at this new development.

The story of the past 13.77 billion years of the universe (since BB) is that the "stuff" called matter keeps recycling itself in different forms, with more and more mass and "staying power" to borrow a phrase from deodorant commercials. Not only that, the mass of the universe seems to be increasingly organized into long-lived "structures"—ranging from galaxies and planets to elements, chemical compounds and biological species—that can also be thought of as stocks of information or negentropy. These complex structures, especially those embodied in Planet Earth, both living and non-living, constitute natural wealth.

I said that the Second Law appears contradictory at first. Why? At first sight, the Second Law says that the entropy of the universe keeps on increasing, which implies increasing disorganization. But how does that fit with creating structures? The answer, in brief, is that evolutionary "progress", both in biology, ecology and economics, is based on "dissipative structures" that involve maximizing local entropy creation while simultaneously maximizing *negentropy* (order, structure) (Schrödinger 1945; Prigogine et al. 1972; Nicolis and Prigogine 1977). So how does maximizing local entropy creation to create local order fit with minimizing the *rate* of entropy increase in the universe as a whole? It is puzzling, at first sight. But there is no actual contradiction.

There is more to be said about the detailed mechanisms driving evolution and complexity. Darwin identified competition as the key element of natural selection, but he was careful to say that there are other means of modification. Symmetry-breaking in the sense of divergence and compartmentalization is involved at every stage. The deeper question is what causes speciation and diversification in biology. The same question will arise later in connection with human evolution and economics.

Ilya Prigogine has said (Prigogine 1991):

"Irreversible processes have likely played an essential role in inscribing time, so to speak, into matter at the early stage of the universe and producing the information carried by basic biochemical compounds such as DNA. Indeed the basic feature of biomolecules is that they are carriers of information. This information is quite similar to a text which has to be read in one direction. Biomolecules represent, in this sense, a broken symmetry ..."

In any case, life as diversity, complexity and organization, is all around us. Evolutionary change has been explicitly interpreted as an entropic process by several modern biologists, notably Brooks and Wiley, Schneider and Kay, and Schneider and Sagan (Brooks and Wiley 1986; Schneider and Kay 1994; Schneider and Sagan 2005). At the systems level, Holling and his colleagues have pointed out there is an adaptive cycle in ecosystems that actually depends on "creative destruction" for continued growth in new forms (Holling 1973, 1986). Today we oversimplify agriculture and other natural systems to increase "productivity" at great hazard.

The *rate* at which entropy increases on the macro-scale, whether in star formation or planet formation or in living organisms, does not seem to be the maximum theoretically possible. On the contrary, the universe keeps finding ways to slow down the approach to "heat death". As I said a few paragraphs back, and earlier in Sect. 3.9, the maximum rate of entropy increase after the BB would seem to have been for all that heat to simply escape, as an instantaneous blast of radiation, into the void. But the laws of physics (or God's will) seem to be making things happen in a more complicated way that enables us to exist and enjoy our existence however long it may be.

Part II

Chapter 8
Energy and Technology

8.1 The Enhancement of Skin

A million years ago, in the mild environment of the inter-glacial periods, upright 2-legged proto-humans (*hominids*) began to roam the fields and forests of Africa, or East Asia, as foragers and collectors of seeds, nuts and fruit, with an occasional meat supplement when they encountered a recently killed beast and could scare away its killer. At some point the proto-humans learned to hunt larger animals, always in groups (because they were physically slower and weaker than any of the other predators).

But it seems that (thanks to organization) they developed a capacity to run for longer distances than the faster four legged creatures (like antelopes) that they hunted. The antelope would run away when the hunting party came close, but only far enough to be out of sight. Then the antelope would pause to rest until the hunters came again. And so it continued until the chosen antelope was exhausted and could run no further.

For purposes of long-distance running, heat dissipation by evaporation is crucial—as any long distance runner will tell you—and bare skin is far more efficient at heat dissipation than fur. Bare skin and widely distributed sweat glands became an evolutionary advantage. It enabled hairless hominids to hunt (and travel) during the heat of the day when the large predators with inefficient heat dissipation mechanisms were inactive and the large animals like rhinos, hippos and elephants lounged in the water. (N.B. have you noticed that your dog can only sweat through his tongue?)

There is a question about whether hairlessness was encouraged by a pre-existing habit of running for long distances, or whether the long-distance running habit followed the hairlessness. What was cause and what was the effect? As it happens, there is another theory that would account for the hairlessness, namely that human ancestors lived in the water for a few million years. This theory seems to explain a few other anatomical features of humans, mainly the shape of the human foot,

© Springer International Publishing Switzerland 2016
R. Ayres, *Energy, Complexity and Wealth Maximization*, The Frontiers Collection,
DOI 10.1007/978-3-319-30545-5_8

which is a little better suited for paddling than for gripping. However, there is no direct evidence to support this theory.

It is true that we humans, like other animals, come into the world naked. But we do not stay that way for long. At some later time back in the ice ages, our increasingly hairless ancestors, being cold when they were not running, had the bright idea of covering their bare skins with the skins of other animals. Thus clothing was invented. I know not when this happened, but the link between hairlessness and the use of animal skins seems obvious.

The caves where hairless hominids lived during the glacial epoch were, of course, also another form of protection from large predators of the night. When the ice melted, people learned how to create artificial caves, or huts. Groups of huts became villages, which soon acquired walls. Skin by another name, also thanks to organization and communication.

8.2   The Taming of Fire

At some point, primitive proto-humans made a huge leap forward by capturing and taming fire. This seems to have occurred about a million years ago. We can imagine the bold leader of a family group or tribe such as *Sinanthropus Pekinensis* grabbing a burning branch from the flames that engulfed a tree ignited by a lightning bolt. He bravely carries it to his cave or hut. After getting over their initial fear, the members of his tribe begin to guard and nourish the fire, to keep it alive.

Fire was an important element in ancient religion, especially Zoroastrianism, which was the main religion in Persia before the Muslims. For Zoroastrians Baku— and several other places in the Caucasus where there were "eternal flames"—were holy places. Marco Polo remarked on this in his journeys. According to the Iranian Ferdowski's epic poem Shahnameh (c. 1000 AD):

"One day Hushang reached a mountain with his men
And saw afar a long swift dusky form
With eyes like pools of blood and
The smoke from whose jaws bedimmed the world
Hushang the wary seized a stone
Advanced toward the beast and hurled it royally.
The world-consuming snake escaped
But the stone struck on a larger
And, as both shivered,
Sparks issued and a dried bush was set ablaze.
Fire came from its stony hiding place
And again, when iron knocked.
For such a radiant gift
The lord of the world thanked God.
This lustre is divine said he to his people
And you, if wise, will hold it sacred.

Making fire a cynosure for all to behold ..." [Wikipedia, "Zoroaster"] For hunters and gatherers, fire provided protection at night, even from large fierce predators like tigers, bears or wolves. (We now realize that neither bears nor wolves actually hunted humans for food—they were competitors, not predators, notwithstanding the occasional conflict.) But fire was especially effective at scaring off night-hunting members of the cat family.

People learned more and more ways to use the heat from fire (and how to start one). Protection against cold was surely one of them. Fire encouraged family groups or tribes to stay in one location, which was beneficial for the safety of young children and nursing mothers. Roasting meat (to soften it and make it more digestible) and, later, making bread from the edible seeds of certain grasses, were early uses. Some foods had to be cooked to be eaten at all. Cooking added to the availability of some nutrients.

It has been suggested that before cooking, humans had to do a lot more chewing—up to 50 % of waking time—whereas after cooking, the need for chewing declined drastically. This had the further benefit of reducing the need for big teeth for grinding and large guts for digestion, which paid off in other ways.

Fire was also useful for hardening the points of spears and later still, for making clay pots to hold water or other fluids. Fire was also the only source of light after the sun had set. More and more tribes and villages kept an "eternal flame" going. Gradually, the benefits these innovations concerning the use of fire spread among all the tribes of men.

William Nordhaus has estimated that, at the time of Peking man (500,000 years ago), 1 h of labor would suffice to gather and trim 10 pounds (4.5 kg) of firewood—after drying (Nordhaus 1998 Appendix). Thus 200 h of labor for wood-foraging would have yielded a ton of dry wood yielding five million Btu of thermal energy (heat) when burned. At 0.69 lumen-hours per thousand Btu-based on actual measurements by Nordhaus (ibid) with a Minolta TL-1 illuminance meter—burning this amount of firewood in a crude fireplace would have produced 3460 lumen-hours of light. That is the amount of light produced by a standard 100 W incandescent light in 2 h and 52 min. In other words, 200 h of labor, more or less, would have been needed to provide enough fire-light to read this book. Sorry for the confusing un-metric units, but please blame Nordhaus.

Some of those prehistoric experiences have been passed down as myths and legends. In Greek mythology a man named Prometheus stole the fire from Mount Olympus, the residence of the gods, and brought it to the humans on Earth. Zeus, the Godfather, punished Prometheus cruelly by chaining him to a rock, where birds pecked at his liver (which seemed to grow back every night). But Prometheus' gift gave humans power to resist the angry and fickle Gods and saved them from doom. Here is a quotation from Hesiod, 700 BCE: "Fire is the basis of all craft. Without it, Man could not persist."

8.3   Transmission of Knowledge: Writing and Replication

The story of language, writing, paper and—finally—printing needs to be introduced at this point. Some would argue that the invention and development of spoken language was the crucial difference between our species and the other great apes. Transmission between generations did occur, partly by means of explicit training father-to-son, mother-to-daughter, in techniques pertaining to everyday life. Fathers taught their sons how to use weapons and tools, and how to tame and train animals. Mothers taught their daughters domestic skills and medicine. Community history was passed on in songs, legends and ceremonies. But this mode of knowledge transmission was limited.

Modern historians are inclined to think that written language was the most important of all inventions because it greatly enhanced the transmission and accumulation of knowledge. But writing itself took a long time to evolve. The earliest hieroglyphs on stone may have been records of the storage of crops, or other day-to-day housekeeping, but writing for ceremonial purposes, especially the burials of kings, came soon after (if not before). More recently, inscriptions on stone, such as the Rosetta stone, now in the British Museum, were used to transmit royal decrees.[1] Most such inscriptions were in the form of hieroglyphics or pictograms.

There is no need for me to recapitulate the history of written language (fascinating though it is) except perhaps to note that the Greek alphabet was an enormous leap forward. Countries, notably China, Korea and Japan, relying on pictographic characters, have run into difficulties in adapting it to express abstract scientific ideas and to educate large numbers of people capable of communicating such ideas. The idea of an alphabet (a word constructed from the first two Greek letters) was revolutionary, because all words are constructed from a few characters (26 letters in the current English version).

Similarly, the Hindu-Arabic notation for numerals (0,1,2,3,4,5,6,7,8,9)—and especially the zero and expression of large numbers in powers of 10—have turned out to be a revolutionary improvement over so-called "Roman" numerals for arithmetic and mathematics. Can you imagine how to subtract DXXXIII (533) from MCDXLII (1492)? It is no wonder that keeping track of money, not to mention other things, was a very rare skill in Europe until the Arabic notation arrived. Pope Sylvester II, whose previous name was Gerbert of Aurillac, had studied in Barcelona (then under Arab domination) and had learned about the Arabic numbering system. As Pope in the 1080s he promoted its use, primarily by the clergy. Later, **Leonardo Fibonacci** of Pisa, who studied in Algeria as a boy, did even more to promote the use of Arabic numbers, with his book *Liber Abaci* (1202). By the fifteenth century the Arabic numerals were widely used. After 1450, the printing press did the rest.

---

[1] The text on the Rosetta stone, in three languages (hieroglyphic, demotic and ancient Greek), was a decree by Ptolemy V, issued at Memphis and dated 196 BCE.

Accumulation and communication of facts and ideas also requires a storage medium. The earliest examples were inscriptions on stones (such as the famous Rosetta stone). But, apart from the labor intensiveness of writing on stone, storage and transmission limits soon dictated another solution. Papyrus (made from the pith of a wetland sedge plant) and parchment (from sheepskin) were the first—very costly—solutions, going back about 5000 years (Egypt).

Paper made from wood-pulp was the next. It was the first industrial material to be made in large quantities according to a recipe.[2] The "inventor" according to Chinese history was Cai Dun, an official of the Han dynasty who offered his recipe, based on crushed mulberry wood, to the emperor in 105 CE. (However recent discoveries in China suggest that paper of some sort existed at least 100 years earlier.) Paper came to Europe via the "Silk Road", possibly thanks to **Marco Polo**, and was adapted first by the Muslims in Spain. The first paper mill in Europe was built in Leiria, Portugal (1411). The first in England was built in 1490.

Printing by spreading ink over carved wood-blocks was first done in China, also during the Han dynasty (100 CE), about the same time that paper was "invented" there. This set the stage for printing of books. But the first known printed book, a Buddhist Sutra, was dated 868 CE, 750 years after the prerequisites became available. Printing by means of movable (carved wooden) types was first done by Bi Sheng, in China (about 1000 CE). The first printer to use movable type made from metal (lead) castings was a nameless Korean (c. 1375 CE).

But the Oriental printers were still dependent on pictographic characters. Around 1450 **Johannes Gutenberg** used the Greco-Roman alphabet to form words and added some other inventions of his own. Together, his innovations made printing (replication) of books into a large-scale business. Gutenberg's own press could print 3000 pages per working day, and popular authors like Martin Luther and Desiderius Erasmus sold hundreds of thousands of copies of their books. The Bible, in various languages, was the major best-seller, of course. Printing on paper cut the cost of books enormously. It is not unreasonable to suggest that it kicked off the Protestant Reformation. Obviously Martin Luther's "Ninety-five theses" (in Latin) was not only nailed to the door of the Castle Church in Wittenberg; it was printed on paper and widely distributed to the other universities, and cities, especially in Paris. The new technology of the printing press enabled many people other than ecclesiastics to debate the issues. The Bible was rapidly translated into the vernacular languages. Martin Luther himself was the first to translate it into German.

The lack of standardization of materials was certainly true for wood, ceramics and "pigs" of lead, tin, copper or iron. The latter had to be further treated by rolling, stamping, hammering (forging) or some other way. The first to partially solve the problem for a mechanical product, it seems, was an Italian gunsmith named Bartolomeo Beretta. He filled an order for 500 harquebus (musket) barrels for the Venetian armory in 1526 CE. (The Beretta firm is still family owned, but is now

---

[2] The quantities were quite large. One Chinese province paid its annual tax by shipping 1.5 million sheets of paper to the central government.

producing pistols, rifles, shotguns and other things in the millions.) Evidently, the technology of "widget" manufacturing, even 300 years later, was still very primitive, due to what has been called the "complexity-variety barrier". The more complex the product, the more design variants there can be. True mass production had to wait until the late 19th Century.

8.4   The Dawn of Civilization and the Beginning of Agriculture

The upright hairless bipedal hominids of genus *Homo* divided into several branches during the last million years. *Homo sapiens* seems to have evolved about 400,000 years ago. The proto humans also learned to devise traps and simple weapons, made from bones or stones, such as flint. This period is known as the *Paleolithic* age. (The word "lithic" refers to stone.) Soon after that they also learned how to make spears for defense against larger predators and also for purposes of hunting.

Human hunter-gatherers needed a lot of land, from 20 to 100 ha per capita, depending on location. Our ancestors used dogs for hunting at least 32,000 years ago. Hunting was evidently responsible for the extinction of some species, such as the giant mammoth (whose remains are still being uncovered, from time to time, in northern Siberia). And, as they became more efficient at hunting, they also took advantage of the warmer inter-glacial periods to migrate north into Europe and Asia, and eventually across the Bering Straits "land bridge" from Siberia into North America. As far as we can estimate, the exergy budget of those early human hunter-gatherers was about two kilowatt-hours (2 kWh) per day, per capita.

The bipedal hominids competed with each other for living space. Finally, our sub-species (*Homo Sapiens*) eventually conquered or outbred its chief rival, *Homo Neanderthalensis*. The details of that competition are unknown, but it seems to have ended around 40,000 years ago. It seems that the Neanderthals did not use dogs, whereas *H. Sapiens* did. They had bigger brains—and bigger skulls—than ours, which may have been a factor negatively affecting child-bearing. Moreover, Neanderthal males needed more Calories per day than *H. Sapiens*. By the time of the Neanderthals final disappearance, the average energy consumption per human had risen to around six kilowatt-hours (6 kWh) per day.

The oldest artifacts that we can date with some confidence are the cave paintings of large animals, in the Vézères Valley of south central France. A large number of inhabited caves have been found in the area, some of them decorated. The famous paintings at Lascaux have been dated at 17,300 years ago. That period is known as "Upper Paleolithic".

Human "civilization", consisting of centralized governments over large regions, wars of conquest, and large towns, probably began either in the so-called "fertile crescent" between the Euphrates and Tigris rivers in what is now Iraq, or along the

Nile River. Between 12,000 and 10,000 years ago the last glaciers were receding and the average temperature of the Earth was in the process of rising by about 5 °C to the level we have enjoyed for the past several millennia.

In this new, warmer, environment—which must have seemed like paradise compared to the preceding ice age—*Homo Sapiens* took another great leap forward: the *"Neolithic* revolution". Of course "Neolithic" is another word for "stone age" and it is interesting that a factory for manufacturing, storing and exporting flint tools existed in central France (the Claise Valley) in the third millennium BCE. Farming (of grain) and cattle breeding were literally invented. Instead of just hunting and gathering food from wild plants and animals that live in grasslands, forests, and surface waters, our ancestors learned how to harvest solar energy more systematically, by cultivation of plants for food.

We can imagine a Neolithic woman—call her "Eve"—who collected edible seeds from nearby clusters of wild grasses. (Not the kind of grass we see in lawns or on golf courses; these were precursors of the grains we now call wheat and corn, or maize). She naturally collected the biggest and ripest seeds. Her family ate most of the grain she collected. But some seeds got scattered around the hut or cave. Those seeds grew and produced more of the big seeds. She realized, perhaps for the first time, that plants grow from seeds. It was one of the great *Eureka* moments in human history.

Fire also helped, for land clearing. In fact early agriculture was mostly of the kind what we now call "slash and burn". A tribe would clear a small area by burning it (which also released nutrients in the ash), and then would move on when the fertility "spike" from the fire was used up. This type of agriculture is still practiced in a few remote areas. However, that type of agriculture was still inefficient in land use: it needed 2–10 ha per capita. The floodplains of the Yellow River in China, or the Tigris and Euphrates in Mesopotamia or along the Nile took care of the need for annual renewal of fertility, and the land requirement dropped to 0.5–1.5 ha per capita.

Meanwhile the hunters—like her man, call him "Adam"—made a comparable great discovery. They discovered that when a baby animal orphan of some (not all) species was raised by humans in a human environment, it became tame and biddable. This probably worked first for dogs. The dogs (prehistoric wolves?) probably started by following human hunting parties to scavenge. But in the course of time the dog-packs occasionally left behind orphan babies that were adopted by human families and tribes. (Who can resist cuddling a puppy?)

Over the centuries, dogs that had been raised by humans became "different" from the wild dogs. They became guardians and helped with tracking and hunting of other animals. The cynical saying about human relationships, that "if you want a friend, get a dog", is also saying something about dogs. There are numerous documented cases of pet dogs dying of grief after the death of a master, or giving up their lives to protect, or save, a master.

The ancient relationship between humans and dogs (and wolves) is the basis of the legend of *Romulus and Remus*. They were unwanted by their father, King *Amulius* (who was afraid they would grow up and depose him). So, the twins were set adrift in the Tiber river—a form of infanticide. But the babies were found and suckled by a she-wolf. They later grew up and founded Rome.

Humans learned how to control the animal breeding process. Later, cattle, sheep, goats, camels, horses, donkeys, oxen, pigs, geese, ducks, chickens and some other large animals (e.g. elephants) were gradually domesticated. Some humans became sheep or cattle herders. Domestic animals provided skins, fur and hair (wool), meat, milk, eggs, manure for fertilizing the land and muscle for tilling the soil. Horses, donkeys and camels also provided mobility and carrying capacity.

Not every wild animal can be tamed. In fact, most of the animals now seen in zoos, from zebras, antelopes, giraffes, rhinos, hippos, and large members of the cat family are difficult or impossible to domesticate. Small cats seem to have made themselves useful—probably by catching rodents—without ever becoming teachable or "biddable" in the way that dogs and horses are.

It seems that most of the animals suitable for domestication were originally found in Eurasia, not in Africa or the Americas. Furthermore, agricultural innovations diffused much more easily along Eurasia's east-west axis than along Africa's and the America's north-south axis, broken up by geographic and climatic obstacles. Around those two axes, it has been noted, turned the fortunes of history (Diamond 1998).

Since hunting for large animals, such as antelopes, is an easy way to feed a group of people for several days, the invention of farming seems likely to have been triggered by game (and firewood) scarcity. Probably agriculture, as a primary occupation, was a desperate response to over-hunting, land clearing and over-crowding.

It began about 9000 to 10,000 BCE, both in the Nile Valley, the Tigris–Euphrates, the Indus valley, the Mekong, the Yangtze and the Yellow Rivers of China, where human population had grown too great to live on the proceeds of hunting. The simultaneity of this shift may conceivably be explained by climate changes related to the crustal displacement hypothesis (Hapgood 1958). In any case, land clearing, for purposes of cultivation, drove the wild animals further away. Early agriculture was *not* very productive, as can be seen today by comparing the skeletons of prehistoric farmers with the skeletons of the hunters who had previously lived off the same land. The hunters were as much as 6 in. taller and they lived longer. The agricultural population of Neolithic times was considerably less well-nourished (and had to work much harder) than their earlier hunting-and gathering ancestors (Diamond 1991 Chap. 10). For the people struggling for a bare survival, it must have seemed as though they had been kicked out of a paradise of plenty. This historic experience has been described in stories handed down through the generations, such as the expulsion of Adam and Eve from the Garden of Eden.

I mentioned the limits of fire as a source of light. But during the Paleolithic era, from 15,000 to 40,000 years ago, some tribes learned how to make stone dishes, with shallow depressions, capable of retaining melted animal fat. These were the first lamps, of which several hundred have been found by archeologists. It is estimated that such a lamp, burning animal fat, produced about the same amount of light as a candle (Nordhaus 1998). Terracotta lamps burning oil (probably sesame oil) existed in Babylon 5000 years ago, and were used (at least) in the temples.

As an indication of the price, it seems a common laborer at the time earned about one shekel per month, which was the price of 10 L of sesame oil. Nordhaus says that a quarter cup of oil (with a wick) burned for 17 h in a Roman copy of such a lamp, producing 28.6 lumen-hours, or 17.5 lumen-hours per thousand Btus. From then until 1992, Nordhaus estimates that the labor-cost of illumination declined by a factor of 350,000 (ibid). It is still declining, thanks to light emitting diodes (LEDs) that have been commercialized in recent years.

Other consequences of primitive agriculture include the need for means of storage, and the need for protection of the immobile farmers and their crops from mobile predators, both animal and human. The need for protection was probably the cause of walled cities and armies, not to mention class differences, epidemics, starvation and gender discrimination (Diamond 1991 Chap. 10).

Despite their impressive cultural achievements, agrarian civilizations have always been handicapped in their development by the limitations of the physical work that can be derived from human or animal muscles. Wheels, levers, inclined planes, pulley blocks, water mills, windmills and sails have provided limited help to surmount human biological limitations. The earliest farmers had to do the cultivation of land themselves—with the assistance of children. This is why agriculture began in river valleys where there was a reliable annual deposition of fresh silt that was easy to plow and not already occupied by trees or deep-rooted grasses and shrubs. Of course plowing, back then, was probably nothing more elaborate than dragging a sharp stick through the damp silt to make a groove in which to drop the seeds.

The taming of animals to do the plowing, probably oxen at first, eventually enabled farmers to leave the river valleys and cultivate higher grasslands. This task was very challenging, and required much more brute strength than humans could provide. Several inventions were required to make this possible. One was a more elaborate plow, initially of wood, but later of crude metal. Another was some form of harness to enable the animals to pull the plow. Horses were more difficult to tame and harness than oxen, but they were naturally more at home in grasslands. They were also quicker than oxen and could also be tamed (and ridden). Later, horses were also used to pull wheeled vehicles. Horses gradually replaced oxen for plowing in most of the upper-story world outside the river valleys.

It is tempting, but quite wrong, to think of horses or oxen as workers capable of feeding themselves, i.e. as "free lunches" so to speak. It is true that these animals can feed themselves by munching on grass; they do it in their natural habitat. But if a horse is working, harnessed to a plow (or a cart) all day, it must be fed some more concentrated form of nutrition. Moreover, hay (dried grass) is not sufficient as feed for working animals. It will not provide enough either of energy (Calories) or of protein. A working horse needs grain, usually oats.

To put it in terms of modern units, a working horse generates roughly one "horsepower" which is equivalent to about 700 W or 0.7 kW of power. By comparison, a working human can generate only 50–100 W of power (Smil

2008). In terms of work output, a horse can produce around ten times the output of a human worker. Not only that, a colt can start working after 2 years or so, whereas a human child cannot do hard physical work beyond sorting seeds or tubers or feeding chickens, until he or she is at least 12 or 13 years old.

The energy (Calorie) content of the feed (for a horse) or the food (for a human) must be at least seven times as great as the energy equivalent of the food produced. This is because the energy conversion efficiency of the animal metabolism is not more than about 14–15 % at most (ibid). This is a biological limit that has not changed in the last 5000 years since animals were first domesticated. Among other things, the need to provide high-energy feed (not just hay) for working animals means that about 15–20 % of all cultivated land has always been needed for that purpose.

The quantity and plant composition of crops depend on two properties of the biosphere: One is weather and climate, i.e. the short term and longer term fluctuations of sunshine and rain. The other factor is the natural fertility of the land. This means its content of humus, nitrogen, phosphorus (and other useful trace elements), on the one hand, and the absence of noxious salts (and pests), on the other. Technical progress in tilling the soil, like the transition from the wooden pickaxe to the animal-drawn iron plough, cut the labor (and energy) inputs needed for tillage. Also, crop rotation and the use of animal manure for retaining soil nutrients have increased crop yields enormously—if it were not so, China and India could not feed themselves today.

Domesticated grazing animals, especially cattle, sheep and goats, convert cellulosic biomass into food available to humans. However, the thermodynamic efficiency of producing animal biomass from grass is fairly low. For instance, about 8 tons of dry vegetable matter is converted into a ton of dry milk by cows. Chickens make eggs somewhat more efficiently, which accounts for their wide exploitation. Fish are even more efficient convertors because, being cold-blooded, they do not need energy to maintain body heat. Thus, in times of population growth and food scarcity, shifting from pasture to cultivation can increase the output of edible food energy by something like a factor of 5. Estimates for different agricultural technologies arrive at the following annual yields of food energy per hectare ($=10,000$ m^2), including the fallow periods:

- Rice with fire clearing (Iban, Borneo): 236 kWh
- Horticulture (Papua, New Guinea): 386 kWh
- Wheat (India): 3111 kWh
- Maize (Mexico): 8167 kWh
- Intensive farming (China) 78,056 kWh

Hunters and gatherers only obtain between 0.2 and 1.7 kWh per hectare per year. Thus, as compared to Chinese intensive farming, as many as 50,000 times more people can live on the produce of a given area today than under pre-agricultural conditions.

There are people today (mostly living in cities) who imagine that it would be a great idea to return to an economy of small self-sufficient farms, such as envisioned by Thomas Jefferson, the author of the Declaration of Independence of the 13 American colonies. This was possible, even in his native Virginia, when he was alive. But it is possible no longer. To feed the current global population of the Earth (7.5 billion or so by now), intensive (and somewhat industrialized) agriculture is unavoidable.

8.5 Agricultural Surplus and Cities

As I have mentioned above, thanks to the domestication of animals and later inventions like the harness and the metal plow, more and more land came under cultivation. It seems to have started about 10,000 years ago, as the glaciers were melting, in the highlands of present-day Kurdistan, where it rained enough for crops year-round. But 6000 years ago (4000 BCE) a band of farmers moved south into the valley of the Tigris-Euphrates. They made it into the "fertile crescent" by digging a ditch—then many ditches—to irrigate the land during dry seasons. Similar migrations occurred in the Valleys of the upper Nile, the Indus, the Ganges, the Mekong, and the Yangtze and Yellow rivers in China. The richer irrigated farmlands created a food surplus. By 3000 BCE the agricultural surplus in these valleys was enough to provide food and feed output equivalent to 14 kWh per person per day (although much of it was consumed by the animals).

An accompanying development was urbanization: villages, towns and cities. The oldest (or, at least the most thoroughly studied) human settlement was Jericho. It was settled about 9000 BCE, as the last glaciers were melting. The first occupants were Natufian hunter-gatherers. A stone wall 3.6 m high and 1.8 m broad at the base surrounded 70 dwellings. A tower inside the walls stood 8 m high, with 22 stone steps. This was not needed to protect against wild predators. It was needed to protect the villagers from forays by other humans on "hunting" expeditions.

Over the next 7500 years, nearly 20 settlements came and went. By the time Joshua arrived with his Israelites, c. 1550 BCE, the stone walls were much higher and there was a brick wall on top. The collapse of the walls is hard to explain unless there was a fortuitous earthquake. Large clay pots containing grain have been found in the ruins.

Larger cities, like Memphis in Egypt and Babylon in Sumeria, probably also began as defensible forts which served as storehouses for the surplus grain, as well as protection for the population. The vulnerable farmers, who were unable to leave their land and herds, needed defense, so tribes needed protectors. The would-be protectors also fought each other, human nature being what it is. The winners gradually increased their domains and eventually became warlords and kings. The kings lived in the cities, and collected taxes (often in the form of grain) from their subjects. These taxes financed other activities, including the support of increasingly powerful priesthoods, and the manufacture of other goods such as pottery, textiles, and primitive metal objects (from horse-shoes to swords and plowshares).

On this agrarian base rested the first "high" civilizations of China, India, the Middle East, and Egypt. In the agrarian societies, economic and political power rested with the landowners, because they controlled the useful energy embodied in herds of animals and the products of photo-synthesis. In Latin, the original expression for cattle, "pecunia", assumed the meaning of "money" and hence of portable wealth. Even today, in parts of Africa, such as the Masai of Kenya, wealth is measured in terms of cattle ownership. In due course, the cattle-owning tribal leaders became land-owning nobility and accumulated political power. Feudalism in some version became the dominating political system of most agrarian societies. It gained strength with the increasing energy needs of these societies, as they advanced technologically, commercially, and militarily.

8.6   Slavery and Conquest

Human history from 3000 BCE to the Industrial Revolution is normally presented in the Western World as a sequence of kings and conquerors, like the Assyrians and Babylonians, the Greco-Persian wars, the fabulous Alexander the Great, the rise of Rome, Hannibal, the defeat and conquest of Carthage and Gaul, the eventual downfall of the western Roman empire, Charlemagne, the rise of the Ottoman empire and the fall of Constantinople, the sieges of Vienna and the endless wars of religion. The history of China, too, is presented as a sequence of dynastic changes, as the "mandate of heaven" passed from one emperor to the next. Too much attention is awarded to Great Men (mainly prophets, kings, generals, Emperors and Popes) plus a few women (Joan of Arc, Queen Elizabeth I, Catherine the Great, and the legendary "four great beauties" of China).

By the same token, historians have focused far too little to the role of culture, technology, resources and institutions. I suspect that the evolution of social institutions deserves far more credit from historians than it has yet received, but I am a physicist, by training and inclination, so I will stick—so to speak—with my knitting. The rest of this book deals especially with the major inventions and innovations in history, including ways to exploit the sources of useful energy (exergy) to perform useful work in human society. I cannot hope to compensate for the historical distortions. However, I can focus on a few key innovations while ignoring the self-important men on horseback.

Imagine how human history would have appeared to a very long-lived observer from some Tibetan "Shangri-La" where occasional travelers, like Marco Polo, would stop in occasionally to gossip. He (she?) would have noticed that, around 4000 years ago, slavery became the dominant mode of socio-economic organization. The first empires, East and West, were slave states organized by the winners of territorial conflicts (who became the landowners) to keep themselves in power and the rest in permanent subjection. The pyramids of Egypt and the Great Wall of China were built by slaves or land-less "guest workers". So was the system of irrigation canals in Mesopotamia (Postel 1989). All of those immense undertakings required a lot of ditch digging, hauling and lifting and there were no machines to do it. Nor, at that time, were there any animals like water buffaloes yet tamed and able to pull heavy weights. The work had to be done by humans.

Later, around 500 BCE, armies had become professionalized. Slaves still did almost all of the work required to support the life-style of the (relatively) few "citizens" of Rome (or the mandarins of Han China), and their armies. Slaves, or landowning peasants, plowed the fields, delved in the mines, made the great roads and the aqueducts, and propelled the ships against the wind. The ordinary soldiers, who fought the battles, captured the prisoners and kept them in their chains—literally or figuratively—were not much better off than the peasants, even if they were legally "free" men, even volunteers.

What may not be obvious, at first sight, is that slavery in Rome, or in Han China, was a step up, in terms of modern values, as compared to its predecessor. The invading Hebrews, having escaped from Egypt, killed all the inhabitants of Jericho, except for one girl. The fate of Troy was the same. (If the defenders had won, of course, the attackers would have received the same treatment.) In prehistoric times (and as recently as Genghis Khan and Tamerlane) it was normal for an invading army to kill all the inhabitants of any conquered city that had offered resistance. Later, the winners kept the women and girls, but killed the men and boys.

But as time passed, the armies became increasingly hierarchical, with a professionalized officer class of armored cavalry, a middle class of bowmen, and a lower class of spear-carriers. Some, like the Hoplites of Athens, were trained. When a military force surrendered, the leaders sometimes had cash value and could be ransomed. The mercenary professional soldiers, under contract, just changed sides. Draftees died or disappeared into the woods after a battle. (Many lived on as vagabonds or outlaws).

But as regards the untrained peasant soldier, he could be disarmed, restrained by iron chains, and put to work in an irrigation ditch, a mine or galley or a construction project. In ancient Greece, as reported by Homer's Iliad, one slave was worth four oxen. Slaves of that worth were well treated. But things changed when lots of prisoners of war were thrown on the slave market at the same time. This must have happened during the Alexandrian wars of conquest during the fourth century BCE and the Roman-Carthaginian wars. The market value of slaves crashed and their treatment deteriorated correspondingly. Slave insurrections like those of Spartacus (73–71 BCE) were the result.

The population of the world, and of the territory controlled by Rome in the West and Han China in the East, grew exponentially. Hindu culture in India had already become stratified by adopting a rigid caste system. Buddhist culture spread throughout south Asia, and China, where it lived in harmony with Confucianism and Taoism and never became stratified or militarized. China developed during its early history in the fertile central plain between the two great rivers, the Yellow River in the north and the Yangtze in the south, where agriculture prospered. The horse-breeding Mongols and the Manchus north of the Wall lived primarily by raiding. South of the Yangtze the climate was tropical and (like India) covered by jungle and difficult to cultivate.

The Han dynasty (206 BC–220 AD) is known as the "golden age" of Chinese history. The emperor Wu Di conducted a census in the year 2 CE, which set the population of China at 57,671,000. At its peak, around 200 AD, China encompassed all the territory north of Viet Nam and south of Mongolia, including part of Korea. The capital of China, An Fang (a few km from present-day Xi'an)—the site of the recently discovered terracotta soldiers—was a trading city, with a population estimated at two million. It was the eastern end of the "silk road" which linked China to the Mediterranean coast in Syria. Why did the Han dynasty collapse? (And why did its successors collapse?) It is a good question that can only be answered by hints. Corruption, misgovernment and military revolts were always involved. But famine due to periodic floods or over-population was the most common factor.

This history raises an interesting question: if famines due to population pressure repeatedly caused serious social unrest and revolts, which are documented, how is it that today's China, which has a population of 1.2 billion, living in roughly the same territory as Han China, can feed itself at a standard probably considerably higher than 2000 years ago? The radical increase in agricultural productivity, noted in Sect. 8.4 a few pages back, was responsible, of course. China currently imports some rice and some grain, but most of its food is home grown and it also exports. True, some of the former jungle south of the Yangtze has been cleared and converted into wet-rice paddies. But, the rest of the answer is that, during every one of its dynastic changes, there was a period of "creative destruction", resulting in changes and improvements in the technology of agriculture. Agricultural productivity in China today is many times higher than it was during that "golden age".

To feed Rome and its armies, new large estates or "latifundia" in North Africa, using slave labor, were introduced. For a time, they produced food at much lower cost than the free peasants of Greece or Italy could do. The population of Rome was kept happy—a relative term—by free grain and "pane et circenses" (bread and circuses). The bread was made from imported grain from North Africa and the circuses were mostly contests between (slave) gladiators or dangerous animals. Ruined economically, the farmers of Rome became the permanent Roman army (whose business was to acquire more slaves) and the Roman proletariat. The Roman Empire was fed for several hundred years by the grain-lands of North Africa, Egypt and Gaul. But the eastern regions around the Black Sea became more important, and Diocletian split the government into two regions, keeping the eastern one for himself, which he ruled from Croatia (Split). The Emperor Constantine, who followed, moved the eastern capital to Byzantium (renamed Constantinople) which dominated East-West trade for the next thousand years and more.

The western Roman empire finally perished because it had become a military society that had no room to expand except to the north, where there were no more cities to loot and where the land was heavily forested, easy for barbarians to defend, and difficult to farm. The Christian Church, having been legalized by Constantine, spread rapidly throughout both halves of the empire after 325 CE. This process was accelerated after the key doctrines—the "Nicene creed"—were agreed to by the Council of Nicaea (now Iznik, Turkey). There was one key difference, having to do with the nature of the "Trinity" and the divinity of Christ. It is said, tongue in cheek, that the Eastern and Western branches of the Catholic Church were separated by the placement of a single letter, hence the phrase "an iota of difference".

The (western) Roman Empire finally collapsed under the onslaught of barbarian Germanic tribes, such as the Huns and the Visigoths, probably due to inability to pay their own armies or protect the farmers who fed the cities. New feudal societies evolved in Europe, with parallel and competing secular and clerical (Church) hierarchies. The slow-motion collapse of Rome left a vacuum briefly filled by the Emperor Charlemagne. But his successors, the Holy Roman Empire (neither Holy nor Roman) were weakened for several hundred years by the unresolvable conflict between ecclesiastical and secular power centers, which meant control over resources. Ecclesiastical power peaked when Pope Alexander III forced the Holy Roman Emperor, Frederick I ("Barbarossa") to acknowledge his authority over the Church and to kiss the Pope's foot in Venice (1178). Barbarossa died during the third crusade (1190) leaving no obvious successor. Thereafter, the Papal power was supreme in Western Europe for a century.

Its signal "achievement" was the fourth crusade, initiated by Pope Innocent II, in 1198, who wanted to end the split between the Eastern and Western branches of the Catholic Church, under his own control. The avowed purpose was to support and defend the Christian communities in the Holy Land. The Crusaders agreed to pay the Venetians 85,000 silver marks for the ships and personnel to transport the huge Crusader army to Cairo. The Holy See, having no money of its own, reneged on most of this debt. This left Venice in a severe financial difficulty.

To recover their investment the Venetians, under doge Dandolo, diverted the bulk of the Crusaders to Constantinople. They were welcomed as protectors at first, but they proceeded to sack it in 1204. Crusaders destroyed the Hagia Sophia and the great library, among other desecrations. The value of gold and other precious goods stolen was about 900,000 marks, of which the Venetians recovered about 200,000 marks, making a profit. (The marble face and interior walls of St Mark's Cathedral in Venice were all brought from Constantinople.) The rest of the loot was divided among the Christian Princes who led the Crusade and the Venetian, Florentine and Genoan merchants who financed it. Byzantium itself was subsequently split into three parts, but the Mongols, Tatars, Turks, Kurds, Turcomans and Persians gradually swallowed the hinterland. Byzantium survived in Greek enclaves like Trebizond and Jaffa around the Black Sea for another 250 years, but finally fell to the Ottoman Turks in 1453 CE. (Georgia and Armenia remained independent).

The peasant farmers of Europe, both east and west of the "pale"—roughly the boundary between the two branches of Catholicism (the eastern border of Poland)—were not bought and sold. But they were taxed, drafted into military service (for which they had to provide their own weapons) and tied to the land. Most were no better off than slaves. Jews were expelled at times by England, Spain and Portugal and Russia "beyond the Pale". They moved, when possible, to other places, such as the Low Countries, Austria, Poland, the Ukraine (especially Odessa) and Turkey. Residence in towns in Western Europe was very tightly restricted as membership in the guilds of artisans (not to mention musicians, like the Meistersingers of Nuremberg) was even more tightly controlled.

Slavery and the near slavery of feudalism were abolished in Western Europe in stages, beginning in the fourteenth century and finishing early in the nineteenth century. Western European feudalism started to break down during the "Black Plague "of the fourteenth century, which created a labor shortage and enabled towns to become more independent of the landed aristocracy. Slavery in Europe was finally swept away by the French Revolution and the Industrial Revolution. But slavery was still a big business for European ship-owners supplying the plantation economies of America. Over ten million African slaves were transported to North and South America during the seventeenth and eighteenth centuries. Slavery existed in the Arab countries of Africa until the middle of the nineteenth century. It still survives in parts of India, where children are sometimes sold by their parents for debt.

Slavery in European-American civilization finally ended with the American Civil War, and Abraham Lincoln's Emancipation Proclamation of January 1, 1863. That was followed by several amendments to the US Constitution. In that first modern war the industrialized, "abolitionist" Northern States crushingly defeated the agrarian, semi-feudal, slave-based southern Confederacy, even though the latter had quite a bit of help from the English, whose Manchester-based textile industry depended on cotton imports. The American Civil War probably demonstrated, once and for all, that a slave society cannot compete economically with a free one.

8.7   Money and Coinage

In the last section, I mentioned money as a hidden driver of great affairs, such as the Crusades, even though religious faith was ostensibly the cause of some of them. Money—in some form—was a necessary precondition for trade (beyond simple barter) and markets. Markets were necessary to provide goods and services not provided by slaves (even when some of the goods were produced with slave labor.) And even slave markets—like most markets—required a medium of exchange.

The medium of exchange in primitive markets could have been some familiar and standard commodity, like cattle skins, olive oil, cacao beans, coconuts, ivory, salt, or even cowrie shells (Weatherford 1997). There were debts and credits, recorded in a variety of ways from stone tablets to papyri, long before money appeared. But the medium of exchange had to be transportable and storable (long-lived).

When you think about it, the ancient world had few storable commodities of general use that had long enough lifetimes and that could be accurately quantified. Measurement of weight was a major problem in ancient times. The discovery that made **Archimedes** most famous was the law of specific weight, which enabled him to determine that the gold crown of the tyrant Hieron II (of Syracuse) had been adulterated with silver. According to legend he was in the bath when he realized how to solve the problem without destroying the crown, and ran naked into the street shouting "Eureka" ("I have found it!").

To make a long story short (long in the historical sense) "markets" before money were very local, amateurish, and inefficient for thousands of years. This began to change when gold, silver, copper or iron coins were invented. Gold was probably the first, and especially important, metal for coinage. The metal is malleable and can be melted at a reasonable temperature. It was occasionally found in nature in relatively pure form, as nuggets or dust. Jason was a character in Greek mythology. But the "golden fleece" he allegedly sought was an early method of recovering fine particles of gold from a flowing stream using a sheepskin. The particles of gold stick to the lanolin in the wool. Of course gold was also decorative, and gold jewelry was a sign of wealth. In India it is the principal mode of inter-generational transfer of wealth (as brides' dowries) to this day.

Like gold, both silver and copper were occasionally found as recognizable nuggets. But as demand increased, the metals had to be mined and smelted. They were found in sulfide ores at first, where heat alone was sufficient to drive of the sulfur and purify the metal. It seems that copper was first found in Egypt, but was carried by traders from many other places, especially along the "ring of fire". In any case, the spread of copper tools and ornaments occurred roughly at the times of various documented migrations of tribes. The origin of the more sophisticated carbo-thermic smelting technology for oxide ores can only be guessed. But that technology, when it did appear, led to the "iron age". Archeological evidence (consisting of rust associated with decorative beads) suggests that iron for decorative purposes was known before 4000 BCE. Other uses, e.g. for spear tips, were becoming widespread around 1300 BCE. Ultra-pure iron, as exemplified by the so called "Ashoka's tower" in Delhi, India, was produced in small quantities in a few places around the world before the Christian Era (BCE). In those days pure iron from meteorites was perhaps almost as precious as gold.

Mines were strategically important, since the first "industrial metals", notably bronze and brass, could be made into weapons for armies and the armies could be paid in coins. The coins had to have fixed weights and fixed purity to be acceptable. The legendary inventor of metal coinage was King **Croesus** of Lydia, a small but prosperous Greek city-state in western Anatolia (Turkey). This happened about 560 BCE.[3] The problem of standardization was solved by monopolizing the process and stamping an image of a lion's face on the two sides of the coin and forbidding anyone but the king to produce coins. Coinage (and trade) spread rapidly after Croesus' landmark innovation, and it was accompanied by a huge increase in regional trade. Lydia, itself, was briefly an important center for cosmetics production. Lydia's wealth made the city a target for the Persians, who conquered the city in 547 BCE and adopted its coinage system.

---

[3] The first silver coins were minted by King Pheidos of Argos, about 700 BCE. Croesus' coins were made from electrum, an alloy of gold and silver. The Persians subsequently eliminated the silver and produced pure gold coins.

Curiously, coins also seem to have been invented and used, around the same time, in India and in China. We don't know the names of the innovators, but the coincidence of dates is interesting. David Graeber suggests that one driver of that innovation was the advent of mercenary soldiers, who needed to be paid in some way less destructive than by allowing them to loot (Graeber 2011). ("Cry havoc" was the traditional command for unpaid soldiers to rape and pillage.)

Credit, as an institution, seems to be as old as agriculture. If there was a bad year, many of the farmers needed to borrow money to plant their spring crops, and the records of those credit transactions go way back. But cash was not necessarily involved during the heyday of self-sufficient empires with central administrations, like China, Babylon and Egypt. It was the need to pay soldiers, above all, that made coinage important. The Greco-Roman era of (mostly) cash transactions and coinage (roughly 800 BCE–600 AD) ended in the West during the "dark ages".

Credit was revived by the Crusades. In particular, it was provided by the Pope's private army, the Knights of the Temple of Solomon ("Templars") during the twelfth and thirteenth centuries. Credit was offered to members of the feudal nobility (and royalty) who needed cash to buy arms and transport their soldiers to the Holy Land. The Templars' own funds consisted of gifts and endowments from their initiates, at the beginning, and later from the lands and castles of the Crusaders who did not return with enough loot to repay their loans, or did not return at all. In fact, the debt burden from financing the Crusades triggered a shift away from ecclesiastical to secular power in Europe.

That power shift was very sudden. Secular power first prevailed over ecclesiastical power when Philip "the fair" of France (he was a blonde) destroyed the Templars in a single stroke: it was Friday the 13th of October in 1307. (Friday the 13th is still regarded as an omen of ill-luck.) Along with most of the French aristocracy, he was heavily in debt to the Templars. King Philip destroyed the Templars because he couldn't pay his debts to them. He erroneously expected to find a literal pot of gold. What his agents probably found was titles to a variety of castles and villages held as security for loans outstanding. Frustrated and furious, he took it out on the prisoners. Captured Templars were accused of horrible crimes, tortured until they confessed, and then burned at the stake. Philip also kept title to those lands and castles, which he used to reward supporters and centralize power.

Philip then took control over the Papacy itself, and moved the Vatican from Rome to Avignon, where it stayed under the French King's thumb for over a century. But without a central secular authority, Europe disintegrated into a number of competing but essentially independent principalities ruled by barons and a few bishops. The change in the role of money beginning at that time is nicely illustrated by the following quote: Here speaks a fictional English Franciscan Monk, William of Casterbridge (student of Roger Bacon) explaining money, in Italian cities c. 1326, to his German scribe, the Benedictine novice, Adso.

> "The cities are like ... so many kingdoms ... And the kings are the merchants. And their weaponry is money. Money, in Italy, has a different function from what it is in your country, or in mine. Money circulates everywhere, but much of life everywhere is still dominated and regulated by the bartering of goods, chickens, or sheaves of wheat, or a

scythe, or a wagon, and money serves only to procure these goods. In the Italian city, on the contrary, you must have notices that goods serve to procure money. And even priests, bishops, even religious orders have to take money into account." (Eco 1984 p. 126).

Both Venice and Genoa were already great maritime and financial powers, and had trading colonies all over the Mediterranean and Black Seas. Other Italian city-states, especially Florence and Siena, were important centers and financiers of commerce. The first banks, created in the fourteenth century to fill the vacuum left by the departure of the Templars in 1307, were family affairs: the Bardi, Peruzzi and Medici of Florence, the Grimaldi, Spinola and Pallavicino of Genoa, the Fuggers and Welsers of Augsburg, and so forth. When the Templars disappeared, the need for their financial services did not. These banks financed the age of discovery and the Italian Renaissance. When the Popes moved back to Rome in the fifteenth century, after leaving Avignon, the Renaissance was under way and unstoppable.

International banking, with currency exchanges, developed first in Italy (Florence, Venice, Genoa) and spread in the sixteenth century to other European cities including Augsburg, Frankfurt, Paris, Antwerp, and London. For international transactions metal coins were not much used. Purchases were financed by claims on another bank or goldsmith or a promissory note from a King. A "bill of exchange" issued by a reputable banker could be "cashed" in a branch of that bank in another currency in another city. The heyday of the Medici's banks was the fifteenth and sixteenth century. By the middle of the sixteenth century, the Queen of France (and mother of three kings) was Catherine de Medici. The Fuggers and Welsers of Augsburg bloomed in the sixteenth and seventeenth centuries. Those banks and many others were destroyed by lending (under duress) to monarchs who lost wars and could not pay.

The Rothschild bank was started in 1760 by Mayer Rothschild in the free city of Frankfurt. But he wisely sent four of his five sons to create branches in other cities (London, Paris, Vienna and Naples), thereby protecting the family's wealth from the consequences of wars and revolutions. It is said that the Rothschilds were the financiers behind J.P. Morgan and Henry Rockefeller.

For more on money, banks and economics, see Sect. 12.4.

8.8 Productive Technology

There is much more to be said about technological evolution during the pre-Roman period, especially in Greece and Alexandria. Invention and innovation were not lacking during that period. Several sorts of looms were in use in Mycenaean Greece, more than 1000 years BCE. **Archimedes** is justifiably regarded as one of the great inventors of history, but he was only the most outstanding one of a considerable list of inventors, including Aristotle, Eudoxos, Hephaestus, Heron, Hipparchus, Ktesibios, Philon, Plato, Ptolemy, Pythagoras, Timosthenes and other nameless ones (Kotsanas 2011). Kotsanas has described and reproduced over 200 of the inventions of Ancient Greece and its satellite territories (Alexandria, Byzantium, etc.). Most of the mechanical inventions of modern times were anticipated by the Greeks.

These inventions ranged from automata, sundials and hydraulic clocks, oil presses, screw pumps, compound pulleys, astronomical instruments, cranes, ships, siege machines, cross-bows and catapults. Many had military uses, while others were primarily scientific instruments, time-keeping devices or amusements. But the Greeks had screw-cutters (both male and female), nuts and bolts, hemp ropes and cables, pulleys for lifting heavy weights,[4] hydraulic and pneumatic machines, chains and sprockets. With this technology, they built marvelous temples, free-standing statues, massive fortifications and monumental tombs. These were individual creations. Some were reproducible, by skilled artisans. What they did not have was structural materials apart from wood and stone, or manufacturing (replication) systems—other than human slaves—in the modern sense. I will come back to this point later.

While manufacturing technology was slow to develop, the Dutch made a huge advance in hydraulic engineering during the Middle Ages. The reasons for colonization of the waterlogged marshlands in the Rhine River delta by farmers are lost in history. But the likely reason was for protection against the raids by mounted or seaborne raiding parties. But the land of the delta was marshy at the best of times and frequently subject to floods, both from storms off the North Sea and from spring floods down the Rhine. (Some of those floods were very large. One notable Rhine flood in 1784 raised the water level more than 20 m above normal levels at Cologne. Most of the cities along the Rhine were totally destroyed by that flood.)

What the Dutch did, starting around 1000 CE was to build dikes (to keep the river within its banks), drainage canals and windmills. Those highly sophisticated windmills were built primarily to pump water continuously, by means of an Archimedes screw, from canals and ditches to reservoirs at higher levels, and thence into the river. (Some of them are now UNESCO "heritage" sites). Much of the land of Holland is now permanently several meters below sea level (and the land is still sinking). One of the interesting features of the Dutch water management system is that it is an essentially public enterprise in which every citizen has a direct interest.

It is evident that the technological innovations before 1500 CE did not increase manufacturing productivity in any significant way. The only materials that were produced in bulk before 1500 CE were paper and silk (in China), cotton fabrics (in India), linen and wool cloth, plus wood, glass, bricks and tiles. There are two categories. The first category consists of materials that were processed

---

[4] One of the tourist sites in Mycenae (supposedly the home of Agamemnon) is an ancient royal tomb, still in nearly perfect condition. Above the entrance is a lintel stone resting across two walls that weighs 120 ton (est.) How did it get there? The crane—made from wood—that lifted that stone had to be colossal.

mechanically. Thus paper was the final product of a sequence starting with cutting mulberry trees, crushing the wood, bleaching the fiber, dewatering the mush and pressing it into a sheet that could be cut into pieces. (One district in China paid its annual tax to the central government in the form of a million and a half sheets of paper.)

Cotton, linen and wool fabrics are the end products of a long sequence starting with harvesting the plants or shearing a sheep, washing the fiber, bleaching, combing, spinning into yarn, dyeing, color-fixing (with alum) and weaving on looms—not necessarily in that order. Bricks and ceramics are the end product starting from clay and sand, mixing with water, dewatering and pressing into molds, followed by baking to initiate a chemical reaction at high temperatures. Leather is produced from the skin of an animal, after washing, hair removal, and chemical treatment (tanning) to fix the protein in the skin and stop it from decay.

The second category consists of products involving a biochemical transformation depending on anaerobic organisms (e.g. yeast) that convert sugar by fermentation into alcohol and carbon dioxide. This applies to bread, beer, wine and other beverages. The sequence starts with harvesting grain, potatoes or fruit (e.g. grapes), threshing (in the case of grain), crushing, grinding, mixing with water (in the case of beer) and confinement in an enclosed anaerobic volume for the fermentation to take place. Grain for beer may be roasted. Bread is baked after the dough "rises" thanks to fermentation by the yeast. The baking process for bread breaks up large molecules and makes them more digestible.

In all the above cases there is a considerable need for energy in useful form (the technical term is exergy) beyond the energy-content of the raw original material. Energy is needed especially for the washing, bleaching and baking steps. Note that all of these are "batch" processes, based on a recipe, but that no two batches were ever identical. Technological improvements over time have made some of the mechanical steps (like crushing, spinning and weaving) more efficient, but usually by increasing the mechanical energy required.

Bread, beer and wine are consumed as such. But more complex objects were made from wood, paper, fabrics, and leather by craftsmen, possibly with help from apprentices. Clothes for the rich were made by tailors or dress-makers. Shoemakers made shoes and boots. Horse-shoes and other farm implements were made by blacksmiths, knives and swords were made by armorers. Books were copied individually (by monks). Items of furniture were made by carpenters and cabinet-makers.

The "division of labor", observed by Adam Smith in 1776, had barely begun by 1500 CE except in textiles, construction, ship-building, sailing long distances and military operations. Most things, like shoes or guns, were made by craftsmen one at a time. The problem of replication, even of simple mechanical designs, had not yet been recognized, still less solved.

The problems of standardization and replication (especially of interchangeable metal parts), became more difficult as the objects to be manufactured became more complex, with different shapes requiring different materials (ibid). For instance, a Jerome clock c. 1830 might have had 100 different metal parts, ranging from threaded nuts and bolts, to axles, washers, rivets, sleeve bushings and various stampings, plus several complex components such as escapements and gear wheels that had to be machined more accurately (Ayres 1991). The nuts and bolts and washers were probably purchased from another firm. A sewing machine (c. 1840) had 100–150 parts, including some product-specific castings and machined parts like gear-wheels, plus a number of stampings and screw-type items that could have been mass-produced elsewhere [ibid].

The Springfield rifle of 1860 had 140 parts, many of which were specialized castings. Next in increasing complexity, and later in time, were typewriters, bicycles, automobiles and airplanes. IBM's "Selectric" typewriter (no longer in production) had 2700 parts [ibid]. Very complex objects such as Boeing's 747 jet aircraft may contain several million individual parts (nobody has counted them). A firm like AMP that only makes electrical connectors, has over 100,000 different types in its catalog. Nowadays, only 10–15 % of the parts of a complex modern product are typically made in the same factory as the final assembly.

The technical problem of standardization still vexed manufacturers until late in the nineteenth century. It was the Connecticut hand-gun manufacturers, notably Colt and Remington, who were the pioneers of the so-called "American system of manufacturing" (Rosenberg 1969; Hounshell 1984). The solution of the interchangeability problem, for guns, clocks, and more complex products, was basically a question of material uniformity and mechanical machine tool precision and accuracy. It began with artillery manufacturing, under the rubric "système Gribeauval" for the French artillery general who introduced it in 1765. Gribeauval's system was further developed in France by Honoré Blanc, to achieve uniformity of musket parts at an armory in Vincennes. Thomas Jefferson, who was in France at the time, learned about Blanc's work. After the Revolution, it was quickly adopted and imitated in the US by **Eli Whitney** (c. 1798). Whitney's efforts were obstructed by the lack of machine tools. The general purpose milling machine was his invention (1818).

The most important pioneer of mass production was Samuel Colt, who invented a pistol with a revolving breech (the "Colt 44") that could fire multiple shots without reloading. This was a huge advantage over muskets. It was said that a Comanche Indian warrior could fire six arrows in the time needed to reload a musket. Colt's factory was the first to mass produce a complex metal object. In fact the British Parliament sent a delegation to Connecticut to visit Colt's factory in 1854 and later invited him to build a similar factory in England (Hounshell 1984).

The "systems" concept, in which a manufacturing plant was designed for each product or family of products was truly revolutionary. It enabled American manufacturers in the nineteenth century to take the lead in a variety of manufacturing sectors, such as machine tools (Pratt & Whitney), watches and clocks (Waltham, Hamilton), sewing machines (Singer), typewriters (Underwood, Remington), calculators (Burroughs), bicycles (Schwinn), motorcycles (Harley-Davidson), agricultural machinery (International Harvester, John Deere, Caterpillar), automobiles (Chevrolet, Oldsmobile, Buick, Ford, Chrysler) and finally aircraft.

There were other important innovations later in the nineteenth century. The "division of labor" was advanced especially by **Frederick Winslow Taylor**, the pioneer of time and motion studies, who advocated scientific management, based on the application of engineering principles on the factory floor ("Taylorism"). Taylor's ideas strongly influenced **Henry Ford**, who adapted them to the mass production of automobiles starting in 1908. Ford is credited with innovating the moving assembly line.

Mechanization of industry yielded enormous productivity gains between 1836 and 1895, according to a massive study by the US government (Commissioner of Labor 1898). A few examples make the point. Comparing the number of man-hours required in between the year of earliest data and 1896 the productivity multiplier for cheap men's shoes was 932 (1859–1896), for axle nuts 148 (1850–1895), for cotton sheets 106 (1860–1896), for tire bolts 46.9 (1856–1896), for brass watch movements 35.5 (1850–1896), for rifle barrels 26.2 (1856–1896), for horseshoe nails 23.8 (1864–1896), and for welded iron pipe 17.6 (1835–1895). While these are just the most spectacular examples, they make the point. Mechanization drove US industrial output in the latter half of the nineteenth century. Other countries, especially Germany, were close behind.

The outstanding change in manufacturing technology in the twentieth century has been programmable automation, initiated during the Second World War.

In summary, the invention of printing, followed by large-scale manufacturing of goods with division of labor, standardization and interchangeable parts, is somewhat analogous to the "invention" of cellular (RNA and DNA) replication in biological evolution. The analogy is rough, but the emphasis in both cases was on copying accuracy and error minimization.

Chapter 9
The New World: And Science

9.1 The Discovery of the New World

The year 1500 was a major turning point in human history. In the year 1500 CE China was technologically more advanced than Europe in most arenas, including paper and printing, not to mention gunpowder and the compass. Chinese society was as hierarchical and bureaucratic, but far more introverted, than Western society. The demise of the Silk Road was probably triggered by the rise of the Ottoman Turks, who managed to capture Constantinople and defeat the Byzantine Empire in a few years. Soon after, they started to tax and otherwise interfere with the caravans on the traditional "silk road" to China. The western demand for spices, dyes and raw silk did not abate. New trade routes became highly desirable.

After the fall of Constantinople (1456) the "age of exploration" began to get serious in the West, led by Portuguese sailing ships, under Prince Henry "the Navigator". In 1477 the Duchy of Burgundy was split after the death of Duke Charles "the Bold", with the southern part merging with France under King Louis XI and the northern parts passing to Austrian (Hapsburg) control. In 1478 Ferdinand and Isabella had merged their kingdoms (Aragon and Castille) to create Spain. They began a campaign to expel the Moors. In 1487 the Portuguese explorer Bartolomeo Dias reached the southern tip of Africa (Cape of Good Hope). In 1492, the Kingdom of Granada surrendered to Ferdinand and Isabella, and the three ships of Christopher Columbus had reached America (for Spain). In 1498 Vasco de Gama reached India, and later expeditions established Portuguese trade in the region. In 1496 Pope Alexander VI divided the "new world" into Spanish and Portuguese parts, leaving the Dutch and the English out in the proverbial cold.

By 1500 CE the Ottoman Turks had cut the "silk road" completely. It was the year of the birth of Charles V, later emperor of the Holy Roman Empire (from 1519 to 1556). Charles V and his son, Philip II fought the Ottoman Turks along their eastern frontiers more or less continuously. By mid-century Spain under Charles V had found great sources of gold and silver in Mexico, Bolivia and Peru. This gold

© Springer International Publishing Switzerland 2016
R. Ayres, *Energy, Complexity and Wealth Maximization*, The Frontiers Collection,
DOI 10.1007/978-3-319-30545-5_9

enabled Spain (with the Hapsburg territories) to become the dominant power in Europe, at least for the next century until the end of the 30 Years' War (1648). The Spanish monarchy paid for its wars with Aztec gold, Inca silver and silver "thalers" from a mine in Bohemia. The Turks conquered much of the Balkans (Bosnia-Herzegovina, Montenegro, Albania) but were stopped in Hungary, Croatia and Serbia. That line of demarcation, which also roughly follows the line between Roman and Eastern Orthodox Catholicism, was a sore spot in European history until the twentieth century.

Ever since 1450, the printing press had greatly increased the spread of news and new ideas, while sharply decreasing the centralized authority and prestige of the Church of Rome and its prelates. But back in Rome, the Popes needed (and wanted) more money than the church tithes brought in. Having nothing tangible to sell, the Popes started to sell lucrative Church offices (including Bishoprics and Cardinal's hats) and certificates promising blanket forgiveness of all sins, known as "indulgences".

In 1517, Pope Leo X authorized a new round of sale of indulgences in Germany to finance the building of the basilica of St. Peter's in Rome. Martin Luther, a priest and a monk who was also Professor of Theology in Wittenberg, Saxony, objected strongly to this money-raising stunt (along with many others). He nailed a paper listing "95 theses" to the door of the Castle Church. Specifically it was a challenge to debate (in Latin): *Dusputatio pro declaratione vertutis indulgentiarum*. This was a custom at the university, comparable to publication of a journal article nowadays. He also sent handwritten copies to some clerics, including the Archbishop of Mainz who was in charge of the sale of indulgences locally.

Luther's challenge "went viral" in modern terms. The handwritten paper was translated into German and printed by a Luther supporter, Christoph von Scheurl. It had spread throughout Germany within 2 months and throughout Europe in 2 years. In 1520, the Pope issued a rebuttal, *Exsurge domine* (Arise O Lord). Luther was told, on pain of excommunication, to retract his theses. He refused. The Pope then called him to account at the Diet of Worms (1621). Again, Luther refused to retract any of his theses unless he could be shown refutations in scripture. There were none. He finished by saying *"Here I stand. I can do no other"*. Martin Luther was duly excommunicated from the Roman Catholic Church and declared an *"Outlaw"* by the Emperor (Charles V). Yet he was physically protected by admirers in Germany. Within 5 years he had translated the Bible into German, where it became a best-seller. In 5 years Lutheranism was already an important alternative religion.

A similar scenario was shortly enacted in France, where Jean Cauvin (John Calvin) founded another branch of Protestantism (soon moved to Geneva) and others followed in several countries, including Switzerland and Scotland. King Henry VIII's successful "nationalization" of the English Church in order to divorce his Spanish Queen and marry Anne Boleyn (accompanied by privatization of much of the Church's wealth in gold and silver) could not have happened without the German precedent and the accompanying turmoil. The confiscated gold and silver was mostly converted to coinage (a lot of it to used pay mercenary soldiers, on both sides).

Seaborne trade was expanding rapidly, not only to the Far East. In 1519–1522 another Portuguese explorer, Ferdinand Magellan led the first expedition to circumnavigate the globe from east to west, via the stormy straits of Magellan. Francis Drake, from England, followed Magellan's route about 50 years later. The continuous wars between Christian nations and the Ottoman Turks, before and after the Protestant reformation, stimulated technological development. This was particularly notable in metallurgy, ship design and weaponry. (Curiously, China had discovered gunpowder much earlier than the West, but never found a way to use it safely in guns as a propellant for bullets.)

The sixteenth century was also more or less the beginning of capitalism as a social phenomenon, albeit not yet explicitly recognized as a social philosophy. The phenomenon was essentially a form of organized long-distance trade, which required modes of organization and finance that had no obvious precursors in the village marketplace. One was a unique cross-national alliance. The Hanseatic League spread among trading cities of northern Europe, mainly the North Sea and Baltic ports, from Novgorod (Russia) and Riga at the eastern end to Bruges, London and Cologne at the western end with Lübeck and Hamburg in the middle. The league traded in timber, resins, wax, furs, amber and other local products. It also had its own navy and legal system and engaged in ship-building, The Hanseatic League was effectively a cartel, which negotiated for special privileges including local compounds for its resident traders. At its peak in the fifteenth century it protected free trade among its members and raised prices for others. That cartel was broken, eventually, by traders backed by nation-states, starting with the Dutch.

The first recognizable example of shareholder capitalism was the foundation of the Company of Merchant Adventurers by Richard Chancellor, Sebastian Cabot and Hugh Willoughby in 1551. It was chartered in 1553 for the purpose of finding a northwest passage to India (China) and capitalized by the sale of 240 shares at £25 each. Three ships started but they were separated and sailed (by mistake?) around the North Cape and into Russian waters. Two of the ships were lost but Chancellor's ship anchored at the mouth of the Dvina River and met some Russians. In the end, Chancellor traveled overland to Moscow, met with Tsar Ivan ("the terrible") and returned to his ship and back to London with a letter from the Tsar approving trade relations.

The company of Merchant Adventurers was subsequently rechartered by Queen Mary as the Muscovy Company (1555). It was followed by the Levant Company (1580), the East India Company (1600), the Virginia Company (1609) and the Hudson Bay Company (1620). Some of these companies were profitable, some were not, but with their charter privileges they were able to negotiate deals with local rulers (in India, especially). This created the pattern that led directly to colonialism.

There is a continuing dispute among scholars as to the relationship between the rise of capitalism and the transformation of religion. One view, espoused by Max Weber, was that the Protestant "work-ethic" (a term he coined) was crucial to the success of industrial capitalism (Weber 1904–1905). Historian R.H. Tawney took the contrary view, that capitalism, with its emphasis on material wealth, was responsible for the decline of Christian values (Tawney 1926). There is some evidence for both views.

9.2   From Charcoal and Iron to Coal and Steel

As mentioned, agrarian societies (Rome was agrarian, as were China, India and all others before the nineteenth century) obtained heat almost entirely by burning wood or charcoal or even dried dung. There was, as yet, no technology to transform the heat of combustion, whether of wood, peat or coal, into mechanical work. Rotary motion was generated only by wind (after windmills were invented), water (after water wheels were invented), or by horses or human slaves on a treadmill. The few machines like winches, or grinding mills for grain or gunpowder, were limited in size and power by the available structural materials (e.g. wood beams with iron bolts) and the available energy source (wind or flowing water).

Pure iron or steel were very expensive in ancient times, because their properties were very valuable but smelting iron from ore was extremely difficult. (The only natural source was nickel-iron meteorites which were very rare and very hard to melt. One of the Egyptian Pharaohs had an iron dagger made from a meteorite, as did some Eskimo tribes. It was called "sky-metal"). The fact that smelting technology was known over 2000 years ago is confirmed by the existence of "Rajah Dhava's pillar"—also known as King Ashoka's pillar—in Delhi. It is a 6-ton object, 22 ft high with a diameter ranging from 16.5 in. at the bottom to 12.5 in. at the top. It is rust-free and looks like stainless steel, but it is actually pure iron (Friend 1926 Chap. XI). The Muslim conqueror Nadir Shah fired his cannon at the pillar, leaving dents, but otherwise doing no damage (ibid).

Pure iron melts at a temperature of 1536 °C, which was very hard to achieve in Europe, until Huntsman's "crucible" process in the mid-eighteenth century. The cost of steel was still extremely high until the Bessemer process was invented in the mid-nineteenth century. High quality iron continued to be expensive until well after the industrial revolution because of technological constraints on the size and temperature of blast furnaces and lack of an effective technology for the removal of carbon and sulfur from the metal. Greater use of iron was frustrated by the high (energy) cost associated with the smelting of iron ore using charcoal as fuel.

In case you have forgotten (or never knew) most iron ore is hematite, an oxide (Fe_2O_3 or Fe_3O_4). It was laid down in ancient times by specialized bacteria called "stromatolites" (Chap. 5). Those bacteria were oxidizing ferrous iron (FeO) dissolved in the oceans for their own metabolic energy. The smelting process in a blast furnace is a way of detaching oxygen atoms from the iron and moving them to the carbon, producing CO_2. It sounds simple. But the process in reality is quite complicated, since it involves several stages.

The first stage is incomplete combustion of fuel (charcoal or coal) to obtain high temperature heat, and carbon monoxide (CO, which is the actual reducing agent). In the second stage the carbon monoxide reacts with the iron oxides, creating carbon dioxide and pure iron. That's the big picture. However, at a temperature of around 1250 °C some of the carbon from the charcoal or coke actually dissolves in the molten iron. But if the fuel was coal, and there was some sulfur in the coal, the sulfur would also be dissolved in the liquid iron. The higher the temperature, the more carbon dissolves. Pig iron is typically 5 % carbon by weight (but 22.5 % by volume). Unfortunately, cast "pig iron" is brittle and not much use for most practical purposes. So the metallurgical problem was to get rid of the carbon without adding any sulfur or other contaminants.

Unfortunately, metallurgists at the time had no idea about the effects of any of those impurities. Swedish chemists in the early eighteenth century knew that sulfur in the ore (or the coal) produced "hot shortness", an undesirable condition. It was mostly trial and error. The long-term solution was coking, to remove the contaminants in the coal by preheating and burning off the volatiles. The first to use coke was **Abraham Darby** of Coalbrookdale. He built the first coke-burning blast furnace in 1709 and supplied the cylinders for Newcomen's steam engines (pumps) and, later, the iron for the Severn River Bridge. He died in 1717 but his family tried to keep his methods secret. Some other iron smelters began using coke in the mid-eighteenth century, but adoption of coke was very slow.

Pure (wrought) iron was made mainly in Sweden, in the early eighteenth century by heating pig iron (from good quality ore) with charcoal in a "finery forge" where further oxidation removed carbon and other impurities. The melting point of pure iron is higher than steel, so almost all operations took place with solids. Wrought iron is very durable, forgeable when hot and ductile when cold; it can be welded by hammering two pieces together at white heat (Smith 1967).

The next innovation in iron-making was **Henry Cort**'s "puddling and rolling" process (1783–1784) which was a tricky way of using coal rather than coke to get rid of silicon without adding sulfur to the metal, based on the different melting temperatures of the pure iron vs. the dross. The manufacturing process involved physically separating the crystals of pure iron from the molten slag. In a second step, the remaining slag left with the iron was expelled by hammering (forging). It was very labor-intensive, but it did yield pig iron as good as Swedish charcoal iron. This made it possible to produce pig-iron pure enough to be forged and rolled. Cort also introduced rolling mills to replace hammer mills for "working" the metal.

By 1784 there were 106 furnaces in the UK, producing pig iron, of which 81 used coke. Furnaces sizes were up to 17 tons/wk. Annual output in that year was about 68,000 tons of pig iron. The puddling process speeded manufacturing, and cut prices. Output boomed. Thanks to the demands of war, English production of iron jumped to 250,000 tons by 1800.

To make steel, which is much harder than wrought iron, and consequently far more useful and valuable, most of the carbon has to be removed from the "pig", but (as we now know) not quite all of it. (Low carbon steel still has 0.25 % carbon content, by weight. High carbon steel may have 1.5 % carbon content). But the problem was that steel does not melt at 1250 °C, which was about the limit for a blast furnace at the time. It needs to be heated to between 1425 and 1475 °C. to liquefy, depending on the carbon content. (The melting point of pure iron, carbon-free, is 1535 °C.) That is the reason why—until the industrial revolution—steel was mainly reserved for swords and knives that needed a hard edge. But steel blades had to be forged by an extremely labor-intensive process of repetitive hammering and folding a red-hot solid iron slab, burning the excess carbon off the surface, and repeating the process many times. The high quality steel swords made in Japan and Damascus were produced this way (Wertime 1962).

Steel was produced from wrought iron (usually from Sweden) before the eighteenth century, albeit in small quantities. Iron bars and charcoal were heated together for a long period (a week) in a stone box. The resulting metal was steel. But the steel rods had tiny blisters on the surface (hence the name "blister steel"). The red-hot metal was then hammered, drawn and folded before reheating, as in a blacksmith's forge. Blister steel sold for around £3500 to £4000 per metric ton in the mid-eighteenth century. The next step was to bundle some rods together and further re-heat, followed by more hammering. This was called "cementation" This homogenized material was called "sheer steel". It was made mostly in Germany.

The first key innovation in steel-making was by Benjamin Huntsman, a clockmaker in Sheffield who wanted high quality steel for springs (c. 1740). After much experimentation, his solution was "crucible steel". Starting from blister steel, broken into small chunks, he put about 25 kg of chunks into each of 10 or 12 white-hot clay crucibles. The crucibles were re-heated, with a flux, at a temperature of 1600 °C to for 3 h. The resulting crucible steel was used for watch and clock springs, scissors, axes and swords.

Before Huntsman, steel production in Sheffield was about 200 metric tons per annum. In 1860 Sheffield was producing about 80,000 tonnes, and all of Europe steel output was perhaps 250,000 tons by all processes. By 1850 steel still cost about five times as much as wrought iron for rails, but it was far superior. In 1862 a steel rail was installed between two iron rails in a London rail yard for testing purposes. In 2 years about 20 million rail-car wheels ran over the steel rail. The iron rails at either end had to be replaced seven times, but the steel rail was "hardly worn" (Morrison 1966) p. 123. That explains why steel was demanded for engineering purposes, despite its cost.

The Kelley-Bessemer process was the game-changer. Elting Morrison called it "almost the greatest invention" with good reason (ibid). The solution was found independently by William Kelly in the United States (1851), and **Henry Bessemer** in Wales (1856) (Wertime 1962). It was absurdly simple: the idea was to blow cold air through the pot of molten pig iron from the bottom. The oxygen in the air combines with the carbon in the pig iron and produces enough heat to keep the liquid molten. It took some time to solve a practical difficulty (controlled recarburization) by adding a compound of iron, carbon and manganese, called spiegeleisen. But when the first Bessemer converter was finally operational the result was a spectacular fireworks display—and 2.5 tons of molten steel—in just 20 min.

The new process "went viral" to use a modern phrase. Bessemer, with help from investors, did more than Kelley to make the process practical and to overcome some technical difficulties along the way. So he deservedly got his name on the process, which was patented and licensed widely. It was rapidly adopted around the world because it brought the price of steel down to a point where it could be used for almost anything structural, from rails to high-rise structures, machinery, ships, barbed wire, wire cable (for suspension bridges), and gas pipe.

As it happens, the Kelly-Bessemer process was fairly quickly replaced by the "open hearth" process, for two reasons. First, the Bessemer process required iron from ore that was very low in phosphorus and sulfur. Such ores were rare. In fact the best iron ore in France (Lorraine) was high in phosphorus. That problem was eventually solved by the "basic process" of Thomas and Gilchrist (1877). They lined the convertor with bricks of a basic material such as dolomite, which bound the phosphorus and sulfur into the slag. The other problem was that blowing cold air through the bottom of the convertor left some of the nitrogen from the air dissolved in the molten iron. Even tiny amounts of dissolved nitrogen weakened the steel. The final answer was the Siemens-Martin "open hearth" regenerative process in 1864, which accomplished the decarbonization by a less spectacular but more controllable means that produced a higher quality product.

The Open Hearth process dominated the steel industry until after World War II when it was replaced by the "basic oxygen furnace" or BOF. The BOF is essentially the same as the Bessemer process, except that it uses pure oxygen instead of air. Since the advent of scientific metallurgy (c. 1870), it has been found that small amounts of other less common metals, including chromium, manganese, nickel, silicon, tungsten and vanadium can provide useful special properties. "Stainless" steel (with nickel and chromium) is a particularly important example today.

The year 1870 was the advent of the "age of steel". In 1860 US steel output from five companies, by all known processes, was a mere 4000 tons. By the end of the nineteenth century US steel production was ten million tons. US production peaked around 100 million tons in the 1990s. Global production in 2013 was 1.6 billion tons, of which China was by far the biggest producer (779 million tons).

Energy requirements per ton to make iron and steel have declined enormously since the eighteenth century, but the quantity of iron and steel produced—and consumed—in the world today has increased even faster, thanks to the creation of new markets for the metal, and to economic growth in general. Neilson's fuel-saving innovation was one of several (Bessemer steel was another) that brought prices down sufficiently to create new uses for iron and steel. The price of steel dropped by a factor of 2 between 1856 and 1870, while output expanded by 60 %. This created new markets for steel, such as steel plates for ships, steel pipe and steel wire and cable for suspension bridges and elevators. The result was to consume more energy overall in steel production than was saved per ton at the furnace.

Because of the new uses of coal and coke, English coal production grew spectacularly, from a mere 6.4 million tonnes in 1780 to 21 million tonnes in 1826 and 44 million tonnes in 1846. The increasing demand for coal (and coke) as fuel for steam engines and blast furnaces was partly due to the expansion of the textile industry and other manufacturing enterprises. But to a large extent, it was from the rapidly growing railroad system, which revolutionized transportation in the early industrial era.

9.3 Gunpowder and Cannons

The Chinese are thought to have been the first to discover gunpowder. They claim it as one of their "four great inventions". Since the ninth century they used it in fireworks. But they never succeeded in casting metal gun barrels that could withstand the sudden release of energy of a gunpowder explosion. About 500 years later, in the fourteenth century, Roger Bacon (an English monk) or Berthold Schwarz in Germany (also a friar), or possibly an unknown Arab, rediscovered gunpowder, possibly based on information brought back from China by the Venetian traveler, **Marco Polo** (1254–1324). Thereafter, metallurgical progress in bronze and iron casting enabled the Europeans and the Ottoman Turks, to build cannons that could resist the high pressure of the exploding gases.

Guns did not replace swords overnight. In Japan, the Samurai cultural tradition, based on skill and bravery, rejected the possibility of being beaten in warfare by the chemical energy of gunpowder. Two Portuguese adventurers brought iron (in the form of primitive harquebuses) to Japan in 1543. They were copied, and half a century later Japan was well equipped with firearms. But the government, controlled by Samurai warriors, preferred swords and, over a period of years, guns were prohibited and eliminated in Japan. That ended in 1853 when Commodore **Matthew Perry** arrived in Tokyo Bay for a visit, with four steamships armed with cannons. At first the Japanese thought the ships were dragons puffing smoke. Perry carried a letter to the Emperor from President Millard Fillmore who came later to open trade relations. But his visit also convinced Japan to resume gun manufacture.

Even so, demand for iron was growing rapidly, not only for guns and stewpots but for plows, axles for wagons, and for the steam engines produced by **James Watt** and his business partner (and financier) **Matthew Boulton**. The first substitution of wind power for muscles was by sailing ships. We know that the sailing ships from Tarsi brought "gold, silver, ivory, apes, and guinea-fowls" to King Solomon. Later, sails were employed by Phoenician, Greek, Carthaginian, and Roman ships that cruised the Mediterranean, even though much of the motive power was provided by human galley-slaves chained to oars and encouraged by whips. Sails with steering oars and rudders, of a sort, were also employed by Chinese junks and Arab dhows navigating along the shores of the Pacific and the Indian Ocean.

For a long time the galleys were needed because sailing ships can't sail directly into the wind, so they must "tack" back and forth at an angle to the wind. That maneuver requires an efficient rudder attached along its length from a vertical sternpost. This invention was first introduced around 1156, but it was adopted only very slowly because the engineering details were invisible to observers. The Portuguese caravels were the first ships really capable of sailing against the wind, and it was this development (plus maps) that enabled them to undertake long-distance voyages across the Atlantic and around Africa. The ships capable of sailing closest to the wind were the fastest and most maneuverable, which made ship design a very important military technology.

Galleys were still used in some parts of the world as late as the eighteenth century to provide additional power, especially when the wind was in the wrong direction. However, the "age of discovery" in the fifteenth century was driven by "full rigged" sailing ships capable of sailing entirely without galleys, and thus requiring a rather small crew and carrying food and water for much longer periods out of the sight of land. Full-rigged schooners, some with steel hulls and 2000 tons of displacement, operated competitively over some ocean routes (especially from Australia to Europe) well into the twentieth century.

The sailing ships, armed with bronze or cast-iron cannons, became the instrument of Europe's rise to global military power. Europe's global dominance between the sixteenth and twentieth century was owed (at first) to the ever more efficient maritime use of wind power by ocean-going sailing ships (starting with Portugal). Later, during the heyday of the British Empire, it was due to "gunboat diplomacy" and the military use of chemical energy in firearms.

Guns were the first weapons whose destructive impact did not depend directly or indirectly on muscle-power (The counterweight of a trubuchet had to be loaded). Rather, they transformed the chemical energy of gunpowder into the kinetic energy of bullets, cannon balls, and grenades. Since the end of the fourteenth century Venetian, Genoese, Portuguese, Spanish, English, Dutch, and French sailing ships with cannons and soldiers armed with matchlock (later flintlock) muzzle-loaders, carried the European conquerors to the Americas, Africa, Asia, and Australia. By the end of the nineteenth century much of the world was divided up into European colonies or "protectorates", all conquered and defended by men with guns, carried on "ironclad" ships.

For example, Pope Alexander VI divided the world between Portugal and Spain by a papal bull in 1493, awarding to Spain all newly discovered lands west of any of the Azores or Cape Verde islands, and to Portugal all newly discovered lands to the east (Africa, India and Indonesia.) Except for eastern Brazil (Recife) this gave South America to Spain. Needless to say, the rest of Europe did not agree that the Pope had any right to make such dispositions, and by 1512 the lands around the Indian Ocean were being colonized by the Dutch, French and English. Spain, on the other hand, hung on to its Papal benefice until the nineteenth century.

It is ironic that fifteenth century China, then the world leader in technology, missed the chance of beating Europe to colonize the globe. This can be traced to Chinese politics in the fifteenth century. In the words of Jared Diamond (Diamond 1998):

> "In the early 15th century it [China] sent treasure fleets, each consisting of hundreds of ships up to 400 feet long and with total crews of up to 28 000, across the Indian Ocean as far as the coast of Africa, decades before Columbus's three puny ships crossed the narrow Atlantic Ocean to the Americas' east coast. ... Seven of those fleets sailed from China between 1405 and 1433 CE. They were then suspended as a result of a typical aberration of local politics that could happen anywhere in the world: a power struggle between two factions at the Chinese court (the eunuchs and their opponents). The former faction had been identified with sending and captaining the fleets. Hence when the latter faction gained the upper hand in a power struggle, it stopped sending fleets, eventually dismantled the shipyards, and forbade ocean going shipping ..."

9.4 Steam Power

1776 was a memorable year in human history for several reasons. Most of all, it was because the "Declaration of Independence" was approved on July 4 by the Second Continental Congress in Philadelphia. It started the history of the United States of America with the noble words written by **Thomas Jefferson** :

> "We hold these truths to be self-evident, that all men are created equal, that they are endowed, by their Creator, with certain unalienable Rights, that among these are Life, Liberty, and the pursuit of Happiness."

It is interesting that Jefferson's original draft of the Declaration of Independence included a denunciation of the slave trade, which was later edited out by southerners in Congress. It seems that only after industrialization had provided enough "energy slaves", could the words of the Declaration be finally put into practice—even then, not without great suffering.

As it happens, 1776 was also the year of publication of a book entitled *"The Wealth of Nations"* written by a moral philosopher named **Adam Smith** (Smith 1759 [2013]). This book is widely credited for laying the foundation for economic science, although the French physiocrats also deserve credit. In any case, Smith's book was the source of the powerful ideas like "division of labor", "free trade" and the beneficent "invisible hand" of the market. (To be accurate, Adam Smith never suggested that the "invisible hand" was always beneficent. We will return to this important point later in Chap. 10.) But Adam Smith's thesis was focused mainly on trade as an engine of economic growth and prosperity. That was a new idea at the time, but it came to dominate economic thought, and policy, in the two centuries following.

But Human Rights, as proclaimed by the Declaration of Independence, and market economics, as established by *"The Wealth of Nations"*, could not have become ruling principles of free societies without a revolutionary technology.

We could mention the fact that in 1776, coke-based iron-making overtook charcoal-based iron-making in England, as mentioned in Sect. 9.2. And 1776 was also the year that the first steam engines designed by James Watt, and manufactured by the newly created firm of Watt and Boulton, were installed and working in commercial enterprises. This constituted the second "kick" of the Industrial Revolution and the start of the "age of steam". It is not too outlandish to suggest that steam engines, and later developments, created the preconditions for human freedom from the hardest kinds of muscular work such as pumping water from a well, grinding grain, cutting trees, digging for coal, quarrying stone, plowing and harrowing the soil, "puddling" iron, and carrying bricks.

James Watt was not the first to employ steam power in place of muscles. That honor probably belongs to Thomas Savery or **Thomas Newcomen**. Newcomen's atmospheric pumping engine represented tremendous technical progress as compared to earlier muscle-powered pumps. His first engine, in 1711, was able to replace a team of horses that had been harnessed to a wheel to pump the water out of a flooded coal mine. This was the origin of the "horse-power" unit. The invention solved a serious problem. By 1776 coal was increasingly needed as a substitute for scarce charcoal.[1] But as coal near the surface was removed, the mines got deeper, and as they got deeper they were usually flooded by water. Getting that water out of the mines was essential.

Newcomen's clever trick was to use the energy from burning the coal itself to produce steam. In the Newcomen engine, low pressure steam—barely above the boiling point of water—entered the bottom of the cylinder, while the piston was at its highest position, and pushed the air out. Then cold water was injected to condense the steam, creating a partial vacuum. Then the air pressure above the piston forced it back down, lifting the other end of a rocker-arm enabling it to operate the pump. (It was a reciprocating pump similar to the ones still found in some oil wells.) The steam was at atmospheric pressure. In Newcomen's engine steam was not a source of power to drive the piston itself. During the next 50 years, 75 of these engines were installed in coal mines all over England. Very few changes were made to the basic design until James Watt came along.

James Watt was a Scottish instrument maker. He had been given the job of repairing a model of the Newcomen steam pump for the University of Glasgow. In so doing he became aware of the extreme inefficiency of this engine. (Considering that thermodynamics was not yet a science, this realization on his part required some insight.) Moreover, Watt had a simple and better idea. Watt realized that this arrangement wasted a lot of heat unnecessarily. By saving the condensed steam (which was still hot) in a separate chamber, much less fuel was needed to turn it back into steam.

Watt's first backer was John Roebuck (the man who first industrialized sulfuric acid). But Roebuck went bankrupt and sold his share to **Mathew Boulton** who became Watt's business partner. For a while, progress was frustratingly slow and several times Watt almost gave up on the project. But Boulton persuaded him to continue. The main problem was that they had to work with crude iron castings that couldn't be shaped to a satisfactory degree of precision by the tools they had to work with. They needed to manufacture a large cylinder with a tightly fitting piston to prevent the slippery steam from leaking out.

The solution to that problem was an invention by another engineer, **John Wilkinson**, who owned an iron works and had developed the first practical machine for boring cannon barrels—a fairly big business in England at the time. Wilkinson's boring machine worked just as well for boring the cylinders of Watt's steam engines. The first customer of Watt and Boulton was John Wilkinson, in 1775. He used the engine to drive the boring machine. The next engines were used as pumps in coal mines, like the Newcomen engines. Orders began to pour in, and for the next 5 years Watt was very busy installing his engines, mostly in Cornwall, for pumping water out of tin and copper mines. The efficiency improvement, in terms of lower fuel requirements, was dramatic: The addition of an external condenser

---

[1] From 1850 till 1950 coal commanded more than 90 % of British commercial energy use, wood having completely disappeared by 1870 (McMullen et al. 1983).

saved about 75 % of the fuel used by a similar Newcomen engine. Since the initial changes were fairly easy to make, Boulton and Watt began to license the idea to existing Newcomen engine owners, taking a share of the cost of fuel saved by their improvement.

Watt kept on inventing. He smoothed the movement of the piston by injecting steam alternately on the two sides. He invented the "sun and planet" gear system to transform the natural reciprocating motion of the piston to rotary motion. He gradually made his engine smaller and faster. By 1786 it had become a flexible multi-purpose steam engine. This engine became the main source of mechanical work for the Industrial Revolution—apart from existing water wheels—at least in its earlier years.

In 1806 the first self-propelled traction vehicles on rails, called "locomotives", were built for coal mines using a new generation of steam engines developed by **Richard Trevithick**, a Cornish engineer. Trevithick's "strong" steam engines used steam pressure, in addition to the air pressure, to move the pistons. Since the steam pressure could be several times greater than the air pressure, Trevithick's engines were more compact and more powerful than the Watt and Boulton engines, but also more dangerous. (Remember, both Newcomen and Watt were still using steam at atmospheric pressure, partly because boilers at the time were not capable of safely containing steam at higher pressures).

Engine development did not cease with Watt or Trevithick, of course. Reciprocating steam engines (with pistons) got bigger, faster, more efficient, and better over time. They were used for pumping water from mines, and for powering steamships, railroad locomotives and factories.

The basic components of railways—iron locomotives and iron rails—were mostly accomplished before 1820, thus opening the way to rapid and widespread railway-building, not only in Britain but in other countries. **George Stevenson** built an 8 mile railway for Hetton Colliery in 1819. Stevenson's Stockton-Darlington line opened to passengers in 1825. The 29 mile Liverpool and Manchester line, which cost £820,000 was completed by Stephenson in 1830, and was an immediate success. Railways opened for public traffic before 1830 in the United States, Austria, and France, and very soon afterwards in many other countries. The first steam locomotive used in the U.S. was the "Stourbridge Lion", purchased from England for the Delaware and Hudson in (1827). Peter Cooper built his "Tom Thumb" for the Baltimore & Ohio Railroad in 1830. The first major railway-building boom in Britain occurred in 1835–1837, when many companies were formed, most locally, and a number of disconnected point-to-point lines were built.

These early railway investments produced high financial returns, commonly returning 10 % or more per annum on their capital (Taylor 1942). This attracted more capital to the industry. In the second "boom" period (1844–1846) new railway companies were formed with an aggregate capital of 180 million pounds. Indeed, this boom virtually consumed all the capital available for investment at the time (Taylor 1942). Returns began to fall when the main lines were finished and railway service was being extended into smaller towns by 1860 (in the UK) and by 1890 in the US.

Another direct consequence of the mid-nineteenth century railway-building boom was the very rapid introduction of telegraphy. Cooke and Wheatstone's first practical (5-needle) telegraph system was constructed for the Great Western Railway, from Paddington Station (London) to W. Drayton, a distance of 13 miles (1838). Four years later, it was extended to Slough (Garratt 1958 p. 657). Thereafter, virtually all newly built railway lines were accompanied by telegraph lines. Wheatstone and Cooke formed the Electric Telegraph Co. in 1846; 4000 miles of line had been built in Britain by 1852 (ibid).[2] While telegraphic communication soon developed its own *raison d'être*, the needs of railways provided a strong initial impetus and created a demand for still better means of communication.

Steamboats actually preceded railways. There were a number of experimental prototypes during the eighteenth century: experiments began as early as 1763 (Henry), 1785 (Fitch), 1787 (Rumsey). Symington's *Charlotte Dundas* towed barges on the Forth and Clyde canal to Glasgow in 1803. Commercial service began in 1807 when Robert Fulton's *Clermont* carried passengers 150 miles up the Hudson River from New York to Albany in 32 h, and back in 8 h (Briggs 1982 p. 127). Fulton's rival, John Stevens, is credited with the first sea voyage by steamship (1809), from Hoboken to Philadelphia around Cape May, N.J. (Briggs 1982).

Riverboats became quite common thereafter, especially in the US. Fulton and Livingston started steamboat service from Pittsburgh to the Mississippi in 1811 and on the lower Mississippi River in 1814. These boats all used paddle-wheels and high pressure steam engines evolved from Trevithick's design. Boiler explosions were a regular feature of Mississippi steam boats; at least 500 explosions occurred with more than 4000 fatalities, mostly in the first half of the nineteenth century. However the problem of containing high pressure steam was finally solved by confining the steam to welded tubes and confining the tubes within a stronger outer container, as in the famous Stanley Steamer automobile.

Steamships began to displace sailing ships in the second decade of the nineteenth century. By 1820 the U.S. merchant marine had only 22,000 tons of steam powered shipping, 1.7 % of the total fleet (United States Bureau of the Census 1975). By 1850 this fraction had increased to 15 % ibid. The pace of substitution slowed in the 1850s, in part due to the development of large full-rigged schooners with small crews and power-assisted rigging, but increased thereafter. The year 1863 seems to have been a turning point: it was the year the first steel (as opposed to iron-clad) ship was built, as well as the year of the first steel locomotive (Forbes and Dijksterhuis 1963). Nevertheless, only 33 % of the fleet was steam powered as

---

[2] By contrast, telegraphy got a slower start in the U.S., even though the system developed by Morse (with assistance from Joseph Henry and others) was technically superior and was eventually adopted in most countries. Morse's system did not achieve recognition until the U.S. Congress appropriated money for a demonstration line between Washington D.C. and Baltimore. The demonstration took place successfully in 1844. Again, it was railroads that were the early adopters.

late as 1880 ibid. Thereafter, penetration was very rapid, even though the last sailing ships were still working in the 1930s.

The days of the great three-, four- and five-masted sailing schooners that carried tea from China and grain and wool from Australia to England and passengers in the other direction for much of the nineteenth and part of the twentieth century were the essence of romance. They were also an effective way of projecting European military and economic power, in their day. But those days were cut short by a new and dirtier kind of power. The change is symbolized by Turner's painting "The Fighting Temeraire", which depicts a huge, pale, three-masted battleship with tied up sails. A small, black tugboat, its chimney belching flames and smoke, is seen towing her up the River Thames to her last berth, where she is to be broken up. We know, of course, that the tugboat is powered by a coal-burning steam engine. And we understand that Turner's painting from 1839 signals the triumph of coal and steam.

Most early ocean-going steamships also carried sails to keep fuel costs down. Iron-clads started replacing wooden hulls in the 1840s, as the price of timber kept rising, whereas the price of iron-plates kept falling as iron-works got more and more efficient. The screw propeller (1849) put too much strain on the structure of wooden hulls. But screw propellers and iron-clad ships became standard by the 1860s. Ships began to get much larger after that time.

Unlike sails and wooden hulls, the steam engine did not die. It merely got smaller (in size) but faster and more powerful. It became a turbine. The conversion from reciprocating to turbine steam engines after 1900 was remarkably fast. In fact the last and biggest reciprocating steam engine, which produced 10,000 hp, was completed in 1899 to power the growing New York City subway (metro) system. The pistons needed a tower 40 ft high. But by 1902 it was obsolete and had to be junked and replaced by a steam turbine only one tenth of its size (Forbes and Dijksterhuis 1963) p. 453.

Undoubtedly the most important early application of steam power after railways and ships was in the textile industry, as a supplement (and later, substitute) for water power. Yet in 1800, when Watt's master patent expired, it is estimated that there were fewer than 1000 stationary steam engines in Britain, totaling perhaps 10,000 hp (Landes 1969 p. 104). By 1815, however, the total was apparently 20 times greater (210,000 hp), all in mines or mills. By mid-century the total of stationary engines had increased to 500,000 hp, in addition to nearly 800,000 hp in locomotives and ships (Landes 1969). At that point (1850), the cotton industry was still using 11,000 hp of water power, but 71,000 hp of steam. The woolen industry, which was slower to mechanize, used 6800 hp of water power, as against 12,600 hp steam power (Landes 1969). Mechanization of textile manufacturing was one of the few examples of a purely labor-saving technology introduced during the industrial revolution.

Turbines powered by steam or (later) by high temperature combustion products (gas turbine) deserve special mention. This is because they have to operate at very high speeds and very high temperatures, constituting a major technological challenge. The first effective multi-rotor steam turbines were built in 1884 by Sir

Charles Parsons. His first axial-flow design, based on earlier hydraulic turbines,[3] was quickly followed by a prototype high speed turbo-generator.[4] The first installation for electric power generation was a 75 kW unit for the Newcastle and District Electric Lighting Company (1888). By 1900 Parsons installed a pair of 1000 kW turbo-generators for the City of Elberfeld, in Germany. Radial flow designs and compounding designs soon followed. Other successful inventors in the field included C.G.P. de Laval (Sweden), C.G. Curtis (USA) and A.C.E. Rateau (France).

The steam turbine rotor wheels utilize high quality carbon steel with some alloying elements such as molybdenum for machinability. They now operate at steam temperatures above 500 °C. The first effective gas turbines (for aircraft) were built in 1937 by Frank Whittle, and adapted to electric power generation by Siemens and GE in the 1950s and 1960s. Gas turbine wheels must now withstand much higher temperatures, as high as 1200 °C. To do so, they must be made from so-called "superalloys" usually based on nickel, cobalt, molybdenum, chromium and various minor components. These metals are very hard to machine or mold accurately, as well as being costly to begin with, which is why gas turbines are used only by high performance aircraft and have never been successfully used for motorcars, buses or trucks.

Steam turbines now supply 90 % of the electric power in the US and most of the world. Most of the rest is from hydraulic turbines. Steam turbine technology today is extremely sophisticated because steam moves faster and is much "slipperier" than liquid water. Combustion products from a gas turbine move even faster. Therefore turbine wheels need to turn much faster, too, requiring very good bearings and very precise machining. This need had a great influence on metallurgy, machining technology and ball bearing design and production in the early twentieth century.

9.5 Town Gas, Coal Tar, Aniline Dyes and Ammonia Synthesis

Coal was the energy basis of the first industrial revolution, but it is bulky, inconvenient and dirty. Yet it was the primary fuel for heating buildings and driving trains and ships for the over a century (in the case of trains and ships) and for two

---

[3] The first hydraulic motors were undershot or overshot water wheels, going back to Roman times and perhaps even earlier. A French engineer, Bernard Forrest de Belidor wrote a monograph *Architecture Hydraulique* in 1757, describing alternative hydraulic motor design possibilities. The first enclosed hydraulic turbines, using a screw-type propeller, were built by Benoit Fourneyron in the 1830s. An improved design by Uriah Boyden and James Francis of Lowell, Massachusetts, the Francis reaction turbine in (1848) achieved 90 % conversion efficiency (Hawke 1988) pp. 197–198. This design (not the turbine) is still in use today. Many factories, especially textile plants near fast-flowing rivers like the Merrimack River in Connecticut, used water-wheels for power well into the nineteenth century.

[4] Before Parsons the maximum rotational speed of dynamos was 1200 rpm. Parsons increased this in his first generator (1889) to 18,000 rpm.

centuries in the case of buildings. The famous "pea soup" London fogs were a feature of every major European and American industrial city, not restricted to London. They were a fact of urban life as late as the last half of the twentieth century when natural gas finally replaced coal as a domestic fuel in Europe and America. They remain so in China and elsewhere in the developing world. Not only did the smoke leave its black imprint on every building and every gentleman's collar and every lady's handkerchief, it caused asthma, emphysema, lung cancer and heart disease. Life expectancy in northern Chinese cities today lags behind industrial countries by several years.

Dirty coal has kept its choke-hold on the urban home-heating market so long, despite the availability of alternatives, for one reason only: it was "dirt cheap". Coal fires could heat a room, a coal-burning stove could cook food, and a coal-burning furnace could also heat water and provide central heating. But there was one important domestic energy service that coal (or its predecessor, wood) could not provide, at least in raw form. That service gap was illumination. Lighting is so important to people, even the poorest, that in the eighteenth century people in England spent as much as a quarter of their incomes on lighting, mostly by candles or whale-oil lamps. (Oil for lamps from sperm whales, was the basis of New England's prosperity in the late eighteenth and early nineteenth century. The species was nearly hunted to extinction for its oil, and is now protected). But candles and oil lamps were not suitable for street lighting, and the need for street lighting was growing even faster than the cities themselves.

The first street lighting solution came about as a result of the demand for coke by the iron industry. Coking produces a gaseous by-product, called coke-oven gas, which nobody wanted at the time because it was so smelly (due to the presence of sulfur dioxide and ammonia). Coke-oven gas, in those days, was generally flared. It was **William Murdoch**, one of the employees of Watt & Boulton, who first produced a cleaner gaseous fuel from the coke. He used a process now called "steam reforming", in which red-hot coke reacts with steam to produce a mixture of hydrogen and carbon monoxide. This gas became known as "town gas" because it rapidly became the fuel for street lighting in London, Paris, New York, and other big cities. It also served for interior lighting (replacing candles) wherever the expanding network of gas pipes made it possible to deliver the fuel inside a house. Town gas was widely distributed in cities around the world by the end of the nineteenth century, and continued to be used until after WW II. (Some of those old pipes now deliver natural gas).

Of course, the iron (and steel) industry required more and more coke, eventually resulting in a glut of the by-product coke-oven gas. Cleaning it for domestic use (in combination with town gas) was technically possible. The sulfur dioxide could be removed by passing it through a solution of quicklime, similar to what is done today in large coal-burning electric power plants. But getting the ammonia out of the coke-oven gas was too expensive until German chemists realized that it could be converted into ammonium sulfate, a valuable fertilizer.

Nitrogenous fertilizers were increasingly necessary to increase the grain yield of European farmlands, as the population was growing fast. Coke oven gas became a primary source of nitrogen fertilizers for several decades in the late nineteenth century. Of course, extracting the ammonia and the sulfur left the gas much cleaner, and less smelly. This improvement soon allowed other uses. One man who found such a use was **Nikolaus Otto**, of Cologne. But that begins another story, the development of the gas-burning stationary (and later, liquid burning mobile) internal combustion engine (discussed in Sect. 9.7).

The gas-light industry produced another nasty by-product: coal tar. In Britain this product was mostly used, at first, to caulk ships-bottoms and to seal leaky cellars and roofs. However, those German chemists recognized an opportunity for developing products of higher value and began to support fundamental research on coal-tar chemistry at several institutions. Nevertheless, the first important discovery occurred in England in 1858. A chemistry student, **W.H. Perkin**, accidentally synthesized a brilliant mauve color from aniline, a derivative of coal-tar, while he was searching for a way to synthesize quinine. Perkin saw the value of synthetic dye materials for the British cotton industry, and began to manufacture the mauve dye commercially. In the following year Perkin also succeeded in synthesizing another dye color, alizarin, the coloring agent in "rose madder."

However, Perkin's early lead was trumped by German chemists, who also synthesized alizarin and began production. The firms Badische Anilin und Soda Fabrik (BASF), Bayer, and Hoechst were all in business by 1870. Thereafter the Germans gradually forged ahead, by investing heavily in research. More and more aniline-based dyes were introduced in the 1870s and 80s, culminating in 1897 with a major triumph: BASFs successful synthesis of indigo (which the British imported from India). This firm also developed the successful "contact" process for manufacturing sulfuric acid in the 1890s.

BASF began work around the turn of the century on the most important single industrial chemical process of all time: the synthesis of ammonia from atmospheric nitrogen. Synthetic ammonia was needed to manufacture nitrate fertilizers to replace the natural sodium nitrates that were then being imported from Chile. The basic research was done by a university chemist, by **Fritz Haber**, while the process design was accomplished by BASF chemist **Karl Bosch**. The first laboratory-scale demonstration of the Haber-Bosch process took place in 1909. The process was shelved for a while due to the (temporarily) low price of a competing nitrogenous fertilizer, calcium cyanamid, made from calcium carbide. (Calcium carbide was being mass-produced by the Union Carbide Company and others for acetylene lamps). Finally, under the pressure of war-time shortages, a full-sized ammonia plant did go into production in 1916 to supply nitrates for fertilizers and munitions in Germany.

Still, the German chemical industry was predominantly based on synthetic dye manufacturing through the 1920s.[5] Indeed, it was experience with color chemistry

---

[5] When the three biggest firms, Bayer, Hoechst and BASF consolidated (c. 1925) they took the name I.G. Farbenindustrie. Farben is the German word for "color".

that led to several major pharmaceutical breakthrough's (including the anti-syphilis drug "salvarsan"), and helped the Germans take a leading role in color photography.

The gas-light industry had been created primarily to serve a growing demand for illumination. It could only do this, however, in large cities where a central gas-distribution system could be economically justified. However, in the 1850s most of the population of the world still lived in small towns or rural areas where gas light from coking coal was not an option. The alternative illuminants at that time were liquid fuels: whale oil and kerosene. (Acetylene came later). Whale oil was cleaner burning and generally preferable, but the supply of whales from the oceans of the world was being rapidly depleted. Animal fats were a possible substitute, but the oil had a bad smell. Kerosene could be produced in small quantities from coal, but not in useful amounts. Petroleum—known for centuries as "rock oil"—was the other alternative.

9.6   Petroleum

The 1850s saw the beginnings of the global petroleum industry. The raw material ('rock oil') was known, thanks to seepages in a number of places, and the decline of whaling (due to the scarcity of whales), which led to increased interest in alternatives for lighting purposes. These discoveries were prompted by the introduction of distilled coal oil (Young, 1850–1851) and kerosene (1853). Interest was further stimulated by **Benjamin Silliman's** pioneering work on fractional distillation of petroleum (1855), and the rising price of liquid fuels such as kerosene (due to the scarcity of whale oil) for household illumination. It was Silliman's report to an investor group, that later became the Pennsylvania Rock Oil Company, that kicked off the search for petroleum in western Pennsylvania.

The year 1857 marked the beginning of commercial petroleum production in Rumania, followed in 1859 by the discovery of oil in Pennsylvania. Petroleum was found by the drillers under "Colonel" Drake, and crude refineries were soon built to separate the fractions. The most important fraction was kerosene, known as "illuminating oil", which rapidly became the dominant global fuel for lamps (replacing whale oil). In the early days, refinery output was about 50 % kerosene, 10 % gasoline, 10 % lubricating oil, 10–15 % fuel oil and the rest consisting of losses and miscellaneous by-products like tar.

In fact, an expert commission of geologists had concluded (back in 1871) that the Ottoman territory between the Tigris and Euphrates rivers was likely to be a good source of high quality (easy to refine) petroleum. But after Pennsylvania, the next major source of petroleum discovered was near the town of Baku, in Azerbaijan, where Robert Nobel (older brother of Alfred) started developing an oil industry. That was in 1873. By the end of the century, the Rothschilds had joined the Nobels in Azerbaijan, still an Ottoman province. By 1900 it was the primary oil producer in the world, albeit landlocked and increasingly dominated by neighboring Czarist Russia.

In 1892 Marcus Samuel, a London merchant, persuaded the Rothschild bank to finance a new venture, to sell "Russian" kerosene in the Far East in competition with Standard Oil of N.J. It is referred to as the "Coup of 1892", and it created Shell Oil Company. Shell later merged with Royal Dutch to create the basis of the modern behemoth, Royal Dutch Shell.

As I said, the early petroleum refining industry was focused on manufacturing "illuminating oil" (i.e. kerosene). The more volatile hydrocarbon liquids (notably natural gasoline) were only used in those days for dry-cleaning or as a solvent. But natural gasoline was crucial for the adaptation, by **Gottlieb Daimler**, of **Nikolaus Otto**'s stationary high compression gas-burning internal combustion engine to mobile transport purposes. I discuss that development in the next section.

The growth of the petroleum industry in the United States was extremely rapid. Production in 1860 was about 0.5 million barrels (bbl). Output in the first decade of production multiplied tenfold to 5 million bbl per year. Production nearly doubled again by 1874 and once again (to 20 million bbl) in 1879. Thereafter, output of crude oil reached 24 million bbl in 1884, 35 million bbl in 1889, 49 million bbl in 1894 and 57 million bbl in 1899. And then came the colossal "Spindletop" gusher in January 1901 which created the Gulf Oil Company and put Texas on the oil map.[6]

From 1857 to 1910 or so, the major product of the U.S. petroleum industry was kerosene or "illuminating oil, much of which was exported to Europe. Other products, in order of importance, included "naphtha-benzene-gasoline", fuel oil (in the 1890s), lubricating oils, paraffin wax, and residuum (tar and bitumen). The gross value of refined products was about $43.7 million in 1880, $85 million in 1889, and $124 million in 1899. (Only the steel industry was bigger at that time). Employment in U.S. refinery operations averaged 12,200 workers in 1899. In 1911 demand for gasoline (for automobiles) overtook demand for kerosene. That event kicked off another technology competition, namely to increase the yield of gasoline by "cracking" heavy petroleum fractions.

It is important to bear in mind that this was already a very large industry, based primarily on human demand for illumination, long before demand for motor gasoline became significant in the years after 1905. It was, of course, the existence of a petroleum industry that made the automobile and the automobile industry (not to mention aircraft) possible, and revolutionized transportation. For about 50 years it was the most important manufactured export to Europe and the basis of **John D. Rockefeller**'s Standard Oil Company of New Jersey. Rockefeller was originally financed by the National City Bank of Cleveland, a Rothschild bank. The Rothschilds were major backers of the global oil industry in Baku (Azerbaijan) as well as Ohio.

---

[6] The original producer was the Guffey Petroleum Company, later merged with the Gulf Refining Company, both financed largely by the Mellon bank in Pittsburgh. By coincidence, the Carnegie Steel Company was sold to JP Morgan (to create US Steel Co.) about that time, and a number of Carnegie's employees became instant millionaires. Several of them invested that money in Spindletop.

A 1901 German report said that Mesopotamia sat upon a "veritable lake of petroleum" constituting an inexhaustible supply. However, transportation was poor, and this fact was one of the incentives for Germany to move into the region. German railway engineers were hired to build railways in Turkey. The Deutsche Bank and Württembergische Vereinsbank formed the Anatolian Railway Company in 1888. The rail line from Istanbul to Konya was complete by 1896. In 1898 the Ottoman government awarded a contract to complete the line from Konya to Baghdad. The winner was Deutsche Bank, with French financing, and in 1903 the Ottoman government gave permission for construction of the last sections of the Berlin-to-Baghdad railway to proceed. (It was still incomplete by 1914, and was finished finally by the Iraqi government in 1940) [ibid].

After 1903 the British government took notice of the Berlin-to-Baghdad railroad project. On completion it would have given Germany rail access—by-passing the Suez Canal—to the Persian Gulf and the Indian Ocean, and its colonies in East Africa (present day Ruanda, Burundi and Tanzania), as well as German Southwest Africa (Namibia). Quite apart from this, it would have given Germany access to oil that had recently been discovered in Persia, where more discoveries followed rapidly.

Meanwhile, in 1901, after a long search, oil was found in Persia (Iran) by British geologists, and the Anglo-Iranian Oil Company—with an exploration license— controlled it. That company later became British Petroleum. The Iranian resource assumed greater importance when the British Navy, urged by First Sea Lord Winston Churchill, switched from coal to oil as fuel for its ships after 1911. The advantages were that oil-burning ships had longer range, and needed to carry fewer men (no stokers). Also, there was no source of coal to be found in the Middle East, whereas there was plenty of oil (from Persia) at the time. The possibility of a German port on the Persian Gulf, and possibly German warships in the neighborhood, was also seen by some in London as a potential threat to British interests in India. The British played hardball. One consequence was the 1914 seminationalization of Anglo-Iranian Oil and the forced marriage of Shell with Royal Dutch, in both cases due to British admiralty concerns about assuring fuel oil supplies for the navy and merchant marine. The Rothschilds and Nobels were driven out of Russia during the 1905 anti-Czarist revolution. Azerbaijan was briefly independent after 1918 but it became part of the USSR in 1920.

The next great oil discovery (1910) was in Mexico, which became the world's second largest producer by 1921 (Yergin 1991). And in 1930 leadership was back to Texas again when "Dad" Joiner and "Doc" Lloyd hit another huge gusher.

A revolutionary invention in 1909 greatly increased the depth of wells that could be drilled. The original drilling technology, adapted from water well drilling, was to drop a heavy "bit" onto the rock. The rock would shatter and be periodically removed by a scoop. This method is slow, inefficient and not feasible below a thousand meters or so. The far more efficient continuous hydraulic rotary drilling system used today was invented by Hughes Tool Co. c. 1900. It consisted of a rotary cutting tool[7] at the bottom end of a pipe, which was rotated as a whole by an engine at the top. A lubricant was pumped down from above, both to cool the drill and to carry the cuttings back to the surface outside the pipe. This enabled the drill to operate continuously, except for pauses to add new sections of pipe at the top. Depth was virtually unlimited.

As mentioned, gasoline overtook kerosene as the primary product of the industry in 1911. About that time the need for lighter fractions of the petroleum had outrun the output of "natural" gasoline. The technology for thermal "cracking" of heavy fractions was originally invented and patented by Vladimir Shukhov in Russia (Baku) in 1891, but the process was not needed at the time. William Burton of Standard Oil of Indiana[8] invented an effective batch process that became functional in 1913. The Burton process doubled the yield of natural gasoline (to 25 %). License fees generated enormous profits for Standard Oil of Indiana, perhaps half of the total profits of the company during the rest of that decade (Enos 1962). A number of continuous processes were patented, starting in 1920, too many to list. Two are worth mentioning: the Dubbs process by Universal Oil Products Co. (UOP) and the "tube and tank process" (Enos, ibid). The details don't matter here; these processes were designed by chemical engineers, rather than by chemists like Burton.

The next major step was catalytic cracking. Eugene Houdry was the prime mover. In 1927, by trial and error (like Edison) he finally discovered an effective cracking catalyst consisting of oxides of aluminum and silicon (Enos 1962 p. 136). With support from Vacuum Oil Co. (later Socony-Vacuum and finally Sun Oil Co). Houdry got the first catalytic cracking process operational in 1937. It processed 12,000 bbl/day and doubled again the yield of gasoline by the Burton process to 50 %. By 1940 there were 14 Houdry fixed-bed catalytic plants in operation, processing 140,000 bbl/day. Shortly after that a moving bed "Thermofor catalytic cracking" process (or **TCC**) was put into operation in 1943 and by 1945 it was processing 300,000 bbl/day into high octane gasoline for the war effort.

The so-called fluid catalytic cracking process (**FCC**) in use today was developed by a consortium of companies led by Exxon, primarily to by-pass the Houdry patents. Success was achieved quite rapidly, thanks to a suggestion by two MIT professors, Warren K. Lewis and Edwin Gilliland, for fluidizing the catalyst itself. This turned out to be successful, and now all refineries use some version of the FCC process.

---

[7] The tool was described as "a rolling bit with two cones with unevenly spaced teeth" (Rundell 2000). It was patented by Howard Hughes Sr. and Walter Sharp. When Sharp died in 1913 Hughes bought his share of their business and renamed it Hughes Tool Co. in 1915. The Hughes Tool Company was the source of profits that created (among other things) Hughes Helicopter and Aircraft Co. A later merger created Baker-Hughes, the oil service company of today and, subsequently spun off a number of aerospace enterprises.

[8] The Standard Oil Trust was broken up by order of the US Supreme Court in 1911, creating a number of newly independent "children", of which Standard Oil of Indiana was one.

On October 3, 1930 wildcatter "Dad" Joiner and self-educated geologist "Doc" Lloyd (with financial help from H.L. Hunt) discovered the huge East Texas field. This discovery was disastrously ill-timed. During 1931 an average of eight wells per day were being drilled in East Texas, resulting in a huge glut. The surplus was exacerbated by declining demand due to the depression. The price dropped to an all-time low of 10 cents per barrel, far below cost. During that year the Texas Railroad Commission tried to limit output to 160,000 bbl/day, but actual output at the time was 500,000 bbl/day. This restriction was toothless and was ignored. However the governor of neighboring Oklahoma, William Murray—who had the same problem—put all the wells in the state under martial law, from August 11 1931 until April 1933.

Thereafter the Texas Railroad Commission (**TRC**) took over the task of regulating output, bringing it down to 225 bbl/day per well. The price gradually recovered as the economy recovered, but the TRC set world prices until **OPEC** took over in 1972. It is worthy of note that the energy content of the oil produced, in those distant days, was well over 100 times the energy required to drill the wells and transport the oil. Today the return on energy invested in oil drilling is around one fifth of that, taking a world-wide average, and the energy return on some of the newer "alternatives" (such as ethanol from corn) is around one twentieth, or less, of what it was in 1930.

The year 1933 was also when Socal (now Chevron) got the first license to explore for oil in Saudi Arabia. California-Arabia Standard Oil Co (Casoc) was created, and Texaco joined Casoc in 1936. The first strike was at Dharan in 1938. The Ras Tanura refinery (world's largest) started operations in 1945. In 1948 Esso and Mobil bought into Casoc and the name was changed to Arabian-American Oil Company (Aramco). That was when the world's largest oil field, Ghawar (still not exhausted) was discovered. It was the peak year for discovery. In 1950 King Abdul Aziz threatened nationalization, and agreed to a 50–50 split of the profits. The US government gave the oil companies a tax break called the "Golden Gimmick" equal to the amount they had to give Saudi Arabia. The Trans-Arabian pipeline (to Lebanon) began operations. Aramco confirmed the size of Ghawar and Safaniya (biggest offshore field) in 1957. The whole story of the search for "black gold" is told very well in "The Prize" (Yergin 1991).

9.7   The Internal (Infernal) Combustion Engine

The steam engine, as applied to vehicles, had three fundamental disadvantages, which were gradually recognized. The first, and most obvious, was that a steam engine needed either a large water tank and frequent water refills, or an external condenser to recycle the steam. The first option worked fairly well on the railways and for large ships, but not for smaller applications, especially private cars. The famous "Stanley Steamer" won a lot of races in the early days because of its high torque, but it needed frequent drinks of water. It is said that the demise of the early steam-powered automobiles was because the horse troughs—where horses used to drink along the roads—disappeared along with the horses themselves. The Stanley steamers, which had no condenser, were left (so to speak) "high and dry". The later "Doble", built in the 1930s, had a condenser that worked fine, but only a few were produced. It was too expensive and too late.

The second fundamental disadvantage of steam was that high efficiency requires high temperatures (the **Carnot law**). It follows that a really efficient steam engine needs to use special steel alloys for the boiler that do not lose their strength at high temperatures. Plain old carbon steel is not strong enough. In modern steam-electric generating plants it is necessary to use more expensive alloys containing more exotic metals like manganese, molybdenum and nickel. These special steels might be economically justified in large central station power plants, but not in engines for cars or trucks. (Jet aircraft are another story.) The third disadvantage, related to the second, is that the condenser makes the engine considerably heavier and bulkier than a comparable gasoline powered internal combustion engine.

The reciprocating (piston) internal combustion engine (ICE) avoids all these problems. Instead of making high pressure steam to drive the pistons (or turbine wheels) it makes use of the high temperature exhaust gases from the combustion itself.[9] There is no need for a condenser because the exhaust gas returns to the atmosphere (as air pollution) but that is another problem. Problem #1 is thereby avoided. Problem #2 is avoided also, because the combustion process in an ICE is intermittent, not continuous. After each explosion of the compressed fuel-air mixture, the temperature drops rapidly as the gaseous combustion products expand and energy is transferred to the piston.

The temperature of the metal inside the cylinder and the piston itself never gets too hot because the metal is a good heat conductor that conducts heat away from the surface where the combustion takes place, between explosions. So, it turns out that a cast iron engine block works just fine. In the case of gas turbines, stronger alloys are needed, which is why gas turbines came along much later and are not nearly as cheap (or mass-producible) as ICE piston engines.

Of course, the stationary steam engine did have one great advantage: It could use cheap coal, or coke, available everywhere (by the mid-nineteenth century), as a fuel. However, the rapid expansion of coking to feed the new blast furnaces of the Ruhr valley in Germany created plenty of by-product coke-oven gas which needed a market. Moreover the use of "town gas" for street lighting and interior lighting in offices, hotels and restaurants had already created a system of gas pipelines in the major cities. Widespread availability of coke-oven gas, a by-product of the steel industry, enabled (and possibly stimulated) the development of internal combustion engines. The early driving force behind his innovation was the need for more efficient and cheaper prime movers. This was especially true for the smaller

---

[9] The power stroke of an ICE is driven by the expansion of hot gases released by the explosive combustion of fuel. Since neither temperature not pressure is constant, the resulting free energy is some compromise between the Gibbs and Helmholtz versions.

machine shops and manufactories that were springing up along the Rhine River and its tributary, the Ruhr.

Cologne-based entrepreneur-inventor **Nikolaus Otto**'s commercially successful high compression gas engine, the "Silent Otto" (1876) was the culmination of a series of earlier inventions. Some of the most noteworthy were the prototype "explosion engines" of Street (1794), and Cecil (1820) and the first commercial stationary gas engines built by Samuel Brown in the 1820s and 1830s. Key steps forward were taken by Wright in (1833) and Barnett in (1838), who was the first to try compression. Lenoir (1860) built and commercialized a double-acting gas engine modeled on the double-acting steam engine invented by James Watt. Like Watt's engine, it did not compress the fuel-air mix. These engines were functional, despite being very inefficient (about 4 %). The so-called "free-piston" engine, invented by Barsanti and Matteucci in (1859), was briefly commercialized by Otto and Langen in (1867). Siegfried Marcus built and operated the first self-propelled vehicle in Vienna (1864–1868). It was shown at the Vienna Exposition in 1873. But it was a one-of-a-kind, dead-end, project.

The need to compress the fuel-air mixture prior to ignition had been recognized already (by Barnett) in the 1830s. But the first to work out a way for the pistons to perform the compression stage before the ignition—now known as the "4 stroke" cycle—was Beau de Rochas in (1862). Not for another 14 years, however, was this revolutionary cycle embodied in Nikolaus Otto's revolutionary commercial engine (1876). The "Silent Otto" rapidly achieved commercial success as a stationary power source for small establishments throughout Europe, burning illuminating gas, or coke-oven gas, as a fuel. Otto's engine produced 3 hp at 180 rpm and weighed around 1500 lb. It was very bulky by today's standards, but much more compact than any comparable stationary steam engine. By 1900 there were about 200,000 of these gas-fired engines in Europe. They still had to be attached to long rotating shafts that drove machines by means of belts. Those shaft-and-belt combinations were replaced, after 1900, by electric motors.

The next challenge was to make the engine smaller and lighter, as well as more powerful. The key to increasing the power-to-weight ratio was to increase the speed. The prototype automobile engine was probably a 1.5 hp (600 rpm) model weighing a mere 110 lb. It was built by **Gottlieb Daimler** (who had once worked for Otto) and his partner, **Wilhelm Maybach** (1885). They built four experimental vehicles during the years 1885–1889. These were not the first self-propelled vehicles—"automobiles"—but they were the first ones with a commercial future.

The first practical car that did not resemble a horse-drawn carriage without the horse was Krebs' Panhard in (1894). Benz did introduce the spark-plug, however, a significant advance over Otto's glow-tube ignition. A large number of subsidiary inventions followed, ranging from the carburetor (Maybach in 1893), the expanding brake (Duryea in 1898), the steering wheel (1901), the steering knuckle (Eliot in 1902), the head-lamp, the self-starter (1912), and so on.

The success of the spark-ignition high compression "Otto-cycle" engine created enormous interest in the technical possibilities of compression engines. **Rudolf Diesel** realized that higher compression made the engine more efficient. He also discovered that, above a certain compression (about 15:1), there was no need for spark ignition, as the compressed fuel-air mixture becomes hot enough to self-ignite. And the fuel need not be as volatile as gasoline; heavier oils work OK. The compression-ignition internal combustion engine was patented by Rudolf Diesel in 1892. It was first commercialized for stationary power in 1898, but development was very slow because the need for very high compression (more than 15:1) resulted in technical and manufacturing difficulties. Diesel power was adopted for the first time for railway use by the Prussian State Railways (1912), and for marine use shortly thereafter. Diesel-electric locomotives for railway use were finally introduced in the 1930s by General Motors. The first automobile diesel engine was introduced in the 1930s by Mercedes Benz, but penetration of the automobile market was negligible until the 1980s, when the turbo-diesel was introduced. Diesel power dominates the heavy truck, bus, rail and off-road machinery fields today, and is rapidly penetrating the automobile market in Europe (until 2015).

The single most important technological barrier in the way of practical self-powered road vehicles (until Daimler) was the lack of a prime mover with sufficient power in a small enough "package". The same barrier applied even more strictly to heavier-than-air craft. The key variable is power-to-weight ratio. In retrospect it is clear that the minimum feasible level for road vehicles was about 50 kg/hp, or about 0.02 hp/kg. The Daimler-Maybach engine achieved 35 kg/hp or 0.0275 hp/kg in 1886. Cars did not become truly practical until further developments brought the engine weight down (or the power up) to around 7 kg/hp or roughly 0.15 hp/kg. See Fig. 9.1.

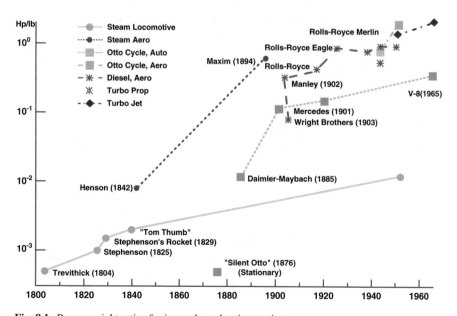

Fig. 9.1 Power-weight ratios for internal combustion engines

Actually, the 1901 Mercedes Benz engine produced 35 hp with an engine weighing 215 kg (6.2 kg/hp). (The "Mercedes" car was named after the 10 year-old daughter of an early investor, Emil Jelinek.) But the Manly engine, designed specifically for Langley's "Aerodrome" (1903), achieved 52 hp in a package weighing only 68 kg. or about 1.3 kg/hp. This was an improvement over the original Daimler-Maybach engines by a factor of 38! The early aircraft engines obtained part of their "punch" by the use of special high octane fuels (e.g. benzene) that permitted higher compression ratios, and hence greater power, but which could not be produced in large enough quantities for normal automotive use. Obviously practical air transportation (which came much later) required substantial further progress in increasing the power-to-weight of the engines. Progress in gasoline refining technology in the 1930s and 1940s played a major role in this.

It is worth mentioning that the performance improvements of gasoline engines since 1886 also resulted partly from thermodynamic efficiency improvements somewhere in the neighborhood of a factor of 2 or 3. The point is that higher efficiencies did not save energy. On the contrary, energy consumption for automotive and air transportation has "sky-rocketed". (No pun intended.) It is a good example of the so-called "rebound effect", where an efficiency gain results in lower prices and that, in turn, drives higher demand (Herring and Sorrell 2009).

Actually the greatest efficiency improvement came about in the 1920s as a result of the discovery of tetra-ethyl lead as an octane additive. All engines before that time were effectively limited to a compression ratio of 4:1, because they were forced to burn natural gasoline. Higher compression ratios led to explosive pre-combustion before the cylinder completes its stroke. This had an effect known as "knocking", that sharply reduced efficiency and damaged the machinery.

The solution, tetraethyl lead, was discovered in 1921 by **Thomas Midgeley**. He was a chemist working in a tiny firm called the Dayton Engineering Laboratory in Dayton Ohio, under Charles Kettering. (Kettering's firm merged with GM where it became the Delco division). GM and Standard Oil of N.J. commercialized tetraethyl lead under the trade name "Ethyl". Later, it is said that the giant German chemical firm IG FarbenIndustrie, spent many millions of Deutschmarks trying to find a way to avoid the Ethyl patents, but without success.

The first effective gas turbines (for aircraft) were built in 1937 by Frank Whittle, and adapted to electric power generation independently by Siemens in Germany and GE in the US during the 1950s and 1960s. Gas turbine wheels must withstand much higher temperatures, as high as 1500 °C. To do they must be made from so-called "superalloys" usually based on nickel, cobalt, molybdenum, chromium and various minor components. These metals are very hard to machine or mold accurately, as well as being costly to begin with, which is why gas turbines are used only by high performance aircraft and have never been successfully used for motorcars, buses or trucks.

9.8 Electrification and Communications

Electricity is the purest form of energy, since it can be converted to any other form, from heat to light to mechanical motion with practically no loss. However, apart from the occasional lightning strike, there is no natural source of electric power, as such. In fact, electric power is derived today mostly from kinetic energy of motion, originally from reciprocating engines based on Watts' rotary design, and since 1900 in the form of a turbine wheel. Early contributions to the science that enabled the development of an electrical industry included discoveries by Andre-Marie **Ampere**, Charles-Augustin **Coulomb**, Humphrey **Davy**, Michael **Faraday**, Benjamin **Franklin**, Luigi **Galvani**, Joseph **Henry**, Hans **Oersted**, Georg Simon **Ohm**, Alessandro **Volta**, and many others.

The first two practical applications of electricity were the electric telegraph and the light-house (using arc-lights). The other major component of an electric power system is the generator or dynamo. Early key inventions were the efficient DC dynamo (and its alter ego, the DC motor). The dynamo-and-motor evolved in parallel over a period of many years, following Michael Faraday's initial discovery of electromagnetic induction in 1831. The major milestones were European, with significant contributions by Woolrich (1842), Holmes (1857), Wheatstone (1845, 1857), Siemens (1856), Pacinotti (1860), Siemens (1866), Gramme (1870) and Von Hefner-Altaneck (1872) (Sharlin 1961). The most widely-used commercial dynamo before Edison (by Zénobe Gramme) achieved an efficiency of barely 40 % in terms of converting mechanical energy to electrical energy. Gramme's machine was the first capable of producing a relatively continuous current.

The prevailing engineering practice was to produce large DC currents for arc-lights at low voltages. Edison reversed this practice. Edison's breakthrough, the "Jumbo" generator (1879) came late in the game because he started from a "systems" perspective. He was intending to make and sell electric power systems to supply whole regions, from central generating plants. Edison's generator took advantage of the fact that higher voltages greatly reduced resistance losses in the wiring, thereby effectively doubling the efficiency of the generator. Edison was also the first to build an integrated system of electric power, as applied originally to incandescent lighting in buildings. The efficiency progression is shown in Fig. 9.2.

Once electric power became available in the 1850s and 60s, new applications—and new industries—quickly emerged. The first application of DC power was for electric lighting. Arc lights were already known, and a number of practical arc-lighting systems for public purposes were developed in the late 1870s, pre-eminently those of Charles **Brush** (1876) and Elihu **Thomson**-Edwin **Houston** (1878). These systems were suitable for outside use and large public rooms, though not for more intimate indoor quarters. But soon they were being produced in significant numbers. By 1881 an estimated 6000 arc lights were in service in the U.S., each supplied by its own dynamo. See Fig. 9.3.

Thomas Edison's decision in 1877 to develop a practical incandescent lamp suitable for household and office use (in competition with the gas-light), was

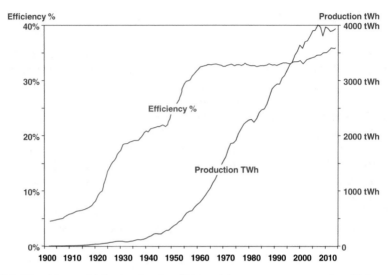

Fig. 9.2 Electricity production by electric utilities and average energy conversion efficiency US

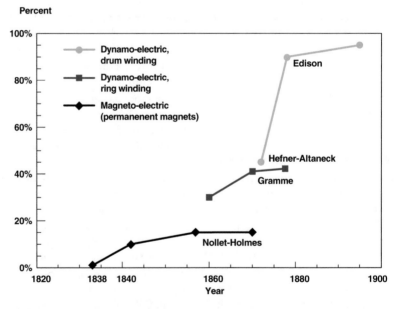

Fig. 9.3 Generator efficiency: electric power output per unit mechanical power output

historically momentous, albeit not especially interesting, or original, technologically. In fact, Edison was by no means the first to work on the idea of incandescent light. At least 22 inventors, starting with Humphrey Davey (1802) demonstrated incandescent electric light. The first to be patented was in 1841, in England. It was a design based on a coiled platinum filament in an evacuated glass bulb. The first patent on a carbon filament in a glass bulb was granted to an American in 1845. Neither was actually produced commercially.

Nor was Edison alone in the field when he began his campaign; he had several active competitors, some with important patents on features of the incandescent light. Joseph Swan, in England, was actually the first to light his house by electricity, and the first to go into commercial production. Swan's lightbulbs lit the Savoy Theater in 1881.

However Edison was the first to install a complete electrical lighting system in a commercial building, in the fall of 1880. It, too, went into production in 1881, both by Edison's company and several competitors, including Swan in England. Both incandescent and arc-lighting systems met rapidly growing demand. By 1885 the number of arc lights in service in the US was 96,000 and the number of incandescent lights had already reached 250,000. Five years later those numbers had risen to 235,000 and three million, respectively.[10] By 1891 there were 1300 electric light generating stations in the USA. See Fig. 9.4.

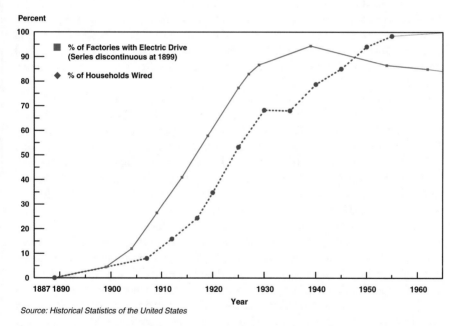

Source: Historical Statistics of the United States

Fig. 9.4 Rapid growth of electrification of factories and households in the US. *Source: Historical Statistics of the United States*

---

[10] Edison's success was bad news for the gas-light industry. After a long period of prosperous growth, gas shares slumped sharply in the late 1880s, which may have contributed to the recession of the early 1890s. As it happened, the gas-light system was temporarily reprieved by the invention of the "gas mantle" by Carl von Welsbach (1885). The incandescent thoria-ceria mantle increased the luminosity of a gas flame by a factor of a little more than 6 (Passer 1953 p. 196), and its adoption kept gas lighting competitive for another 20 years. Actually 1910 was the peak year for

The main difference between Edison and his competitors was that Edison was business-oriented, and had the patience and the financial backing (from J.P. Morgan) to make it work. (It was during the development phase that he remarked to someone that *"genius is one percent inspiration and ninety-nine percent perspiration."*) The first problem he faced was to develop a filament and a way of mass producing it. The constraints were that the bulb would last a reasonable number of hours before burning up, and that it could be manufactured in quantity. His carbon filament, in an evacuated bulb (an idea borrowed from several predecessors) solved that problem temporarily, though the carbon filament was eventually replaced by tungsten filaments. The key was to get as much as possible of the air out of the light-bulb. Sprengel's improved vacuum pump was an important prerequisite of Edison's success.

It is of some interest to mention why electric light was so revolutionary. When the sun is shining, the light level on a clear day is about ten thousand *lux* (lumens per square meter). The light level in a contemporary home is only around a hundred lux. (The iris of the human eye is able to adjust for the difference automatically by opening the pupil, so the difference doesn't seem so great.) But the light supplied by a wax candle, which was the only source of indoor light available before Edison and Swan, produces only 13 lumens. A 100 W incandescent lamp produces about 1200 lumens, or nearly 100 times as much. And recent developments have gone much further.

William Nordhaus has calculated the price of illumination in terms of labor-hours (Nordhaus 1998). In 1850, when people still used candles, the price of light was 3 labor hours per 1000 lumen hours, the equivalent of about 10 candles.[11] By 1890 the price of lighting was down to 0.133 labor hours per 1000 lumen hours (a factor of 21) and by 1992 (the last year of Nordhaus' calculations), it was down to 0.00012 labor hours per 1000 lumen hours, or a factor of about 24,000. Today we take light for granted and barely give a thought to its price. But in terms of human well-being, the difference is as between night and day.

Non-lighting applications of DC power soon followed. The most notable in terms of immediate impact was the electric street-railway (or trolley). Street railways sprang up more or less independently in a number of cities. The contributions of Van De Poele (especially the carbon-brush, 1888) and Sprague deserve particular mention. The building of urban transit systems not only employed a good many people, but also permanently influenced urban geography, permitting much higher density development than had previously been possible, as well as creating many new "suburbs" along the trolley lines. A remarkable number of street railways were built in the following decade. By 1910 it was said to be possible to

---

gas-light. The electric light was adopted much more rapidly in the U.S. than it was in the U.K., primarily because gas was much cheaper in Britain, making electricity much less competitive for purposes of lighting in the 1890s. This, together with restrictive legislation, seriously impeded the British in developing a competitive electrical industry.

[11] Assuming a candle weight 1/6 of a pound producing 13 lumens burns for 7 h.

travel by trolley—albeit with many interchanges—from southern Maine to Milwaukee, Wisconsin.

To supply electricity in large amounts economically it was (and is) necessary to take advantage of economies of scale in production, whether from available hydropower sources or from large steam-powered generating plants. In the eastern USA the obvious initial location for a large generating plant was Niagara Falls. It was some distance from the major centers of industry (although some energy-intensive industry soon moved to the neighborhood). In Europe, the first hydroelectric generating plants were naturally located on tributaries of the Rhone, Rhine and Danube rivers, mostly in remote Alpine Valleys also far from industrial towns.

To serve a dispersed population of users from a central source, an efficient means of long-distance transmission was needed. But efficient (i.e. low loss) transmission inherently requires much higher voltages than lighting. DC power cannot easily be converted from low to high or high to low voltages. By contrast, AC power can be transformed from low to high and from high to low voltages easily. (It's what electrical *transformers* do.)

This simple technical fact dictated the eventual outcome. All that was needed was the requisite technology for generating and transmitting AC power at high voltages. Elihu Thomson (inventor of the volt-meter) had already developed a practical AC generator (1881), as the arc-light business of Thomson-Houston went into high gear. Gaulard and Gibbs, in Europe, and Stanley in the U.S. developed prototype transmission and distribution systems for AC incandescent light by 1885. **Nicola Tesla** introduced the polyphase system, the AC induction ("squirrel cage") motor and the transformer by 1888. Tesla was one of the great inventive geniuses of all time.

George Westinghouse was the entrepreneur who saw the potential of AC power, acquired licenses to all the patents of Gaulard, Gibbs, Stanley and (most of all) Tesla and "put it all together". His original agreement with Tesla called for a royalty payment of $2.50 for every horsepower of AC electrical equipment based on his patents. If those royalties had been paid Tesla would soon have been a billionaire, but the royalty was wildly unrealistic. It would have made the equipment so costly that it could never have competed with Edison's DC equipment. (Tesla generously tore up his contract, but failed to negotiate a more realistic one.)

Sadly, Thomas Edison is regarded as great genius whereas Tesla, probably the greater inventor, is now regarded as a failure and mostly forgotten. Westinghouse's success was assured by a successful bid for the Niagara Falls generating plant (1891) and, subsequently, the great Chicago Exhibition (1893). It was Nikola Tesla, incidentally, who persuaded Westinghouse to establish the 60 cycle standard for AC, which ultimately became universal in the USA.

Yet Edison, the "father" of DC power, strongly resisted the development and use of AC systems (as Watt, before him, had resisted the use of high-pressure steam). During the 1880s both the Edison companies and the Thomson-Houston companies had been growing rapidly and swallowing up competing (or complementary) smaller firms, such as Brush Electric, Van De Poele and Sprague Electric. In 1892 the two electric lighting and traction conglomerates merged—thanks to

J.P. Morgan's influence—to form the General Electric Co. (GE), with combined sales of $21 million, as compared to only $5 million for competitor Westinghouse. Yet, in the end, AC displaced DC for most purposes, as it was bound to do.[12]

By 1920 all US cities and towns also had electric power generating and distribution systems. Growth was facilitated thanks to the introduction of regulated public utilities with local monopoly over distribution. This innovation facilitated large-scale central power generating plants capable of exploiting economies of scale. In the 1920s and 1930s the electrical industry expanded to embrace a variety of electrical appliances and "white goods" such as refrigerators, vacuum cleaners and washing machines.

From 1910 to 1960 the average efficiency of electric power generation in the industrialized countries increased from under 10 % to around 33 %, resulting in sharp cost reductions and price cuts. Today the electric power generating and distribution industry is, by capital invested, the biggest industry in the world. Unfortunately, the efficiency of the US electric power system has not increased since 1957, due primarily to perverse regulations (Casten and Ayres 2007).

Two new and important applications of electric power also appeared during the 1890s, viz. electro-metallurgy and electro-chemistry. The high temperature electric arc furnace, invented by William Siemens in (1878) was first used commercially to manufacture aluminum alloys (Cowles in 1886). In 1892 Moissan proposed the use of the high energy gas, acetylene (C_2H_2) as an illuminating gas. In the same year (1892) Thomas L. Willson demonstrated a method of producing acetylene from calcium carbide, made from limestone and coke in an electric furnace. The carbide is a solid that can be transported and stored safely. Acetylene can be made from the calcium carbide by simply adding water.

This made the use of acetylene practical for many purposes, from welding to illumination, and quickly led to the creation of a large industry, of which Union Carbide was the eventual survivor. Acetylene was rapidly adopted as an illuminant for farms and homes in towns without electricity. By 1899 there were a million acetylene gas jets, fed by 8000 acetylene plants, in Germany alone (Burke 1978 p. 209). The acetylene boom collapsed almost as rapidly as it grew, due to the impact of the Welsbach mantle on the one hand and cheaper electricity on the other. Acetylene continued as a very important industrial chemical feedstock, however, until its displacement by ethylene (made from natural gas) in the 1930s.

Another early application of electric furnaces was the discovery and production of synthetic silicon carbide ("carborundum") (Acheson in 1891). This was the hardest material known at the time, apart from diamonds. It found immediate use as an abrasive used by the rapidly growing metal-working industry, especially in so-called production grinding machines introduced by Charles Norton in (1900).

---

[12] However, in the mid-1890s the two firms were fighting hundreds of lawsuits over patent rights; this problem was finally resolved by creation of a patent pool which gave GE 63 % and Westinghouse 37 % of the combined royalties of the two firms for 15 years. It worked. The main benefit of this arrangement was that neither incompatible standards nor patent restrictions held back further technological progress in applications of electricity, as might otherwise have happened.

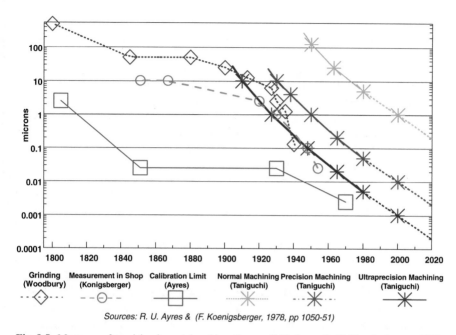

Fig. 9.5 Measures of precision in metalworking. *Source*: R.U. Ayres & (F. Koenigsberger, 1978, pp. 1050–51)

High-speed grinding machines turned out to be essential to the mass production of automobile engines, especially the complex shape of the crankshaft.

The research of Henri Moissan in the late 1890s also led to the use of the electric furnace for melting metals with high melting points, such as chromium, molybdenum, nickel, platinum, tungsten and vanadium. Heroult further developed the electric furnace for industrial purposes and his work was instrumental in permitting its use for the production of ferroalloys and special steels, beginning in Canada after 1903. The importance of abrasives (for grinding) and special tool steels and "hard" materials, such as tungsten carbide, is illustrated by Fig. 9.5 in terms of metalworking rates.

Electrochemistry—the application of electrolysis—also became industrially practical in the 1880s. The first, and most important, industrial electrolytic process was discovered in 1887 independently by Paul **Heroult** and France and Charles M. **Hall** in the U.S. It was a practical way of producing aluminum from aluminum oxide (alumina). To avoid the classic difficulty that defeated earlier efforts (that electrolysis of aluminum salts dissolved in water produced aluminum hydroxide, not aluminum metal), both inventors came up with the same solution. They dissolved aluminum oxide in molten cryolite a mineral (sodium-aluminum fluoride) found originally in Greenland and later made synthetically. This process was commercially exploited in both countries within 2 years. The price of metallic aluminum dropped rapidly, from $2/lb in the late 1880s, to $0.30/lb by 1897. Not surprisingly, many new applications of aluminum emerged, starting with pots and pans for the kitchen, which were previously made from cast iron and didn't conduct heat very well.

Curiously, the availability of metallic aluminum had no real impact on the infant aircraft industry in its first three decades, during which air-frames were made from wood and wire and surfaces were often made of fabric. It was 1927 when the first all-metal plane (Ford Tri-motor) was built. Needless to say, the commercial airline industry, in its present form, could not exist without aluminum. The next generation Boeing 787 "Dreamliner" will be the first to utilize significant quantities of carbon-fiber composites materials in place of aluminum.

The second important electrolytic process (1888) was Castner's system for manufacturing metallic sodium or sodium hydroxide (and chlorine) from fused sodium chloride, better known as common salt. At first, it was sodium hydroxide that was the important product. It was used in soap-making and for "whitening" illuminating oil for kerosene lamps, and later for converting bauxite to alumina. At first, chlorine was a cheap by-product, used initially—and still today—as a disinfectant for swimming pools and municipal water treatment. However, chlorine was soon in demand for a variety of chemical purposes, as well as for bleaching paper.

But the German discovery of a host of uses of chlorine in organic chemistry was the game-changer. Today, chlorine is one of the most important basic industrial materials, with a host of important uses from plastics (e.g. polyvinyl chloride) to insecticides (beginning with DDT). The cumulative economic impact of electro-metallurgy and electrochemistry has clearly been enormous, although most of it was not felt until many years after the key manufacturing innovations.

9.9 Telecommunication and Digital Technology

Communication since prehistoric times until the mid-nineteenth century was a major challenge, especially for ships at sea, or military commanders. Ships were alerted by light-houses on coasts; lookout towers warned villagers that pirate ships were approaching. Postal services, like the "pony express" in the American west depended on messages carried by couriers on horseback, between stations where fresh horses were provided. In some military situations messages were conveyed by semaphores or flags. One legendary financial coup (by the Rothschilds?) was accomplished by sending a false message to London that the Battle of Waterloo had been lost, resulting in a sharp sell-off of British government bonds and a huge profit for the buyers of those bonds when the true state of affairs was revealed. The message in question was almost certainly sent by means of semaphores between agents placed along the roads, and possibly between ships in the Channel.

Be that as it may, everything changed when the railroads, accompanied by the telegraph, came. The telegraph was commercially developed and patented in 1837 by Charles **Wheatstone** (England) and Samuel **Morse** (USA). At first, the messages, consisting of dots and dashes, were carried between railroad stations by electric currents (from batteries) conducted by the iron rails. But copper wires were both better conductors and not restricted to the rail line. By the 1860s commercial services, operated by national post offices, or by Western Union in the USA, had connected most towns and cities. Messages to be sent were written or dictated to telegraphers who coded them. At the other end, they were received and decoded by trained operators who typed the message on paper and had it delivered by a messenger. Later, the coding and decoding was done by machines.

By some accounts, the most valuable invention of all time, was the first device to transmit the sound of a human voice over a copper wire. It was, to some extent, accidental. **Alexander Graham Bell**'s backers were merely seeking to increase the capacity of the telegraph system, which was having difficulty expanding its network fast enough to accommodate growing demand. The original invention was the practical implementation of the notion that speech could be transmitted and reproduced as an "undulatory electric current of varying frequency".

To be sure, the telegraph was already a well-established technology by the 1860s, having grown up with the railroads. But the telephone was disconnected from the railroads from the beginning, and it needed its own power supply. Bell's first telephone in 1876 actually preceded Edison's breakthrough in DC power generation, and the first units utilized battery power. The entire telephone system, as well as radio, TV, and now the Internet are all essentially developments for the transmission of electro-magnetic signals as "undulatory currents of varying frequency".

In any case, Bell's invention was soon followed by the creation of a business enterprise (American Bell Telephone Co., later AT&T) which grew with incredible rapidity. Manufacturing began under license within a year and 5600 phones had been produced by the end of 1877. In 1878 the first commercial switchboard was placed in operation in New Haven, Connecticut, with 21 subscribers. Telephone companies sprang up in almost every town and city, not only in the U.S. but also in Western Europe. The annual rate of U.S. telephone production rose to 67,000 in the year 1880 (of which 16,000 were exported), and the number of units in the hands of licensed Bell agents in the U.S. alone reached 132,692 as of Feb. 20 1881 (Smith 1985 p. 161). The number in service nearly doubled 2 years later.[13]

Bell's original patent was the subject of a long, drawn-out lawsuit due to the near-simultaneous patent submission by another inventor, Elisha Gray, a co-founder of Western Electric Co.[14] The dispute was based on whether or not

---

[13] The Bell licensees gradually evolved into local operating companies. Western Electric Co. was acquired by American Bell in 1881 to become its manufacturing arm, and AT&T was formed to operate "long lines" interconnecting the local companies. It gradually acquired stock in many of the operating companies and subsequently exchanged its stock with that of American Bell, becoming the "parent". AT&T was broken up by the US Department of Justice in 1984, although the "long distance" component of the original conglomerate still exists as a separate telecom company, having first been swallowed up by one of its former "children". After the breakup, Western Electric became Lucent, which later merged with the French firm Alcatel.

[14] Western Electric was later acquired by AT&T and became its manufacturing arm.

Bell had stolen Gray's idea of a liquid (mercury) conductor in the transmitter. It was not decided in Bell's favor until 1888. Bell's invention was only the first step in a massive technological enterprise. Telephony has spawned literally tens of thousands of inventions, some of them very important in their own right. One of the first and most important was the carbon microphone invented by Edison and Berliner (1877). Elisha Gray, a lifetime inventor with 70 patents (including a method of transmitting musical notes) is credited with inventing the music synthesizer.

Many of the key inventions in sound reproduction, electronic circuitry and radio were by-products of an intensive exploration of ways to reduce costs and improve the effectiveness of telephone service. There were 47,900 phones actually in service at the end of 1880, 155,800 by the end of 1885, 227,900 by the end of 1890, and 339,500 by the end of 1895. The original Bell patent expired in 1894, releasing a new burst of activity. U.S. demand skyrocketed: a million phones were added to the system in the next 5 years, 2.8 million in the period 1905–1910, and 3.5 million in the period 1910–1915. The U.S. industry grew much faster than its European counterparts after 1895 (having lagged somewhat behind Sweden and Switzerland previously).

Employment generated by the telephone system in all its ramifications has not been estimated, but was probably not less than one employee per hundred telephones in service. By 1900, the principal manufacturer, Western Electric Co. (later acquired by AT&T) alone employed 8500 people and had sales of $16 million; by 1912 it was the third largest manufacturing firm in the world, with annual sales of $66 million. It need scarcely be pointed out that the telephone system could not have grown nearly so fast or so large without the concomitant availability of electric power.

An offshoot of the telephone (and another application of electricity) was radio-telegraphy, which has subsequently morphed into radio, TV and the Internet. It all started in 1867 when James Clerk Maxwell published his comprehensive theory of electromagnetism, which predicted the existence of electromagnetic waves. This was one of the greatest intellectual achievements in human history, although it was barely noticed except by a few physicists at the time. Twenty years later **Heinrich Hertz** demonstrated the existence of radio waves. The measure of frequency is named for him.

The man who saw a business opportunity for radio-telegraphy and developed the technology was Guglielmo Marconi, starting in 1896. However, the technology was very limited at first and its applications remained few and specialized—mainly to ships—for the next two decades. But electronic technology progressed very rapidly, thanks to Alexander Fleming's invention of the thermionic diode, or "valve" in 1904. It got that name because, like a valve, it could open and shut like a switch. (He was working for Marconi at the time, taking time off from his professorship.) This "valve" was the predecessor of the ubiquitous "vacuum tube", which has by no mean been completely replaced by solid-state transistors.

Fleming's invention was followed by a clever "add on" 2 years later by the American inventor Lee De Forest (no slouch—he had 180 patents). De Forest added a third filament—creating a triode—and observed that the "valve" had become an amplifier of sorts. De Forest and Fleming fought for years over patent rights, while a young genius named Edwin Armstrong proceeded to invent several of the most important circuitry innovations in the radio business, starting with the regenerative amplifier, the super-regenerative circuit for transmitters, and the super-heterodyne circuit for receivers. (Armstrong had 70 patents.)

Broadcast radio took off in the 1920s. By 1929 Radio Corporation of America (RCA) was the hottest stock on Wall Street. In the 1930s Armstrong pioneered frequency modulation (FM)—previously invented (1922) by John Renshaw Carson as single sideband modulation (SSM), at Bell Laboratories—as a substitute for amplitude modulation (AM) for radio broadcasting. The superiority of FM is perfectly evident today. But it was fought tooth and nail in the courts and the Federal Communications Commission (FCC) by RCA, under its unscrupulous chief, David Sarnoff (himself a former Marconi inspector). RCA's legal victories impoverished Armstrong—who had been a wealthy man in the 1930s, thanks to his radio patents—and he committed suicide in 1954. Armstrong's widow finally succeeded in reversing RCA's legal victories, but too late for him.

The underlying ideas for TV were developed gradually in several countries over the decades after 1897, when British physicist J.J. Thomson demonstrated the ability to control cathode rays in a cathode ray tube (CRT). The first to patent a complete system including "charge storage" for scanning and displaying an image was an Hungarian engineer, Kalman Tihanyi in 1926. Tihanyi's patents (including patents not yet issued), were bought by RCA. RCA also bought Vladimir Zworykin's 1923 "imaging tube" patent from Westinghouse, further developed by Zworykin into a color version in 1928 and incorporated in RCA's "iconoscope" camera in 1933.

Meanwhile Diekmann and Hell patented an "image dissector" camera in Germany in 1925. That later became the basic design for European TV systems. Meanwhile, Philo Farnsworth in the US had independently invented his "image dissector" camera which he patented (and demonstrated in 1927). (RCA sued Farnsworth for infringement on Zworykin's 1923 patent, but lost and had to pay him $1 million starting in 1939). Farnsworth demonstrated his improved system at the Franklin Institute in 1934. During the early 1930s other imaging systems for CRTs were being developed and demonstrated in Berlin (Manfred von Ardenne), Mexico (Gonzales Camarena) and London (Isaac Schoenberg at EMI). However, the signal-to-noise ratios (sensitivity) of the RCA iconoscope, and the first EMI "emitron" were both still much too low. In fact, under Schoenberg's supervision the signal to noise ratio of first "emitron" was radically improved (by a factor of 10–15 or more) as a "super-emitron". That breakthrough became the basis for the world's first 405-line "high-definition" TV broadcasting service, by the BBC, in 1937.

Television using vacuum tubes and cathode ray tubes (CRTs) for pictures was further developed in the 1940s and spread rapidly in the 1950s. Bell Telephone Laboratories demonstrated a mechanically scanned color TV in 1929. Camarena's patent on a "trichromatic field sequential system" in 1940 was one of the enablers of color TV. There were a number of problems, and a great many firms in different countries were working on the problems. The first color TV broadcast in the US was in 1963 from the Rose Bowl parade. The "color transition" in the US started in 1965 when all the TV networks started transmitting more than half of their programming in color. By 1972 the transition was complete. But by 1970, thanks to transistors, RCA had lost its technological edge and the TV industry had moved to Japan (Sony, Matsushita).

Japan pioneered digital high density TV (HDTV) and began to penetrate the global market in 1990. The digital TV transition started in 2007 or so. Analog broadcasting was rapidly phased out, and was expected to be finished by 2015 or so.

The flat screen "plasma display" system was first proposed in 1936, by the same Hungarian engineer, Kalman Tihanyi, whose TV patents from 1926 had been acquired by RCA. Flat screen systems, mostly using liquid crystal displays (LCDs) using solid-state light-emitting diodes (LEDs) took a long time to be commercially viable, in competition with CRTs. But by the 1990s, flat screens were replacing CRTs and by 2007 LCD displays outsold CRTs for the first time.

By 2013, 79 % of the households in the world had TV sets.

After WW II electricity demand was driven by the spread of telephones, radios, TV, freezers, air-conditioning and—of course—more lighting. The rising electrical demands of the rapidly expanding telephone system in the US, which depended entirely on electromechanical switching systems, prompted the executives of the Bell Telephone Laboratories to initiate a new R&D project in 1945. It was aimed at finding cheaper and more economical solid state switching devices to replace vacuum tube diodes and triodes.

The outcome of that project was the transistor, a solid-state electronic device based on semi-conductors, a newly discovered class of materials, and the so-called "semi-conductor" effect: to alter the conductivity of a material by means of an external field, discovered previously by Shockley. The Bell Labs project stalled at first, when the experiments got nowhere. Bardeen found (and published) the explanation, in terms of quantum-mechanical surface effects. That idea led to the first point-contact transistors, which were demonstrated in December 1947 at Bell Telephone Laboratories and announced publically in June 1948. This breakthrough by three physicists (**John Bardeen, Walter Brattain and William Shockley**) earned a Nobel Prize in 1956.

The problem of patenting the transistor was complicated by the fact that the first patents for a solid-state triode were filed by Julius Edgar Lilienfeld, in Canada and the US in 1925 and 1926. In 1934 Oskar Heil patented such a device in Germany, but neither Lilienfeld nor Heil published research articles or demonstrated actual prototypes. However the lawyers decided to patent "around" the Lilienfeld patents. They submitted four narrow patent applications, none of which contained Shockley's name. This omission angered him, and broke up the group. In 1951 Shockley—working in secret—independently invented the bipolar junction transistor, which is the basis of most later devices. Publicity for this infuriated Bardeen and Brattain. Bardeen left Bell Labs and moved to the University of Illinois in 1951 to work on the theory of superconductivity (for which he received another Nobel Prize in 1972).

In 1956 Shockley (still angry) moved to California to run Shockley Semiconductor Lab, a division of Beckman Instruments. But because of his decision to terminate R&D on silicon (and his unilateral management style) eight of his top employees left as a group to form Fairchild Semiconductor. A few years later, several of those men departed again, to form Intel.

Because of a "consent agreement" by AT&T with the Anti-Trust Division of the US Department of Justice, that fundamental patent was licensed by AT&T to all comers, for a modest fee ($50,000). One of the earliest licensees was Sony, which used transistors in its successful portable "transistor radios", starting in 1957. The original black-and-white TV sets, using vacuum tubes in their innards, were replaced by color as transistors replaced vacuum tubes, in the 1960s and 70s. In the 1970s and 1980s the TV manufacturing industry had moved to Japan and South Korea.

The early stored program digital computers, starting with ENIAC in 1948, used vacuum tubes for switching. The next generation of computers by IBM and UNIVAC in the early 1950s was using transistors instead of vacuum tubes. At that time one of the chief executives of IBM opined that there was room in the world for a dozen or so of those "main-frame" computers. (I worked on one in the summer of 1953.) Other entrepreneurs saw it differently. The next major breakthrough was the "integrated circuit" invented, independently, by **Jack Kilby** at Texas Instruments and **Robert Noyce** at Fairchild Semiconductor in 1958–1959. The circuit-on-a-silicon-chip technology took a decade to scale up, but the dynamic random-access memory (D-RAM) on a chip arrived in 1970, followed by the first micro-processor (1971), both by Intel.

Those technologies, in turn, enabled computers to become faster and more powerful as they shrank in size. The first personal computer by Olivetti in (1965) cost $3200. Hewett-Packard began manufacturing small computers soon after that. To make a long and interesting story short, the merger of telecommunications and computer technology resulted in the internet, the mobile phone and the digital revolution. Since 1980 information technology devices, starting with personal computers (PCs) and more recently cellphones, and so-called tablets have spread rapidly.

In retrospect, it can be argued that the transistor and the integrated circuit *did* open up a new resource, albeit one that is hard to compare with natural resources like coal or copper. The new resource can be called "bandwidth". It revolutionized the radio and TV industries and led to the ubiquitous digital information and information technology that dominates the world economy today.

Technically, the useful bandwidth (of an electromagnetic transmission) is the part of the frequency spectrum that can be detected, amplified, and converted into other frequencies. As higher frequencies allow greater bandwidth, there is a direct relationship both between bandwidth and energy, and also between bandwidth and information. In fact, if an elementary particle can be thought of as a packet of condensed energy, with a corresponding frequency, it can also be thought of as a packet of condensed information. Perhaps at the instant of the Big Bang, the universe was "pure information". A few theoretical physicists have been thinking along these lines. But I only mention it as a possible link between the subject of Sect. 6.1 and this chapter.

9.10   The Demographic Transition: The Final Disproof of Malthus or a Prelude?

The early demographic history of *Homo Sapiens* was covered, very briefly, in Chap. 8, Sects. 8.1–8.4. Human population in the Old Stone Age cannot be estimated with any accuracy but given the scarcity of evidence of habitation the numbers must have been small. The transition from hunting-gathering to settled agriculture began in the valleys of the Nile, the Tigris-Euphrates, and presumably the big rivers of South Asia and China, about 10,000 years ago. That was an indication of scarcity of game, probably due to population density. By 3000 years ago (1000 BCE) it had probably increased to 150 million or so (Fig. 9.6). At the peak of the Roman Empire it might have doubled, as suggested by the several waves of "barbarians" seeking new lands to settle. A few moderately large cities existed in that time. But when Rome collapsed, the urban population also declined.

Growth was slow, barely keeping up with land clearing (the fertile river valleys having been settled long since) until the devastating bubonic plague in the fourteenth century, which killed between a third and a half of the population of Europe and presumably elsewhere. The labor shortage following that plague resulted in

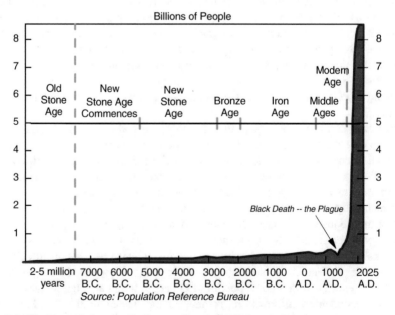

Fig. 9.6 World population growth. *Source: Population Reference Bureau*

higher wages and food prices in Europe. It also gave the towns more leverage against feudal landowners, which accelerated economic activity and trade—and urbanization. The fall of Constantinople to the Ottoman Turks (in 1453) was a watershed, not only in Europe, because it cut the traditional "silk road" for trade with China and stimulated navigation, exploration and long-distance maritime trade instead.

Birth rates (children per woman) increased during the fifteenth and sixteenth centuries thanks to progress in reducing the death rates of young women giving birth in the absence of medical facilities or antiseptics. Early childhood death rates fell due to increased medical knowledge (e.g. about disease transmission) and simple public health measures. So population growth rates accelerated while death rates fell dramatically. The decline in death rates in the nineteenth century was mostly due to further progress in public health (e.g. water treatment and sewage treatment) and more effective antiseptics.

In the twentieth century the main factor bringing death rates down has been the conquest of infectious disease (like flu, smallpox and measles) by segregation, vaccination or (for bacterial infections) by sulfa drugs and antibiotics (penicillin and streptomycin). During the Napoleonic wars and the US Civil War, the vast majority of wounded soldiers died. Death rates among wounded soldiers fell somewhat during the Crimean War (thanks to nursing) and still more during World War I. The death rate for wounded soldiers fell dramatically during and after WW II, as medical help became available near the front lines (as illustrated by the TV series M*A*S*H.) Today doctors travel with the soldiers and the death rate among wounded soldiers is negligible.

So the declining death rate preceded the declining birth rate. Only recently have the two rates begun to converge in the most advanced countries. Indeed, birth rates are now below death rates in Germany, Italy, Japan and Spain, and Eastern Europe including Russia. Figure 9.7 shows the current declining population growth rates in the world as a whole, as compared to the rates in less developed vs. more developed countries, while Fig. 9.8 shows the global population growth since 1700.

Death rates have continued to decline thanks to medial progress, but birth rates have also begun to decline, at least in the industrialized countries. The weakening of religious opposition to birth control and the availability of cheap contraceptives are contributing factors. The declining birth rate is due, very largely, to the introduction of birth control pills ("The Pill") starting in 1960 in the US and now widely available.[15] The Pill has been given credit for the so-called "sexual revolution" (giving women control over progenerative behavior) and the decline of the Catholic Church's influence around the world.

---

[15] There are a number of names associated with this development, but the most outstanding (in retrospect) is the work of Carl Djerassi, who co-founded Syntex Corporation with Luis Miramontes and George Rosenkrantz. Djerassi was one of the outstanding chemists of the twentieth century, with many other contributions.

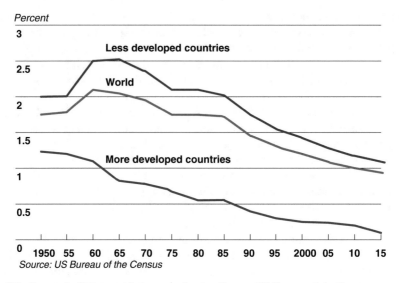

Fig. 9.7 Current declining population growth rates. *Source: US Bureau of the Census*

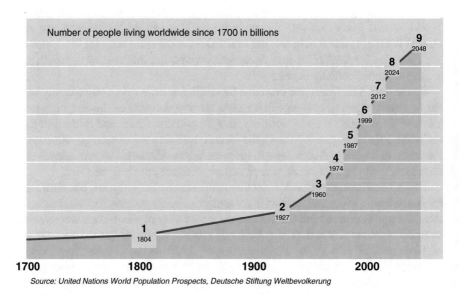

Fig. 9.8 Population of the Earth. *Source: United Nations World Population Prospects, Deutsche Stiftung Weltbevolkerung*

Other factors include urbanization and the high cost of raising children in cities. (Children are useful labor on small farms, but much less so for households in cities.) However, it is increasingly clear that the critical variable is female education. Females with primary education (or none) have many more children than females

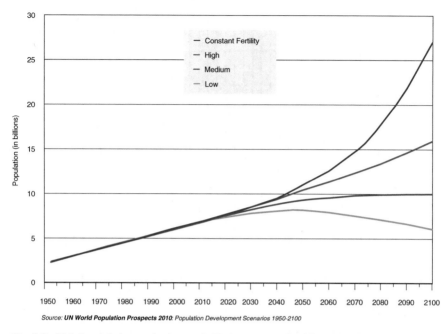

Source: **UN World Population Prospects 2010**: Population Development Scenarios 1950-2100

Fig. 9.9 Global population projections to 2100. *Source*: **UN World Population Prospects 2010**: *Population Development Scenario 1950–2100*

with primary or secondary education. Females with tertiary education have even fewer children, less than one child per woman. Yet a number of European countries (and Japan) are now subsidizing children, under the (false) impression that there will be a demographic crisis in the future: too few working people to support the elderly by means of their social security taxes. This child subsidy policy makes absolutely no sense for countries with youth unemployment rates of 20 % or more. The global economy does not need more unskilled or semi-skilled labor supply, and even college-educated boys and girls are having trouble finding jobs. The social security problem needs to be addressed in a different way, perhaps by means of a guaranteed income or negative income tax. (Or it may not be solved at all.).

A few years ago it was thought that the global maximum will be between 9 and 9.5 billion, sometime after 2050 e.g. (Lutz et al. 2004). Recent revisions have suggested that the global population will reach 11 billion by 2100 and that peak will be later and higher, if there is a peak at all e.g. (Gerland et al. 2014). It is not clear what may have contributed to the shift in expectations. See Fig. 9.9.

However it is very clear that urbanization is one of the main factors contributing to lower birth rates. Lack of space, less demand for child labor (sadly, not zero) and the high cost of raising and educating children in cities are the primary reasons why birth rates are so much lower in the highly urbanized (developed) countries than in the more rural, less developed parts of the world. Urbanization for different parts of the world, with extrapolations to 2050, is plotted in Fig. 9.10.

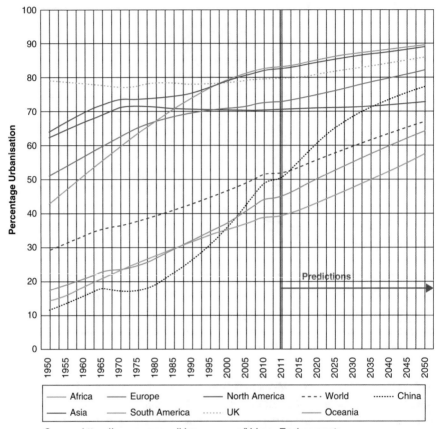

Fig. 9.10 Graph showing world urbanization for different world areas 1950–2050

Declining birth rates, in turn, have a major impact on the social structure of countries, as shown in Fig. 9.11. Countries with high birth-rates (especially Muslim countries like Egypt, Indonesia and Pakistan), have very young populations. This young population is sometimes portrayed as an economic advantage, since it implies that there are many people of working age (yellow) to support the elderly non-working population (red).

But the young population (blue) guarantees a lot of future growth, even if birth rates slow down. Moreover, it imposes a heavy load on the educational system. (Many of the young boys in poor Muslim countries like Egypt and Pakistan end up in "madrassas", which essentially teach nothing except memorization of the Koran.) It tends to lead to an under-educated class of young people with no job skills. These children become easy prey for religious fanatics, jihadists or criminal gangs. On the other hand, China's unpopular "one child" policy allowed that country to educate its young people, enabling China to become the manufacturing hub of the world. (To be fair, it seems possible that the education of women may

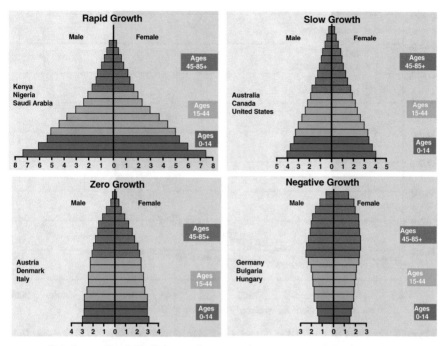

Data Source: Population Reference Bureau. x-axis represents population (percent)

Fig. 9.11 Population age structure. *Data Source: Population Reference Bureau. x-axis represents population (percent)*

have been more important than the 1-child policy.) By contrast, countries like Japan and Germany with old populations have a different problem: namely too few tax-paying workers with jobs to support the large number of elderly pensioners (many with health problems).

Urbanization, in turn, is driven by the economics of specialization (Adam Smith called it "division of labor") and economies of scale. The small self-sufficient farmer may be a social ideal for Jeffersonians who dislike the encroachment of "big government". But that kind of farmer needs to be a "jack of all trades" whereas a large factory can hire specialists for many different functions. And in most enterprises, increased scale allows fixed costs to be spread over more output. It is an old story that productivity and incomes are higher in cities—hence in urbanized areas—than in the surrounding rural areas. The very sharp increase in GDP per capita after 1960, in almost all regions of the world, is partly attributable to rapid urbanization (Fig. 9.12). Cities also offer greater variety of ideas, educational opportunities, job opportunities, housing and entertainment opportunities, not to mention choice of products for shoppers to buy.

The other reason for sharp increases in GDP per capita after 1960 is partly attributable to increased per capita energy consumption, especially of petroleum products in Asia (Fig. 9.13). This was mostly liquid fuels for aircraft, trucks, buses

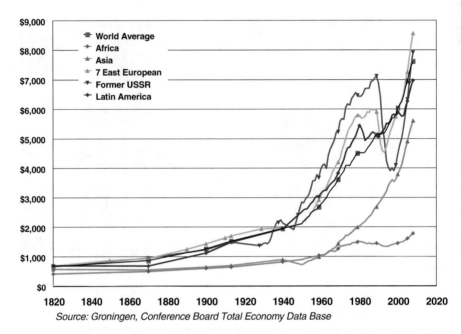

Fig. 9.12 Per capita GDP by region. *Source: Groningen, Conference Board Total Economy Data Base*

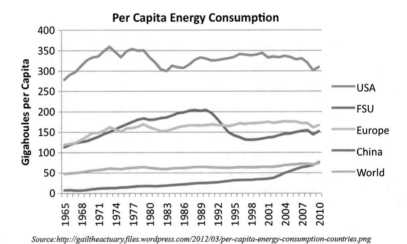

Fig. 9.13 Per capita energy use for five regions, 1965–2010. *Source*: http://gailtheactuary.files. wordpress.com/2012/03/per-capita-energy-consumption-countries.png

and private automobiles. The other big increase in per capita energy consumption since 1960 was that of natural gas for a variety of purposes including electric power, space heating and petrochemicals (e.g. plastics).

Use of biofuels, mostly wood and charcoal (blue), is declining everywhere due to deforestation, and most of what remains is in less developed countries that do not have hydrocarbons. Coal consumption per capita is roughly constant (which means it is still increasing globally) primarily because of its importance for the iron and steel industry, the cement industry, and for electric power generation. Overall, there is no indication of declining energy consumption per capita.

It is worth mentioning (as shown in Fig. 9.13) that per capita energy use in the US (and Canada) is much higher than in any other region: it is double the level in Germany and Japan, and five times larger than the world average. This fact (among others) suggests that there must be large opportunities for energy efficiency and renewables in the US and Canada. I will discuss some of them in Chap. 9.

Chapter 10
Energy, Technology and the Future

10.1 This Time Is Different

That trite phrase has been used too many times to explain why disasters, ranging from great storms to financial bubbles, couldn't have been anticipated. Some gurus characterized the financial collapse of 2008 as a "black swan", implying that it was an unpredictable phenomenon that might occur only once in a thousand years (Taleb 2007). It was, of course, no such thing, and it was predicted by some stock traders (Lewis 2010) and at least one academic economist, Steve Keen, see (Keen 2011a). The phrase has also been used ironically to make the point that lessons from the past are not being learned, still less acted upon (Reinhart and Rogoff 2010). That is clearly true.

However, what I mean by "this time is different" is that this time in human history is truly different: the economic growth paradigm that has been in place since the industrial revolution is now changing. I tried to make this point in a book in 1998 entitled "Turning Point: The end of the growth paradigm" (Ayres 1998). (It was not a best-seller.) This chapter is, in some ways, a recap (and revision) of that thesis to focus more on energy availability as the catalyst of change. The point remains the same.

In a book I co-wrote in 2009 with my brother Ed, we noted the harbingers of three disruptive and imminent energy-related challenges, as follows: "First is the aftermath of 'peak oil', that fateful moment when the public is finally convinced of what experts will have known for some time: that global oil production has begun its final decline" (Ayres and Ayres 2010 p. 19). Today, 6 years later, the subject of "peak oil" has fallen out of favor, and some experts are touting "Saudi America". The current position and prospects for oil are discussed in Sects. 10.2 and 10.3.

The second disruptive challenge we saw in 2009 was "the conspicuous 'old age' of several key technologies on which the fossil fuel-based economy has depended. Neither the internal combustion engines produced today nor the steam turbines we use to generate our electric power are much more efficient (at converting fuel to

© Springer International Publishing Switzerland 2016   303
R. Ayres, *Energy, Complexity and Wealth Maximization*, The Frontiers Collection,
DOI 10.1007/978-3-319-30545-5_10

useful energy) than they were in the 1960s. Our 'rust belt' is rusting for a *reason*". This topic is considered further in Sect. 10.4.

"The third disruptive challenge is the acceleration of climate change and a range of associated catastrophes like storms, droughts, and ocean acidification. This challenge will compound the global economic and energy crisis by increasing the urgency both to achieve oil independence—not only for the US—and to fully develop new, non-fossil fuel technologies . . . .[that are] not yet nearly ready to take over on a large scale" (ibid).

This topic will be discussed from a current perspective in Sect. 10.5.

There is another reason why "this time it is different" cuts ice. The point is that there are cycles in politics and economics that mirror cycles in ecology. In ecology those cycles of rapid growth, maturity, aging and collapse are most evident in ecosystems—such as forests—that are subject to occasional fires, insect infestations, storms or clear-cutting (Holling 1973). In human affairs the analogous cycle starts with a financial collapse, an epidemic, a revolution or a war. Recovery tends to be faster, at first, than growth in the previous "mature" stage. But later the economy becomes increasingly "sclerotic" as institutions and special interest groups proliferate (Olson 1982). The following disaster—war or revolution—constitutes a time of "creative destruction when many old habits and institutions are destroyed." After each such cycle things are likely to be different, and re-growth is less inhibited. However I leave it for Chap. 11 to discuss the mechanisms involved.

Some readers may be wondering why this book does not emphasize the great digital revolution that has made Apple (rather than IBM, General Motors or Exxon) the most valuable company in the world (as measured by market capitalization) with Google, Microsoft, Amazon and Facebook close behind. It is true that the information technology industry, combining telecommunications, and computer technology, centered in "Silicon Valley" is undoubtedly the primary driver of economic growth in the US (software) and in parts of East Asia (hardware manufacturing) at present.

There is great excitement among market analysts, currently, about such questions as how soon cars will be automated and whether people will do their banking from their iPhones or Apple watches. Other topics in the background concern Big Data and the "cloud" (Amazon, Google, Microsoft), the feasibility of automated driving (Google, Apple, Tesla, everybody), future applications of drones for delivering goods (Amazon), the feasibility of brick-laying or housecleaning robots, and the future of artificial intelligence (AI). A few people, so far, wonder about the social consequences of these technologies. (See Chap. 12 for a little more on these issues.)

From an historical perspective, however, what has happened since 1950 or so (Sect. 8.8) was very much like the discovery of a new resource (e.g. a gold rush). The new resource was, in the first instance, semi-conductors that made it possible to replace expensive and short-lived vacuum tubes by solid-state devices—now called "chips". Since 1990 the real new resource for the economy has been the "stretching" (as it were) of the electro-magnetic spectrum by miniaturization of electronic circuitry. This revolution led on to the communications revolution, the

Internet, and its avatars (like Apple and Google) that dominate the growth center of the world economy today.

In short, I think this time in history is really different, in the sense that several 'old faithful' assumptions and obsolete paradigms must now be discarded. New economic thinking and new paradigms are badly needed. I will make my modest contribution in Chap. 12.

10.2 "Peak Oil"

In the past few decades, many petroleum geologists have become convinced that global output of petroleum (and of natural gas soon after) is about to peak, or may have peaked already (Hubbert 1956, 1962; Ivanhoe 1996; Hatfield 1997; Campbell 1997, 2003; Campbell and Laherrère 1998; Deffeyes 2005). From the perspective of 2016 it appears that production of oil from "conventional" sources—so to speak—may indeed have peaked, although there is still some oil to be found in deep water (e.g. off the coast of Brazil), in very remote places (e.g. the Arctic), or in tar sands and shale. Unconventional oil is much more expensive to find and recover than oil from the Middle East, but with new technology it is still profitable at high enough prices. Investment in unconventional oil depends on prices.

This geological scarcity, together with rising demand from China and India, and turmoil in the Middle East, seemed to be reflected in pre-2014 prices. But short term oil price movements have several drivers, including futures speculation, OPEC politics, the Sunni-Shia rivalry in Islam, Russian and Iranian affairs, the slowing of the Chinese economy and declining US demand for gas guzzlers. Notwithstanding the turmoil, an upward long-term price trend, seems sure to follow eventually. The deeper the bottom the faster (and higher) the price recovery is likely to be. Geology does drive the recovery price of oil in the longer term.

The fact that discoveries have lagged behind consumption has been known for a long time (Figs. 10.1 and 10.2). The reality of peak oil has been obscured up to now by sudden but unexplained increases in officially reported reserves in the Middle East in the late 1980s (Fig. 10.3), and uncritical forecasts by industry figures and government agencies. These optimistic forecasts are strongly influenced by mainstream economists who still argue—as they did in their response to the "Limits to Growth" book (Meadows et al. 1972)—that there is still plenty of oil in the ground and that rising prices will automatically trigger more discovery and more efficient methods of recovery.

Discovery of conventional oil peaked in the US in 1930 and globally in 1965. Total discoveries in a given year have not kept up with depletion since 1982 (except for 1992) and the ratio of discovery to depletion is continuously declining. An influential oil consultancy, Wood-MacKenzie, noted in 2004 that oil companies were discovering an average of 20 million bbl/day, while global consumption was up to 75 million bbl/day. The discrepancy is even worse now that daily global consumption is around 95 million bbl.

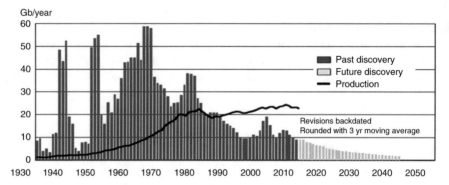

Fig. 10.1 Gap between oil production and oil discovery. *Source: Colin Campbell ASPO*

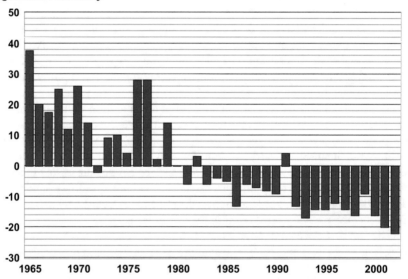

Fig. 10.2 Global oil discoveries vs. global oil production 1986–2002. *Source*: Heinberg 2004, "Powerdown", Figure 5 page 43. *Until well into the 1970s, new global oil discoveries were more than sufficient to offset production every year. Since 1981, the amount of new oil discovered each year has been less than the amount extracted and used*

The implications for total Saudi output in the coming years are ominous, as Figs. 10.4 and 10.5 show. To what extent has the longer term been compromised by near term political considerations? Nobody knows outside of Riyadh.

The average size of new discoveries has fallen precipitously since the year 2000 although this may be partly due to the war in Iraq and the chaos in Libya. The simple facts are that new 'oil provinces' are not being discovered. No 'super-giant' field has been discovered since the Alaska North Coast. The North Sea has passed

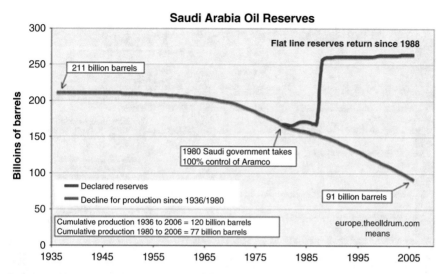

Fig. 10.3 Official Saudi oil reserves 1935–2008. *Source*: http://3.bp.blogspot.com/_Jx78YcF-F8U/TVLD6NTh9KI/AAAAAAAAEKE/eei6WpDlDoM/s1600/07-06-19b_Saudi_reserves.png

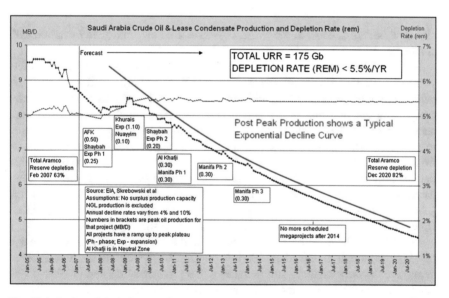

Fig. 10.4 Projected depletion rate for Saudi Aramco. *Source*: http://www.theoildrum.com/files/Production.jpg

its peak and—except for the Barents Sea (where Statoil is exploring), Kashagan in Kazakhstan (under the north Caspian Sea), Brazil's new deep-water field (at a depth of more than 10 km) and the one just announced in central Java—the western oil companies are now mostly frozen out of the only regions where significantly more

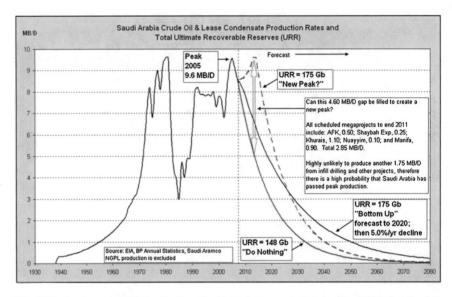

Fig. 10.5 Another miracle? Saudi oil production rate forecast. *Source*: http://i129.photobucket. com/albums/p237/1ace11/200705fig10.jpg

oil might be found. Most of the world's reserves are now controlled by national oil companies.[1]

So-called 'proved resources' (90 % certain) were still increasing (barely) a few years ago because formerly 'proved and probable' resources (50 % certain) are being converted to 'proved' as existing fields are fully explored. But the latter category is the one that best predicts future supplies—and the two curves are converging (Fig. 10.6). Big publicly traded oil companies are showing increased reserves, but what they do not mention is that this appearance of growth is mostly from "drilling on Wall Street"—i.e. buying existing smaller companies—rather than drilling in the earth. Because share values supposedly reflect reserves, companies that did not adopt this strategy—like Shell—faced strong pressures a few years ago to meddle with their reserve statistics in order to reassure stockholders. Now Shell seems to have switched to the acquisition strategy.

On-shore production, before shale, peaked in the 1970s. Combined on-shore and off-shore production peaked in 2005. 'Gas liquids', which are an indication of aging, are the only category still increasing. US production peaked in 1969–1970, just before Prudhoe Bay came on stream, and even that discovery did not raise US output above the 1970 level. The Cantarell field in Mexico, the second largest ever found, is well beyond its peak and output is declining fast. North Sea output has peaked and is now declining. The giant Kashagan field, discovered in 2000 (with an estimated 13 billion barrels of recoverable, but high sulfur, oil) is under the northern

---

[1] The best source on all this is (Strahan 2007). For the geological background see (Campbell 1997, 2003; Campbell and Laherrère 1998; Deffeyes 2001).

Billion Barrels

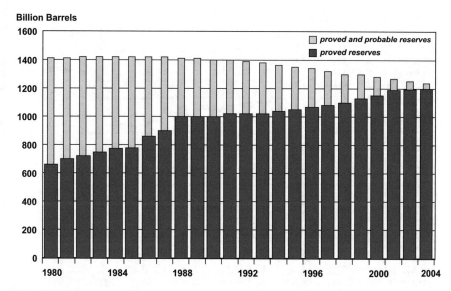

Fig. 10.6 The category "Proved and Probable" is vanishing. *Source*: Strahan 2007, *"The Last Oil Shock"*, Figure 13 page 71. *Global "proved reserves" (wide bars) give the reassuring appearance of continuing growth, but the more relevant "proved and probable reserves" (thin bars) have been falling since the mid-1980s*

part of the Caspian Sea. It has been called "the last pearl in the crown of the world oil industry". Kashagan is now under development by a consortium of companies but has proved to be very complex and expensive ($135 billion spent so far), and harmful to wildlife. Commercial production only started in 2013 (full production expected in 2017). The oil will be going to China via a new pipeline.

In 2007 the Brazilian nationalized company Petrobras announced the discovery of a 5 billion barrel "giant" field off the coast south of Rio de Janeiro. It may presage a larger ("super-giant") field in the Santos Basin with up to 50 billion barrels of recoverable oil. However, it is located under dense layers of salt at a depth of 10 km below the ocean surface, so getting the oil out will not be easy. The energy return on energy invested (EROI) is unlikely to be better than 10 (typical of deep-water fields) which means that 5 billion of the 50 billion barrels of possible output will be used—as exergy—for the recovery, refining and distribution. Corruption in Petrobras has destabilized the Brazilian government.

Russian output looks like it will peak very soon, if it has not done so already. The director of one of the largest Russian firms says that his country will never produce more than 10 Mb/d. The CEO of *Total* (the French oil major) has said that 100 Mb/d is the absolute maximum for world production. According to the British Petroleum *(BP) Statistical Review*, of 54 producing nations only 14 still show increasing production, 30 are past peak output, while output *rates* are declining in 10 (Fig. 10.7).

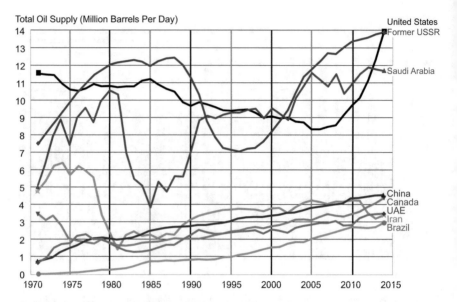

Fig. 10.7 Oil producing countries 1971–2014. *Source*: http://www.eia.gov/cfapps/ipdbproject/
IEDIndex3.cfm

Most non-OPEC oil producing countries (except the US) are now in decline, and
very few significant countries, mostly in the Middle East, claim to be able to
increase output. The "end of oil" may not be an immediate threat (thanks to other
factors). But the halfway point, corresponding to peak output, is likely to be just
around the corner, i.e. probably 5–15 years in the future.

But as the discovery rate has slowed, the energy and other costs of replacing oil
that has already been consumed are now rising fast. The energy return on energy
invested (EROI) of oil discovered in the 1930s and 1940s was about 110, but for the
oil produced in the 1970s it has been estimated as 23, while for new oil discovered
in that decade it was only 8 (Cleveland et al. 1984). This was still tolerable since
only 12 % of the oil discovered was needed to discover, drill, refine and distribute
it. In the case of deep-water oil, and for heavy oil, the EROI is currently estimated to
be about 10.

The end of near total dependence on fossil fuels is already in sight, although a
little more distant than was thought a decade ago. Some years ago the oil geologist
M. King Hubbert, taking into account discovery rates and depletion rates of known
fields, predicted that global oil output would peak between 2000 and 2010 (Hubbert
1973, 1980). Others, notably Colin Campbell and Jean Laherrère came to the same
conclusion (Campbell and Laherrère 1998).

From the perspective of nine years later (2014) it can be seen that global oil
production did not actually peak in the period 2005–2013. Data from EIA for 2003
through 2015 (Fig. 10.8) show clearly that production from standard "legacy"
sources of crude oil, such as the OPEC countries, Africa and Russia, did not

Total Oil Supply (Million Barrels Per Day)

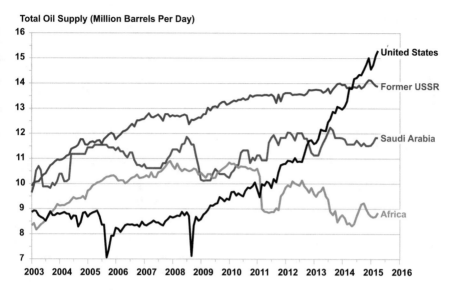

Fig. 10.8 Convergence of US and Saudi Oil production. *Source*: http://www.eia.gov/cfapps/ipdbproject/IEDIndex3.cfm

increase. The small increase in total output since 2005 has been entirely due to US shale oil and Canadian tar sands production, known as "unconventional" oil. However, whereas a number of influential reports in the 2009–2019 era predicted a gap by 2015 of up to ten million bbl/day between demand and supply, the opposite situation actually occurred. The ten million bbl/day gap has become a two million bbl/day glut.

It is worthwhile to look at oil prices over time. Figure 10.9 tracks the price through 2015. The two dramatic "spikes" in 1979, 1979–1980 and 2008 (and a small one in 1991) have all been followed by recessions (indicated by the gray shade.) The oil price since 2010 is shown in more detail in Fig. 10.10. The question causing such volatility in the stock-market in the winter of 2016, was mostly whether the current low oil price will cause another recession, despite the (fairly) strong US economic data.

Evidently a small change in output after 2008 was followed by a very large change in price, although the indirect effects on the global economy, resulting in sharply reduced global demand, were mainly responsible. But when the economic downturn "bottomed" in 2009, the price went up again and stayed fairly high for several years. The price peaked over $100/bbl in 2011 and remained near or above $90/bbl until midsummer 2014. John Watson, CEO of Chevron, has said in interviews during that period that $100/bbl is "the new reality" in the crude oil business and that "costs have caught up to revenues for many classes of projects." In 2013 or so it was being said around the executive suites that $100/bbl had reached OPEC's "threshold of pain". This is because many oil-producing countries were,

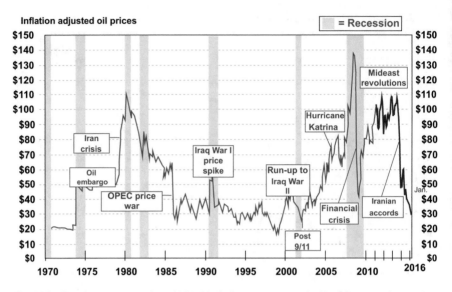

Fig. 10.9 Oil prices and recessions 1971–2015. *Source*: www.peak-oil-crisis.com using production data from www.chartoftheday.com (Accessed 27 August 2013) updated by Ayres September 2013–January 2016

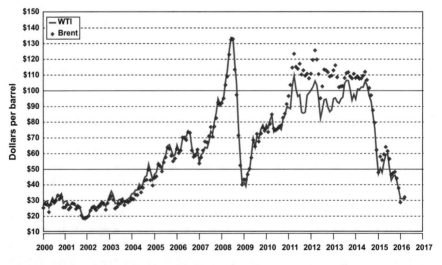

Fig. 10.10 Crude oil; monthly spot prices Jan. 2000–Jan. 2016. *Source*: http://www.eia.doe.gov/dnav/pet/TblDefs/pet_pri_sbt_tbldef2.asp

even then, earning less from petroleum sales than their national budgetary requirements.

The weighted average of OPEC member's receipts in 2013 was $106/bbl. But as of late September 2014, the average prices received by OPEC members was down

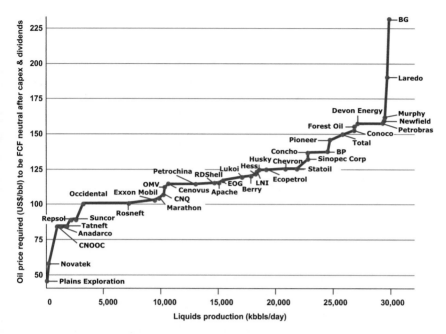

Fig. 10.11 Hurdle rate for many oil companies since 2012. *Source*: Kopits (2013)

to $98.63/bbl. Iran, Iraq and Nigeria were all earning less than they were spending. By winter 2015, with prices at $50 or less, the budgetary pain was much more severe. The price of oil is back to around $50/bbl at the time of writing (June 2015–2016).

Yet, the costs of maintaining existing wells and discovering new ones continues to rise faster than revenues (Fig. 10.11) (Kopits 2013). In fact, revenues from the legacy wells began declining after 2005, long before the price decline in 2014–2016. From July 2, 2014 through January 2016 crude oil prices fell from a peak above $110/bbl (Brent) or $108/bbl for West Texas International (WTI) to around $28/bbl (Fig. 10.10). This decline was totally unpredicted, although numerous pundits have unhesitatingly explained it retroactively. The source of the increased output was a huge boom in hydraulic fracturing ("fracking") of shale in the US, especially Texas and North Dakota, with some contribution from Canadian tar sands.

Even before the 2014–2016 price decline, most oil companies needed to recover more than $100 per barrel for new wells to achieve positive free cash flow, according to Kopitz, (op cit). Half of the oil industry needed $120/bbl, and the bottom quartile in profitability terms requires $130/bbl. See Fig. 10.11. In retrospect, the median breakeven point seems to have been considerably lower than $120/bbl, probably around $70 or $80/bbl. But by winter 2015–2016 virtually all producers except Saudi Arabia and Kuwait were unprofitable, and most are living

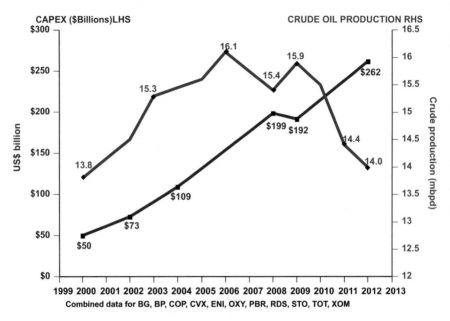

Fig. 10.12 Mismatch between capital expenditures and demand. Combined data for BG, BP, COX, CVX, ENI, OXY, PBR, RDS, STO, TOT, XOM. *Source*: Kopits 2013

off cash reserves, insurance (hedging) or credit. Many companies will either be swallowed up by the deep pocketed majors, or shut down by bankruptcy.

However, it is important to recognize that free cash flow is a time dependent measure. It can be low during periods of exploration and development of new fields when CAPEX is high, and high during periods when already developed resources are being drawn down (see Fig. 10.12). The present (2016) seems to be such a period. Hence Fig. 10.12 is not necessarily a reliable indication of the true financial condition of the industry.

Low prices do seem to be here to stay for another year or so. But, in that scenario, the global economy—already stagnant—could fall into a deeper recession. It is probably fear of some such outcome that is driving the stock market sell-off and keeping the US Federal Reserve from raising interest rates as fast as other data suggests it should. However, if efficiency gains in the auto industry causes continuing drop in domestic US demand—as seems to be happening already—or if Chinese demand slows further, it could keep oil prices low enough, for long enough, to drive a lot of producers, especially the shale plays, out of the business. This is important for explaining fluctuations in the stock market. But what does it mean for long-term global economic growth? How much of the existing reserves can never be burned and must, therefore, be regarded by shareholders as "stranded assets"? These questions are still moot.

To summarize a few points: global output from conventional ("legacy") terrestrial and offshore sources of crude oil, plus natural gas liquids (NGL) peaked in

2005 at 84.5 Mb/d. The increase in total liquids (including NGL) since then was 7.5 Mb/d. *All of that increase was from unconventional sources in the US and Canada.* The low prices as this is being written (2016) will probably continue for some time to come, probably until global production finally declines as the high price producers (with high debt-loads) finally fall by the wayside. Yet that day is still in the future, because most will continue to produce, even at a loss, just to keep paying their debts.

If global liquids supply had increased at the previous historical rate (75 % of the GDP growth rate) then liquids supply from 2005 through 2013 would have grown by between 12 and 18 Mb/d, depending on the assumed rate of efficiency increase. A 2.5 % gain in efficiency would be consistent with a total increase of 12 Mb/d; an efficiency increase of half that amount would be consistent with a supply increase of 18 Mb/d. The actual supply increase was just 7.5 Mb/d. (This implies that was one or more of the assumptions is wrong.)

Capital expenditure (capex) by the oil industry to maintain (and increase) output from those "legacy" sources from 1998 to 2005 was $1.5 trillion. Output increased by 8.6 Mb/d during those years. Capital expenditure (capex) from 2005 through 2013 was $4 trillion (of which $1 trillion was spent to maintain and upgrade the gas distribution system). Of that, capex on non-conventional oil and gas, including tar sands, fracking, liquefied natural gas (LNG) and gas-to-liquids (GTL) amounted to about $500 million. *There was no increase (actually a slight decrease) in output from the legacy sources, which remained near the peak 85 Mb/d level.* In fact legacy crude oil production is down 1 Mb/d since 2005. All of the increase in total liquids output was in the US and Canada, and all of it was unconventional (and most of that was financed by "junk bonds", depending on low interest rates from the Fed.).

As of 2016, global demand seems to be about 1.5 Mb/d below supply. That is the accepted explanation of the recent (2014) price decline. Political factors, notably response to Russian behavior in the Ukraine, cannot be ruled out. The nuclear accord with Iran will bring Iranian production back to the market this year. Obviously there is a chance that the political uncertainties in North Africa (and Iraq) will go away, resulting in still higher output during the coming decade. Evidently several countries are producing less than they once did and probably could again. Summarizing the main points:

- Global output from conventional ("legacy") terrestrial and offshore sources of crude oil, plus natural gas liquids (NGL) peaked in 2005 at 84.5 Mb/d. The increase in total liquids (including NGL) since then was 7.5 Mb/d. *All of that increase was from unconventional sources in the US and Canada.*
- If global liquids supply had increased at the previous historical rate (75 % of the GDP growth rate) then liquids supply from 2005 through 2013 would have grown by between 12 and 18 Mb/d, depending on the assumed rate of efficiency increase. A 2.5 % gain in efficiency would be consistent with a total increase of 12 Mb/d; an efficiency increase of half that amount would be consistent with a supply increase of 18 Mb/d. The actual supply increase was just 7.5 Mb/d. (Does that mean there was no increase in efficiency?)

- Capital expenditure (capex) by the oil industry to maintain (and increase) output from those "legacy" sources from 1998 to 2005 was $1.5 trillion. Output increased by 8.6 Mb/d during those years.
- Capital expenditure (capex) from 2005 through 2013 was $4 trillion (of which $1 trillion was spent to maintain and upgrade the gas distribution system). Of that, capex on non-conventional oil and gas, including tar sands, fracking, liquefied natural gas (LNG) and gas-to-liquids (GTL) amounted to about $500 million. There was *actually a slight decrease* in output from the legacy sources, which remained near the peak 85 Mb/d level. In fact, by 2013, legacy crude oil production was down 1 Mb/d since 2005. *All of the increase in total liquids output was in the US and Canada, and all of it was unconventional.*
- Costs of maintaining the legacy production system are rising rapidly. Many fracking (and other) projects outside the US have been cancelled or shelved due to cost overruns.

The fracking boom (discussed below) is not a new phenomenon. Peace has not broken out in the Middle East. Instability is greater, if anything, in Libya and especially in Syria and Iraq. However, whereas the downfall of the Libyan regime and the Iraq war interfered with exports, it is likely that continuing instability (the rise of the Islamic State, ISIS) may have actually spurred current output from existing wells by producers worried about potential pipeline shut-downs due to fighting. On the other hand, those worries have undoubtedly cut down on exploration and capital expenditure (capex) in the unstable regions (Fig. 10.12).

As regards the unpredicted demand shortfall, apart from too much economic austerity in Europe, the other plausible candidate is China's growth slowdown. Chinese growth has certainly slowed (see Fig. 10.13). From 2003 through 2007 it was above 10 %, and in 2007 it exceeded 14 % p.a. But from that time the trend has

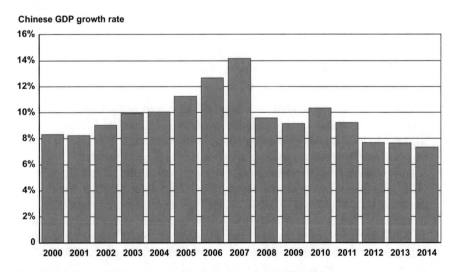

Fig. 10.13 Chinese GDP growth 21st Century. *Data Source*: *World Bank*

been down, except for 2010 when it was slightly above 10 % once more. From 9 % in 2011 it dropped below 8 % in 2012 and 2013, to an annual rate of 7.3 % in the third quarter of 2014, and 6.8 % in the third quarter of 2015. A similar decline in GDP growth (at a much lower level) has occurred among the Mediterranean countries of the European Union.

The next candidate for blame (by some pundits) for the drop in oil prices is the end of the Fed's QE (quantitative easing) policy and the first rise in interest rates. The correspondence is interesting if not remarkable. During QE1 the price of West Texas Intermediate (WTI) rose from $38 to $85, then fell back to the $70s. But during QE2 the price rose from $74 to $110, before falling back to $98. But the drop continued until the Fed's "Twist" program began. There was another lower peak in 2012. Followed by another decline. But when QE3 started the price of WTI was about $86. During QE3 it rose again to more than $100. But when QE3 ended it was back to $86, and since then it has fallen through the floor. Is there a causal connection? I think it is possible, but the link is not a strong one; who knows?

Finally, some older folks among us will recall a period of very low oil prices starting after 1983, with a bottom in 1986–1987 (see Fig. 10.9). (That bottom corresponded with the year of lowest unemployment during the Reagan Administration.) OPEC oil production was rising during those years, thanks to West African investments induced by the high prices of 1979–1982. Meanwhile Soviet output, which had been around 4–4.5 million barrels per day in the early 1980s declined precipitously after 1987. That combination of events cut Soviet hard currency income drastically. It forced the USSR to cut back its overseas activities in Cuba, Angola, Afghanistan, Syria and elsewhere. This was followed by the political collapse of the USSR itself in 1988–1990. There is natural speculation that something of the same sort may be happening now, vis a vis Russia. I think it is unlikely.

More likely, in my opinion, the refusal of Saudi Arabia and the other Gulf producers to cut output is a commercial strategy aimed at killing off as many high cost competitors as possible, with US frackers among the main targets. The pressure to cut production will come mostly from the governments of OPEC countries that depend on oil revenues to balance their national budgets. Nigeria (2.6 % of global supply) needs $122.70 to balance its budget, and the fight against the Islamist Boko Haram will raise the cost further. Venezuela (2.7 % of global output) needs $117.50 to balance its shaky budget, which probably means the end of subsidies for Cuba. Those two countries have no foreign currency reserves to speak of, so financial collapse is a real possibility. Saudi Arabia (13.6 % of global output) needs a price of $86.10 to balance the national budget. Iran needs $130.50; Iraq needs $109.40; UAR needs $74.30; Kuwait is in the best shape among the Gulf States, needing only $52.30 to balance its budget. With its massive financial reserves, the Saudis may prefer to keep up the pressure for quite a while. Russia and the US frackers may be its primary targets, but Sunni Saudi Arabia's Shia rivals, Iraq and Iran, are also on the list.

In 1932, overproduction and ultra-low petroleum prices led to the creation of the Texas Railroad Commission, a regulatory agency empowered to control production and keep prices above costs. OPEC (with 39.3 % of global output) still aspires to play this role. But without coercive power over its members, it is a "paper tiger". For that reason, OPEC has been unable (or unwilling) to cut production, at least so far. Only Saudi Arabia could break the stalemate. The two biggest non-OEC producers, the US (14.7 % of global output) and Russia (11.6 %) will not help.

In the US, the oil-gas industry is entirely private and cartels are illegal. Since the demise of the Texas Railroad Commission, there is no longer any government agency with the power to restrict output. Moreover, the US is still a net importer of oil, so low prices are beneficial (on balance) to the rest of the economy. In the Russian economy, by comparison, exports of oil and gas are its life-blood. The low ruble means that imports are suddenly much more expensive, driving inflation. But there is an obvious way for Russia to escape from this bind, at least in the medium term: to increase sales to neighboring China. This requires a pipeline. The first steps toward a major bargain along these lines may have been taken in the last months.

Low prices will cut the hard currency income of both ISIS and Russia, as well as providing relief for European, South Asian, Japanese and Chinese consumers. China and India are the major beneficiaries. Russia has been hurt the most. The Russian ruble has depreciated by more than 50 % against the dollar, and inflation in Russia is now approaching 12 % p.a.

The low prices will cut the profits of the big oil companies, but none will suffer greatly. It will also hurt some of the smaller energy companies, such as shale "frackers", especially those with heavy debt loads at high interest rates. The OPEC countries themselves will lose a lot of money but, except for Nigeria and Venezuela they can afford it. But, some analysts expect oil prices to be back over $80/bbl by next fall (2016), simply because the major OPEC exporters need the high prices to balance their budgets—and because the costs of drilling and fracking are very high and rising all the time. If OPEC cuts output to raise prices, as many hope, but few expect, the "plot" (if it is one) will probably have failed.

10.3   More on Fracking: Is It a Game Changer?

What are the possible substitutes for "conventional" light crude oil? Since 2009 there has been a boom in hydraulic fracturing ("fracking") of shale for gas, and now oil, in the US. The fracking boom is partly a creation of the Energy Research and Administration (ERDA) arm of the US Department of Energy, based on research and development carried out over several decades. ERDA financed the research in horizontal drilling technology, polycrystalline diamond compact bits and hydraulic fracturing (fracking). The system, in its original version, consisted of a drilling site or "pad" covering about 7 acres. The well had to be connected to power sources, pumps for the fracking fluids and other infrastructure. Recent developments are mentioned later.

From 1973, during the oil embargo, to 1984, the price of natural gas (NG) rose 12-fold, while US output declined by 15 %. From 2001 through 2007 the price of NG rose 50 %. This provided the financial incentive for innovation. The entrepreneurial pioneer of fracking in Texas was Mitchell Energy and Development Company, which applied the technology on the Barnett Shale. During those years fracking was very profitable.

Then, rather suddenly, NG prices dropped dramatically in 2008 (due to oversupply), and have since remained below the cost of drilling and operating new wells. Since then, demand from the electric power industry, has helped NG prices recover somewhat, but in summer 2014 they were still below breakeven, despite near-zero interest rates. But the low prices have given an economic boost to the US economy, badly needed in the post-2008 era.

There has been a lot of hype. One fracking optimist is Ed Morse, head of commodity research for Citigroup, frequently interviewed on CNBC and quoted in the Wall Street Journal (WSJ). Others include Porter Stansbury ("The Oil Report"), and the notorious "activist investor"—some would say "corporate raider"—H. Boone Pickens, who also gets interviewed on TV a lot.[2] Pickens is a cheerleader for "Saudi America". Morse thinks US oil production will surpass that of Saudi Arabia by 2020, based on continued growth of the shale "fracking" sector (and oil from Canadian tar sands). He foresaw declining US demand (due to increasing efficiency) and a declining rate of economic growth rate in China, resulting in an "oil glut" by 2020 with prices down by 20 % from 2011 levels. (His forecast was too conservative.) The US Energy Information Agency (USEIA) and the International Energy Agency (IEA) both accepted this view. Nothing has happened to change the "official" view since then.

The profitability of fracking for gas is now in question, partly because several of the early leaders were losing money even when oil prices were higher. Four of the main shale gas and oil producers (Chesapeake, Devon, Southwestern and EOG) lost a cumulative total of $42 million during the 5-year period from 2008-through 2012. (Chesapeake cancelled its dividend in July 2015). Of course, they still have assets, but many firms have had to write-down their claimed gas reserves. Contrarian energy analyst Bill Powers has noted that some of the shale startups booked reserves based on claimed well lifetimes of up to 40–65 years (Powers 2013). Yet most shale wells exhibit rapid decline in production. Typical Bakken shale oil wells, starting at 500 bbl/day, were down to 150 bbl/day at the end of 12 months, 90 bbl/day after 24 months 50 bbl/day after 48 months and continuing to nearly zero after 5 years (Hughes 2010).

The natural gas price collapse in late 2008 was partly due to the global collapse of oil prices in that year and partly to the recession that followed the financial storm. This prompted a massive shift from shale gas to shale oil, partly based on the sharp recovery in the price of crude oil after the deep dip in 2008-9. But the problem of rapid depletion of shale wells remains. There are "sweet spots" but no gushers to counterbalance the large number of dry wells. Very few wells in a typical field are

---

[2] Pickens was a wildcatter and founder of Mesa Oil Co. He merged several times with larger companies, yet retained control. His financial engineering forced Gulf Oil into the arms of Chevron in the mid-80s and originated the phrase "drilling for oil on Wall Street".

highly productive. For instance, about 30 % of wells yield less than 250 bbl/day at a given time, while 30 % produced more than 500 bbl/day and perhaps 3 % produced more than 1000 bbl/day (Hughes *op cit*). This depletion rate means that just to maintain current output requires far more drilling than conventional oil or gas fields.

What of the future? On the positive side for the frackers, since 2010 or so is the evident success of multi-well pad drilling ("octopus") technology, whereby a single pad supports a mobile drill rig that can crawl around from place to place on the pad in as little as 2 h. Each pad can support up to 18 (or more) wells, drilling horizontally in different directions, thus assuring that a significant area is covered. It is expected that a pad of 7 acres may be able to "frack" 2000 underground acres of shale reserves. This technology was pioneered in the Piceance formation in Colorado. In 2011 Devon Energy (the successor of Mitchell Energy) drilled 36 wells from a single pad in Marcellus shale. Shortly after that, Encana drilled 51 wells, covering 640 acres underground, from a 4.6 acre pad. The multi-well pad drilling technology is now operating on the Bakken shale. This technology cuts drilling costs dramatically and (it is argued) also reduces the environmental impact.

One negative indication for shale oil is that the very low interest rates that financed a lot of the drilling after 2003 and again after 2009 are presumably a temporary phenomenon due to the actions of the Federal Reserve Bank (FRB). As the US economy recovers from the debacle of 2008, it is assumed that interest rates will eventually rise. In that case, debt-financed drilling for shale gas (or oil) will get even more expensive than it is now. But, as of 2015, the conventional wisdom about rising interest rates has proven premature, if not false, and interest rates the commercial banks have to pay the Fed have remained at near zero.

A second negative indication is that oil shale has a very low energy return on energy investment (EROI). See Figs. 10.14 and 10.16. In a study commissioned by the Western Resource Advocates (an NGO) Prof. Cutler Cleveland of Boston University found that EROI for shale oil at the well-head is about 1.7. After refining, the EROI for shale oil falls to 1.5, making it, at best, a very marginal resource (about the same as ethanol) (Cleveland 2010). However, that could increase as the technology improves. See Figs. 10.15 and 10.16. Figure 10.15 refers to gas wells, but the depletion rate for oil is comparable.

Another negative indication for shale gas is that the USGS has cut the original estimates of 800 trillion cubic feet (tcf) (supposedly a 40 year supply for the US) to 400–500 tcf. Several early shale plays are already peaking. This applies to the shale plays of Barnett, Fayetteville, Haynesville, and Arkoma Woodford. Only Marcellus is still growing. Powers predicts that total recovery will be 125–150 trillion cubic feet (tcf), only an 8 year supply (op cit).

Of course the price of oil from early 2015 onwards was significantly lower than it had been in 2011 (Fig. 10.10). Clearly, global economic growth has not recovered in the aftermath of the economic crisis, as economists expected. The widespread adoption of austerity policies around the world is undoubtedly part of the reason for lagging economic growth. Optimistic projections of economic growth have assumed a continuing significant increase in unconventional supplies, overcoming

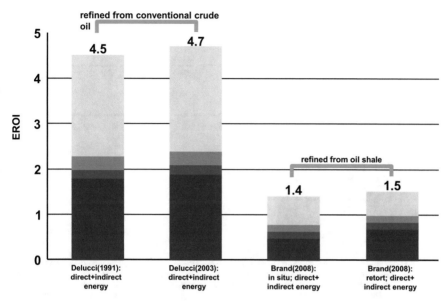

Fig. 10.14 Refined fuel produced: Comparison of estimates of *Energy Return on Energy Investment* (EROI). *Source*: Cleveland, Cutler J. 2010. *An assessment of the energy return on investment (EROI) of oil shale*

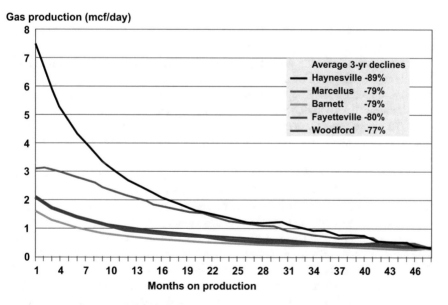

Fig. 10.15 Average production profiles for shale gas wells (mcf/year). *Data source: DrillingInfo/HDPI, March 2013*

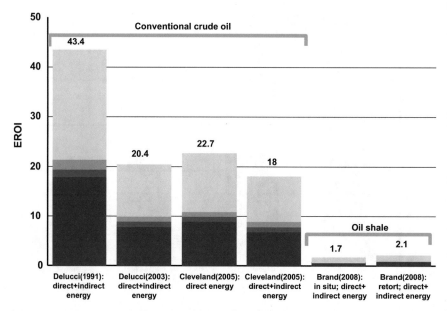

Fig. 10.16 Crude oil & shale: Comparison of estimates of *Energy Return on Energy Investment* (EROI). *Source*: Cleveland, Cutler J. 2010. *An assessment of the energy return on investment (EROI) of oil shale*

expected declines in conventional crude output. But that increase may have been overestimated.

The mismatch between declining Chinese economic growth rates and the rise in US oil output (triggered by a huge investment in new hydraulic fracturing technology) has altered the market power of OPEC. Not surprisingly, OPEC spare capacity rose dramatically in 2009, thanks to declining demand, and it stayed high throughout 2010. Spare capacity declined sharply for a while, but rebounded again in 2013. It still remains well above 2003–2006 levels. That (partially) explains the low current price of oil.

The glut will not go away quickly. One reason is that the cost of production from major established fields is low, so even reduced output is profitable. Producing firms in the private sector have no incentive to cut output from existing wells. They will simply slow down their exploration and development activities. However, US frackers are in a more difficult position, because their wells depreciate much faster than conventional wells (typically more than 40 % per annum and up to 60 % in the first year). The fracking sector depends on drilling new wells at an increasing rate in order to grow. However when prices are below breakeven cost (as now) drilling stops, as it has. This hurts the drillers, of course, but production from existing wells is not affected for at least six months. Only after that will output start to fall.

One common measure of current activity is the "drill rig count". In 2011 the drill rig count peaked at about 2000 active rigs drilling for shale oil in the US. On May

16, 2015 that number was down to 668, and it has dropped further since then. However, the "drill rig" count is an imperfect proxy for production, because the time required to drill a well has declined about 50 % since 2011, while the output per well has also roughly doubled in that time, suggesting that each rig today is four times as productive as a rig in 2011. [http://nocalledstrikes.com/2015/02/05/obser vation-on-the-rig-count/]. Based on this proxy, US shale oil output may not have fallen at all.

In any event, the glut will end before 2018 *unless there is a global economic recession*. That possibility is "the elephant in the room". It depends on factors beyond my insight. But, if there is no repeat of 2008, I think prices will rise rapidly to early 2014 levels and more slowly after that.

10.4 The Inevitable Decline of the Internal Combustion Engine

From a macro-perspective, it is clear that fossil fuels have been utilized in one of four ways. The first, historically, was to produce mechanical power (by means of steam engines) for factories, ships and railway locomotives. Second came the internal combustion (IC) engine for motor vehicles and aircraft. The third is to produce mechanical power (from steam turbines) devoted to generating electricity. The fourth was for chemical uses, starting with iron smelting and other metallurgy, then street lighting, synthetic dyes, synthetic fibers (rayon, nylon, dacron), synthetic rubber and plastics.

These categories are getting fuzzier. The first, steam power for mechanical drive in factories and railroad locomotives, has now been completely replaced by the IC for mobile activities or by the steam or gas turbine producing electric power. Electrification is now in the process of encroaching on the domain of the IC. - Short-haul passenger trains and trams (trolleys) have been electrified for decades, and that process is continuing as more and more long-haul railroads are being electrified. Now, automotive transport is beginning to electrify. It began with the Toyota and other hybrids, now selling in the million plus numbers per year. Now every large automobile manufacturer produces both hybrids and all electric EVs.

Still, the vast majority of motor vehicles being sold today depend on spark ignition IC engines (although more efficient diesel engines are increasing their penetration), while trucks and buses are almost exclusively diesel-powered. Apart from electric power generation, transportation is the biggest producer (by category) of Greenhouse Gases, especially carbon dioxide. This is a temporary situation. By "temporary" I mean one or two decades at most. There are three big reasons for the coming change:

First, almost all transportation modes in the industrialized world (excepting bicycles, electric cars and electrified railways and trams) depend upon mobile power sources. And all the portable mobile power sources, with the minor exception of electric batteries and fuel cells, are internal combustion engines fueled by liquids, derived from petroleum.[3] Yet petroleum is geologically scarce. That means prices for liquid hydrocarbons will certainly rise, as long as economic growth continues, perhaps not very sharply during the next decade (thanks to fracking) but certainly before 2030.

The second factor working against the IC is the greater energy efficiency (and lower operating cost) of hybrids and EVs. Details vary from model to model, but on average, current EVs consume only about half of the energy consumed by piston engine vehicles, per km traveled, even taking into account the inefficiency of centralized electric power generation. This translates into operating cost savings, over the life of an EV, notwithstanding the relatively high price of batteries today. As battery prices come down in the next few years (and they will) those operating cost advantages will increase.

The third factor working against the IC is that, despite complex and expensive exhaust gas treatment systems, motor vehicles with piston engines are major polluters. Even though leaded gasoline is on the way out and gasoline engines are now quite "clean" as compared to their ancestors 50 or a 100 years ago, there are now 700 million automobiles in the world and their numbers are growing inexorably. These motor-vehicles are concentrated in cities. The unburned hydrocarbons, carbon monoxide, nitrogen oxides and sulfur dioxide that are still emitted from cars and trucks, added to diesel emissions including micro-particles of soot, are still enough to cause smog in the air above large cities in the US and Europe. Air pollution in Paris, London, Chicago or Los Angeles is mainly from vehicle emissions, since emissions from heavy industry and power plants is mostly outside the city proper, and fairly well controlled. The air above Asian cities—notwithstanding fewer vehicles per capita—is currently much worse. Apart from "smog" pollutants, internal combustion engines produce at least 20 % of the total carbon dioxide generated by all fossil fuels.

Today the liquid fuels we use to move ourselves and our goods are mostly derived from petroleum: gasoline, kerosene or diesel oil. Two thirds of the petroleum consumed in the US is used for transportation purposes, and the fraction is increasing. It was in the US that oil was first extracted in large quantities, and it was in the US that it first found its major use as fuel for internal combustion engines to power horseless carriages, now known as automobiles.[4] But US oil consumption has outrun domestic production since the late 1960s and (notwithstanding shale fracking) the US is still one of the world's major importers, along with China, primarily for use by cars and trucks.

---

[3] Steam engines and other so-called external combustion engines require large and inherently bulky heat exchangers (condensers).

[4] The internal combustion engine was a German invention (by Claus Otto), but Otto's stationary engines used coke oven gas as fuel. The first application to an automobile, using gasoline as fuel, was also in Germany (Daimler-Benz). But mass production (by Ford) first occurred in the U.S.

The substitution of automobile and truck transport for rail transport, especially in the US, has re-organized the urban landscape in ways that are not only inefficient from an energy perspective, but also from a land-use and capital-investment perspective. About 50 % of urban land, even in Europe, is now devoted to roads, parking lots and auto-related service facilities, even though this is a very unproductive use of both land and capital. Compared to bus or rail-based transport systems, the existing system is very inefficient as a means for moving people and goods.

This market distortion, in turn, has serious geo-political implications. One of them is the continuing need by the industrialized countries for reliable foreign sources of supply. The war in Iraq and continuing US military involvement in the middle-east is unquestionably oil related. The US military presence in Islamic exporting countries is increasingly resented by, and will eventually be intolerable to the local populations. (It was the explicit motivation for Osama Bin Laden's Al Qaeda.) Another international consequence of imported oil back in the 1970s was the huge transfer of dollars to oil exporters, especially around the Persian Gulf. Those "petrodollars" were, in turn, deposited in US banks and re-loaned at low interest rates to Latin American, South Asian and African countries who also needed dollars to import oil. But when the persistent US inflation in the 1960s (due to the Viet Nam war) became a drag on economic growth ("stagflation") in the late 1970s, it prompted the US Federal Reserve under Paul Volker to raise interest rates dramatically in 1981–1982. That move was catastrophic in 1982–1983 for the countries that had borrowed dollars and needed to repay in dollars at the new high interest rates. It led to the first major international financial crisis. The sharp cut in oil prices starting in the summer of 2014 has generated a reverse version of that crisis, temporarily helping the major importers like India and China but hurting the OPEC countries, Iran and Russia.

Clearly, the US—and global—addiction to imported foreign oil is unsustainable in the long run. The place where oil consumption must be cut most urgently is, of course, the transportation sector. The possibilities are very limited. They are: (1) increased fuel efficiency, (2) alternative fuels (biofuels, hydrogen), or (3) alternative modes of transportation.

Following the dramatic oil price increase in 1972, so-called corporate average fuel economy (CAFE) standards for new cars were set in 1978. Overall US fleet average consumption rose from about 13.5 mpg in 1978 to a little over 21 mpg by 1992. Since then the average actually decreased until the year 2000 (as SUVs increased their share of the market) and increased slightly to 20.6 mpg in 2006 or 11.6 L per 100 km. There has been some slight decline in gross consumption (in the US) since 2000, due to design changes, though efficiency gains for cars have been largely compensated by increased numbers of light trucks and SUVs.[5] The lesson from CAFE experience is that most of the early gains were due to weight reduction,

---

[5] There are no standards for vehicles over 8500 lb (about 4 metric tons) which leaves the GM Hummer and the Ford Excursion unregulated. The Hummer apparently averages about 11 mpg.

with minor gains from other design changes. However, the reduced compression ratio for gasoline engines (due to the elimination of tetra ethyl lead, TEL) has had the opposite effect, increasing fuel consumption per km.

Followers of Formula One racing and BBC's "Top Gear" show will find this message hard to believe, but I think the spark-ignition Otto cycle internal combustion piston engine, born in the workshop of Klaus Otto in 1876, together with its cousin Rudolph Diesel's compression-ignition piston engine (1892), will soon be obsolescent, as marked by declining sales, perhaps by 2030. The next step will be either a hybrid-electric or an all-electric battery powered car. The hybrids, such as the *Prius*—pioneered by Toyota (1997) and the Honda *Insight* (1999)—have been a success story in the marketplace. The hybrid allows higher fuel efficiency and cuts emissions but it requires 30 % more components than a conventional ICE car, and will never be cheaper to build or sell. Plug-in EVs, like the Nissan *Leaf* and the GM *Volt* are still expensive compared to conventional vehicles, yet inferior in performance.

However, these first-generation electrified cars are made of steel, with small gasoline engines assisted by batteries, or they carry big batteries but with limited range. The problem is that the hybrids and plug-in EVs currently in production (except the Tesla) use a standard vehicle 'platform' shared by other models, and based on the mass-produced steel body and frame. They are standard mass-produced cars, modified slightly to accommodate electric batteries and motors.

But much more radical design changes are now on the way (Lovins 1996; Lovins et al. 1996). They are as follows: (1) whole system design, (2) advanced composite (carbon fiber) materials and (3) electrified power trains. Weight is the primary variable at the disposal of designers. What Lovins called "whole system design" was pioneered in the 1990s by a group at Lockheed–Martin led by D.F. Taggart (Lovins 2011) pp. 24–31. The team re-designed the F-35 Joint Strike Fighter (JSF), which was 72 % metal, to cut its weight by a third. Their design used carbon fiber composites, which are many times more expensive than aluminum or steel. Yet the resulting design would have been cheaper to build because it involved far fewer parts, light enough to be hand-carried, and utilizing snap-together joints that could be bonded at a later stage.[6]

Such a lightweight carbon fiber body can cut the weight of the vehicle—and with it, the weight of the power plant needed—by a factor of 2 or more. However at present, such bodies are mostly custom produced, for race cars, power boats, or military aircraft. (The Boeing 'Dreamliner' uses a lot of composites). GM's carbon-fiber ultra-lite 4-seat 'concept car' in 1991 had a curb mass of 635 kg. It got 26.4 km per liter (3.8 L per 100 km) without a hybrid drive, and presumably would have achieved somewhere in the neighborhood of 50 km/L (2 L/100 km) with hybrid drive (Lovins 1996) (ibid).

The second new design "enabler" is the carbon-fiber composite materials. Twenty years ago such materials were not only expensive to manufacture, but

---

[6] It was not built, but that was for other reasons.

extremely difficult to form into parts. They are used for Formula 1 racing cars, but every such car is handmade. The Fiberforge team of Taggart and Cramer have patented a process for high speed forming of carbon-fiber parts. Some of the manufacturing problems have been solved by Boeing for the 787 "Dreamliner". The strength and weight advantages of fiber composites are tremendous. Moreover, the material does not rust or corrode, and the color can be incorporated into the molding process, eliminating the paint-shop.

It is now estimated that manufacturing auto bodies with composites can cut fixed costs of manufacturing by 80 % and variable non-materials cost by 25 %. Many steps in the conventional manufacturing process are not necessary for the light-weight carbon-fiber or thermoplastic hypercar. There is no need for shipping heavy materials like steel, no need for heavy lifting equipment on site, since all the parts (except the battery) are light enough to be carried by hand. No need for a paint shop, since pigments can be incorporated directly into the material. No need for welding, since parts are snapped together and bonded by special glues, etc. The manufacturing facilities for both batteries and vehicles can probably be built as fast as the growth of demand. But that depends, in turn, on the rate at which production costs can be brought down to affordable levels and a sales and service network for the vehicles can be created.

The future of composites in mass production now depends on reducing the raw material cost itself. This was around $16 per pound in 2010, compared to $1 per pound for carbon steel sheet. At that time Oak Ridge National Laboratory (ORNL) estimated that material costs could be cut by 90 % (Lovins 2011) p. 28. As of mid-2015, the costs were down to $8–10 per pound, and the official goal of the ORNL program, in partnership with Ford, GM and Chrysler-Fiat, was to achieve $3–5 per pound. At that price large automakers could afford to use about 300 lb of composites per car, cutting vehicle weight by 60 % and fuel consumption by 30 %, while also cutting global carbon emissions to the atmosphere by 10–20 % [http://www.reliableplant.com/Read/850/carbon-fiber-cars-could-put-us-on-road-to-efficiency].

The third "design enabler" for EVs is electrification of the drive train. That includes the electric motors and controls, of course, but those are already mass produced as components. One important benefit of integrated electrification is that up to 70 % of the energy normally wasted as heat in brakes, can be recovered by so-called "regenerative brakes" (hybrids already do this). These act like generators on a downhill, or during a slowdown, and send the energy back to recharge the battery. The other main source of progress is the rechargeable electric battery itself. As will be pointed out below battery costs are declining, partly because of continuing improvements and partly due to economies of scale.[7]

---

[7] See the book entitled *Alternatives to the Internal Combustion Engine* that I co-authored (Ayres and McKenna 1972). That book discussed the future of electric cars and dozens of the plausible electro-chemical rechargeable battery combinations, some still being considered today.

I co-authored a book on alternatives to the internal combustion engine in 1972 (Ayres and McKenna 1972). The most interesting possibility mentioned in our book was Ford's sodium sulfur battery (1966), operating at a temperature of 360 °C, with a molten sulfur cathode, a molten sodium anode and a solid electrolyte of β-aluminum, which allows sodium ions to pass but not sodium or sulfur atoms.[8] It would have stored 15 times as much energy per kg as the familiar lead-acid batteries. That battery, or a descendant of it, is now available commercially in large sizes (but not for vehicles).

General Motors was developing a lightweight lithium-chlorine battery at that time, with even higher energy density. It had a molten lithium anode, a molten lithium chloride electrolyte and a chlorine (gas) cathode, with an operating temperature of 615 °C. Needless to say the application of high temperature batteries such as these to ordinary cars was not attractive; both sodium and lithium, as metal, will burn if exposed to air and chlorine gas is highly toxic. The auto companies stopped battery development in the 1980s. That might have been a mistake.

In our 1972 book we did not discuss the lithium-ion battery, which had not yet been invented, although the electrochemical combination had been known for a long time. The invention that made the lithium-ion battery possible (and rechargeable) at reasonable temperatures was the lithium-cobalt-oxide cathode, invented by **John Goodenough** (1980). Goodenough's invention was arguably as important as the invention of the transistor back in 1947, though he has never received any monetary reward. Goodenough (aged 93) is still working on a new invention he thinks will supplant the current version. Don't bet against him.

The rechargeable lithium-ion battery was commercialized (with a carbon anode) by Sony in 1991. There have been a few variants, notably the lithium-iron-phosphate material invented by Goodenough's students Padhi and Tokada. Soon after Sony's commercialization, lithium-ion batteries were utilized in a variety of portable electronic devices from razors to cameras, to cell-phones to laptop computers. Prices for consumer applications have fallen along a classic experience curve from nearly $2000/kWh in 1995 to under $300/kWh by 2005, and the trend continues.

Li-ion batteries in larger sizes are already in use for many applications, including aircraft (in the Boeing 287) and in electric automobiles. Prices for these applications have fallen 40 % from over $1000/kWh in 2010 to $338/kWh in 2012 and (expected) $300/kWh in 2015. Moreover, prices are expected to decline considerably in the coming decades, from the present range to somewhere less than $100/kWh by 2030. See Figs. 10.17 and 10.18. As vehicle designs have been getting better, an outsider, Tesla, has opened a new high performance sports car market that is expected to broaden, shortly, into a mid-range SUV market. Tesla's cars already have a range in excess of 220 km. Tesla's sports models were based on a Lotus body, with some carbon-fiber composites to cut weight. It has an advanced motor-invertor developed by GM for its Volt, together with a patented battery pack consisting of 6831 individual lithium cells designed for consumer electronics.

---

[8] The inventors of record were Joseph T. Kummer and Neill Weber.

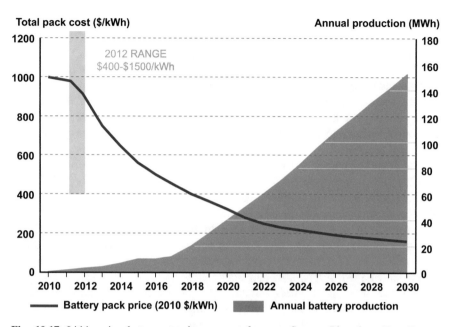

Fig. 10.17 Lithium ion battery experience curve forecast. *Source*: *Bloomberg New Energy Finance*

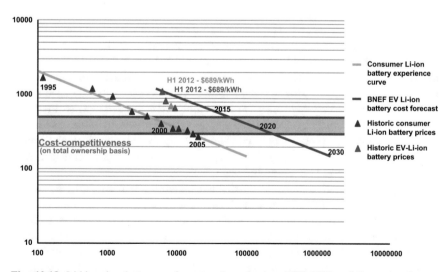

Fig. 10.18 Lithium-ion battery pack cost and production 1995–2005 and forecasts. *Source*: Battery University, MIT, IIT, Bloomberg New Energy Finance, January 2013

The major innovation, thanks to chief engineer J.B. Straubel, was to eliminate the original 2-speed gearbox by redesigning the motor-invertor to get 40 extra horsepower, 10 miles longer range and 14 lb less weight (Lovins 2011) p. 31.

The next generation of Tesla SUVs is expected to be competitive in both performance and price to most conventional SUVs while polluting far less and costing less to run. It will be a new design, making more use of light metals (aluminum, titanium) and composites. The main remaining drawback to the EV is the longer recharging time, and there are several schemes under development to solve that problem.

The Tesla GigaFactory #1 in Nevada is expected to be fully operational in 2020. It will have a capacity of 35 GWh/y of lithium cells and 50 GWh/y of lithium battery packs, enough to supply 500,000 T cars per year. Cost is estimated at $5 billion, of which Tesla's partner Panasonic will invest between $1.5 and $2 billion. A new product *PowerWall* (for short-term storage at rooftop PV sites) is now being marketed. In fact, during the first week of opening for order-taking in June 2015, $804 million in orders were received, of which $179 million were for PowerWall [https://en.wikipedia.org/wiki/Gigafactory_1]. Tesla CEO, Elon Musk, says that several more new Gigafactories will have to be built soon to accommodate soaring demand. The next one would probably be in Japan.

There is no apparent obstacle to mass production of EVs today except for the long time-scale for designing and building any highly automated manufacturing facility. A lot of people outside the auto industry imagine that the conversion could be done in a matter of a year or 2, based on the World War II experience when the US auto industry rapidly switched from making cars to airplanes and tanks. But the manufacturing technology for cars at the time was much simpler and much more flexible (and labor intensive) than today's steel-based technology. In those days, machine tools were controlled entirely by the skilled machinist-operator, and the tools were not computer programmed and not linked to each other by specialized transfer devices, as they are today. The assembly process itself was still entirely manual, not done mostly by specialized single-purpose robots, as is the case today. Typically the design process for a conventional highly automated plant to build steel frame, steel body cars (or engines) requires 5 years. Another 5 years is needed to build the heavy equipment and the plant.

As regards alternative fuels, there are three categories, namely liquefied natural gas (LNG), bio-fuels (ethanol, methanol, bio-diesel), and hydrogen. Actually LNG can only be regarded as an alternative fuel for buses or other large vehicles in places where natural gas is abundant, either near a source or a terminal. It is fairly easy to convert an internal combustion engine (ICE) to burn gas, but the combined problems of compression and decompression, as well as storage, make LNG an unlikely long-term solution. It is almost always more efficient to burn the natural gas in a combined cycle power plant (which can achieve something like 60 % efficiency) or to produce methanol as a hydrogen carrier for an on-board fuel cell.

For political reasons that need not be spelled out here, the favorite candidate in the U.S. at present is ethanol from corn. It is it is sold in a mixture of the two fuels, known as E-85, in which the ethanol fraction is 15 %. While alcohols contain less useful energy per kg of weight,[9] they have higher octane numbers than gasoline and therefore permit higher compression ratios and slightly greater efficiency. The US process starts from corn (maize). The current process—similar to that for making beer—uses only the corn starch in the kernels (66 % by weight), not the oil (3.9 %) or the gluten feed and meal (29.7 %). These by-products have economic value on their own as animal feed or other food products. They also have some energy value, which has to be accounted for in the calculation. But producing bio-fuels also consumes exergy for nitrogen fertilizer for the crop, processing and transportation. The fossil energy inputs to corn-ethanol production are nearly as great as the energy content of the grain, or even greater, depending on the study and its assumptions (Natural Resources Defense Council (NRDC) 2006).

However, grain production requires lots of water, several thousand gallons per gallon of ethanol (Brown 2006a, b). There is no way to justify using so much of this valuable resource for producing a gasoline additive. The case for corn-based ethanol *per se* is even weaker than the above argument suggests, since it ignores substantial irrigation requirements, nitrogen pollution (from excess fertilizer use) and the fact that subsidies also create incentives to cut trees in developing countries (notably Brazil) to create more cropland. In fact, when other uses of land are considered, it turns out that, under reasonable assumptions, the reverse strategy (converting cropland to forest and thus sequestering carbon in biomass) would be more effective at reducing global GHG concentrations in the atmosphere than converting forest to cropland used to grow corn or soya beans (Righelato and Spracklen 2007). Ethanol is not a long-term answer.

Today Brazil is the only country using ethanol on a large scale, and probably the only place where it makes sense economically. In Brazil the ethanol is made from sugar cane by fermentation with brewer's yeast. The yield is about 70 L of ethanol per dry ton of sugar cane.

Starting from cellulose rather than corn or sugar cane makes more sense. By contrast, a metric ton of dry cellulose such as rice straw (or cornstalks) now yields 330 L of ethanol by the simultaneous saccharification and fermentation (SSF) process developed by the National Renewable Energy Laboratory (NREL) in Colorado (Wyman et al. 1993; Wingren et al. 2008). That yield implies a cellulose to ethanol conversion efficiency of 40.6 %. Improvements in the technology over the coming years might bring this up to 470 L/ton or 56 % conversion efficiency (which is the target).

Annual world production of biomass is estimated at 146 billion metric tons. Of course only a small fraction of this would be available for conversion to ethanol or methanol. On the other hand, it is likely that the total tonnage of biomass produced from existing land could be increased significantly if that were desirable (e.g. by growing algae in ponds). In any case, based on the rice straw data cited above, the global annual output of biomass could potentially yield 60 trillion liters of ethanol. Suppose future cars drive 10,000 km per year and consume 6 L of ethanol per

---

[9] One liter of ethanol has the energy content of 0.65 L of gasoline.

100 km (achievable years ago by the Toyota Prius, the Peugeot 207 and others) or 600 L per year. There would be enough ethanol to support 100 *billion* automobiles. Obviously a small fraction—around 1 %—of the global biomass production would suffice for the much smaller number of cars—perhaps 1 or 1.5 billion—that will actually be choking the world's highways by 2050. Assuming most of those cars will actually be plug-in hybrids, using electricity (possibly from wind turbines) the biomass requirement would be even smaller. This would *not* affect food production and should be technically feasible by 2035 or so.

So, whereas ethanol from corn can be expected to produce (at most) about 15 billion gallons a year in the US by 2012—around 6 % of projected U.S. gasoline demand—the theoretical potential for ethanol from woody plants (and municipal refuse) is at least ten times greater, or 150 billion gallons a year according to the Natural Resources Defense Council (NRDC). That would be over two thirds of current consumption. Together with other measures to cut consumption, ethanol from biomass could conceivably replace US gasoline consumption altogether by 2050 or so.

The economics are another story. I have no crystal ball to forecast the costs. If it were only a question of converting rice straw or corn-stalks (stover) into ethanol, the organisms in the rumens of cows, sheep, goats (and termites) would provide a straightforward solution. One problem is that simple cellulose—a long chain of sugar molecules—constitutes only a half of the mass of wood. The remainder consists of more complex carbohydrates, hemi-cellulose and lignin, which most ruminants cannot digest. Therefore, a future industrial cellulosic ethanol process will have to start with a chemical process to break down the complex tree-structures that 'glue' the cellulose chains together in wood and woody plants, into component sugar molecules.

One promising approach is the integrated bio-refinery system under development since 1980 at the Tennessee Valley Authority (TVA) (Broder and Barrier 1990). The idea is to convert raw crop feedstocks into a range of useful products, including animal feeds and industrial feedstocks. The starting point is an air separator to segregate leaves from stems. In the case of alfalfa, for instance, a metric ton yields 400 kg of high protein leaf material for animal feed, and 600 kg of stems for further processing. The next step is first stage hydrolysis to convert most of the hemi-cellulose and cellulose to sugars. Second stage hydrolysis (by concentrated sulfuric acid) converts the remaining cellulose to glucose. In the final step, the glucose and some other 6-carbon sugars (sugars with six carbon atoms per molecule) are then converted to alcohol by brewer's yeast, just as ethanol is currently made from corn or sugar-cane.

However, hydrolysis of hemi-cellulose and lignin yields a lot of 5-carbon sugars—notably xylose, and arabinose—that brewer's yeast cannot convert into ethanol. This results in a significant un-reacted residue. Also, the purified synthetic enzymes now used for enzymatic hydrolysis are expensive: they currently account for up to 60 % of total costs. Also, the chemical hydrolysis step can produce by-products (such as furfural) that interfere with the subsequent fermentation step, and must therefore be removed. Unconverted solids (lignin) can be dried and burned as fuel. These problems may be overcome in time, but probably not in the very near future. There is virtually no commercial production of cellulosic ethanol today, nor will there likely to be for several more years.

A possible breakthrough of major importance has recently been reported. It is known as the 'carabao paradigm'. The carabao is a ruminant (a domesticated sub-species of water buffalo) common in the Philippines, Malaysia and southern China. It not only digests grass as cattle and sheep do, but it also digests woody shrubs. Researchers recently discovered that the micro-organisms in the carabao rumen can hydrolyze complex carbo-hydrates, including hemi-cellulose and lignin, and apparently convert 5-carbon sugars to 6-carbon sugars (glucose) that support life processes. In short, the micro-organisms in the carabao rumen converts woody materials into stuff that brewer's yeast can subsequently ferment into ethanol. The research on this process has barely begun. However, preliminary calculations suggest that 1000 kg of woody biomass can yield 117 L of ethanol (Agriculture Business Week 2008). Scaling up would obviously be a major problem, so this is a long-term solution at best. If all goes well, the "carabao paradigm" (as it is called) could be of great importance, at least in the Philippines, Indonesia and South Asia.

Like ethanol, methanol can be burned in a slightly modified gasoline engine, although to take advantage of the high octane number a higher compression ratio is needed. (Race cars have used methanol for a long time.) Methanol is more toxic than ethanol and more corrosive. It also has slightly less energy content per unit volume. Nevertheless, methanol is also a conceivable future substitute for gasoline. Like ethanol, it can be produced from biomass.

As it happens, a thermo-chemical process developed by Battelle Columbus Laboratories can produce methanol from cellulosic biomass (Spath and Dayton 2003). The BCL process consists of a high temperature gasifier to produce 'synthesis gas', a mixture of carbon monoxide (CO) and hydrogen that was formerly used to produce 'town gas' (from coal) in the old days before natural gas became available. Syngas, or synthesis gas, is a variable mixture of carbon monoxide and hydrogen. It can be produced by pyrolysis or incomplete combustion of coal, oil or biomass. However there is not enough hydrogen in the biomass to convert the synthesis gas into methanol. To obtain additional hydrogen, the BCL process includes a steam reforming step, which uses heat to break the water molecules in the steam into hydrogen and oxygen, and uses the oxygen to convert some of the carbon monoxide into carbon dioxide, yielding more heat to drive the process and extra hydrogen. This process requires extra equipment for the steam-reforming step and to remove the resulting carbon dioxide.

A still newer process called Hynol gets around this difficulty by adding some natural gas and steam as co-inputs to the gasifier, along with the biomass feedstock, providing just enough hydrogen to make the methanol *without* the need for a separate stage of steam reforming and carbon dioxide removal. The Hynol process can achieve 66.6 % conversion efficiency, even higher than the 63 % efficiency achieved by the conventional natural gas to methanol process (Steinberg and Dong 1993; Borgwardt 1999). A slightly less efficient but simpler scheme starts from the gaseous BCL product and feeds it, with some natural gas, into the steam reformer.

Lower temperatures and less equipment are required. The combined biomass plus natural gas process is up to 12.5 % more efficient than either of the processes starting from biomass or natural gas considered separately.[10] However, it is in the early stages of engineering. The scaling problem still remains.

In principle, a significant quantity of methanol could be achieved by pyrolysis of municipal solid wastes (which are largely paper and plastics) and adding some natural gas to avoid the steam reforming step (Bullis 2007a, b). The technologies are fairly well known and costs seem to be potentially competitive. The only barrier would probably be to sort and screen the inputs to remove metals and PVC, and the need to treat the resulting gases to eliminate toxics. But this problem applies equally to the incinerators now used in many cities. The problems are as much political as technical. However, right now the economics depends a lot on the conversion efficiency and, hence, on the cost of capital per unit output.

The economics of methanol also depend on the cost of feedstock (agricultural wastes and/or woody materials) which have to be transported to the gasifier. Feedstock should be considerably cheaper per ton than corn or soybeans. In the case of municipal wastes, feedstock would be free and the bigger cities might even pay to get rid of it. So, when the hydrolysis problem can be solved economically, both capital and operating costs will drop. Thus, under not unreasonable assumptions, cellulosic methanol could be cheaper than gasoline or diesel oil by 2050.

Bio-diesel is a vegetable oil that has been refined and processed so that it can be mixed with regular petroleum-based diesel or (in some cases) used directly as an alternative. In principle, many crops would provide suitable oil-seeds. However the oils are not chemically identical. At present, 90 % of US bio-diesel is produced from soya beans, even though the diesel output (40–50 gallons/acre) is not particularly high. The reason is simply that soya is the only oilseed crop now grown in the US in large enough quantities to provide reliable feedstock for a refinery. However, other crops are much more promising on a gallon per-acre basis: rapeseed (110–145), mustard (140), Jetropha (175) and palm oil (650). (Unfortunately, the soaring demand for palm oil today is causing extensive deforestation in places like Indonesia.)

With one exception (jojoba, see below) all vegetable oils must be refined before they can be used as a diesel fuel. The current bio-diesel refinery process yields glycerin as a by-product, and the recent boom in bio-diesel has produced a glut of glycerin, which was formerly synthesized from natural gas. Some of the new supply can be used in the manufacture of epoxy resins, but it remains to be seen whether the market can absorb much more.

A potentially promising new approach starts from algae, grown either in fresh water ponds or in salt water. In contrast to agricultural crops, algal ponds can, in principle, produce as much as 40,000–80,000 L of bio-diesel per acre (Briggs 2004). The refining process would first extract the oil and then process the solid residue into ethanol. Practice is another story. Enclosed ultra-clean climate

---

[10] The Hynol process can be adjusted to produce hydrogen rather than methanol.

controlled ponds would be expensive to construct and cost a lot to maintain, while open ponds are subject to invasion by other species. Up to now it has also been difficult to control the algae growth process adequately to produce a continuous output of feedstock for a refinery. It is not clear whether these difficulties are fundamental or temporary. A number of algae companies have failed. However, in early 2008, PetroSun Inc. announced a joint venture to build a bio-refinery for algae in Arizona, with a projected capacity of 30 million gallons per year (PetroSun Inc. 2008).

Another approach has been developed by Solazyme Corp. Their system does not require ponds or external sunlight. Instead, micro-algae are fed with biomass (switchgrass), similar to the use of yeasts to convert carbohydrates into ethanol. In the Solazyme process, algae convert the carbohydrates into oil. The company claims the oil is equivalent to #2 diesel fuel and can be burned in pure form, without blending, year round.

The potential for bio-diesel from jojoba beans (a sub-tropical tree crop) appears promising, because the jojoba oil can also be used in diesel engines without further refining (New Scientist 2003). This is a significant advantage over bio-diesel from soya beans, rapeseed or sunflower seeds. However the economics are unclear, since the trees take some time to grow and the land they will require currently has other uses, such as citrus. Even under the most optimistic assumptions, jojoba diesel cannot make a significant contribution to the energy picture within the next two decades.

A few years ago, there was a lot of hype about the "hydrogen economy" and hydrogen fuel cells in particular. The attraction of hydrogen fuel cells arises from the fact that, in principle, they could roughly double the fuel economy of a hybrid-electric car and treble the fuel economy of a comparable gasoline or diesel vehicle. The hydrogen fuel cell was seen as an intermediate term solution by both the automobile industry and the oil/gas industry. The most familiar scenario was that hydrogen would be produced from natural gas, compressed, shipped by pipeline to service stations where some of the energy of compression could be recovered by means of a back-pressure turbo-generator, and (still partially compressed) used to fill the storage tanks in vehicles.

However, the necessary special-purpose infrastructure investment now appears rather daunting, to say the least. This scenario is unlikely because there are much better uses for the supply of natural gas, for example in combined-cycle electric power plants or the production of cellulosic methanol via the Hynol process, as discussed above.

To summarize, bio-fuels are a long-range possibility, but major breakthroughs are needed, and short-term benefits are very unlikely. Small amounts of bio-diesel are currently derived from soya beans (in the US), and from sunflower seeds or rapeseed in Europe. Output can increase in some locations, but the competition with food is likely to impose a strict upper limit. One future hope is for an efficient process to convert ligno-cellulosic waste from paper mills, agricultural wastes and municipal wastes into sugars that can then be fermented into ethanol. This should be possible: after all, termites and wood-boring worms do it routinely, with the help of specialized bacteria in their guts. However, wood-based fuel is not on the immediate horizon, and it may be several decades away, depending on where the R&D money is spent and how the land-use problems are resolved.

The easiest way to save money and energy (exergy) for transportation, at the same time, is to reduce the amount of driving we have to do. The simplest way to do this is to provide more and (much) better bus service on existing routes and to expand the area coverage. Underground rail transit is energy efficient in densely populated cities, but very costly to create or extend once the city above is already built. To be sure, the world's largest cities will doubtless go on building metro-rail systems, and there may be some integration of the underground systems with rail systems on the surface, but this will have at most a very modest—and long delayed—impact on energy use.

The most cost-effective solution for big cities in the near term can be called Bus Rapid Transit, or BRT (Levinson 2003; Vincent and Jeram 2006). BRT is an integrated system, not just an upgrade of existing bus lines. Tickets are prepaid by electronic fare-cards; no need for conductors. It operates along a designated corridor between purpose-built stations. For optimum performance, it requires dedicated (confined) bus lanes, and station stops spaced several blocks apart, ideally in the center of the corridor between two opposite traffic lanes. It usually involves articulated vehicles (similar to a tram) but current vehicle designs require no rails or overhead wires.

As compared to metro-rail, it is much less expensive—as little as one twentieth. Mexico City has a new BRT system along its major artery (*Insurgentes*) that cost as much as just two underground metro stations. BRT began in Curitiba, Brazil in 1966. Thanks to support from the World Bank, there were 49 BRT systems operating around the world as of 2009, of which 16 were in Brazil (only 3 are in the US). As of 2009, another 26 systems were being planned, mostly in Asia and Latin America. Curitiba's BRT system carries 2,190,000 passengers per day, and it is used regularly by over a third of the population.

It is a little puzzling that so few BRT systems have been started in recent years. One articulated BRT vehicle of current design can carry 160 people, equivalent to 100 cars averaging 1.6 passengers per car. Suppose that bus were allowed to travel on a reserved lane, on existing highways, stopping only at stations from 2 to 4 km apart in the outer suburbs. A BRT system with confined bus lanes is capable of as much as 40–50 % higher average speeds than multi-use lanes. (Such a reserved lane might be shared during rush hours, for instance, by cars carrying at least three people.)

The electronic fare-card system described above is nothing new. Exactly the same system has been in operation on the Washington DC metro-rail system since it opened in 1976. Essentially the same system is now used by the Bay Area Rapid Transit (BART) system around San Francisco and more recently in the London underground system ("oyster cards") and in all the European metro systems. The capital costs of creating this system would depend a lot on the geography, the existing highway network and the cost of land, but any place where a rail system can even be contemplated should first consider a surface BRT network.

Ideally, any large city should also impose a road-pricing or congestion charge of some sort on private vehicles (including trucks) entering the central area. A wide variety of such schemes have been implemented around the world since Singapore introduced the first congestion charge in 1975. The first version was controlled by police. Now it is entirely electronically controlled and variable pricing schemes (according to traffic conditions and time of day) are operational. Bergen, in Norway, introduced a charge in 1986; today there are six such schemes operating in Norway with a national coordinating scheme called AutoPASS. Stockholm has had a congestion charge in effect for the central city since 2007. Durham was the first in the UK (2002). London has had a congestion charge in place for the central districts since 2003; it was extended in 2007. Austria and Germany now have vehicle pollution charges for trucks. Austria has a comparatively simple scheme, operational since 2004. The German scheme, operational since 2005, is more elaborate, depending on the Galileo satellite system and other advanced technologies. It charges trucks on a per kilometer basis depending on their emission levels and the number of axles. Milan introduced a vehicle pollution charge in 2008 for all vehicles, but with a lot of discounts and exemptions. New York seems to be the only US city to consider the possibility seriously, but has been prevented from doing so by the State legislature.

The funds from such a congestion charge, as well as parking and other fees, should be used to subsidize any revenue shortfall in the public transportation systems, with some preference for BRT. The funds transfers from vehicle congestion charges to public transport users should be revenue-neutral in financial terms. BRT users along any reasonably well-traveled route would save money, perhaps quite a lot. The overall energy savings and pollution reduction could be quite large, again depending on the specific details. Incidentally, it should be pointed out that BRT buses need *not* be noisy and need not spew black smoke. In fact they need not depend on diesel engines, even though virtually all buses today do so. In fact, buses are excellent candidates for LNG or battery-assisted fuel cell operation.

Car-sharing is another way to cut down on street parking. The Uber system for ride-hailing recently introduced in many cities (but banned in some) is one approach. There are others (Britton 2009).

Finally, I come to the humble bicycle. Bicycling is good exercise, and it is the national sport of France, so should be encouraged. In fact, there are many more bicycles in the world than automobiles. There are more than 450 million ordinary bicycles in China, for instance, as compared to a 100 million or so private cars. But in China bicycles are being squeezed off the roads by the cars. The car-owners in China are far more influential than the cyclists. Moreover, they are supported by a belief in high places that it is necessary for China to develop an automobile industry in order to industrialize. The car owners of China are driving the government of China to build freeways where rice paddies used to be. These expensive highways certainly cut travel times between major centers of activity, including airports, for the lucky few, but simultaneously they spill automobile traffic onto all the connecting streets and roads. This increases congestion, pollutes the air and makes it much harder for local cyclists to compete for road space. (The conversion of rice paddies to highways also drives up the price of land and rice.)

The obvious answer for China to reduce congestion, pollution and save energy (exergy) would be to introduce BRT on a large scale, using reserved lanes on the recently-built highways. Congestion and pollution may turn out to provide the incentives leading to this transition.

In some European countries like the Netherlands, bicycles are already the dominant mode of transportation in towns and cities. Nearly half of all urban trips in the Netherlands were by bicycle as long ago as 1987, when gasoline prices were very low. Dutch commuters use bicycles to go to school, to university classes, to shops and offices, sports facilities, and to the nearest railroad station (which is never very far). In all these places they will find convenient bicycle parking spaces. The Dutch climate is by no means sunny and warm for much of the year, but it doesn't stop the cyclists.

To be sure, the bicycle solution is most suitable for the young, especially the students, and those commuters and shoppers who don't mind a little damp in case it is raining. But older folks ride too. Bicycles are not really suitable for commuters wearing business suits (though some do and others change clothes at the office), for mothers with little children in tow, shoppers with large or awkward packages, and even less so for the elderly and infirm. But many person-trips in Amsterdam and other Dutch cities do not require automobiles, which saves energy, cuts emissions and leaves more room for the buses and taxis (and some private cars) to move.

In big cities there are enough people able to make use of a bicycle for short trips to make a significant dent on urban traffic congestion, if (and only if) the automobile-oriented traffic authorities would cooperate. The first step must be to prohibit on-street parking by cars in downtown areas absolutely, except for delivery vehicles during non-rush hours. Segregated bicycle lanes, often in parks or along rivers or canals, are features of most European cities, though there are only a few in North America and China. Segregation of buses, taxis and bicycles into special lanes (as in Paris) is another approach. (The idea is that professional drivers can share road space with cyclists without endangering them.) Starting in Amsterdam in the 1960s, several large European cities have experimented with free 'bicycle sharing'. Most of these plans were inadequately thought through or too small in scale to succeed, But the schemes are getting more sophisticated. Paris is the most exciting example. On-street parking has been banned on avenues and boulevards for many years, and the no-parking zones are spreading.

All new buildings in central cities should have underground parking facilities, and parking must never be free. In July 2007 the city of Paris inaugurated a program called 'Velib" (a contraction of "vélo libre", or "free bike") with an initial endowment of 10,600 bicycles, of uniform design (Wikipedia 2008). They were paid for by a subsidiary of the big advertising firm JCDecaux and allocated to 750 reserved parking racks around the central city. The number of bicycles was increased to 20,600 within the next year, and the number of parking stations increased to 1450. The bikes are activated by an electronic credit card and electronically monitored. There is a small annual fee to belong to the 'Velib club'. The first half hour is free of charge, with a nominal hourly rental fee thereafter. Typically, each bike is used several times each day, and 95 % of the trips are free. Most uses of the trips are point-to-point, between one reserved parking place and another.

It is difficult to obtain good data on the impact of these shared-bicycles on other modes of travel. Hopefully they are replacing at least some private auto trips, although probably not very many (because very few central city residents use cars for short trips inside the city.) The Velib shared bicycles mainly replace walking trips, or bus and metro trips, reducing congestion on those modes. The main benefit to users is speed. The bicycles are faster between most pairs of destinations in the central city private cars or buses. So far there is very little direct evidence of energy conservation or cost saving, but since bicycles do not use fuel, the savings must exist. They will grow as the system becomes more widely accepted.

The shared bicycle programs in Paris, Vienna and elsewhere may be stepping stones to a much more significant future program combining arterial BRT routes with shared electric vehicles aimed at reducing commuter trips from within the city or the inner suburbs, and later from the outer suburbs, where most commuters travel by private car. At present the average time/distance for an average urban bicycle is less than half-an-hour or 5 km, enough for some lucky commuters, and enough for many others to get to a bus or tram station or a train station. Bicycle-parking facilities at railroad stations (as in the Netherlands) would sharply increase the utility of this option. Obviously a great many young men and women can go much faster and farther, at least where the terrain is flat, as in most cities.

The next step beyond the human-powered 'push-bike' is the battery-powered 'e-bike'. Such bikes are capable of an average speed of 15–20 km/h, depending on traffic. By the end of 2007 there were already at least 30 million of these e-bikes in China, out of a total bicycle population of 450 million (Weinert et al. 2007). The market for e-bikes in China was 40,000 units in 1998, but it exploded to an estimated 16–18 million units in 2006, produced by over 2000 firms, mostly small and local.[11] There are two types, 'bicycle-style' (perhaps with pedals for supplementary muscle power) and 'scooter style'. The former carry storage batteries, with a capacity ranging from 0.4 to 0.6 kWh, while the latter carry batteries with about twice as much capacity (around 1 kWh).

The dominant battery type today is the valve-regulated lead-acid battery (VRLA), which accounts for 95 % of Chinese production, and almost all of domestic consumption, since the other battery types, especially the bikes with nickel hydride and lithium-ion batteries, are mostly exported. The lead-acid batteries have a lifetime of only about 2 years. However, the more advanced battery types are lighter, more powerful, and have a much longer expected lifetime. They are rapidly becoming more popular, and optimists expect them to account for 40 % of Chinese

---

[11] The local firms don't really manufacture anything; they convert ordinary bikes into e-bikes by adding a battery pack and replacing a conventional wheel by a wheel with a motor. The battery packs and motorized wheels are made by a much smaller number of suppliers.

e-bike output by 2020. In 10 years or so the lead-acid batteries for bikes are likely to go the way of the dodo.

Electric bicycles and scooters are still rare in Europe and America, but they have the potential for changing commuter behavior radically, even in a spread-out American city such as Los Angeles. All it takes is serious effort on the part of municipal authorities to discourage cars—especially through congestion charges and parking restrictions—and to make it easier for bicycles to use the roads and to find safe parking facilities at bus stations, tram stations, railroad stations and so on.

Obviously, gasoline powered motor-cycles are already a feature of the highway. However, most of us over the age of 30 who don't ride them ourselves, see them as noisy, brutish machines—called 'hogs' by the aficionados—driven much too fast by crazy young men like Jimmy Dean, Marlon Brando, Evel Knievel, Hell's Angels and so on. That image is not very accurate today, but it persists. A much more accurate picture is the one seen by commuters stuck in the frequent traffic jams during rush hours on the main auto-routes into large European cities. The car-bound folks curse in frustration at the steady stream of motor-cycles that pass freely between the stalled lanes of cars, without even slowing down. These are not recreational drivers; they are people going to or coming from work, consuming much less gasoline and producing much less carbon dioxide (but much more noise) at much less cost per km traveled, than the 4-wheel vehicles they pass. The 2-wheelers are also easier to park. Why doesn't everybody do it? Well, increasing numbers of people are doing it, because it is so practical. But there are problems and barriers.

One barrier for many is simple fear. Motorcycles are known to be dangerous. Accidents are much more frequent than for 4-wheelers. Remember the opening scene from "Lawrence of Arabia" where T.E. Lawrence dies in a crash? Quite a lot of people seem to have chosen to drive big, heavy SUV's—rather than small, more fuel efficient cars—precisely because (encouraged by auto company marketing) they believe big cars are safer. Such people are not likely to switch to motorcycles anytime soon. Then, there is the need for a special license to ride a high powered Suzuki, Honda or Harley-Davidson motor cycle on the highway. To obtain that license you must obtain special training, registration and insurance. In Europe the training is fairly expensive, yet millions take it. In the US the motorcycle safety foundation (MSF) offers free or inexpensive classes. In many states, proof of passing this course is enough to get a waiver of the state road test and written test.

The main problem for the gasoline powered 2-wheelers is noise and pollution. These vehicles are supposed to be equipped with mufflers, but the mufflers often are not functional, and there is still no legal constraint on emissions. (In fact, I suspect that for some users, noise is part of the attraction). To put emissions control on a motorcycle adds significantly to weight and cost, and detracts from performance. Up to now, many 'gear-heads' simply disconnect the unwanted pollution-control equipment.

But as the number of motorcycles on the roads increases, the pressure to eliminate this loophole in the anti-pollution and anti-noise laws will grow. The likely answer is the electric motorcycle. Of course, the electric bikes now on the market are more expensive than human-powered bikes but they are already cheaper than gasoline powered motorbikes. They are quiet and have no tailpipe emissions whatsoever, and operating costs (electricity) are considerably lower than gasoline powered vehicles. Even if the electric power is generated by burning coal or natural gas, the electric version will be at least twice as efficient (in life-cycle terms) as the gasoline-powered version.

The average middle-aged suburban commuter in North America, Europe or even China—male or female—is not likely to buy and use either a powerful gasoline-powered motorcycle or an electric scooter to ride to work. But the costs of the e-scooter have already dropped radically and will fall further in the next few years, both because the lithium-ion battery technology is still being improved rapidly, and because mass production has barely begun. Moreover, those e-bikes will be able to use the bike paths or reserved lanes, from which motor-bikes will be excluded, that many cities are going to build into their traffic plans for the coming decades.

10.5   On Opportunities for Energy Efficiency Gains by Systems Integration

Before discussing specific opportunities, it is important to clarify a persistent confusion in the literature. I refer to two ways of defining energy efficiency, based on the two laws of thermodynamics. A study in 1975, sponsored by the American Physical Society (APS) defined "first law" efficiency (e_1) and "second law" efficiency (e_2) (Carnahan et al. 1975). "First law" efficiency e_1 was defined in the APS study as a simple ratio of "useful" output to total input. By this definition, a boiler can have an efficiency of 80 % if 80 % of the heat of the flame goes into the water and only 20 % of the heat goes up the stack.

The essential point is that a high efficiency furnace or boiler in the sense of "useful output" as a fraction of total energy input doesn't mean that the heating *system* is efficient compared to other ways of achieving the desired result. To put it another way, the "first law" efficiency takes account of the energy losses applicable to specific pieces of equipment, but when applied to a system, it implicitly assumes (erroneously) that the particular choices of equipment in the existing system are optimal. It does *not* reflect potential gains that might be achieved by different configurations of equipment or (for instance) by using more insulation.

The so-called "second law efficiency", or e_2, as defined in the APS study is a different ratio (Carnahan et al. 1975). The numerator of this ratio is *the amount of useful work actually done* by the energy input (taking into account the second law of thermodynamics, or entropy law). The denominator of the ratio is defined as *the theoretical amount of work that could be done by the same actual energy input* to the process. This theoretical maximum is technically denoted "exergy", a term familiar to engineers but not to most other people. Second law efficiency is usually called exergy efficiency in the engineering literature.

As the American Physical Society (APS) Summer Study in 1975 showed, the second law efficiency e_2 of a conventional boiler in a hot water central heating system for a building is quite low—typically in the range 5–8 %—as compared to the much higher theoretical potential. The crucial point is that the conventional hot water or steam heating system is by no means optimum. There are much better ways to achieve the desired temperature of the air in the room. The reason is that the combustion flame produces heat at a very high temperature—say 1500 °C—which *could* be used (in principle) to do quite a lot of useful work, for instance by driving a gas turbine, the exhaust heat of which could drive a steam turbine (as in a combined cycle), the waste heat from which could heat the room. Or the fuel itself could be used to generate electric power, via a fuel cell, a small part of which could then be used to drive a heat pump to heat the room, the rest being available for other kinds of work. (In fact, the simplest and easiest way to reduce energy losses from heating would be to adopt district heating systems, as in much of the world—but not the US—using the low temperature "waste" heat from electric power generation.)

More generally, by drawing the boundary for the analysis too narrowly around a specific equipment configuration (the boiler) completely misses the most obvious way to reduce energy losses (apart from district heating). That is to improve insulation, install double glazed windows and LED lights and make the whole boiler unnecessary. In fact, the so-called "passive houses", now being built in Germany and Austria, cut heating and cooling requirements by around 95 %, by using solar heat, high quality insulation and a special heat exchanger that heats incoming air by taking heat from outgoing air (Elswijk and Henk Kaan 2008). Whereas the conventional heating system has a second-law (exergy) efficiency of 5–8 %, at best, the passive house design has an exergy efficiency approaching 90 %. In fact, some recent house designs go beyond that and actually permit surplus energy exports.

Another illustration of the boundary problem is automobiles. Conventional efficiency analysis draws the boundary around the vehicle itself, including the engine, drive-train, and tires. But the easiest way to improve the efficiency of automobile transport *as a system* (after weight reduction) would be to induce drivers to increase vehicle occupancy from the current average of 1.5–2 (or 2.5) perhaps by creating stronger incentives such as special lanes for multiple-occupancy cars. Beyond that, car-sharing offers further gains (Bryner 2008; Orsato 2009) and bicycles even more. All of these are outside the domain of the vehicle *per se* (engine + drive-train) which is where conventional energy-efficiency analysis stops (see Sect. 10.4). Evidently there is plenty of room for further efficiency gains in the transport system taken as a whole (Britton 2015).

The technology for increasing overall system efficiency by utilizing waste process heat is known as combined heat and power (CHP) or "combined cycle" or "co-gen". This technology is already well developed, and scalable (Casten and Collins 2002, 2006). The basic idea is shown in Fig. 10.19. If applied widely, CHP could cut waste heat and carbon emissions from the electric power sector alone by as much as 30 %, by producing electricity more efficiently, thus cutting demand for coal or natural gas.

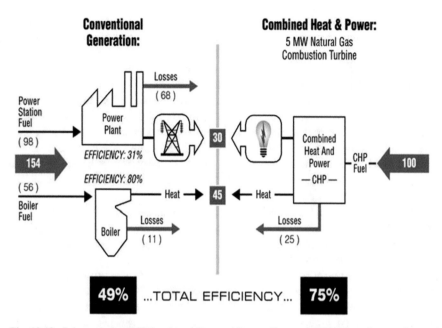

Fig. 10.19   Schematic for CHP Combined Heat and Power. *Source*: *US EPA* http://www.c2es.org/
technology/factsheet/CogenerationCHP

For example, Arcelor-Mittal Steel (formerly Inland Steel), in Gary Indiana, has a
coking plant (Cokenergy) that supplies its neighboring steel-making facility. The
heat from the coking process drove a CHP unit that generated 90 megawatts
(MW) in 2005. That electricity, which produced no carbon emissions, was utilized
by the neighboring steel plant (on the same property), thus using 90 MW less power
from the local utility. A neighboring US Steel plant nearby in Gary also generated
100 MW from CHP, using waste blast-furnace gas. Together those two plants
produced 190 MW in 2005 (Ayres and Ayres 2010).

It is worth recalling that the discovery of coking in the eighteenth century played
a central role in the history of technology. It enabled the production of high quality
iron (and later, steel) without using charcoal from trees for the smelting process.
The gaseous by-products of coking enabled gas lighting, and later the first internal
combustion engines. Aniline dyes and aspirin were still later by-products. My point
is that the application of CHP at the Cokenergy plant revealed yet another important
by-product of coking, namely *emissions-free electric power*. I'm not saying that
coking generates no pollution. It does. I am saying that utilizing that hot exhaust
through a heat-exchanger to make steam to make electricity, *does not generate any
additional pollution*. (In fact, by cooling the exhaust, it probably makes pollution
control, by standard techniques, slightly more efficient.) CHP attached to an
existing high temperature heat source generates effectively emissions-free
electricity.

There are a lot of industrial facilities around the world that produce hot exhaust streams that could be diverted through heat exchangers to make steam and electric power, as in the Cokenergy plant. Petroleum refineries are obvious examples, as are smelters, glass fabricators, cement plants, carbon black plants (they make the stuff that makes tires black), silicon refineries, and so on. Steam systems, which account for 35 % of US (and global) manufacturing energy consumption, lose 45 % of the input heat before reaching the point of use. (Some of this loss is due to inefficient pumping). By far the best way to reduce the energy loss in steam generation (apart from plugging leaks) is to install combined heat and power (CHP) systems in the plant. Where the same plant can use the electric power, this is being done. But where the excess power would have to be sold to other users in order to finance the CHP system, there is a problem.

The implementation problem is that the law that allowed the formation centralized electric utilities back in the 1920s (based on a theory of "natural monopoly") says that by-product electric power cannot be sold to any users located on the other side of a public road (or other public land). Under current law, any by-product electric power can only be sold to the local monopoly electric utility, at whatever price the utility offers. The two examples mentioned above got around this law because their properties were adjacent along the shore of the lake. The utilities don't want to outsource electricity generation, because the law allows them to recover the full cost of borrowed capital by charging it to users, whereas any innovation that cuts their costs has to be passed on to consumers as lower prices. So they have no incentive to innovate. It is why the efficiency of electric power generation has barely increased since the late 1950s. Consequently a lot of hot exhaust gases from high temperature industrial processes that use coal, oil or natural gas are being lost, literally, "up the spout". Changing that law would unlock a very large potential (Casten and Collins 2006; Casten and Schewe 2009; Ayres and Ayres 2010). Most industrialized countries have similar laws, though some (notably the European Union) have begun to change the laws to accommodate CHP.

Actually, from 2000 to 2010 applications of CHP in the US nearly doubled. In principle, industrial CHP could be tripled by 2030 or so, resulting in a substantial decrease in carbon emissions as well as an increase in the overall efficiency of electric power generation at the national level (Casten and Ayres 2007). But it would take market share from the centralized electric utilities that depend on their monopoly business model. If the utilities were to give up their centralized monopoly model and focus on decentralized power generation, using combined head and power (CHP) for space heating, they could also supply the total demand for decentralized space heating. There are technical problems associated with decentralization, of course. One of them is that support for the grid would shrink. But the main one is that demand for space heating is seasonal—mostly in the winter—whereas in the summer demand is for air-conditioning (more accurately, cool air and cold water). There is an advanced heat-pump technology, called pumped heat energy storage (PHES), now being tested in England, that could solve this problem. I discuss it later (in the next section) in connection with storage for the grid.

Needless to say, it is not possible to cover all the opportunities for efficiency gains in this section of a single chapter. I can only touch on the high points and point to the sources. In the US the DOE has estimated that energy used in manufacturing is allocated to (direct) process heating (39 %), process steam (35 %), electric motor systems (12 %), facilities (8 %) electro-chemical systems (2 %), process cooling (1 %) and "other" including lighting and computers (4 %) (Energetics 2004).

Globally, motor systems account for an even greater share of manufacturing energy (15 %) according to the IEA, and a considerably larger share of electric power consumption by industry, probably more than 50 % (IEA 2007). More interesting, studies indicate that 55 % of the electric power used by motor systems is lost before accomplishing any work. Compressors, e.g. for gas transmission, are particularly inefficient, losing 80 % of the mechanical work done by the motors, in the process (Energetics and E3M 2004). Pumping systems, which are extensively used in mines, oil wells, geothermal power plants and petro-chemical plants are often much less efficient than they should be because of over-sizing and badly designed pipe networks. Dramatic improvements are possible by pipe connection redesign e.g. (Lovins 2011).

The US Department of Energy has conducted a number of interesting studies to assess the potential for profitable energy savings in the future. One of the most interesting is a cross-cutting study of the potential for savings by optimization of electric motor systems (Lung, McKane, and Olzewski 2003). The study encompassed 41 optimization projects in 17 industries that were carried out in the US between 1995 and 2001. They were selected from a much larger number, primarily on the basis that all of the ones chosen provided comparable data on energy (as electric power) savings and consumption in kWh, project cost and savings in dollars ($).

Aggregate costs of the projects was $16.8 million, with aggregate savings of $7.4 million—a 23 % reduction in average energy consumption—while the net present value (NPV) of the savings over a 10 year (assumed) project lifetime was $39.6 million, far exceeding the aggregate cost. The simple payback time for the 41 projects was 2.27 years. The savings potential in Europe (as percentage) is larger because motor efficiency (as of 2005) was significantly lower in Europe than in the US and Canada.

The U.S. Department Of Energy started a new program in 2006, entitled "Save Energy Now" (United States Department of Energy 2010). Under this program, DOE provided trained energy auditors, who worked with teams from a firm, to assess process heating, steam generation and compressor use in any manufacturing entity with a total annual energy use in excess of 1.1 Petajoules (PJ). Such firms account for half of the total energy consumed by manufacturing firms in the US. Through 2007, there were 717 collaborative assessments and cost savings investments already implemented amounted to $135 million with $347 million further investments planned. However, the assessors recommended further projects that would save 92 PJ and $937 million in costs, with a concomitant reduction of 7.9 million tons of CO_2 *per year.*

About 40 % of the savings would be from changes such as insulation and cleaning of heat exchanger surfaces, with payback times of less than 9 months. Another 40 % of the potential savings, with a payback time of 9 months to 2 years, will come from feed-water preheating using flue-gas or boiler blow-down heat recovery. Another 20 %, with pay-backs from 2 to 4 years, will come from miscellaneous modifications such as steam turbine modifications, changes in the process steam uses and the use of pure oxygen, rather than air, for combustion. Finally, because of the need to install new equipment and reconfigure existing equipment, installation of CHP systems generally have payback times of more than 4 years.

I illustrate the possibilities with three examples from a top professional in the energy recovery business (Casten 2009).

1. A **blast furnace** generates exhaust gas, containing hydrogen, at perhaps 500–650 °C (having already given up temperature to preheat the new layers of coke and iron ore). To prevent the possibility that any spark might start a fire in the top of the blast furnace, the common industry practice is to quench the blast furnace gas by spraying it with water. The cooled gas then goes through a water separation device to remove the water, but it still has pressure sufficient to drive a gas expander. It emerges as 100 Btu/standard cubic foot flare gas that has 10 % the heat value of natural gas. The flow oscillates with each new layer of coke and iron ore added, so industry practice is to burn part of this gas to make steam for process heat and/or electrical power generation, and then flare the rest.

A more elaborate redesign would eliminate the quench, use the higher temperature and pressure to drive more power generation, and then send exhaust gas to the boiler at higher temperatures, thus increasing the steam production. Alternatively, a thermal oil heat exchanger system could capture the relatively low grade heat and then use it to drive the moisture out of the coke and ore, preheating them to reduce the need for coke. This has rather significant efficiency improvement potential. Finally, it is cost effective to install a bladder that absorbs the oscillation in the gas generation from the blast furnace to smooth out the flow of gas to the boiler, which eliminates the need for flaring.

The electric power generated from the heat in the exhaust gas would be greater than that needed by the steel mill itself, so—in principle—it could be sold to the grid. But present rules in many countries will, at best, allow the utility to pay the short-run marginal cost of all generators supplying the grid, and at worst, it will only displace purchases by the rest of the steel mill itself, which are the lowest rates paid by any user on the system.

2. Plate glass manufacturers burn natural gas to melt silicon dioxide (sand) and scrap glass (cullet) at about 1800 °C. They need an elaborate cooling system to remove the heat at a controlled rate from the formed plate glass. In addition, the exhaust from the glass furnace has a temperature of about 800 °C after the use of a recuperator. One energy-saving scheme is to replace some of the glass cooling, to make low pressure steam and pipe it to some nearby user, such as a dairy processing plant. Another possibility is to use the heat in either a steam or organic Rankine cycle engine (one with a condenser) to generate electricity, yielding about 10 MW.

3. Gypsum wallboard (sheet-rock) production first drives water off of the gypsum, (as it comes from the mine), then adds water back in to make a fine slurry that is sprayed between continuously fed sheets of paper. This goes through rollers to fix the thickness and then through a 100 m tunnel in which hot air at a controlled temperature close to 400 °C dries the slurry and sets the sheetrock. This is typically done with natural gas fired duct burners. An alternative scheme is to burn the natural gas in a gas turbine to generate electricity, and then capture the exhaust heat with thermal oil. The thermal oil is then used to pre-heat the drying air to the precise temperature desired. (If the air is too hot, the moisture cannot move through the face of the wallboard fast enough and causes bubbles; if it is too cool the sheetrock does not set.) The exhaust air from the drying process is barely above the boiling point of water, but it is fully saturated (100 % humidity). It is possible to recover some of the latent heat of vaporization to boil propane that drives a gas expander.

Further insight into the economic potential of efficiency—not limited to the manufacturing sector or to industry more broadly—can be gained from various "least cost" studies that have been carried out since the 1970s. A study by Roger Sant and others at the Mellon Institute for the Department of Energy (Mellon Institute 1983) argued strongly for a "least cost energy policy" that would generate substantial savings; see also (Carhart 1979; Sant 1979; Sant and Carhart 1981). The results are worth a brief recapitulation. In economic terms, the least-cost strategy would have saved $800 per family (17 %) or $43 billion in that year alone (ibid.). Taking the year 1973 as a standard for comparison, such a strategy would have involved a sharp reduction in the use of centrally generated electricity (from 30 to 17 %) and a reduction in petroleum use from 36 to 26 %. The only primary fuel to increase its share would have been natural gas (from 17 to 19 %). Interestingly, the Sant-Carhart study suggested that "conservation services" would have increased their share from 10 to 32 % in the optimal case.

A study carried out by the Italian energy research institute ENEA is typical of the results of engineering surveys; e.g. (D'Errico et al. 1984). It found many technological "fixes" with payback times of 1–3 years—well below the typical threshold for most firms and several times faster than investments in new supplies of petroleum. A 1991 study sponsored by four major US organizations (Union of Concerned Scientists, Alliance to Save Energy, American Council for an Energy Efficient Economy and Natural Resources Defense Council) concluded that a 70 % reduction in CO_2 emissions could be achieved in 40 years, at a cumulative cost of $2.7 trillion but with compensatory savings worth $5 trillion, for a net total gain of $2.3 trillion (ACEEE et al. 1992).

Other examples in the older literature are numerous. Apart from examples mentioned above, I should acknowledge three major international studies of technological potential for energy saving, namely: *Energy for a Sustainable World* (Goldemberg et al. 1988), *Energy Technology Transitions for Industry* (IEA 2009), and *Global Energy Assessment* (Johansson et al. 2012) Chap. 8. These three studies summarize a lot of prior work at CEES, LBNL, and elsewhere, of course.

Also, apart from the LBNL group, a group of researchers under W.C. Turkenburg at Universiteit Utrecht, in the Netherlands, has done many studies of the technology potential for conservation, at the sectoral and technology level, in recent years. A series of Ph.D. theses have collected evidence from a bottom-up perspective of the potential for cost effective energy efficiency savings at the sectoral level, as well as household consumption. A major theme has been the development of bench-marking tools at the process level. Several book length Ph.D. theses have been prepared on iron and steel technology options, plastics, paper and paperboard, oil refineries, chemical processes, cement, food processing and others (primarily with reference to the Dutch economy but clearly applicable to much of the industrialized world). See, for instance (Blok et al. 1993; Luiten 2001; Joosten 2001; Smit et al. 1994; Ramirez 2005; Beer et al. 1994, 1996; Beer 1998; Worrell and Price 2000) and several reports in Dutch cited in these sources.

Finally, it is important to acknowledge the enormous amount of work done at the Rocky Mountain Institute (RMI), under the leadership of Amory Lovins. Much of that work is summarized in his latest book, *Re-inventing Fire* and the studies cited therein (Lovins 2011). See also *Least-cost Energy: Solving the CO_2 Problem* (Lovins et al. 1981) and *Factor Four: Doubling Wealth, Halving Resource Use* (von Weizsaecker et al. 1998). However, Lovins' work has been sharply criticized for over-optimism, e.g. by Smil (2003) and Jacquard (2005).

The evidence, at least from paper studies, that efficiency investments can not only save a lot of energy, but also be profitable, is literally overwhelming. On the other hand, engineering "bottom up" studies tend to be more optimistic than economic "top down" studies. There are various reasons why this happens, none of which need to be explored in detail here. Lovins, in particular, has been criticized by economists as being too optimistic about the potential for energy conservation. Time will tell. I can only say, with regard to who you want to believe, "caveat emptor".

The question for economists is: If energy conservation is so profitable, why don't firms do more of it? This question was addressed indirectly in a recent UNIDO publication (United Nations Industrial Development Organization 2011). Having spent most of a year leading the first version of that study, I offer the following hypothesis: When firms have a choice between investing in energy conservation— no matter how profitable—or investing in new products or expansion into new markets, they almost always make the latter choice.

Why? It seems that CEOs tend to see their world in a Darwinian "eat or be eaten" perspective. A small firm that invests to become more efficient and profitable becomes more attractive as a takeover target. It is better (for the CEO) to be a shark than a minnow. This point is worth mentioning, because it is contrary to received wisdom in economics. Received wisdom says that the CEO (being an agent of the owners) will maximize profits and thus maximize shareholder value. Given that capital is scarce, a rational investor will choose the most profitable ventures first. Thus, a large firm will typically try to rank order the various proposals for capital spending (in order of expected ROI, adjusted for risk) and go down the list until either the available money for investment runs out or the threshold is reached. In principle, government would do the same.

In a quasi-equilibrium economy, there should be enough capital to fund all of the promising projects, i.e. all the projects with expected ROI above the appropriate threshold level. It follows that the really "good" (i.e. very profitable) projects should be funded as soon as they appear on the horizon. In an economy very close to equilibrium, there should be no (certainly very few) investment opportunities capable of yielding annual returns of 30 % or more. Yet many authors have argued for the existence of such opportunities, at least *within* firms. Case study after study has shown that changes in the energy production/consumption system can be identified that would pay for themselves in just a few months or a few years at most.

There are some unresolved questions about what is, or isn't, economically rational in this domain. I must leave it here.

10.6   Renewables for Heat and Electric Power Generation[12]

For reasons discussed in Sect. 10.4, and elsewhere, it is not too unrealistic to say that the future of humanity on Earth depends on ending the use of fossil fuels for all purposes, and doing so within a few decades. There are two broad categories of approaches. One is to utilize solar heat or waste heat, resulting in higher system efficiency. The other is to replace fossil fuels, not only for transportation (already discussed in Sect. 9.7) but for the generation of electric power—the world's largest industry.

Speaking of heat, there is another important heat-utilizing process that will be growing in importance in coming years: solar desalination. There are two main desalination processes at present, viz. thermal (multi-stage flash distillation) and pressure-based (reverse osmosis). The distillation processes differ in detail, but all depend on removing the salt from brine or ocean water by evaporating the water and subsequently condensing the vapor elsewhere, leaving salt (or very salty brine). The salt flats found in many deserts of the world are the result of a natural version of this process. Most of the heat for thermal desalination is now obtained by burning oil or gas, which is incredibly wasteful. However, solar desalination is gaining attention, although it currently accounts for only 1 % of the total desalination market.

The simplest version is illustrated in Fig. 10.20. There are many options for increasing the efficiency by (for example) storing heat, reducing air pressure in the evaporator and so on. One firm in California, WaterFX, has a scheme that multiplies the natural evaporation rate by a factor of 30 [http://cleantechnica.com/2015/07/20/buzz-gets-bigger-tiny-california-solar-desalination-plant/]. The potential in drought-stricken California is very large; much of the irrigated land in the San Joaquin Valley is becoming salt saturated. The WaterFX system might be able to save that land for agriculture. In the more distant future, solar desalination may be

---

[12] Parts of this section are taken chapter from Chap. 8 of my previous book "The Bubble Economy" (with permission). It has been updated and extended.

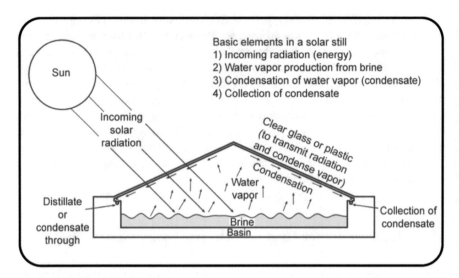

Fig. 10.20 Simple solar desalination system. *Source*: http://www.sswm.info/sites/default/files/ toolbox/MERCHELL%20and%20LESIKAR%202010%20Example%20of%20a%20Solar% 20Desalination%20Process.jpg

applicable in other deserts, like the Atacama, the Gobi, the Sahara and the Middle East.

Reverting to the problem of reducing carbon emissions, the single most important source of emissions is the generation of electric power by burning coal. Increasing efficiency at the power generation stage, and also at the intermediate and end-use stages, are part of the solution. Replacing fossil fuels by renewables is the other part.

In 2001, renewable energy sources (including large hydro) provided 19 % of global electricity supply. The bulk of global energy was supplied by fossil fuels. An estimated 20,000 TWh of electricity was generated in 2014 (due to a slowdown since 2011 in China). Of this amount, renewable energy technologies (RETs) contributed 26 % of the total, but most of this still came from large hydro-electric installations. However, by 2020 RETs should provide close to 34 % of the total and by 2030 this share could increase to 55 %. See Fig. 10.21.

The major renewable energy technologies (RETs), as projected today, are biofuels, wind, and solar PV, with lesser contributions from small-scale hydro, geothermal and solar thermal. Biofuels for electricity production—mainly waste wood and bagasse—are distinct from liquid biofuels for motor vehicles. I also discuss storage, because of its relevance to the intermittency problem.

I have also included a brief discussion of a non-uranium based, nuclear fission technology that does not fit anywhere else. The other fission and fusion nuclear technologies are not considered in this book, probably due to my personal anti-nuclear prejudice. I agree with those who favor banning the current version of nuclear technology. I refer to the technology that uses uranium 235, produces

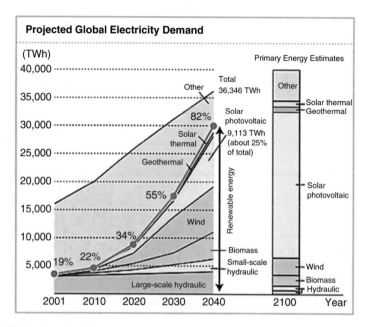

Fig. 10.21 Projected global electricity demand. *Source*: http://pubs.rsc.org/services/images/
RSCpubs.ePlatform.Service.FreeContent.ImageService.svc/ImageService/Articleimage/2012/
NR/c2nr30437f/c2n30437f-f1.gif

plutonium as a by-product, and generates extremely toxic and radioactive wastes
that nobody has yet found a satisfactory disposal solution for. There is a possible
alternative, using thorium, but this is not the place to discuss it (Hargraves 2012).
As regards controlled fusion technology, the day when it becomes cost-effective
and safe for use on Earth, if ever, is too far in the future.

Wind energy technologies are currently the most mature of the RETs. Global
capacity reached 369,553 MW at the end of 2014, with 51,477 MW added in the
year (a 44 % increase over 2013) as seen in Fig. 10.22. The total is increasing
rapidly, although wind still only accounts for a small share of total electricity output
and is heavily concentrated in a few countries, notably China, USA, Germany and
India. A number of global wind resource assessments have demonstrated that global
technical potential exceeds current global electricity production. The great advantage of wind power is that, once built, the cost of operating a wind turbine is nearly
zero (except for maintenance). It is likely that a lot of the maintenance can be
automated, necessitating personal inspections only when something actually
breaks. However, the growth of the industry is limited by the relatively low capacity
factor, which ranges from 12 to 25 %, depending on location. The cost of wind
power, both onshore and off-shore, has been declining at a modest rate, but steadily
(thanks to increasing experience), as shown in Fig. 10.23.

The strong opposition to wind turbines from bird lovers and other environmentalists has largely abated. This is probably due to the fact that the birds seem to be

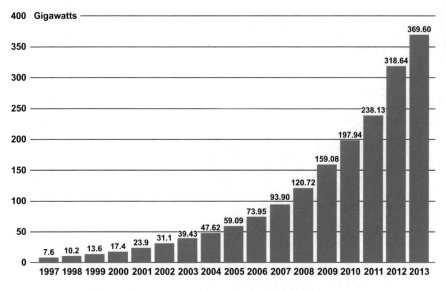

Fig. 10.22 Global wind power capacity. *Source*: *GWEC (Global Wind Energy Council)*

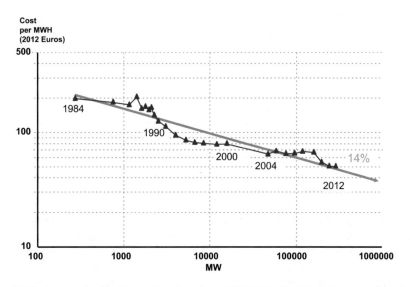

Fig. 10.23 Average levelized cost of wind, onshore 1984–2013. *Note*: Learning curve (*blue line*) is least square regression: $R^2 = 0.88$ and 14 % learning rate. *Source*: Bloomberg New Energy Finance, January 2013

able to avoid the turning blades, and turbines on distant hills are certainly less ugly than overhead high voltage (HV) transmission lines.

Solar thermal heating for buildings is an old technology. However, it is increasingly important for new buildings, especially advanced designs, such as the German "passive house", which relies entirely on solar heat, combined with insulation, double or triple glazed windows, and counter-current heat exchangers to heat incoming ventilation air. In the summer, this procedure can be reversed to conserve air-conditioning (Elswijk and Henk Kaan 2008). In new construction, this technology cuts energy consumption by 90–95 %. However, there is a penalty insofar as capital costs for "top of the line" energy efficient houses are a little higher than conventional designs.

The cost penalty was about 10 % a few years ago, so it took a few years to pay for the extra cost. But in recent times the penalty has declined to less than 1 % according to EEC experts. It would make sense to use governmental licensing authority to enforce advanced thermal management standards for all new housing, starting now. However, the housing stock takes a long time to renew itself, so this action would not have great immediate effect. Moreover, most builders and architects are not yet familiar with the new technology.

The major challenge is to find ways to achieve retrofit some of these efficiency gains in existing buildings, at reasonable costs. Unfortunately, the benefits that can be achieved from fairly obvious leak-reductions like insulating cavity walls, roofs and double glazing windows, and by using LED lights and heat pumps, is far less than can be achieved by new construction with air-exchange. The improvement from insulation add-ons and LED lighting would probably be in the neighborhood of 25 %—but still worthwhile (Mackay 2008: 295).

Another technology with some potential is concentrated solar power (CSP), which is a technology utilizing mirrors to focus sunlight on a spot (traditionally a heat exchanger). The unique feature of this technology is the possibility of obtaining extremely high temperatures at the point where the mirror beams converge. However, up to now, there is no viable scheme to use that very high temperature heat, so it is used to produce steam. At present some installations do employ heat storage, by the use of molten salts, mainly sodium or potassium nitrates. The heat (at a temperature of around 300 °C) can be used to drive a steam turbo-generator. A way of using high temperature heat to drive an industrial batch process, such as lime production, cement production or acetylene synthesis, would be much more valuable. However the engineering would be challenging. The installed capacity of CSP at the end of 2013 was only 3.4 GW, up from 2.5 GW a year earlier. The future of this technology depends largely on molten-salt heat storage.

Photo-voltaic power (PV) is a semi-conductor-based technology that converts the energy of sunlight (photons) directly into electricity. The phenomenon has been known for a long time, but the first commercial applications 30 years ago utilized ultra-pure scrap silicon from computer 'chip' manufacturing. However, the purity requirements for silicon PV cells are much less than for 'chips', and by 2000 the demand for PV justified investment in specialized dedicated fabrication facilities.

Several PV technologies have been developed in parallel. The three major types commercially available now include multi-crystalline silicon, amorphous silicon and thin films on a glass or plastic sheet. Thin film PV compounds in use today include copper indium gallium disulfide(di)selenide (CIGS), cadmium telluride

(CdTe) and amorphous silicon. The panels can be arranged on virtually any scale from a single panel on a rooftop to a multitude of panels organized in arrays that comprise "solar farms". Solar PV power systems can be either off-grid or grid-connected into mini grids or larger national grids. Costs have been declining for a long time, as indicated already in connection with the electric vehicles (EVs).

Many PV systems have been built in China, Japan, Germany, the US, and other OECD countries, both rooftop and large-scale. Costs per kWh are still higher, on average, than for coal-based electricity, but PV costs are now less than costs for coal-based electric power in some locations. The economic advantage of coal is fast disappearing, despite the intermittency problem of solar.

The most practical scheme for matching demand and supply is to extend the grid itself, by means of interconnections. This will enable wind or solar producers to export surplus power to regions where there is pumped storage or where hydro-power can be turned off, and to import from the same countries during times of high local demand and low production. Denmark is already doing this; it has 3.1 GW of wind capacity, with interconnections to Germany, Norway and Sweden capable of carrying up to 2.8 GW (Mackay 2008).

PV costs continue to decline and installed capacity is increasing rapidly. It reached 138 GW at the end of 2013 (see Figs. 10.24 and 10.25), and jumped by 40 GW to 177 GW in 2014. By the end of this year (2015) it is expected to exceed 200 GW. The rapid increase is led by China, as part of a major initiative to cut carbon emissions and air pollution.

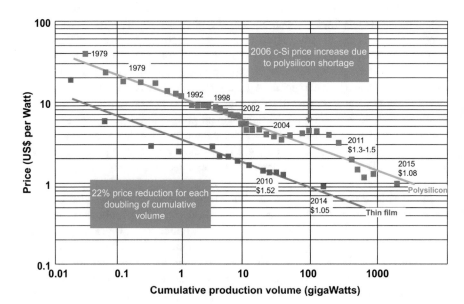

Fig. 10.24 Installed cost of PV electricity. *Source: IRENA(2012), "RENEWABLE ENERGY TECHNOLOGIES: COST ANALYSIS SERIES"; IEA (2011), "SOLAR ENERGY PERSPECTIVES"*

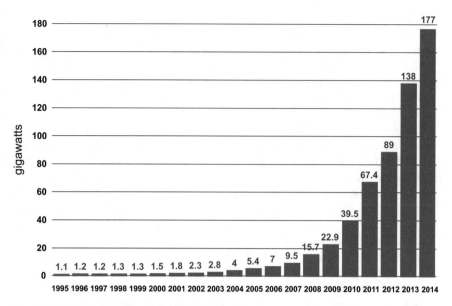

Fig. 10.25 Global solar PV; total capacity 1998–2014

There is an interesting consequence of the coming spread of electric cars (Sect. 10.4) worth considering. In fact the electric utilities in the US are already worrying about it. The centralized electric power utilities are currently experiencing static or even declining demand, partly due to a collection of end-use efficiency gains, especially due to LED lighting, and partly due to the spread of roof-top PV. In fact, the latter not only deprives the central power producers of customers, but—in many countries and states—forces them to pay for excess PV output from consumers, when the sun is shining but the domestic usage is low. This "double whammy" induces utility associations to ask regulators to allow "connection charges" independent of charges for power use.

On the one hand, the case for connection charges seems reasonable, since the utilities do have to provide backup power whenever it is needed, even though inputs from PV are intermittent and unpredictable. The resulting load-balancing problems are very serious and costly.[13] But, on the other hand, if consumers with rooftop PV are forced to pay too much for grid connections, they may choose to buy storage batteries, such as Tesla's PowerWall, for their own load-balancing purposes. So, if connection charges become too onerous, PV users may start to "defect" from the grid (Lovins 2014).

The dilemma for regulated public utilities is that (1) overall electricity demand is now declining and may decline further, (2) reliability of supply is the top priority of

---

[13] For a very good analysis of the problems of providing backup power in a future renewable-dominated world, focusing on future projections for Germany and California to 2050, see (Morgan 2015). The lessons from that analysis are much broader.

regulators, (3) demand management by modifying prices by time-of-day is not sufficiently effective, (4) deficits from solar and wind power must be met by some other dispatchable on-line source, either imports, gas-turbines, or storage. All of the "mature" large-scale storage technologies (pumped storage, compressed air, hydrogen, batteries) are quite costly in relation to the amount of time they are likely to be needed. The bottom line appears to be that a significant amount of nuclear or coal-burning backup power is necessary.

One other possibility now under consideration would utilize the EV fleet as a whole (when it gets large enough) as a storage battery to stabilize the grid. The idea would depend on individual owners being paid to allow their car batteries to feed back into the grid up to 2 kWh per car, (and subsequently be recharged) during short periods when the car is not in use. EVs have a storage capacity ranging from 10 to 50 kWh each A fleet of 3 million EVs could be regarded as a single storage battery with a capacity on the order of 6 million kWh, enough to stabilize most short term supply swings (Mackay 2008).

Intermittency is a major problem for both wind power, which operates when the wind is blowing, and solar PV, which operates only when the sun is shining. Even for grid-connected wind or solar PV, local output varies not only predictably according to the season and the diurnal (night and day) cycle, but unpredictably according to weather conditions. Predictability also varies with location. Wind gusts can increase wind speeds very sharply in a few seconds, with dangerous consequences for both turbines and for the grid. The electrical output of wind turbines varies with fluctuating wind speeds. Predictability of fluctuations in wind speed is an issue because gusts are very difficult to predict, even a few seconds in advance. This variability can, in some instances, have a significant impact on the management and control of local transmission and distribution systems. It may constrain integration of decentralized power sources into the power system.

The penetration of RETs would be much faster and greater if the problem of intermittency, for both wind and solar energy, can be solved. Energy storage is the key problem for both wind energy and PV because it is the only sure way to match demand with supply. The only large-scale storage technology now widely available is pumped storage (of water), using two reservoirs at different altitudes. In these systems water is pumped up from the lower to the higher reservoir to store surplus energy when power supply is higher than demand. Water flows down through a hydraulic turbine to the lower reservoir to generate power during periods when local demand is greater than supply. Round trip efficiency in advanced pumped storage systems is now close to 70 %. The disadvantage is that large amounts of land are needed for the reservoirs and there are few good sites near most large cities.

A new technology called pumped heat energy storage (PHES), now in advanced development by Isentropic Ltd (U.K.) promises to equal, or undercut, the levelized cost of hydraulic pumped storage, i.e. \$35/MWh. The PHES uses a new version of the heat pump technology. See Fig. 10.26. It will operate between 500 °C and −160 °C. with a round-trip efficiency of at least 75 %. It requires very little land, and can even be located underground. In my opinion this technology, which can be utilize either at power sources (wind farms, solar farms) or as backup for power

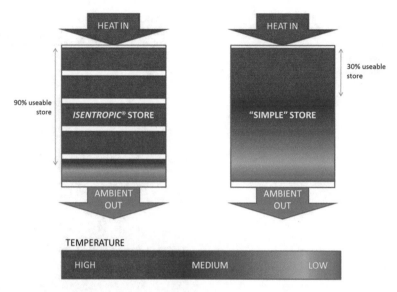

Fig. 10.26 Schematic of Isentropic Ltd.'s PHES system

consumers (factories, hospitals, schools or office buildings) will essentially solve the intermittency problem of wind and PV. It will, in so doing, drastically reduce the need for "backup" power from diesel engines and gas turbines.

A version of this advanced heat pump scheme could also go a long way toward making combined heat and power (CHP) practical for decentralized electricity supply. The problem, there, as mentioned earlier, is that local heat requirements are seasonal (winter) whereas there is a corresponding need for cooling in the summer. Normally that need is met by electric power as applied to air-conditioning systems, but there is no reason I can see why a PHES system could not also be a supplier of "coolth" from the low-temperature reservoir in the summer.

For household rooftop PV, lead-acid batteries can perform a storage function, but for the grid as a whole, they are too slow to recharge after a discharge. For utilities, there are several variants of the lead-acid battery, from several vendors; over 70 MW of these batteries were installed in the US alone by 2010. They are currently cheaper than major competitors, because the technology is so familiar and well understood. But, for several reasons, including the toxicity of lead, I think that is a transitional technology at best.

The high temperature molten sodium-sulfur (NaS) battery, based on a concept developed originally at Ford (for cars), is now operational in fairly large sizes at over 30 sites in Japan, as well as other countries. A schematic of the battery is shown in Fig. 10.27. A 6 MW unit at Presidio Texas stabilizes the power for a long distance transmission line. The biggest application so far seems to be a 51 MW wind power farm at Rokkasho village, Japan, where the NaS system provides 7 h of power (and response time of 1 ms.) The leading manufacturer of NaS batteries is

Fig. 10.27 Sodium-sulfur cell, schematic. *Source*: http://www. twinkletoesengineering. info/battery_sodium_ sulfur_diagram3.jpg

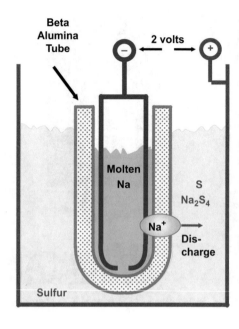

NGK Insulators Ltd. (Japan). The batteries can be installed anywhere, but it is clear that a cheap external source of heat, either in an industrial park or from another storage system, is a potential advantage. The NaS battery could also be a useful "partner" to a PHES system or a solar concentrator with molten salt heat storage.

Another type of battery is the zinc-air cell, which has a great advantage of using cheap and non-toxic materials, but which suffered for a long time from problems with the recharge stage. There are dozens of manufacturers of "button" zinc-air batteries, mainly for hearing aids, but only two manufacturers of rechargeable zinc-air batteries in the MWh storage range. There are two modern versions now being tested by utilities, one from Urban Electric Power and the other from Eos Energy Storage. The latter is being tested by several utilities, including Consolidated Edison of New York, National Grid, Enel (Italy) and GDF-Suez (France). Eos expects that its battery will cost $160 per kWh, for storage, as compared to other battery technologies costing upwards of $400 per kWh. This would be a significant breakthrough, given that zinc is 100 times more available then lithium, and the zinc-air battery requires no rare-earth catalysts.

Geothermal energy is a viable source of power in some locations. The global capacity total today is 12 GW, of which new supply in 2013 added 530 MW. Tide and wave power is another local option in a few locations. The global total amounted to 520 MW at the end of 2013.

There are several technical storage possibilities that have not yet been evaluated in this context. "Super-capacitors" are one. Typical super-capacitors now available can store 6 Wh/kg of weight. A more advanced system promises 280 Wh/kg. Another possibility is flywheels, made of advanced super-strong composite materials. Flywheels now available can store 4.6 Wh/kg of weight. There is no published information on future costs. However, I'll bet that flywheels will find a niche in some locations, especially on railways or trucks operating in hilly country.

The rate of penetration of RETs will be even greater than if (or when) future governments are able to summon the political will to confront the fossil energy lobbies and impose appropriate pollution taxes. (Those taxes could, incidentally, go a long way to pay for public health programs and health-related entitlements that are currently facing increasing deficits.) The increasing rates of return and low risk (due to government guarantees) would make the green energy bonds attractive to insurance companies, pension funds and university endowments, the same customers that bought CDOs created from sub-prime mortgages, because of their (unjustified) AAA ratings.

As mentioned at the beginning of this chapter, there is a "different" nuclear technology that could overcome some of the current objections to solid fuel uranium-based power. I refer to the liquid fluoride thorium reactor or LFTR (Hargraves 2012; Hargraves and Moir 2010). The key feature, as the name suggests, is that the fuel element (thorium or uranium) is dissolved in a liquefied salt (a fluoride for preference) at a high temperature (800 °C) but at atmospheric pressure. The thorium 232 itself is not fissionable, but when it absorbs a neutron from a chain reaction it emits an electron (beta decay) and becomes uranium 233, which is fissionable. Other fission products are generated as the reactions continue, all the way up to plutonium, but in comparatively tiny amounts compared to the uranium chain.

The liquid salt circulates in normal operation. The fluoride is essentially immune to radiation damage, in contract to the fuel rods used in uranium-based power plants.[14] The design includes a "freeze plug" which has to be cooled by a fan. If the power goes off, the fan stops and the liquid salt flows out into a catch-basin where it cools and solidifies. No explosion or "melt-down" such as occurred at Chernobyl or Fukushima is possible. This means that the containment system is far more compact, which means that the LFTR can be made much smaller components, probably small enough to power a jet engine, for example. The high operating temperature permits an electric power plant to achieve 45 % efficiency, as compared to 33 % for a conventional nuclear plant.

Moreover, the quantity of material to be mined and processed and the quantity of waste that must be stored safely for hundreds of generations is far less than with uranium-based reactors with solid fuel rods. The comparison is roughly as follows: 1 ton of thorium is equivalent—in terms of power output—to 250 tons of uranium containing 1.75 tons of U-235. There are no highly radioactive "spent" fuel rods to store. The fission products decay to stability in 300 years, not hundreds of centuries. The fission products, overall, are approximately 10,000 times less toxic than the fission products from uranium-based power plants.

---

[14]. Conventional fuel rods have to be removed after 4–5 years of operation, with only 5 % of the energy extracted. Storing the radioactive fuel rods safely is the major unsolved problem for U-based nuclear power technology.

The technology was actually under development at Oak Ridge National Laboratory (ORNL) from 1965 to 1970, when the program was stopped. It could be re-started. The Indian government is currently doing so. Others may soon follow. The economics of the LFTR are also very attractive, since the enormous containment system and much of the waste recovery and materials processing needed for today's nuclear power plants would be unnecessary.

Part III

Chapter 11
Mainstream Economics and Energy

11.1 Core Ideas in Economic Theory

You readers, having got this far, may wonder why I did not include a section under this heading in Chap. 2, or (on the other hand) why I think that such a section is now relevant. The reason is, simply, that the laws of thermodynamics are central to everything that happens; whereas the core ideas of economic theory (up to now) have never included, or even touched tangentially, on energy or the laws of thermodynamics. One of the purposes of this chapter (and of the whole book, for that matter) is to suggest that this absence needs to be corrected.

The next few pages are an overview of some core ideas of economic theory, and how they have evolved since the eighteenth century. These ideas, and the history behind them, are still important today, sometimes in unexpected ways. Whatever you think about **John Maynard Keynes**, he was unquestionably right when he said: "Practical men, who believe themselves exempt from any intellectual influence, are usually the slaves of some defunct economist" (Keynes 1935, p. 383). Keynes himself is now a defunct economist, but his ideas are still relevant. So, I summarize below a few of the key ideas of other (mostly) defunct economists.

The first comprehensive thinker about economics was **Adam Smith**. Virtually all of the writings of the "classical" economists, beginning with Adam Smith and **David Ricardo**, focused first on markets and trade, and only later on production (Smith 1776 [2007]; Ricardo 1817 [2007]). A major part of the writings of generations of economists since Smith's time has focused on understanding the mechanisms and conditions, such as "comparative advantage", underlying the phenomenon of exchange.

However *self-interest*, as the driver of economic activity, was probably the single most important themes of Adam Smith's most famous work (Smith 1776 [2007]). The term was borrowed from John Hobbes' *Leviathan*, a seventeenth century best-seller. One of the most quoted sentences in *Wealth of Nations* is the following:

... He intends only his own gain and he is, in this as in many other cases, led by an *invisible hand* [italics added] to promote an end that was no part of his intention. Nor is it the worse for the society that it was not a part of it. By pursuing his own interest he frequently promotes that of the society more effectually than when he really means to promote it. Book IV, Chapter II paragraph IX.

Numerous later economists have argued about what Smith meant by some of his words (e.g. "society") and about where he—or others—would have (or should have) recognized exceptions. For instance, in that paragraph, Smith was actually talking about importing versus production at home. Did he really mean that self-interest leads to beneficial outcomes in any and all situations? But there is no doubt that self-interest is central to capitalism.

Adam Smith created a mythic story of how money came about. He postulated "a certain propensity in human nature ... the propensity to truck, barter, and exchange one thing for another." Animals don't do this. "Nobody," Smith observed, "ever saw a dog make a fair and deliberate exchange of one bone for another with another dog." Voila! Smith went on to argue that humans—unlike dogs—are inclined to trade by nature, and that even conversation is a kind of trading. And, in all kinds of trading, Smith asserted that humans seek their own best advantage, i.e. the greatest profit from the exchange. Like Hobbes, before him, he denied all other motivations.

A related topic is "markets". In Adam Smith's day they were village activities, where buyers and sellers knew each other, knew each other's reputations and negotiated individually, keeping tallies of who owed what to whom, and conducting a "reckoning" every six months or so when most debts were cancelled and only the residual was paid off in coin. I suspect that **Robert Solow** would say that "markets" are self-organizing activities involving self-interested, rational traders who—by acting selfishly—will allocate resources efficiently. (If I replaced the word "resources" in that sentence by "exergy" the link to Part I of this book would be obvious.) But markets also allocate resources other than exergy, including labor and capital (money).

The French so-called "physiocrats" in the eighteenth century argued that all traded goods were created by human labor on the produce of the land (mainly agriculture) (Quesnay 1766). It seemed obvious that the monetary value of the goods traded must reflect the labor input. From this assumption the "*labor theory of value*" (Adam Smith, David Ricardo, Karl Marx) emerged. The productive role of physical capital (e.g. tools, machines, buildings and roads or canals) could be accounted for as intermediate outputs, or "frozen" stocks of previous labor. But the value of land itself (and other natural resources) cannot be accounted for in that way.

Karl Marx was the first to try to explain the origin of profit (economic surplus), albeit unsuccessfully. He thought that the origin of surplus arises from the difference between market value of goods (including labor), and their use values. This works for labor, because it produces more value than the minimum (or average) wage. But he mistakenly argued that, for machines, the two must be identical, and equal to depreciation. He—like all before him—failed to notice that useful energy (exergy) plays a role, and that it is the contribution of exogenous energy that accounts for most of the surplus. These facts eventually killed the labor theory of value and led to *utilitarianism* and the "marginalist revolution" in the nineteenth century.

During the time of Adam Smith, the archetypical exchange transaction in the local marketplace was a purchase by an individual customer of some tangible product, such as a loaf of bread or a pair of shoes, from another individual, at a mutually agreed price. The price was usually open to bargaining. (This is still the case in many parts of the world.) The customer could assess the quality of the object by sight, touch or (in the case of food) by smell. The admonishment: *caveat emptor* (buyer beware) was reasonable advice in the larger city markets.

Often the reputation of the seller was sufficient. As time went on, reputation (as in brand-names) has become increasingly important thanks to the complexity of many manufactured products and the impossibility of explicit quality testing of most products by most consumers. Today, reliance on brands and brand names is a substitute for individual assessment of quality for a wide range of manufactured products.

The essential feature of exchange transactions in village markets (or any transaction between two individuals, for that matter) was that both buyer and seller expected to be better off after the exchange. To clarify this point, consider the purchase of a loaf of bread. The baker has more bread than he can eat, whereas with the money he gets for the sale, he can buy flour for the next day and have something left over (profit) that can be spent on dinner. The customer wants the bread for her family's dinner, more than she wants to keep the money for a future rainy day.

So both parties consider themselves better off after the exchange, at the agreed price, than before. *Otherwise no such exchange would occur.* Hence the very existence of trade meant that the people engaging in it were better off than they would have been otherwise. On average, they would all become gradually more prosperous, albeit some faster than others. Adam Smith recognized this feature of exchange transactions and identified trade as the source of "the wealth of nations". The first economist to define precisely the conditions for an exchange was **Carl Menger** (Menger 1871 [1994]). He did it in terms of incremental gains or losses from exchange "at the margin" (the point of indifference).

Of course, you (the reader) knew all that. But saying it in that particular way was very helpful to nineteenth century economic theorists. Notice the terms "increment" and "margin" in Menger's definition. They are important because it turns out that they are the key to introducing differential calculus into economics. That was actually a revolution in the history of economic thought, because it invited quantification of previously intuitive notions like the declining marginal return on investment or consumption.

However trade in which there are two parties, and each party can reasonably expect to come out better off than before, is no longer the norm. In the first place there are too many intermediaries, and the objects of trade are too complex, for both parties to be sure of being "better off". There is a gambling element in every trade, which brings "gaming" into many types of trading. Moreover, trading on a very large scale, as in the currency exchanges, stock and bond market, and the commodity markets, introduces problems that were unknown in Adam Smith's day, namely "financial externalities". See Sect. 11.3.

Here it is appropriate to mention an observation of attributed to Robert Solow, one of the modern gurus of economics: he said that the three central pillars of economics (the "trinity") are "*greed, rationality* and *equilibrium*".[1] By "greed" I suppose that Solow—like Adam Smith—meant self-interest. Nowadays, the textbooks refer to "utility maximization", borrowing the notion of *utility* from **John Stuart Mill** (Mill 1848). Mill used the term in a social context, as "the greatest good for the greatest number". The neo-capitalist mantra "greed is good" is an oversimplified restatement of utilitarian ideas, and the "invisible hand" proposition. In today's economy, it is fairly obvious that greed is not always good for everybody (or even for most people), and that excessive greed has repeatedly led to economic disasters e.g. (Stiglitz 2002, 2012).

"Utility" is analogous to "price" (or "value"), as applied to a product, and it is analogous to "profit" for a firm or "gain" for a consumer engaged in bargaining. Hence it is convenient to assume that *utility maximization* is characteristic of *rational behavior* i.e. that consumers try to maximize utility in all exchange transactions (Debreu 1959).[2] But the *utility* of a good or service is not measurable because it is different for every consumer, and it also varies from situation to situation. Utility maximization, in theory, depends on having perfect knowledge of one's hierarchy of preferences, perfect knowledge of the preferences of other market participants, and perfect knowledge of the future.

As regards *rationality*, Solow seems to have meant the ability to make consistent choices among alternatives, based on such a clear hierarchy of preferences. In economics, a "rational choice" almost always coincides with self-interest, and utility maximization. But it must be said that the assumption of rationality as applied to behavior is one of the weakest in the economic cosmos. It is mainly helpful for writing journal articles, needed for promotion, because it enables economists to use devices and techniques such as Hamiltonians that really are applicable in physics (Mirowski 1984).

"Bounded rationality" is a notion introduced by **Herbert Simon** (Simon 1979). It reflects the existence of limits to knowledge and limits to computational capability. It is very different from *irrationality*. Irrationality generally coincides with human choices made for reasons other than utility or self-interest. Humans are often irrational by economic criteria, meaning that they often do things contrary to self-interest, such as helping victims, feeding the poor, volunteering for the army, or investing on the basis of astrological conjunctions.

However, economic theory generally assumes rational behavior, if only because the other kind of behavior is hard to predict and to explain. Yet irrational behavior by crowds, including "bubbles", revolutions and wars, are obviously very

---

[1] I can't find the original source. Call it "apocryphal".

[2] There is now an alternative microeconomic theory in which utility maximization is replaced by wealth maximization to explain exchange quantities and prices (Ayres and Martinás 2006).

important. Assuming rationality in a world where irrational behavior is rampant, is arguably more dangerous than relying on behavioral rules taken from religion or experience. I will have more to say about this later.

In particular, the assumption that economic agents actually try to maximize utility is now very controversial. It has been sharply challenged by psychologists (Tversky and Kahneman 1974; Kahneman and Tversky 1979), game theorists (Allais and Hagan 1979) and decision theorists (Simon 1955, 1959; Conlisk 1996). In recent decades there has been a lot of experimental work—mainly using games, such as "The Prisoner's Dilemma"—to elicit the behavioral characteristics of people as economic decision-makers. This research demonstrates clearly that real people do not behave like rational optimizers (or maximizers) in many circumstances. In short, the decision criteria of real people are far more complex than standard neo-classical economic theory suggests. The implications of this are still being worked out, with some help from game theory.

Here I think it is important to acknowledge another perspective. Most economists have no trouble thinking that the monetization of human behavior is a good thing, insofar as it implicitly acknowledges the fact that public policy cannot realistically deal with absolute goods or bads. While it is morally repugnant to put a price on pollution, the fact remains that governments do not have infinite resources with which to protect "goods"—including human lives—that are "beyond price". The reality is that, by refusing to attach explicit prices to some things, wildly inconsistent implicit prices can result. Michael Sandel's book "What Money Can't Buy" offers a great many examples of things money cannot buy (like friendship, Nobel Prizes and admissions to Harvard). The same book also provides many examples of actions that are morally repugnant to most people, but that are nevertheless available through gray or black "markets". The sale of child pornography and drugs like heroin are examples of actions that are illegal, but the legal sale of cigarettes to minors and of machine guns to "gun collectors" is arguably worse.

Sandel, like a great many environmentalists (some of them friends of mine), argues against emissions trading and "carbon offsets" on moral grounds, namely that enabling a rich person (or corporation) to continue to pollute by paying a fee to some other person or corporation to reduce emissions creates a bad attitude. Admittedly it would be better if everybody cut emissions voluntarily. But exhortation doesn't seem to accomplish that result, whereas economic incentives may do so. It seems to me that actions taken for profit that cause inadvertent harm to others, especially third parties, are inherently immoral, whereas actions taken voluntarily between two consenting parties leaving both of them better off—by their own standards—are moral. Enough of that.

As regards the third of Solow's "trinity", viz. *equilibrium*, it is usually taken to mean a situation where "supply equals demand". A huge amount of economic analysis to this day revolves around the (false) idea that economies are always in, or near, supply–demand equilibrium and that economies "want" to be in that state. The French economist **Leon Walras** extended the idea to multi-sectoral economies in the 1870s (Walras 1874). He conjectured that a unique equilibrium must exist for such a system. Mathematical proof of the Walras hypothesis (by Arrow and Debreu) waited until the 1950s (Arrow and Debreu 1954). Walras also formalized the idea of *general equilibrium* (but still only for static economies). Almost all large economic models are based on equilibrium conditions, primarily because such models can be solved in a large computer.

Yet a large part of economic theory—and virtually all theories of economic growth—assumes that the economic system is always in equilibrium, even though "growth in equilibrium" is an oxymoron. Equilibrium, after all, whether in physics or economics, is a state where nothing happens. The assumption of equilibrium is mainly for reasons of convenience (i.e. *computability*). However, in reality, this assumption is often dangerously false, especially in terms of analyzing phenomena such as "bubbles".

A few more words are needed here on the problem of aggregation. Clearly, much of economic theory is based on relationships between abstract aggregates (capital, labor, money supply, unemployment, inflation, etc.) and change-inducing actions of those abstract aggregates (GDP, production, consumption, saving, investment, etc.) Abstraction is not, by itself, a problem. As philosopher-mathematician Alfred North Whitehead once pointed out:

> You cannot think without abstractions; accordingly it is of utmost importance to be vigilant in critically revising your modes of abstraction... (Whitehead 1926)

Aggregation is one mode of abstraction that is arguably essential to economic science. But as Whitehead also says, "one must be vigilant in deciding when and how it shall be done."

The notion of aggregation is worth attention, if only because the procedure is taken for granted.[3] Its explicit purpose in economics is to construct variables (say output Y, capital K, labor, L and energy E) suitable for quantitative analysis and mathematical operations. For instance, GDP (Y) (meaning "gross domestic product") is defined as the sum of all the value-added in the economy, which is defined as the total of payments to labor (salaries and wages) plus payments to capital (profits, interests, rents and royalties).

There are deep and well-known problems in the creation and interpretation of heterogeneous aggregates (such as "labor" and "capital stock").[4] Setting aside such problems, the need to do mathematical analysis means that the constructed variables must be smoothly varying, parametrizable and *twice differentiable* (in order to do analysis). The differentiability conditions are non-trivial because they define the conditions for equilibrium: namely that the integrals along any two paths from point **A** to point **B** in "factor space" must be equal, hence *not path-dependent*. (The same condition defines equilibrium in thermodynamics.)

---

[3] The next few paragraphs are lifted from a paper of mine (Ayres 2014a, 2014b).

[4] See, for example, the so-called "Cambridge controversy" involving Joan Robinson, Robert Solow (and others) e.g. (Robinson 1953–1954).

Mathematically, twice-differentiability means that first and second derivatives are continuous and smooth. But there is a more precise condition: the second-order *mixed* derivatives of a function $Y(x, y, z)$ with respect to all its factors (x, y, z) must be equal. This condition can be expressed as three partial differential equations involving the "output elasticities" of the variables, which are the logarithmic derivatives of Y with respect to capital (K), labor (L) and exergy (E) or useful work (U). The third equation becomes trivial if Y satisfies the *Euler condition* ("constant returns to scale"). The Euler condition, a purely mathematical constraint, means that large economies (like the US) are not necessarily more efficient than small ones (like Switzerland), even though this is transparently not true for firms. In mathematical terms, this means that the sum of the output elasticities must be equal to unity. Well, I had to mention that, but you don't need to know more about it. Not today, at least.

The other two equations have generalized solutions that are functions of ratios of the factors $(L/K, K/E, E/L$ or similar ratios involving $U)$. These differential equations can be solved explicitly in some (limiting) cases. The so-called Cobb-Douglas production function is the simplest solution of the differentiability conditions, where each of the output elasticities is a constant. Other solutions with non-constant elasticities are more realistic, however.

The aggregation process involves three different simplifications.[5] One is the assumption that all objects in the economic system belong to a finite number of categories, or sub-categories. Thus all workers may be assumed to belong to the category "labor", or "manual labor" notwithstanding great differences within the category. Capital stock may be subdivided into active components (machines) and passive components (buildings, pipes, infrastructure). There are other ways to subdivide.

The second simplification is the assumption that the objects in the category are characterized in terms of some measurable quantity of interest, such as monetary value, or an input (e.g. of money or labor or exergy) or an output (e.g. work done or tonnes of product). The third simplification is that there is a natural time period—such as an hour or day or month—within which the objects belonging to the category can be regarded as fixed. In other words, it is important to be able to assume that *changes only occur between periods*. It is also assumed that the time periods of no-change are so short that incremental changes between successive periods are very small (i.e. they can be regarded as differentials). Then the quantities and the changes from period to period can be added up or divided by time increments and expressed as *rates* of change. The aggregate is then definable at a given actual (clock) time, and it can then be considered as a variable for analytical purposes.

---

[5]. I am grateful to Katalin Martinàs for calling my attention to these points.

The marginalist revolution of 1870 or so is traditionally attributed to three economists, Carl Menger, William Stanley Jevons[6] and Leon Walras, but should perhaps more accurately include men from a later generation such as Alfred Marshall, Philip Wicksteed, Eugen von Bohm-Bawerk, Knut Wicksell, Vilfredo Pareto and Irving Fisher (and others less well-known. It was the marked by the publication, over a 3 year period, of papers recognizing the declining marginal utility of consumption (Blaug 1997, Chap. 8). This effectively put an end to the "classical" assumption that value is intrinsic to products (or services) and is determined by labor and capital inputs.

More specifically, classical economists (Smith, Ricardo, Mill, Marx) attributed the surplus (land-rentals) over the cost of agricultural labor; wages of labor were determined by the long-run costs of production of consumer goods, and the rate of profit was a residual. The value of land and capital were determined by different principles from the value of products (e.g. labor input at subsistence wage rates). After the marginalist revolution, value was determined by market prices, which were determined by supply and demand by consumers, not by costs of production. In short, prices were determined by scarcity in relation to wants, which are determined by preferences. This was where the marginal utility of consumption entered the picture.

Economics also evolved, in the late nineteenth century, due to the application of differential calculus to marginal relationships. The idea of a demand function was introduced by Cournot (1838). Indifference curves were introduced by Edgeworth (Edgeworth 1881). Calculus, in turn, permitted explicit maximization algorithms, which were adopted by economists around this time. Mirowski argues that this adoption was a deliberate imitation of "energetics" using the analytical techniques from Newtonian mechanics introduced by Lagrange in the 1780s and by Hamilton (in the 1840s). Energy got into economics indirectly through the widespread adoption of Hamiltonians, which were originally mathematical are expressions of the conservation of energy. The idea of optimal growth in economics also arrived with calculus and Hamiltonian dynamics. It was taken a step further by Frank Ramsey (1928).

Profits (in models) are the difference between prices received and costs incurred, in money terms. All of these are measurable, both at the micro scale and the macro-scale. At the beginning of the nineteenth century the productive role of profits was doubted and was debated: Socialists regarded profits as theft, while the importance of credit and debt ("circulating capital") was—and still is—harder to account for theoretically. Profits are now seen by economists as the motor of the economy, but the role of money (and banks) in money creation is still controversial. Hamiltonian dynamics permits profit maximization models.

---

[6] Jevons, one of the pioneers of marginal analysis, was the first economist to discuss energy explicitly (actually he was talking about coal). In fact, he wrote a best-seller called *The coal question: Can Britain survive?* (Jevons 1865 [1974]).

The next useful insight began with the recognition of a conflict (in principle, at least) between *current consumption* and *saving for the future*. In 1920 Arthur Cecil Pigou suggested that people—and societies—might save and invest too little because of a tendency to *discount* the future, due to *myopia* (short-sightedness) (Pigou 1920). It was assumed, at the time, that capital for investment was created by savings from current income by households or businesses. Pigou, who was also one of the first to focus on welfare as such, was the first to mention the possibility of external third party effects of some transactions. These are now called "externalities".

In Pigou's book on welfare economics, he focuses on cases where "marginal social net product" might differ from "marginal private net product". For instance, Pigou noted that an oil driller has no incentive either to maximize total output from the oil field (by pumping slowly) or to minimize consequences of declining pressure due to his private production rate on other drillers in the same field. A similar problem arises in connection with grazing on common property: nobody has any incentive to conserve because the others won't. The most famous paper on this problem is Garrett Hardin's "*The Tragedy of the Commons*" (Hardin 1968).

The mathematical notion of "optimal" growth, was set forth by Frank Ramsey's seminal paper "*A Mathematical Theory of Saving*" (Ramsey 1928). He worked out how much investment would be required each year to maximize the sum total of utility (equated with consumption) over a very long time.[7] Ramsey's optimal growth theory did seem to confirm Pigou's myopia hypothesis.

11.2 On Credit, Collateral, Virtual Money and Banking

Securitization of bank-loans by land (e.g. mortgages) seems to have been a European innovation. In the fourteenth century Italian bankers made securitized loans to long distance traders (for shares in the profits) and to governments (for monopoly privileges), but played no part in local or village commerce. However, in terms of monetary value, most of the trade in most countries took place in village or town market places. The needs of local entrepreneurs or aristocrats with bad luck or other financial problems were met, if at all, by money-lenders under some other name (e.g. pawnshops, goldsmiths). The innovation of fractional reserve banking may have originated with goldsmiths, who stored gold for wealthy families or merchants, and who made short-term loans "backed" by that gold.

Credit has been offered in other countries at other times without security in the form of land or gold. The credit cards offered by banks to almost anybody today are one example. The "security" in this case is the fact that consumers have "credit ratings", kept in a database accessible to the banks. Those borrowers who repay

---

[7] Ramsey did not believe in discounting and he found a clever way to avoid the need to introduce discounting into the calculation. But the mathematical details are irrelevant.

loans get satisfactory credit ratings, but people who do not repay on time (or at all) will be unable to get credit in the future. (This is currently the only constraint on sovereign international borrowers.)

The first bank in China (Rishengchang Bank, founded in 1826 in the town of Pingyao, Shanxi Province) made loans only on the basis of local reputation for personal integrity. There was no securitization. The bank charged interest of 6–8 % per annum (while paying depositors 2–4 %) and it prospered greatly. Pingyao became a financial hub. At peak there were 22 competing banks in the city. By 1900 the Rishengchang bank, alone, had branches in at least 50 cities in and outside of China.

Rishengchang bank (and some others) made the mistake of lending large sums to the last Empress of the Ming (Qing) Dynasty. The bankers did not believe that such a person would default. They were wrong. The bank failed in 1911 when the Empress was deposed by the revolutionaries under Sun Yat Sen, and could not repay. The losses from that huge bank failure must have decimated the rising Chinese middle class, both depositors and entrepreneurs alike.

But there is another question arising from that story. Evidently, the only way a financial system without collateral for security could operate in ancient times, or in nineteenth century China, was by equating debt-repayment with morality. It seems that this idea is still deeply ingrained, both in China and the West. Debts between people in a modern country can be collateralized by property, but only moral obligation (or military force) guarantees repayment of international loans today.

Questions arise: Was the war reparations imposed on Germany by the Treaty of Versailles a moral obligation incumbent on all of the German people? When dictators in Africa or Latin America borrowed money from U.S. banks that were flush with petro-dollars in the 1970s, and then sent the money to private accounts in Swiss banks, should those debts be charged to the ordinary citizens of those countries? Who, exactly, is morally responsible for the current Greek debts, incurred partly by corrupt politicians (with help from Goldman-Sachs), that the German Finance Minister Schauble is so insistent on being a national obligation that absolutely must be repaid in full? The morality of debt and lending is a topic for serious debate. For more on that, see *What Money Can't Buy* (Sandel 2012).

One of the first "laws" of monetary economics—long before Adam Smith—has been attributed to Sir Thomas Gresham (1519–1579), although it was known to others from Aristophanes to Copernicus. Gresham was a financier based in Antwerp who worked for Henry VIII and later for Queen Elizabeth. (He also co-founded the London Exchange.) Gresham's "law", stated formally, is that when a government over-values one type of money and undervalues another, the undervalued money will disappear from circulation. It is usually stated: "bad money drives out good". Gresham wrote about this problem in a letter to Queen Elizabeth on her accession in 1558 to explain why gold coins had disappeared from circulation in England, due to the debasement of silver coinage by Henry VIII. (He didn't mention the export of silver to China.)

"Virtual" money, consisting of a variety of promissory notes and paper curren-cies, were dominant in Europe prior to the influx of silver and gold from W. Africa, the Tirol and the Americas in the early sixteenth century. But taxes in Europe then were levied in silver, and most of the silver from the Americas was being shipped to Ming China, where it was also used for currency, and where the price in gold was higher. This was why silver was actually so scarce in England in the late Middle Ages (Graeber 2011).

The use of shares as virtual money was an innovation by John Law, a Scottish financier living in France. His ingenious idea (c. 1720) was to fund the huge French national debt (left over from the reign of the "Sun King", Louis XIV) by selling shares in a future trading monopoly based on the territory claimed by the French Crown in the Mississippi Valley.[8] The potential profits were hypothetical. But their potential was "hyped" and a large number of worthless shares were issued. The result was the "Mississippi Bubble". A lot of the rising French bourgeoisie lost their savings in that bubble.

A few years later, England adopted the same idea for paying off the British national debt by selling shares in a trading monopoly. The monopoly was for trade with Chile and the west coast of South America. The result was the so-called "South Sea Bubble". Those two financial disasters arguably led (albeit gradually) to the widespread adoption—led by the Bank of England—of the gold standard as a base for the national money supply. At first, large banks were allowed to create paper money based on gold held in their vaults. Later, the Bank of England, the Banque de France, the US Treasury and other central banks took over the function. Paper money was theoretically redeemable in silver, or gold, until 1933, when the US government unilaterally stopped the issue of gold coinage and prohibited private ownership of gold. The "gold standard" finally failed in 1971, when the Bretton Woods treaty collapsed.

Today paper money, from a centralized government mint is only backed by the "faith and credit" of the government (or the central bank). Now that gold is no longer a monetary base, banks have become independent agents of money creation by providing credit not linked to deposits or reserves. But excess leverage and excess debt has also been the cause of a series of financial catastrophes, most recently the financial melt-down of 2008 (Ayres 2014a, b). That mechanism has not yet been well enough understood, still less taken into account, by economists and governments today.

Meanwhile, the importance of credit as the driver of economic growth is only now being fully realized. Portuguese and Spanish colonies and conquests in the fifteenth century were financed by the Crowns of those countries. But it was credit from individual investors, through joint stock companies, that financed the long-distance trading voyages by English, French and Dutch explorers, traders and empire-builders in the sixteenth, seventeenth and eighteenth centuries. British and European banks financed much of the westward expansion of the US in the nineteenth century, and the global race for oil in the twentieth century. It was the

---

[8] The enormous territory in question was later sold by Napoleon Bonaparte to the United States, in 1804 when Thomas Jefferson was President. It was the "Louisiana Purchase".

land redistribution of Japanese farmland during the post-war American occupation that financed Japanese recovery in 1955–1990. The nationalization of Chinese land by the Communist government under Mao Tse-Tung, and it's resale to developers and mortgaging by banks under subsequent governments, financed the real estate boom that drove unprecedented growth in China in the first decade of the twenty-first century.

11.3   On Externalities

Externalities are "third party effects" associated with exchange transactions. In the eighteenth century, externalities were of minor importance. The examples that come to mind are mainly in the realm of public health and the spread of sewage-borne diseases like typhoid fever and cholera, the spread of mosquito borne diseases (malaria, yellow fever) and diseases spread by rats and fleas (plague). These diseases were a constant problem in densely populated cities. They were mostly abated by public investments in the water supply system and the sewage disposal system. The other examples of urban externalities were congestion, crime and fires. The abatement measures involved public investments: roads, bridges, police and fire fighting organizations. But in rural areas, those externalities were negligible or non-existent. Externalities were rare in the eighteenth century, but they are more and more important—even dominant—in economic life today.

The disconnect between private and public welfare with regard to common property resources (like fisheries or grazing land) was generalized in Arrow's "impossibility theorem" and work on "social cost" and "public choice" in the 1950s and 1960s (Arrow 1951; Scitovsky 1954; Coase 1960; Davis and Whinston 1962; Buchanan and Tullock 1962). An underlying theme of several of these papers was that negative incentives resulting from competition and congestion are increasingly important in modern society as compared to the importance of cooperation in a Jeffersonian society. Coase and Scitovsky were especially relaxed about pollution problems like "smoke nuisance", for instance, on the grounds that such problems are rare to begin with or (Coase) they can be resolved by means of regulation or tort law.

Externalities—third party effects—were not explicitly recognized by economists until the twentieth century. The formal definition of an *externality* is a cost (or benefit) incurred by a party who did not participate in the action causing the cost or benefit (a third party). In the older literature (Pigou 1920; Marshall 1930) an example of a negative externality might have been a fire in a farmer's haystack caused by the sparks from a passing steam locomotive. A real positive externality might be the pollination services performed by the bees of a bee-keeper for neighboring farmers. The most familiar externalities today are pollution and climate change (associated with the consumption of fossil fuels), but "busts" and "bubbles"—and subsequent financial crashes—are also major externalities, as pointed out in Ayres (2014a, b).

When a few economic theorists began to think along these lines nearly a century ago, they assumed that the externalities were minor and that any harmful consequences could be compensated by taxes (Pigou 1920), or legal processes (torts), if legal costs were negligible (Coase 1960). But the realities of the legal process make legal remedies for pollution difficult or impossible for an individual to achieve against a giant corporation, whence Pigovian taxes become a "second-best" solution, provided the distortion of market relationships is not too great

But Pigou and Marshall were not thinking in terms of environmental pollution or climate change as externalities. Nor were they thinking about the social consequences of bank failures or financial bubbles, even though the social consequences were huge (e.g. "The Grapes of Wrath"). In the 1960s and 1970s environmental problems like "smog" in Los Angeles, or "Minimata disease" in Japan, gained more attention—and were recognized—as externalities. Economists in the 1960s began to realize that externalities are not necessarily small and unimportant (Kneese et al. 1970)

Some externalities involve very serious health damages to millions of people, and degradation of the natural environment (Kneese 1977). Rachel Carson's "Silent Spring" told the story of how pesticides used by farmers to kill destructive insects and increase food output, also killed off the birds that ate the insect eggs and grubs (by making their egg-shells too thin) (Carson 1962). The additive, tetra-ethyl lead (TEL) that enabled gasoline engines back in the 1920s to operate at much higher compression ratios (increasing both power output and fuel efficiency) turned out to be so harmful to the environment and human health that it had to be banned.

The use of fluorocarbons as propellants in spray cans and as refrigerants has caused an "ozone hole" in the stratosphere. That phenomenon sharply increased the level of dangerous ultra-violet (UV) radiation on the ground, especially in the arctic and Antarctic regions. Fluorocarbons have been sharply restricted for this reason. Oil spills, like the recent one in the Gulf of Mexico, are becoming more dangerous and costly as drilling goes into deeper waters.

The most dangerous externality of all is probably the increasing concentration of so-called "greenhouse gases" (carbon dioxide, methane and nitrous oxide) in the atmosphere.[9] This buildup is mostly due to the combustion of fossil fuels. It has the potential for changing the Earth's climate in ways that would harm people in large parts of the world. Powerful storms hitting coastlines in unexpected places may be only the first hint of what is coming.

In order to make rational decisions about policy, economists argue that there is a need to put a monetary cost (Pigouvian tax) on environment damages because it is not enough to say that a landscape or a coral reef is "priceless" in a world where funds for social purposes as well as environmental protection are limited. More

---

[9] There are several other more powerful greenhouse gases, such as fluorocarbons and sulfur hexafluoride (SF_6), that are released in very small quantities. Altogether they account for about 2 % of total emissions, in terms of carbon equivalent. Ignoring water vapor, carbon dioxide accounts for 84 % of the GHGs and all but 2 % of that is from fossil fuel combustion.

specifically, it would be very helpful for governments to know the external costs of specific kinds of pollution, especially Greenhouse Gas (GHG) emissions. The obvious example is coal combustion. A recent study at the Harvard Medical School gave low, medium and high estimates of the unpaid social costs, mainly health related, of coal combustion in the US. Their "low" cost estimate was $175 billion per year (Epstein et al. 2012).

This calculation has been criticized on the basis that it includes some future climate costs that are controversial because of underlying assumptions about discount rates. I think the "low" estimate is really too low. But another study entitled "*Hidden Costs of Energy*", by the US National Academy of Sciences-National Research Council, did not include any allowance for GHG emissions, mercury emissions, higher food costs or national security issues. It still set the cost of coal combustion on human health at $53 billion, per year, just from sulfur oxide and micro-particulate (smoke) emissions (NAS/NRC 2005).

These numbers, which are conservative, compare with the total *revenues* of the US coal industry, which are around $25 billion per year (Heinberg 2009). If these unpaid costs were added to the price of electricity, even taking the too-low NAS/NRC estimate, they would just about double the price of electricity per kilowatt. Moreover, if the situation is bad in the United States, it is far worse in coal-burning Asian countries, especially China and India. A recent Chinese study pointed out that exposure to coal smoke has cut 2–3 years from the average life-span of workers in North vs. South China. It is clear that negative environmental externalities, are far from small.

Environmental economists usually propose that unpaid social and health costs should be "internalized" by adding them to the price of coal. If that were done (by means of a tax or penalty), coal would cost from 3 to 20 times as much as it does today, and the industries that use it would be far more inclined to support the development of other sources of energy, such as renewables. This has not happened, to date, because of the enormous political power of the coal mining and using industries (e.g. electric power companies with old coal-fired plants). So the market failure continues. The "brass knuckles" of the "invisible hand" are quite clear in this case.

11.4 Economics as Social Science

It was not by accident that I omitted **Thomas Malthus** in Sect. 11.1. In the first place, he was not much interested in economics as a discipline. He was a very narrow specialist. His one and only intellectual concern was population growth, its consequences, and what to do about it. But, in the second place, he was the first economist to worry about the availability of natural resources (agricultural land). He foresaw unlimited (exponential) growth of population, and consequently of demand confronting slow (at best) growth of the food supply. He pointed out that as population approaches the limits of carrying capacity (for food crops), prices

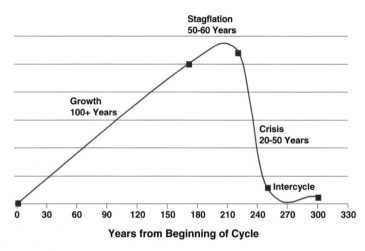

Fig. 11.1 The Malthusian "secular" cycle

rise, real incomes fall, small businesses fail and poverty increases. As a consequence, death rates increase and birth rates decline, so population declines until the
supply and demand for food is balanced again. Moreover, any period of bad
weather—drought or flood—can cause local or regional famines, resulting in civil
unrest, civil wars and widespread destruction. (His gloomy analysis may have
inspired Thomas Carlyle's characterization of economics as the 'dismal science').
Be that as it may, more recent studies by a variety of authors in recent decades,
emphasizing this long-term concern, have been labelled "Malthusian". See
Fig. 11.1. The best-known example of such a study is "Limits to Growth"
(Meadows et al. 1972, 1992).

Before moving on, I should point out that the Malthusian population growth and
decline "model" has actually played out a number of times in human history,
though the statistical evidence of long population "cycles" has only recently been
pulled together by social scientists and economic historians, especially Turchin and
Nefedov in their book *Secular Cycles* (Turchin and Nefedov 2009). Turchin has
introduced mathematical models of population cycles. He also argues that population peaks are often followed by periods of warfare, although the direction of
causation is unclear. Causation may go from warfare to population decline as
well as the converse.

This pattern is most evident in China, which has experienced a series of
dynasties beginning 4000 years ago. Dynastic changes generally occurred because
of uprisings triggered by food shortages. The first unification with central authority
came about in 221 BCE. At the peak of the Han dynasty (140–87 BCE) it's capital
Xi'an had a population estimated at 2 million, active trade with Europe (the silk
road), a highly developed civil service and as much as 60 % of global world
product. But in 220 AD the Han dynasty broke up into three parts. (Dynastic
changes did not occur on precise dates, because they usually involved military

campaigns that lasted some time.) The provinces were gradually reunited from 265 AD to 618 AD under the Jin, Southern and Northern and Sui dynasties, reaching another period of peak prosperity and central government under the Tang dynasty (618–907). But it all came apart again starting with a military revolt in 907. The next three and a half centuries were violent and disunited, to say the least.

Kublai Khan conquered and re-united China in 1271. China continued under centralized and relatively effective governments (Yuan and Ming dynasties) thereafter until 1600. Apparently the "little ice age" affected China, where the population declined by 50 million (30 %) between 1600 and 1644 when the Manchu invasions started.

Europe suffered comparable periods of rising prices and falling populations. There was a period of rising prosperity in the thirteenth century, followed by population decline (due to the black plague) in the fourteenth. There was another period of increasing prosperity in the sixteenth century, followed by decline, or equilibrium (depending on the country) in the seventeenth. The 30 Years War in central Europe (1618–1648) also adversely affected agricultural production, and Germany lost from 15 % to 30 % of its population due to famine and disease during those years. There were extremely bad harvests in France (1693). There was another period of rising prosperity in the eighteenth century, ending with the Napoleonic wars. Harvests were bad in much of the world in 1815 ("the year without a summer"), due to the massive eruption of Tomboro, in Indonesia. There is a lot of debate about what caused what during this period.

This series of crises corresponded to the "little ice age" (1550–1850), during which temperatures were, on average, 0.6 °C below norms, although there were major regional differences. In colder periods growing seasons were shorter, cultivated land declined, wheat prices rose dramatically and malnutrition was widespread (Parker and Smith 1997). The malnutrition has been confirmed by recent studies of skeletons. The nineteenth century saw a return to greater prosperity thanks to the industrial revolution, but the demographic contribution is less clear because other factors were at work.

It is clear from the historical evidence that the periods of prosperity alternated with periods of chaos, during which population pressure, food shortages and high prices played a large part. It is also clear that during the chaotic periods, agricultural innovations of various kinds must have increased the carrying capacity of the land, if only because the population of China in the nineteenth century was far greater than it had been during earlier periods of famine. The historical pattern exemplifies Schumpeterian "creative destruction".

Several philosophers have elaborated this pattern into theories of the rise and fall of civilizations. One of the first was Nikolai Danilevski (1822–1885), one of the promoters of "Pan Slavism". He was a creationist, who disagreed with Darwin and believed that language-cultural "types" are like species that cannot interbreed. Each civilization, in his view, has its own teleological "life cycle", starting with infancy, adolescence, maturity and decadence. In his view, the Slavic culture was destined to overtake the "West". Oswald Spengler, German philosopher and author of *Der*

Untergang des Abendlandes (The Decline of the West), got some of his ideas from Danilevski. Spengler also regarded "civilizations" as species, with natural life-cycles.

While Danilevski and Spengler are no longer taken seriously, they influenced later sociologists, including Vilfredo Pareto (Pareto 1916, 1971), Pitirim Sorokin (Sorokin 1937 [1957]) and perhaps Joseph Tainter. Tainter's view is explicitly focused on complexity, energy economics and diminishing marginal returns on investments (closer to the perspective of this book) (Tainter 1988).

A different perspective on the rise and fall of nations has been put forward by Mancur Olson in two landmark books: *The Logic of Collective Action* (Olson 1965 [1971]) and *The Rise and Decline of Nations* (Olson 1982). Olson's thesis is very similar to that of Holling as regards ecosystem cycles, except for obvious changes in terminology. See Fig. 5.10 in Sect. 5.10. Starting after a disaster of some sort— e.g. a forest fire in the ecological case or a war or depression in the human case—the system grows fast, at first, because most of the barriers that were present during the previous "climax" phase, have been eliminated. But as growth continues, more and more of those growth inhibitors also reappear. In the case of the ecosystem, the growth inhibitors are the longer-lived but slower growing shrubs and deep-rooted trees that capture the most water and the most sunlight. In the case of a firm or a nation, the inhibitors are competitors, government regulations, and long-term investors, such as pension funds and insurance companies, favoring dividends rather than growth.

As time passes, those organizations combine to "lock in" a given development path and prevent or discourage departure from that path (innovation). The same happens in management. It is clear that no manager can deal personally with more than seven or eight immediate subordinates. For this reason a firm employing thousands of people is likely to have at least half a dozen "levels" of management, and an army or a civil service may need ten or more. The more layers of management there are, the longer it takes to make (or approve) any decision.[10]

But bureaucracies do more than interpret and filter commands from the top and advice from the bottom. They protect themselves. Access to the top, in governments and large firms, is fiercely protected by those who have it. The net effect is to restrict and control the information flow both ways, in the interest of those constituting the chain of command.

Similarly, in democratic societies, all sorts of organizations form to promote or prevent laws, regulations or initiatives contrary to the interests of the group. A law may be proposed in a speech, and it may be written by a team of lawyers in a Congressional office, in a month or so. But then it must be subject to comment and modification in committees before being voted on by the whole legislative body. In

---

[10] When IBM was facing a challenge from small "back of the garage" based firms in Silicon Valley c. 1980, it tried to make itself more "agile" by creating a new start-up division reporting directly to the Chairman and by-passing the main organization. It worked, for a while, until the internal bureaucracy fought back. MicroSoft, a small software start-up with only one objective, was the winner.

a bicameral legislature, this process is repeated at a higher level, and then comes a meeting to negotiate a compromise between two versions. After months, if not years, the proposed compromise law is voted on. After that, the chief executive may approve it, or not. If there is a veto, the whole legislative process starts again.

But if the new law is approved, that is only the beginning. Before it can be enforced. rules need to be written covering every contingency and reduced to a set of regulations. Those regulations need to be submitted to "stakeholders" for their comments. This can take many months. Then, the final version is transmitted to regulatory agencies. There, responsibilities for enforcement are allocated and the citizens or other parties affected need to be informed as to what is required, when, where and who to ask in case of questions. Many laws are challenged in the courts after this stage. In the case of international agreements, the process is similar, if not more complex. Cultural differences and linguistic misunderstandings get in the way. Heads of State may threaten to use military force, or they may actually do so.

The bottom line is that as firms get older and larger, and as countries get older (in terms of time elapsed since the last war, revolution or major governmental re-organization), it is harder and harder to make, and enforce, important decisions. This is why small "agile" firms can sometimes win against large, established rivals (as Microsoft and Apple defeated IBM). It is why the losers of a major war may recover faster than the winners (as Japan and Germany recovered faster than Great Britain after WW II), and it is why China is now growing much faster than Europe and America.

Mancur Olson's theory (summarized briefly above) amounts to a cyclic model of organizational and economic behavior, just as Holling's cyclic model explains patterns of ecological succession. These models explain many otherwise puzzling phenomena. They also make the point that change in both of these domains is inherently a cyclic phenomenon. Change is never smooth, gradual or monotonic.

11.5 Economics as an Evolving Complex System

This is the title of a book (Anderson et al. 1988) that deserves more attention than it has received among economists, probably because two of its three editors were physicists. (The third was Kenneth Arrow, a Nobel economist). It is actually the proceedings of a workshop held in August 1986 at the Santa Fe Institute (SFI), in New Mexico, which became volume 5 of a series of studies in the sciences of complexity. The introduction to the volume, by Prof. David Pines, attributes the idea for the workshop to a suggestion by John Reed, Chairman of CitiCorp, to Robert McCormick Adams, Secretary of the Smithsonian Institution and Vice Chairman of the SFI Board of Trustees. Reed suggested that SFI should "examine the potential application of recent developments in the natural sciences and computer technology to the understanding of world capital flow and debt" (Anderson et al. 1988, p. 3).

Topics discussed within the workshop covered a wide range, from pure theory of games, theory of evolution (ontogeny as a constraint on change, e. g. tendency to polarization?), role of foresight in removing positive Lyapunov coefficients (stability, chaos, etc.), meaning or predictability, etc. But there were also workshops on securities markets and excess volatility, the global economy, monetary and real magnitudes in international flows, gold standard vs fixed exchange rates, floating rates, business cycles, leads and lags.

From the perspective of 2016, thirty years after the workshop, it is clear that economic theory, as taught in the universities, and as practiced in the larger world, has not experienced a great epiphany resulting from the work at SFI. However, from the perspective of this book—with the word '*complexity*' in the title—the workshop showcased at least one new and fundamental insight that is very gradually being accepted by the mainstream.

That insight came from Brian Arthur, whose early works were mathematical studies at the International Institute for Applied Systems Analysis (IIASA) with M. Ermoliev, M. Yu and M. Kaniovsky on so-called "urn processes". See Arthur et al. (1987), Arthur (1988) and references therein. The key insight from those mathematical studies was that outcomes can be highly dependent on the path (or trajectory) of a complex dynamical system. This was one of the studies that contributed to "chaos theory" (e.g. Gleick 1987). But from an economics perspective, it led to the observation that *increasing returns to scale*—as contrasted with decreasing returns—plays a crucial role in economic theory. He elaborated on that observation in the first paper of the workshop at SFI in 1986 (Arthur 1988) and his later book, increasing returns and path-dependence in the economy (Arthur 1994).

The key insight of his work (in my opinion) is that self-reinforcing mechanisms, analogous to auto-catalysis, resulting from positive feedbacks in non-linear systems, are common to physics, chemistry and biology as well as economics. In all of these domains they can lead to multiple equilibria (complexity). The selection process by which one possibility emerges from among multiple equilibria is obviously important in all of these domains, but the practical applications are especially apparent in economics. In particular, short-term selection criteria (optimization) can lead to "lock in" of inferior technologies and "lock out" of better ones. This question arises in connection with investment choices: which stock to buy, which start-up to finance, which technology to support.

Conventional economic theory is largely built around declining marginal returns, which are negative feedbacks, resulting in an approach to equilibrium. This is why neoclassical economics cannot explain growth as an endogenous process. Clearly, economic growth does not occur in equilibrium. Hence economists have been forced to assume that growth results from exogenous factors, especially technological innovations (e.g. Schumpeter). Brian Arthur has shown that growth—and other economic phenomena—can be explained by invoking positive feedbacks, such as increasing returns to scale or to learning and experience.

In fact, examples of increasing returns can be demonstrated in many fields of economics, such as trade theory, spatial economics, and industrial organization. It is tempting to assume that the greater the range of choice among multiple equilibria—

many possibilities—the better the outcome. This is not necessarily true. But perhaps the most obvious example is the phenomenon of "lock-in/lock-out" where choice among technologies or competitors or business models becomes irreversible once initial advantages (e.g. learning or experience) kick in (Arrow 1962). This phenomenon can (and almost certainly does) result in sub-optimal or even bad outcomes. An obvious example is the fact that certain English-speaking countries have been unable to adopt the metric system of measurement, despite its demonstrated superiority. The "lock-in" of uranium-based nuclear power technology (and the "lock-out" of a much superior thorium-based technology) is another example.

In fact, "lock-in" phenomenon is applicable to political choices as well as technological ones. The electoral system in the US, which gives excessive power to small states and rural constituencies, is one example. There are more dangerous ones, including the dynamics of arms races and the fact that it is much easier to start a war than to end it. Evidently small provocations can result in disastrous outcomes, if short-term competitive dynamics leads to a cul-de-sac or "over a cliff". The world as a whole is now facing a situation like that, where unrestrained short-term competition to extract scarce natural resources could result in an "Easter Island" scenario.

11.6 Resources and Economics

The purpose of this section is to establish beyond doubt that (1) resources are a legitimate and important topic in economics and (2) that energy (as exergy) is an important resource. More accurately, it is the *ultimate resource*, the substance of every other resource. Mass is energy in a condensed form. Food is a form of useful chemical energy (exergy) i.e. energy that can be metabolized and enable us to perform work. Everybody knows that the economic system cannot function without useful energy to make the wheels turn. Every time the price of oil goes up the GDP goes down, and when the price of oil goes down the GDP goes up. But in theoretical economics (which does matter, as Keynes famously noted) exergy, as the origin of wealth and as the driver of economic growth, is a core issue that has been neglected.

There is a relatively recent branch of economics which deals with physical and biological resources such as metal ores, fossil fuels, land-use, forests, fisheries, and gene pools. Still more recently "ecological economics" includes larger concepts like planetary limits, biodiversity, species extinction and thermodynamics. However, these offshoots of the main trunk of the discipline are not very influential in government and business.

Unfortunately, the word "energy", as now understood in physical science, is awkward because the word is also used in a variety of ways that mean very different things (Sect. 3.9). In ordinary language, energy is often confused with *power* or *work*, two words that also have a variety of confusing and inconsistent meanings. Power, as applied to people or machines, is energy consumed or applied, per unit time.

The word "work" was once associated with physical labor, but today it means activity—physical or mental—associated with earning a living (e.g. farm-work, office-work). However in physics, physical work is energy transferred or consumed in a transformation process. All production processes are transformations of physical materials, either to produce a product or a service. All these processes require physical (thermodynamic) work. Physical work involves an expenditure of energy. In fact, units of work are also units of energy. Yet energy is conserved in all process and actions, meaning it cannot be created or destroyed. But work is not conserved.

Standard economic theory treats energy as an "intermediate good", resulting from some deployment of the two "factors of production", namely "labor" and "capital". The problem is that available useful energy (exergy) is the "prime mover" without which nothing in the economic system, or in the larger world, happens or can happen. Nor can energy be "produced" from something else.

As regards the role of energy in economics, there are just two possibilities: (1) Suppose that energy is available in unlimited quantities, depending only on how much labor and capital can be deployed to find and extract it. Then energy must be an intermediate good—a function of the labor and capital applied—and then energy consumption must be a straightforward output of economic activity. In that case, economic activity (GDP) must be the "prime mover". (2) Suppose that useful energy (exergy) is *not* endlessly available and that both labor and capital themselves depend on the availability of exergy. In that case, exergy itself must be the limiting factor. After all, labor requires exergy, as food or animal feed, to function. Brains depend on exergy inputs as much or more than muscles. (The human brain, only 5 % of body mass, consumes something like 30 % of Caloric intake). Similarly, capital goods without exergy to activate them, like engines without fuel, are inert and unproductive. In this case it must be the energy supply that limits economic activity.

If case (2) is valid, exergy is truly the limiting factor. It must also follow that more exergy and its concomitants—more power, more speed– would be very productive. For instance, if liquid fuel were really as cheap as sea water, and likely to stay so, all intercontinental aircraft would surely be supersonic! Apart from local objections to noise, the only reason civil aircraft are not supersonic today is that the fuel required to fly faster than the speed of sound (Mach 1) is much greater than the fuel required to fly at subsonic speeds.

Half a century ago, the leaders of the airline and aircraft industries were convinced that fuel prices would continue to decline—as they had done since the enormous petroleum reserves in Saudi Arabia were discovered in the 1940s—but they thought that labor costs (hence the value of time and speed) would surely increase. Studies of supersonic passenger aircraft design began in 1954. The supersonic Tupolev Tu-144 first flew in 1968. The prototype British-French

Concorde flew in 1969 and commercial service began in 1976.[11] But the Tupolev retired in 1997. Only 20 Concordes were ever built (including prototypes) and commercial service ceased in 2003. High and rising fuel prices were the reason.

Similarly, suppose for argument's sake that electricity were really "too cheap to meter", as nuclear power enthusiasts promised back in the 1950s and 1960s. In that case, new uses of electric power and new electrical devices would proliferate. In fact, the declining cost of electric power, starting in the 1880s, has already created a lot of new services and new demand for them. Electric light was only the first. The cities of the world today are oases of light at night. Sports activities continue after dark. Business of all kinds continues night and day, as does education and entertainment.

After the turn of the century (1900) washing machines and carpet sweepers were electrified. Refrigerators for the home were introduced. After World War II a host of new consumer products, powered by electricity, appeared: electric stoves, toasters, electric power tools, lawnmowers, hedge-clippers, shavers, tea-kettles, coffee-makers, clothes dryers, dishwashers, freezers, air-conditioners, radios, TVs, PCs—the list goes on and on. A high-performance electric battery-powered car (Tesla) has recently made its appearance. Some would say that the infernal internal combustion engine has finally met its match. Or at least, the day seems to be coming soon.

Since the first electronic computer (ENIAC) appeared in 1946, followed by the solid-state diode (transistor), something similar has happened with computing power (and it still continues). In that case we have seen a similar explosion of new applications of cheap computation. In fact, the field of information technology (IT), which consisted of telegraphs and telephones connected by wires before 1940, or by microwaves up to 1960, has combined with digital machines, microchips and wireless technology and morphed into the internet. The internet itself is just the latest of many new applications that have appeared since large-scale production of electric power began around 1900.

Assuming the existence of electric power "too cheap to meter" (and compact means to store it), there are some other potential applications. To start with, every room in your house would be heated by electricity with individual controls. Many apartment houses built in the 1960s did rely on electric heating (to the regret of later owners); why construct central furnaces and pipes to carry heat around a building in the form of steam or hot air, if a wire would suffice to carry the energy?

If electricity and electric heat were too cheap to meter, the coastal cities of dry countries from Australia to Israel, not to mention the Persian Gulf and southern California, would provide both domestic water and local agriculture with desalinated sea-water, produced by flash distillation. (Other processes currently used, like ion exchange, reverse osmosis, nano-filtration and freezing, are more efficient in energy terms, but more costly. They need to be to save energy.) Further in the future, the deserts of the world would be irrigated by fresh water pumped from

---

[11] The Soviet Tupolev-144 flew a year before the Concorde, but because of two crashes, commercial service ceased after only 55 flights.

coastal desalination facilities. (The cost of pumping is mainly the cost of electricity.)

In the "too cheap to meter" world, electrolytic processes would be much more widespread than they are now. In fact hydrogen from the electrolysis of water would be the fuel of choice, not only for supersonic aircraft, but also for the subsonic variety, as well as for buses and trucks on the ground.[12] Aluminum, magnesium and titanium would be much cheaper than they are now. Automobile frames and bodies would be made from aluminum or titanium.

In that world, ultra-low temperature and high-vacuum processes would also be much more widely available. Superconducting magnets, used today only in the most sophisticated elementary particle accelerators (like the proton super-collider at CERN) would be used routinely to suspend high speed trains through evacuated tunnels, by-passing most short-haul airports. Satellites might even be launched electrically by powerful magnetic "guns".

These applications are pipe dreams today because electricity (pure exergy) is not nearly "too cheap to meter". Yet virtually all neo-classical economists today persist in using a model for economic growth that is based on a completely different proposition. They assume that energy consumption is unlimited and determined by *aggregate demand* and that demand is determined, in turn, by the state of the general economy (i.e. the size of the GDP). Their model implicitly assumes that there is no limit to energy (exergy) supply, either in terms of quantity or price. The standard model, as I mentioned previously, postulates a "production function" in which there are only two important "factors of production", capital and labor.[13] The alternative view, in which energy is also a factor of production just as much (or more) than either of the others, is not taken seriously in textbooks or economics journals today.

I (along with Reiner Kümmel and some colleagues) disagree strongly with the neoclassical orthodoxy expressed in the paragraph above (Ayres and Warr 2009; Kuemmel 2011). Consider one simple question: could there have been an industrial revolution in the early nineteenth century without the availability of vast reserves of cheap hydrocarbons (coal)? Or, from a different perspective, if all the remaining reserves of hydrocarbons were to disappear over the next few decades—without a banking crisis—how much of our present industrial society could survive? The answer to the first question is no. It was all about coal (and coke). The answer to the second is: almost none. I'll get to that.

---

[12] The idea of a hydrogen economy was apparently first suggested by the biologist J. B. S. Haldane at the Heretics Club (1910). Later it enjoyed a brief fad in the 1970s and 1980s initiated by John Bokris at GM Tech. Center (1970).

[13] Some models allow for energy as a factor, but only to the (small) extent determined by energy expenditures as a share of GDP. This is based on a theorem that is taught in all macro-economic textbooks, but that we have recently shown to be inapplicable to the real world. For details see Kümmel (2011) or Kümmel et al. (2010).

Regarding the question: Could large scale industrialization have occurred without coal and other hydrocarbon resources? I answered that it could not. But to dig deeper, does that mean that supply (of cheap coal) then created its own demand[14]— or, more accurately, enabled demand? Or was it demand that created (i.e. induced) supply, as most economists assume today? And, a secondary question arises: if the future supply of useful work (*exergy*) from fossil hydrocarbons is limited by the necessity to curb climate change (caused by the so-called "Greenhouse effect"), can economic growth continue in the future at historical rates?

Imagine an alternate counter-factual eighteenth century universe in which there were no fossil fuels to energize (quite literally) the industrial revolution. Population was increasing, slowly but inexorably. In England (and elsewhere) trees were being cut to provide charcoal and timber for barrels, houses and ships. Forest land was being converted to cropland to feed more people. Common lands in England were being fenced and enclosed to produce wool for export. In that alternate universe, without coal or oil, the economy would be dependent entirely on the energy supplied by food and animal feeds, plus fire-wood. Without coal, the price of firewood would have risen even faster than it did, simply due to population growth. After 1800 when population growth accelerated, the price of firewood must have gone "through the roof".

Actually, without the economic growth that did occur, the price of food would have consumed all the available income of the workers. People would have had to starve or freeze. Malthus' thesis would have been proven correct (Malthus 1798 [2007]). There is no way even a fraction of the current population of the world could have existed, still less with its present level of consumption, without the productive agricultural and other technologies that accompanied the industrial revolution. All of those technologies were enabled by the exploitation of fossil hydrocarbons. There is not nearly enough arable land in the world to support the nearly seven billion people now alive if we all (or most of us) had to be self-sufficient organic farmers. There would be a lot fewer people. Probably no more than one billion humans could survive on the Earth in reasonable health without the benefits of industrialization.

Taking it further, suppose that there had been no population growth after (say) 1700. To start with, almost everybody in that fossil-free world would have to work on the land (as was the case at the time), just as Chinese and Indian peasants have had to do during most of history: People were plowing, planting, and harvesting with horses or oxen (or their children) but without diesel powered tractors and harvesters. Also, without synthetic fertilizers and chemical pesticides, and without refrigeration to preserve meat and dairy products, total crop production *per capita* (and per farmer) would have been much smaller than it is now. Moreover, losses to

---

[14] This phrase is known as "Say's Law", after the French economist J. B. Say, but Say may not have actually said it. An early version of Say's law, in different words, appeared in the works of James Mill (Mill 1808 [1965]). More recently, Keynes used the phrase in his General Theory (some say he coined it) but in a slightly different sense.

insects and rodents would have been greater. (But the cats would have been fat.) That is a plain fact.

It is also a fact that with most people working on the land, the available agricultural surplus to support non-agricultural consumers would have been slim, probably no more than 10 % of gross agricultural output at the very most. (And without ICE powered ships, trains or trucks to transport the food there would be no way to feed very large cities anyway.) Most of that small surplus would be monopolized by the military (necessary to suppress peasant revolts) and the educated elite (since education takes a long time and consumes scarce resources). In practice, the military would likely live off the land—as conquering armies always do—and the educated elite would probably behave like priesthoods always have behaved. But that is somewhat beside the point.

Scarcely any of the tools, appliances and products that we buy from the super-market or DIY store, and take for granted today, could be produced in such a society. The few exceptions would be produced in very small quantities at very high cost. This applies to glass and ceramics, all metals and metal products except whatever iron can be smelted using charcoal. Glass could be produced from sand, or glass scrap, with charcoal, but only in small quantities and crude (but perhaps artistic) forms, as was done hundreds of years ago in Venice. Ceramics were in much the same case.

Iron for horseshoes or stew-pots would be expensive and carbon steel a rare and costly luxury used mainly for cutlery and weapons (as it was). Copper could be smelted from ore and refined with charcoal, but it would be very expensive and not ultra-pure (not good enough to make wire) without electrolytic refining. Local goods transport overland would be by horse-drawn carts, trolleys on crude rails or canal barges. Plastics and synthetic fibers could not be manufactured at all without hydrocarbon feed-stocks. Only wool, linen, cotton or animal skins would be available for most clothing. Long distance transport would have to rely once again on sailing ships, or caravans. Perishables cannot be shipped that way, so tropical fruits (like bananas, lemons and oranges) would not be available in the northern countries. Scurvy would be back.

Electric motors, dynamos and transformers cannot be manufactured without supplies of ultra-pure copper or aluminum wire; nor can they be utilized without electric power. Electric power requires dynamos driven by kinetic energy. The only available sources of kinetic energy in that carbon-free world would be flowing or falling water, or wind. Electrolytic refining (for copper and aluminum) and electric furnaces needed for manufacturing metals and alloys (like stainless steel) with high melting points would be possible only where hydro-electric, concentrated solar or wind power is available.

Electric lights could be manufactured only with great difficulty, if at all. Incan-descent lights, even using carbon or tungsten filaments, require both sophisticated glass-blowing, very high temperature metallurgy and efficient vacuum pumps. The world after dark would have to depend mainly, as it did in 1700, on tallow candles or lamps burning oil from crushed seeds, whale spermaceti or animal fats. It is also possible that some methane could be recovered from local concentrations of decaying organic wastes, as it is now being done in India, but the gas would be difficult to utilize productively without other sophisticated equipment.

The differences between our urban society and the fossil-free society, or our own rural past, is mostly due to the existence of coal, oil and natural gas and their combustion in heat engines, described in Chap. 9. Coal is the easiest to transport and store but the dirtiest and poorest in quality. Petroleum was a great but limited bonanza, whose limits are now beginning to show. It was, of course, a very compact and transportable source of useful energy that is convenient for internal combustion engines. But the cheapest and most convenient sources of petroleum in the Earth's crust are being used up rapidly. Tar sands and "tight oil" from shale are inferior substitutes. They will also be used up in due course (but perhaps not soon enough to stop runaway climate change).

Natural gas is difficult to transport and store. That is why natural gas only became available to consumers in cities in the second half of the twentieth century thanks to pipelines that don't leak (much), refrigerated tankers and efficient compressors. Shale gas is a relatively new source that may extend the global supply of gas for some decades. But all of the fossil fuel reserves are being depleted and, as time goes on, more and more energy will be needed to extract what is left. That means that the energy surplus available to the rest of the economy—the part not involved in digging, drilling, refining and transportation—is declining now and will decline more. Whether the combustion of fossil fuels comes to an end in 30 years, or 50 years, or a 100 years, the end is in sight.

Speaking of the end, it is interesting to think (for the space of a few paragraphs) about that future world where fossil fuels are gone, but viable alternative sources of energy have been developed during the brief "fossil era". That future world will have to be almost entirely electrified, apart from a fraction utilizing hydrogen. The electric power will come from a variety of sources, including flowing or falling water, wind, tidal or wave motion, solar energy, thermal energy from the Earth, and heat from nuclear fission (or fusion). Solar energy may be captured by photosynthetic plants (biomass), by lenses or mirrors (as in solar concentrators) or directly by photo-voltaic (PV) cells.

Some of that electric power may be used to produce hydrogen by electrolysis of water. The hydrogen, in turn, can be converted back into electricity (via fuel cells) for cars or trucks, or used directly as fuel in gas turbines for aircraft. Electrolytic hydrogen will also be needed to manufacture ammonia (for fertilizer), and other chemical intermediates (e.g. alcohols and esters) currently produced from natural gas. But products currently made from ethylene, propylene, butylene, butadiene, benzene, xylene and other petrochemicals (including synthetic rubber, synthetic fibers and most plastics) will largely disappear unless a new chemistry based directly on carbon dioxide and methanol can be developed quickly, as some suggest (Olah et al. 2009).

Some biomass may be utilized as charcoal in a few places, but the need to produce food will impose strict limits on the availability of biomass for other purposes. Bio-products, from wood, natural rubber, cotton, and medicinal plants also require land. Plant-based substitutes for synthetic fibers and plastics will also have to compete for land with food and wood.

The point of the last few paragraphs is that technology now, or likely to be, available *may eventually by-pass the need for fossil fuels*. But the alternatives are likely to be considerably less efficient and significantly more expensive. All of the fossil fuels comprise sources of energy far more concentrated and easier to utilize than the fraction that is captured by photo-synthetic green plants or by wind-powered turbines and most other so-called renewables. Apart from hydroelectric and nuclear electric power, the main source of the energy to do *useful work* today, is the combustion of fossil fuels that were created hundreds of millions of years ago, and the chemical and metallurgical processes those fuels make possible.

The industrial world will eventually have to be electrified. But in some domains, notably transportation and chemistry, the eventual electrification, while technically feasible (and clean), may be less efficient and may consume more of the total energy available, than is now the case. Production of useful work will consume more of the total economic surplus than it does now. That means less of that surplus will be available for investment, debt service or other forms of current consumption.

The original question posed was: which came first, the chicken or the egg? Was industrialization and economic growth (the chicken) the root cause of increased demand for energy? Or was it the other way around? Was industrialization and economic growth an indirect consequence of the availability (declining prices) of energy resources ("the egg")? I think the availability of low-cost energy resources (the egg) definitely came first, back in the eighteenth century.

11.7   Resource Discoveries as Technology Incubators

I think that the logic of Sect. 11.6—that energy availability drove the first industrialization—is very hard to contradict. Energy (exergy) *availability* was the main engine of growth from the start of the industrial revolution. In the future I think that energy (exergy) *efficiency* will be increasingly important. In case you think this is a statement of the obvious, there is a mountain of evidence from the last quarter of the twentieth century that it isn't. Firms, even now, rarely choose to invest in energy efficiency, *if* the other option is to invest in new products or conquest of new markets. Economists, trained to believe that firms always try to minimize costs, have had a lot of difficulty explaining this phenomenon (Taylor et al. 2008).

I know from long experience that most establishment economists will find reasons to object to my heterodox position: that energy (actually exergy) is a factor of production, and that without exergy it cannot grow, or even function. I assume that this is because they have been taught the opposite: namely that the economy "wants to grow" and that energy availability doesn't matter. (After all, the economy has been growing quite steadily for almost three centuries, so why should it stop?) They have also been taught that there is no limit to the availability of energy. Why

390   Mainstream Economics and Energy

no limit? Because "everything has a substitute" and if there is a market demand for energy, it will be met automatically, without (much) higher prices. (If this appears contradictory, you will have to raise your objections with the people who say these things). The professional economists also believe that the economy should continue to grow at near historical rates, even if energy prices no longer decline as they have in the past. I doubt it.

This is a good place to say that the assumption that economic growth "should" continue at the historical (US) rate of 2.5 % or so is almost certainly wrong, for reasons quite apart from declining energy availability (higher prices). Economic growth does not occur in equilibrium, but only after a shock (like a war or a revolution). Growth typically consists of a series of "bubbles", some small, but occasionally a very big one. More on that later.

I and some others have argued, on the contrary, that higher energy (exergy) prices are inevitable, and that they will slow down economic growth. (Unfortunately, I doubt that this will happen soon enough to reverse the atmospheric buildup of greenhouse gases (GHGs) and stop climate change.) In fact, I think that higher energy prices are the reason for the coming "secular slowdown" that Larry Summers suggested at an IMF Forum in 2013, (but did not explain).

Now back to the question; why does mainstream economics ignore energy? Conventional orthodox reasoning goes something like the following: divide all money flows in the economy into two sets, payment of wages and salaries for work done (to "labor") and payments to "capital". The latter category includes interest on loans, dividends, rents and royalties. By definition, those two categories together account for all money flows (expenditures) in the economy. In advanced industrial societies payments to labor typically get 65–70 % of the total while payments to capital get the rest. Recently the labor share has been decreasing and the capital share has been increasing, but that is not relevant to my point about energy.

The point is that there is no third category of payments in the national accounts. There are no payments "to energy" or "resources" or "natural capital", as such, because there is no economic agent to receive that money and then spend it on something else. Payments from people always go to other people wearing different hats. Consumers pay shopkeepers who pay distributors, who pay producers. Producers pay their own workers, or they pay the banks or other producers from which they purchase goods or services. They, in turn, pay the workers. The government collects taxes from both groups, provides services and also pays its employees. Thus energy (and exergy), from this perspective, are "produced" by enterprises employing some combination of labor and capital.

Given this perspective, it makes sense to conceptualize a "production function" that depends only on labor and capital or on labor, capital and "technology". And what makes the economy grow? The conventional answer is twofold: (1) Investment by entrepreneurs because they see growth as the way to increase individual well-being (more profits); and (2) it grows because of steadily improving productivity (due to better technology). In short, there is a maximization process.

Following in the footsteps of mechanics, economists began to look for a way to use the mathematical optimization tools introduced in the eighteenth century by

Euler and LaGrange (see Mirowski 1989). The first attempt to understand economic growth in terms of utility maximization goes back to Frank Ramsay's simple model maximizing a time integral of "utility" (defined as aggregate consumption), subject to discounting (Ramsey 1928),[15] I have no quarrel with the assumption that financial self-interest is a part of the explanation for economic growth. That is not really in doubt.

But I am less satisfied with the idea that technology, or technological progress, is an automatic consequence of human existence. Until the 1990s most economists simply assumed that technology is exogenous—"manna from heaven"—and that it grows automatically thanks to something like "animal spirits" (Keynes) or simple curiosity (the Greeks). Starting with ideas from Paul Romer and Robert Lucas, there have been attempts within neo-classical economics to "endogenize" the technological growth driver, based on the argument that the stock of knowledge, unlike other kinds of capital, has increasing returns-to-scale (like networks and natural monopolies) (Romer 1986). If that is true, then the driver of growth must be investment in education (knowledge) (Romer 1987). Robert Lucas, more or less simultaneously, emphasized "social learning" (Lucas 1988).

Without doubt, education and social learning are both important in today's world. But I doubt that either education or social learning played a significant role in eighteenth century Europe, contrary to the popular idea that the "scientific revolution" initiated by Copernicus, Galileo, Newton et al. (summarized briefly in Chap. 3) somehow triggered the industrial revolution (Beinhocker 2006, p. 259). Instead, I see resource discoveries, in response to scarcities, together with invention (by trial and error) of ways to use the new resources, as the major drivers of technological change until the last decade of the twentieth century.

Science had almost nothing to do with the steam engine, the series of inventions that led up to Bessemer steel, and very little to do with Edison's electrical inventions. Only in the late nineteenth and twentieth centuries did science take a leading role in driving technological change. Moreover, the technological changes that occurred in the eighteenth and early nineteenth centuries were mostly focused on ways to utilize a resource in very specific applications. In short, the discovery of coking of coal was a "technology incubator". Most of the reasons for this view have already been articulated in Chap. 8.

Recall the history of coinage (Sect. 8.7). Money in the distant past, before coins, was any local commodity that could be used as a numeraire for exchange transactions. The most common form of money in many parts of Africa was cattle. In other places and times money has been measured in salt, coconuts, ivory, Cowrie shells, and various other things. Gold and silver were latecomers because of the difficulty of measuring quantities. Metal coins were first introduced in Lydia, using an alluvial mixture of gold and silver ("electrum") about 640–630 BCE. By

---

[15] The maximization (or minimization) technique is now known as "optimal controls", used mainly in orbital calculations for rockets and missiles. (This application may be the origin of the phrase "it's not rocket science")

560 BCE the Lydians had learned how to separate the gold from the silver. King Croesus of Lydia issued both gold and silver coins, with an engraving on them. This innovation greatly reduced the uncertainties inherent in exchange transactions and made Lydia a trading hub. This was why King Croesus became famously rich.

The key point is that for the next two and a half thousand years precious metals (gold and silver) from the ground could be immediately converted into coins, hence money, by anyone with a furnace, a hammer and a mold. The Hapsburg branch of the Spanish empire in the sixteenth and seventeenth centuries was originally financed in part by silver "thalers" minted from the famous mine at Joachimsthaler in Bohemia (the original source of the word "dollar"). The Holy Roman Empire under Charles V (created by a merger of Burgundian and Hapsburg territories with Spain) was partly financed by gold and silver looted by the Spanish "conquistadores" from the Aztec and Incan civilizations. For many years thereafter, Spain obtained silver from the great mine at Potosi in Peru (now Bolivia), which was run like a slave plantation. It is ironic that the Spanish used all that silver (and more) to pay for their vain attempt to subdue the Dutch Protestants from 1550 to 1650 (Kennedy 1989). The silver itself moved gradually to the Far East.

But the influx of Spanish silver (and gold) did not finance much agricultural or industrial development. Instead it financed conspicuous consumption by the Crowns and the Church and (mostly) their wars. The coinage was used to pay the wages of jewelers, dress-makers, mercenary soldiers, and weapon-smiths, not the makers of capital goods. The coinage minted from gold and silver from the New World increased the money supply in Europe without increasing the supply of non-military goods (except for the imports from Asia), and thus increased local inflation. The price index in Europe rose 250 % during the century from 1550 to 1650 or so (Kander et al. 2013, Fig. 4.6). (Admittedly, the influx of Spanish silver was not the whole explanation of high wheat prices in those years.)

Swedish money, in the form of copper coins, was minted from copper from the famous "copper mountain" at Falun in Sweden during the seventeenth century. It helped to finance the military adventures of the Swedish king, Gustavus Adolfus, whose armies contributed a lot to the devastation in Germany during the 30-years war between Catholics and Protestants (1618–1648). So, the discovery of a copper mountain at Falun was mostly wasted on war, not on industrial development.

Money did play a role in the expansion of trade, of course. Henry VIII of England "nationalized" the Roman Catholic Church in the early sixteenth century and confiscated much of the gold from its cathedrals and monasteries. A lot of that gold was subsequently converted into coins, also used to pay soldiers. But some of that coinage financed trading expeditions that imported luxuries: notably spices, tea and textiles (silk, muslin, calico) from India and China. Those luxury commodities could only be purchased for gold or silver, because Europe produced nothing (except arms) of commercial value in India and China. But the investment in ship-building, guns and exploration paid off later.

Neither wool nor coal nor slaves—in contrast to copper, silver or gold—were regarded as inherent "stores of value". They could not be treated as money or converted directly into money. Money is a store of wealth that circulates. Wool had value only when laboriously washed, combed, dyed, spun into yarn, and woven into cloth, then cut and sewed into garments. So, in the late Middle Ages, raw wool from English sheep had to be exported to a city where yarn was spun and cloth was woven and dyed. That city was Bruges in Burgundy (now Belgium). Later the labor-intensive spinning, carding and weaving technologies were mechanized in Manchester and applied first to cotton (imported from India or the southern states of the US). This mechanization, based mostly on steam power, enabled English cotton cloth manufacturers in the nineteenth century to underprice Indian cotton weavers and effectively destroy that industry in Bengal. (Thus did the British Raj create markets for its manufacturing exports.)

Coal had value until the eighteenth century only as a source of heat. It had to be dug up and transported to a user who burned it. The first users of coal were blacksmiths and the great houses of English landowners, which had fireplaces in every room. Other medieval users were brick-works, ceramics, glass and breweries. But metallurgy was not among those early uses of coal. Prior to the middle of the eighteenth century, only **Abraham Darby**'s iron-works in Coalbrookdale, in England, used coke from coal for smelting iron from ore. All other iron works depended on charcoal to make cannons and muskets. Sweden was the dominant iron producer in Europe in the seventeenth and early eighteenth century, mainly because of its plentiful forests for charcoal and the high quality of the Swedish iron ores.

As explained in Sect. 9.2, the reason coal could not be used directly to smelt iron was that all coal contains sulfur (and other contaminants), that get into the molten pig-iron from a blast furnace and spoils the iron castings by making them brittle. Unfortunately nobody really understood the problem at the time. There was no science of metallurgy until much later. So the solutions came about gradually, by trial and error and occasionally by accident.

Abraham Darby's trick was to heat the coal in a closed hearth, driving off the volatiles, before using the solid carbon residue in his blast furnace. This scheme is now called "coking". In fact, most of the world's integrated steel industry today depends upon access to high quality (hard) coking coal. (Coke for a blast furnace must be porous enough to permit combustion gases to pass through, yet retaining enough physical strength to support the iron ore piled above it without collapsing into an impermeable layer. Most coal is not good for coking). Luckily, Darby had access to good coking coal. But his iron ore was too high in silicon, so his iron could not compete successfully with the Swedish iron. (We know this now, but he did not.) However, in the course of the eighteenth century competing English ironworks gradually learned about coking, and abandoned charcoal because of its high, and rising, price. In the seminal year of 1776, coke overtook charcoal as a reducing agent for iron in England.

One major reason for the high price of charcoal, and the substitution of coke, was that firewood to make charcoal is a low value commodity that needs to be used near its point of origin to minimize transport costs. This forced iron-makers into the forests (to be near the charcoal) and put a practical limit on the size of a charcoal-based iron-making furnace. So until the eighteenth century small iron smelters were scattered throughout the woodlands of Europe. But, as mentioned already, forests were being cut back to increase grain production, especially in England, whereas coal was increasingly easy to move from a mine, via the newly built canals. So coking enabled smelters near canals or seaports to grow in size and exploit economies of scale.

Larger blast furnaces, in turn, are more efficient than small ones, and can reach higher temperatures. (They have increasing returns-to-scale). As it happens, "pig iron" containing 4–5 % carbon, by weight, melts at about 1200–1250 °C. But steel, with 0.25–1.5 % carbon, doesn't melt until about 1500 °C. (The purest form of iron melts at 1535 °C.) Molten metal is much easier to form into useful shapes than solid metal rods or ingots. The history of metallurgy (until the electric furnace came along in the 1880s) has been a race to achieve higher temperatures to melt iron and steel in larger volumes.

As the blast furnaces got bigger and demand for coal increased, the coal mines burrowed deeper. Below 10 m or so, they usually encountered water, and had to be continuously drained with pumps. That simple fact, by itself, triggered the invention of steam power, and started the industrial revolution, according to current mythology. The coal mining industry became an important technology "incubator". The first reciprocating steam engines, invented by **Thomas Savery** (1698) and implemented by **Thomas Newcomen** (1711) were specifically designed to pump water out of flooded mineshafts. Coal from the mine was the fuel required to make the steam that drove the piston that made the pump work.

James Watt made the Newcomen pumping engine far more efficient by introducing the separate condenser. There was no science of thermodynamics involved. His first innovation was a separate condenser and a change in the valve arrangements so as to keep the condenser cold and the cylinder and piston hot. Watt and his assistant, **William Murdoch**, subsequently made a number of other improvements. They made the engine more compact, suitable for rotary motion and thus applicable to other industrial uses. Apart from locomotives and steam-boats, the steam engines were used for blowing air into a blast furnace to raise the temperature, powering a boring machine to make cannon barrels (and the cylinders of Watt and Boulton steam engines), turning rotary printing presses for books and newspapers and weaving machines to make cloth from imported cotton.

The growing market for coal and the availability of the high-pressure steam engine after 1800 motivated the construction of canals and rails and locomotives on the rails to carry the coal. The use of coal (as coke) for iron production made iron cheaper. A century of technological progress after Watt's inventions led to the mass production of steel machines and machine-tools, steel rails, steel ships, steel girders, steel bridges, steel pipes, steel barbed wire for ranchers and steel cable for derricks and elevators. As mentioned already, the global steel industry still depends on coke.

Coal was the source of "town gas", first used for street-lighting and household lighting by one of Watt and Boulton's former employees, **William Murdoch**. This application thrived and spread until cheaper natural gas became widely available after the Second World War. Finally, methane-rich combustible gas from coke-ovens powered the first internal combustion engines.

In summary, coking began as a technology for eliminating unwanted sulfur from the fuel used in early blast furnaces. It was the key to making high quality iron and steel. But the gas was inflammable and coke-oven gas turned out to be quite useful in other applications. By the mid-nineteenth century coking was generating quite a lot of coal tar, a material used at first for paving roads, but later for roofing materials (as tar paper), and still later, as a source of nitrogen fertilizers in the form of ammonium sulfate (and explosives for mining and wars). Moreover, the coal tar could be distilled, yielding the aromatics (ring-molecules) benzene, toluene and xylene, which are building blocks for a host of petro-chemical compounds that are important today.

The first of them was aniline, made from nitro-benzene. Aniline was the starting point for the discovery of synthetic mauve (dye) by **W.H. Perkin**. The discovery of other synthetic aniline dyes essentially kick-started the British and German chemical industries. Meanwhile, it was coke-oven gas that provided the fuel for the first practical (stationary) internal combustion engine, the famous "silent Otto" after its inventor **Nikolaus Otto**. Tens of thousands of copies of this engine were made and sold around Cologne and the Ruhr valley in the 1870s and 1880s and later, all over the world, to provide mechanical power for machines in factories.

However, while the first industrial revolution was happening, most classical economists (**Adam Smith**, **Thomas Malthus**, **David Ricardo**, et al.) were still primarily concerned with the productivity of agricultural land, rather than energy or other natural resources. The first economist to call attention specifically to a source of energy as a resource was **William Stanley Jevons** (1865), who recognized the central importance of coal for British industry and British prosperity and worried that known reserves were in danger of exhaustion (Jevons 1865 [1974]). Jevons' book opened with the statement:

> Day by day it becomes more evident that the coal we happily possess in excellent quality and abundance is the mainspring of modern material civilization. As the source of fire, it is the chief agent in almost every improvement or discovery in the arts which the present age brings forth.

That was in 1865, the year of the Bessemer steel invention, a century after the Industrial Revolution began. Oddly enough, hardly anybody worries about coal scarcity these days. Published estimates of reserves by the US Geological Service (USGS), for instance, suggest that the coal we know about could last a 100 or 200 years, even allowing for increasing demand. (More recent reports are less optimistic, but there is certainly no shortage of coal (Rutledge 2007; Zittel and Schindler 2007; Kavalov and Peteves 2007). The worries about future supply are now focused on petroleum and natural gas. Indeed, a major worry for some environmentalists (including me) is that the low-grade coal that remains will be burned and not buried. Some investors are beginning to worry about the risk of "stranded assets" i.e. coal (or oil) that is now counted as an asset of mining companies but that can never be burned for environmental reasons (Leaton et al. 2013; Salzman 2013).

It is ironic that, at the very time Jevons wrote the words quoted above, the US petroleum industry was just getting started, mainly in response to a growing shortage of whale oil for lamp-lighting purposes. The sperm whales were becoming scarce. Animal fats were tried but they smelled bad. The best substitute for whale oil was kerosene made from "rock oil" a moderately commonplace substance found in seepages and exudates from salt wells and sometimes sold for "medicinal" purposes.

The most important new technology for utilizing rock oil was rock drilling. The technology was originally used for drilling salt wells to provide salt for animal husbandry in areas far from the sea. Starting in Kanawha West Virginia, it had spread throughout the Ohio River basin. Drilling had already reached depths of several thousand feet in some places, using copper tubes to bring the brine to the surface. That technology sufficed for oil-well drilling over the rest of the nineteenth century. Walter Sharp and Howard Hughes invented a much more efficient drilling technology in 1909, enabling much deeper oil wells, even offshore. (Hughes Tool Co. was the result.)

The usefulness of petroleum for non-medicinal purposes (e.g. lubrication, petroleum coke, road surfaces) was first suggested in the so-called "**Silliman** report" on *Rock Oil or Petroleum* commissioned by some investors in 1855 (Williamson and Daum 1959, Chap. 4). It concluded that as much as 50 % of the black stuff could be converted by fractional distillation into an illuminant suitable for oil lamps while at least 90 % had some commercial promise. Silliman's report provided a refinery design: a distillation tower. It even hinted at the possibility that the heavier fractions could be "cracked" to increase the output of more volatile fractions. This report triggered the investments that set off the oil boom in Western Pennsylvania. All of Benjamin Silliman's predictions soon turned out to be true, although the first investors lost their money.

The most important product of the oil industry then, and for the next half century, was "illuminating oil" (kerosene), which rapidly became the dominant global fuel for lamps, replacing whale oil. It soon became the most important US manufactured export to Europe and the basis of John D. Rockefeller's Standard Oil Trust. But by far the most important of all the technologies incubated by the discovery of petroleum was unimaginable, at first. It was for use in internal combustion engines. As mentioned in Sect. 9.7, Nikolaus Otto's 4-cycle stationary internal combustion engine (1876) burned coke oven gas and produced 3 hp at 180 rpm. It weighed 1500 lb (670 kg). That was considerably lighter than a comparable steam engine could produce at the time, so it was revolutionary.

The next step was to use liquid fuels instead of coke-oven gas, and to reduce the weight of Otto's 4-cycle engine by increasing its speed. The fuel of choice was natural gasoline, a light fraction of the oil refineries that had no real use at the time except for dry-cleaning. Four prototype gasoline engines, weighing about 50 kg. and producing 1.5 hp at 600 rpm, were built by **Gottlieb Daimler** and **Wilhelm Maybach** during the years 1885–1889. Maybach also invented the carburetor, which was a trick to vaporize the gasoline just before ignition. **Karl Benz**, another inventive automobile pioneer, began manufacturing cars in 1887. The Daimler firm followed 5 years later. (They merged to form Daimler-Benz in 1926).

For the next 15 years automobiles were mainly toys for rich young boys. But after 1900 many other manufacturers were competing in the market and the vehicles looked less like buggies and more like cars. In 1903 **Henry Ford** came on the scene and 5 years later (1908) began the era of mass production with his "Model T" (famously sold in any color so long as it was black). Ford (and "Fordism") revolutionized manufacturing, starting with his moving assembly line. He also revolutionized capitalism by paying his workers more than the market price for labor so that his workers would be able to become his customers.

Petroleum provides all the liquid fuels for the internal combustion engines that power all ground and air transport (except electrified railways), as well as off-road diesel-powered construction and agricultural machines. Petroleum products became a direct substitute for coal in the 1930s, when diesel-electric locomotives began to replace coal-burning locomotives on the railroads.

Air travel, so familiar today, depends entirely on the internal combustion engine in some form. The earliest aircraft, like the earliest cars, were essentially expensive toys for the rich or adventurous. Use during the First World War (1914–1918) was only for reconnaissance over the front lines. But improved engines and metal bodies made aircraft a central element of military power before the Second World War broke out in 1939. The Battle of Britain in 1940 was decided in the air. The kerosene-fired turbojet (gas-turbine), invented in the late 1930s, was a radical improvement in power-to-weight and, hence payload. In improved versions it is the workhorse of today's civil and military aircraft.

The availability of coke oven gas, volatile fractions of petroleum (ethane, butane) and natural gas (originally a by-product of oil drilling) also incubated new technologies for re-combining hydrocarbon molecules, starting in the 1920s. Polyethylene, polypropylene and polystyrene were early innovations. Polyethylene became the insulator of choice for undersea telephone cables. Nylon was used for stockings and parachutes. Synthetic rubber became important for tires during WW II, after the Japanese captured the rubber plantations of Indo-China and Malaysia. Petro-chemicals are the basis for all plastics and synthetic fibers today.

The development of electricity technologies is the major exception to our resource "incubation" thesis. Electrical technologies began with scientific discoveries, notably the discovery of electricity itself (**Ampere, Franklin, Galvani, Ohm, Volta**) and of electro-magnetism (**Faraday, Henry, Oersted**). The voltaic pile (first battery) enabled the telegraph. **Joseph Henry's** electromagnet (1831) and **Faraday's** first generator (1832) were major milestones. **James Clerk Maxwell's** theory of electromagnetism (1867), which predicted the existence of electromagnetic waves, is shown in Fig. 11.2.

The use of electricity to produce arc-light was a significant business long before **Edison's** contribution in the late 1870s. Since then electricity has become the dominant energy carrier and electricity generation is now the world's largest industry.

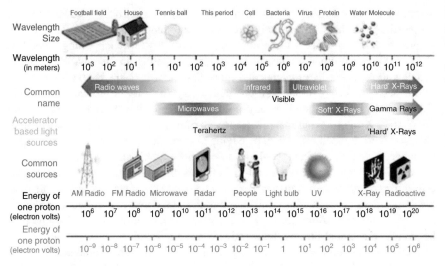

Fig. 11.2 The electromagnetic spectrum as a new resource

The information technology (IT) sector is a new phenomenon. On one hand, it is an evolutionary outgrowth of the telegraph, which accompanied all the early railroads, followed by the telephone, the radio-telephone, broadcast AM radio, FM radio, and TV. All of these technologies, after the radio, involved applications based on some knowledge of science. But the transistor (semi-conductor) that enabled the near-astronomical gains in transmission and data-processing capability in the past half century, was a deliberate application of science to technology. It began from an internal R&D project by AT&T Bell Telephone Laboratories, in response to exponentially increasing demand for telephone service combined with the increasing electric power requirements by the electro-mechanical telephone switchboards that were in use back in the 1940s.

The speed, reliability and mass-produce-ability of transistors enabled electronic computation to take off (with transistors replacing vacuum tubes) in the early 1960s.[16] The development of silicon-based integrated circuits in successively larger sizes, during the 1960s led to the first integrated processor (Intel, 1969), the personal computer (PC) in 1977 and the portable cell-phone soon after. The internet dates from 1992. The demand for computation and information processing via the internet has been so large that the IT industry is already one of the largest electric power consumers. *Most of that power is still generated by burning coal or natural gas.*

---

[16] The use of mechanical devices for computation goes back at least to Blaise Pascal's 8-digit mechanical calculator (1642), which didn't work very well (gears jammed, etc.), Charles Babbage's steam-powered "difference engine" (1820-21)—under construction for 20 years but never finished—and it's successor, his "analytical engine", was described by his friend Ada Byron, Countess Lovelace, who wrote about it. Mechanical calculators were commercially available in the late nineteenth century (e.g. Burroughs' Arithmometer, 1886).

To summarize: it is evident that the discovery of coal and subsequently petroleum both incubated and financed a great many of the technologies upon which contemporary society depends. The mechanism that triggered economic growth was always the same: a technological improvement made the resource itself cheaper, or made a product made from that resource cheaper and better. That triggered increased demand. In some cases, it created whole new products and industries, as exemplified by the examples of cheap steel, internal combustion engines, electrification, electronics and information technology. Those inventions and innovations vastly increased the productivity of our economy, in the literal sense of increasing output per worker-hour.

In the 1950s and 1960s nuclear power—based on the discoveries of nuclear fission ("atom splitting") by Otto Hahn, Lise Meitner and Fritz Strassman followed by **Enrico Fermi**'s controlled self-sustaining nuclear chain reaction[17]—was widely expected to be the next step on the energy (exergy) supply side, after hydrocarbons ran out. The US Atomic Energy Commission (AEC) and the International Atomic Energy Agency (IAEA) were created to promote this outcome. Nuclear power has, in fact, become an established source of base-load electricity in many countries, particularly France. But after several accidents at nuclear power plants, the future of nuclear energy is now in doubt.

It is not too far from the truth to say that the "resource" which will incubate technologies driving future economic growth must be science itself.

11.8   On the Geology of Resources: Scarcity Again?

It is indisputable that the world economy, at least through the first three quarters of the twentieth century, depended essentially on technologies "incubated" by the discovery and exploitation of coal and petroleum, not to mention other natural resources. Consider the words of the most quoted conservative economist, **Friedrich Hayek**. These were written in his best-selling book *The Constitution of Liberty*, which I quote to make a point (Hayek 1960):

> Industrial development would have been greatly retarded if, 50 or 80 years ago, the warnings of the conservationists about the threatening exhaustion of the supply of coal had been heeded; and the internal combustion engine would never have revolutionized transport if its use had been limited to the then known supplies of oil. . ..the conservationist who urges us to make greater provision for the future is, in fact, urging a lesser provision for posterity.

---

[17] Fermi's demonstration of a controlled self-sustaining nuclear reactor at the University of Chicago in December 1942 was part of the US "Manhattan Project", which led to the atomic bombs demonstrated at Alamogordo N.M. and later dropped on Hiroshima and Nagasaki. But all nuclear power plants are now based on some version of Fermi's nuclear chain reaction scheme.

Hayek was one of those economists (Julian Simon was another) who believe that "there is a substitute for everything", hence scarcity is no problem because technology will always come to the rescue; (e.g. Simon 1996). In regard to worries about the coal supply, Hayek was referring to Jevons book "The Coal Question" cited in the last section (Jevons 1865 [1974]). His point about the internal combustion engine is clearly valid, even though he was arguing for minimal government regulation, and especially against conservation by regulation.

Today it is quite evident to governments and business leaders—if not to some orthodox economists in the Hayek camp—that sustainable economic growth in the future still depends a lot on the future availability of high quality liquid and gaseous fuels at low cost. Countries have cabinet-level departments responsible for assuring energy supplies. There is now an international agency (the IEA) tasked with monitoring the global supply situation, especially due to the fact that so much of the petroleum consumed by Europe, India, Japan and China comes from the Middle East and Central Asia, where geo-politics and oil are currently mixed up with religious differences.

Despite Jevons' worries 150 years ago, coal does not appear to be a critical resource today compared to petroleum. The first alarm regarding petroleum occurred just after World War I, following a 10 year period of dramatic increases in both demand and prices. In 1919 both the Director of the US Bureau of Mines and the Director of the US Geological Survey (USGS) predicted imminent scarcity. The first of the two eminent authorities assured the outgoing Wilson administration that peak US output would occur in 2–5 years, while the second authority announced that US oil reserves would be exhausted in a remarkably precise "nine years and three months" (Yergin 1991, p. 194).

Within a very few years, however, worries about scarcity were replaced by worries about glut. In the early 1920s a lot of oil was discovered in Venezuela, Mexico, California, and Oklahoma. Prices dropped again. In 1930, "Dad" Joiner and "Doc" Lloyd discovered the super-giant East Texas field. This discovery resulted in a price crash. At the bottom of the market in1931, oil sold for 13 cents per barrel (bbl), which was actually far below the cost of production at the time (80 cents/bbl). Independent wildcatters went bankrupt in droves. The Texas Railroad Commission was brought in to regulate (i.e. cut) production (Yergin 1991, pp. 248–252). This left excess pumping capacity. From then until 1970, Texas was the low-cost swing producer for the world, and US production had to be limited to maintain prices above costs. That 40 year period ending with the rise of OPEC and the crisis of 1973 may have been the culmination of the demand-driven economic growth paradigm with no supply constraints. The period after 1973 looks very different.

After the Second World War, worries about resource scarcity (not limited to petroleum) emerged again, but only in the academic sense. The first major postwar assessment of resource needs and availabilities was sponsored by the Twentieth Century Fund. It was "America's Needs and Resources" by J. Frederick Dewhurst (Dewhurst 1947, 1955). In 1948 President Truman created the Materials Policy Commission, chaired by CBS Chairman William Paley.

The first comprehensive survey of energy sources *per se*, was a long paper entitled *"Major Sources of Energy"* by Eugene Ayres, a chemical engineer at Gulf Oil Co.[18] It was presented at the annual meeting of the American Petroleum Institute (API), in November 1949 and subsequently expanded into a book, *Energy Sources—The Wealth of the World* (Ayres and Scarlott 1952). The book emphasized the thermo-nuclear origin of all sources of useful energy, ranging from ancient accumulations of "stored sunlight" (coal, oil and gas) from the carboniferous era to atomic energy and "renewables". This book was the first to feature an "energy balance sheet" (op. cit. Chap. 21) with conversion factors for all sorts of fuels and other energy flows. It presented the format that has since been adopted by the US Energy Information Agency (USEIA) and the International Energy Agency (IEA).

The Paley Commission's report, entitled *"Resources for Freedom"* was published in 1952 (Paley 1952). To continue the work of the commission, a think-tank called Resources For the Future (RFF) was created and initially funded by the (new) Ford Foundation, also in 1952. RFF sponsored its first major conference in 1953, resulting in the book *A Nation Looks at its Resources* (Jarrett 1954). Another RFF study built on Ayres & Scarlott's *Energy Sources* was the historical survey *Energy in the American Economy: 1850–1975*, by Sam Schurr and Bruce Netschert, et al. which also included the consumption side of the story (Schurr and Netschert 1960). This work provided the historical background for the landmark *Resources in America's Future* (Landsberg et al. 1962).

In general the RFF studies have been optimistic, in the sense that worries about geological scarcity were generally dismissed in favor of confidence about technological progress. The most optimistic of all was *Scarcity and Growth* by Harold Barnett and Chandler Morse (Barnett and Morse 1963). That study, which has since become regarded as a classic in resource economics, was the first to examine long-term trends in natural resource prices, adjusted for inflation. To the surprise of many, they found that over nine decades, extraction costs and prices had declined almost without interruption. Another RFF study elaborated and confirmed these findings (Potter and Christy 1968). Resource economists, based on this evidence, argued that the only true measure of scarcity must be a rising price trend. I summarize the theoretical arguments more carefully in the next section.

Past experience of energy prices (and GDP) has been summarized by Roger Fouquet, as shown in Fig. 11.3. Evidently prices for iron-heating, land transport, power (from all sources) and sea freight have fallen by factors 20–50 (except for wartime interruptions) since 1700, and mostly since 1825. All of these measures reflect the underlying cost of exergy. Meanwhile the GDP per capita has risen (in real terms) by a factor of 8, with a couple of war-time hiccups (Fouquet and Pearson 2003; Fouquet 2011). If correlation is causation, the driver of GDP growth would seem to be exergy consumption. I will come back to that question in the next-to-final chapter.

---

[18] Confession: He was my uncle.

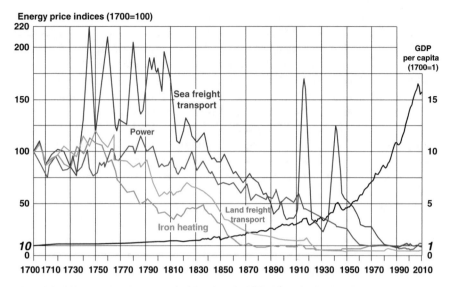

Fig. 11.3 Energy price indexes vs GDP for the UK 1700–2000. *Source*: Fouquet (2011), Broadberry et al. (2013) updated and indexed by Ayres 2000–2010

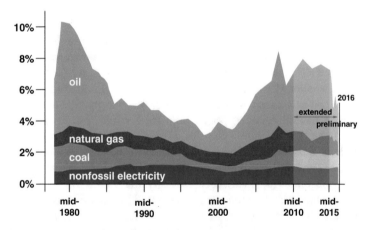

Fig. 11.4 Energy costs and GDP: When energy prices may have reached their lowest point in history. *Source*: King et al. (2015, Fig. 2), updated 2010–2015 by author

As a matter of interest, a recent study by Carey King et al at the University of Texas, using Fouquet's historical price data, has indicated the "real" energy global average prices—as a fraction of GDP—seem to have finally hit their lowest point in 1998. See Fig. 11.4. The original chart only covers the years 1980–2010, so it does not encompass the recent oil price drop. However, we have extended the chart through 2015. Even allowing for the >50 % drop in the oil prices in 2014–2015, it

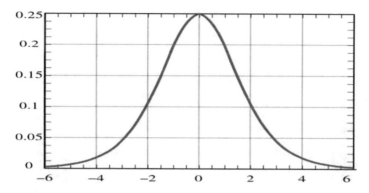

Fig. 11.5 Hubbert Curve

looks as though Dr. King's conclusion will hold. I think prices will start to rise again in 2016 or 2017 at the latest.

A much less optimistic forecast was made in 1956 at a meeting of the American Petroleum Institute in San Antonio, Texas. The author was well-known oil geologist M. King **Hubbert** (Hubbert 1956). He had worked for several oil companies, including Shell International, and had been elected to the US National Academy of Sciences in 1955. He was the first to engage in quantitative forecasting of future energy resources, based on empirical oil drilling statistics and recovery rates. In brief, he argued that the history of every mine or oil-field follows a similar pattern: slow beginning, acceleration to a maximum, followed by a decline (Fig. 11.5). This pattern has been observed in many times and places. The pattern follows a "logistic" probability function (not a Gaussian).

Based on this pattern (which he explained in more detail) Hubbert predicted in 1956 that US oil production would peak in 1969–1970 (Fig. 11.6), contrary to the conventional wisdom in the industry (Hubbert 1962). He wrote:

> The epoch of the fossil fuels as a major source of industrial energy can only be a transitory and ephemeral event—an event, nonetheless which has exercised the most dramatic influence ever experienced by the human species during its entire biological history.

That comment about the past is surprisingly consistent with the Hayek comment quoted earlier, except that they totally disagree about what will happen in the future. Hubbert's prediction that US oil output would peak in 1969–1970, as elaborated in subsequent work (op cit.), turned out to be unnervingly exact. The oil industry and the US Geological Survey did not like Hubbert's conclusion and did not accept his methodology. (Some in the oil industry are still fighting rear-guard battles over the latter issue.) But Hubbert's prediction was correct and its logic is hard to gainsay.

The peak year, 1970, had been preceded by two decades of oil glut and steadily declining inflation-adjusted prices. Prices, adjusted for inflation, fell from nearly $16/bbl in 1948 to about $11/bbl in 1972, in 2000 dollars. This trend encouraged the substitution of oil for coal or coke for home heating, and the proliferation of large gas-guzzling motor vehicles. During the 1950s and 1960s US still had excess

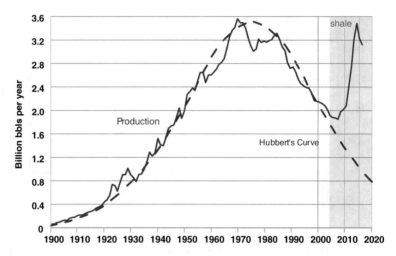

Fig. 11.6 Hubbert Curve imposed on US oil production 1900–2015

pumping capacity, as already mentioned. But US excess pumping capacity had nearly disappeared by 1971, and the US eliminated import quotas in April 1973. This led to a sharp increase in US imports, especially from Venezuela, and correspondingly increased European and Japanese demand for Middle East (especially Saudi) petroleum.

Then came the "Yom Kippur" Arab-Israeli war in October 1973. This, in turn, led to the second Arab oil boycott of 1973–1974 which resulted in sharp real price increases and shifted power over production quotas from Texas to the recently formed Organization of Petroleum Exporting Countries (OPEC). The world price reached $35/bbl in 1974 and it stayed up. It had barely fallen below $30/bbl by 1979 when the Iranian Revolution displaced Shah Reza Pahlavi and installed the ultra-conservative Shiite theocracy that still rules the country today. The price of oil then doubled again, but fell back to the $30 level briefly by 1986. See Fig. 10.10 in Sect. 10.2.

As noted earlier, since the 1970s, Saudi Arabia has been the "swing producer", and OPEC has taken on the former role of the Texas Railroad Commission: to keep prices high enough to discourage investment in alternative energy technologies (and to keep its market share in OPEC), while keeping them low enough to permit modest economic growth by the oil importing countries. Recent volatility (c. 2015) may change the relationships once again.

Geologists in the past have also generally supported optimism about the future availability of natural resources. In particular, the standard rule-of-thumb (sometimes called the "McKelvey rule") in geology says that the lower the ore grade the greater the quantity that can be found in the Earth's crust. History, so far, suggests that technological improvements have indeed made it possible to find and extract lower grade ores at continually declining costs. However the McKelvey rule is not a law of nature (and not even a rule). It clearly does not apply to some of the less common elements, like copper, zinc and lead, which are currently being mined from relatively high grade ores that were formed by particular geological processes (e.g. Skinner 1976, 1987).

Nor does the rule apply to a number of scarce, chemically similar, "hitch-hiker" metals that are found mostly in the ores of other metals (especially copper ores) (Ayres and Talens Peiro 2013a, b). For instance, copper ores were formed (and are found) only on the upper surfaces of certain iron-rich protrusions—analogous to pimples—from the molten magma under the Earth's crust. Copper chemicals, especially sulfides, are not distributed randomly. They were formed and deposited as superheated brines condensed along the outer edges of such protrusions (Kesler 1994). These ancient protrusions are found almost exclusively along the "ring of fire" where most of the world's volcanoes are located. And they are limited in number. The ones that have been exposed by erosion have probably all been discovered. Some others, still underground, or under the sea, may be discovered in the future. But except for these limited ore bodies, most of the copper in the world consists of atomic substitutions—a few atoms per cubic meter—in ordinary rock. That metal is not separable from its matrix by any conceivable process, except in a plasma-arc, requiring extraordinary amounts of exergy (Skinner 1976, 1987).

The origins of hydrocarbons are equally non-random. It is perfectly clear that coal was formed from prehistoric terrestrial vegetation during the so-called carboniferous era (Sect. 5.10). Remnants of the foliage can still be seen under a microscope). Lignite is a lower grade of coal, but not necessarily available in larger quantities than bituminous coal, while the quantity of peat—formed from more recent accumulations of vegetation (mosses) in very wet areas—is even smaller. This fact contradicts the McKelvey rule.

As regards petroleum, the standard theory attributes its formation to accumulations of marine organisms that were covered over by sediments. These were subsequently attacked by anaerobic organisms, releasing carbon dioxide and producing hydrocarbons, while the sedimentary layer hardened into shale and impermeable rock. The gaseous hydrocarbons remain dissolved in the liquid petroleum. They account for the pressure that forces petroleum to the surface in a "gusher" such as "Spindletop". An alternative theory, that petroleum is a sort of condensate from inorganic natural gas, does not alter the fact that hydrocarbons in useful concentrations are geologically scarce.

As applied to hydrocarbons, the McKelvey rule suggests that higher prices alone will suffice to forestall any actual shortage in the foreseeable future. The common sentiment among energy optimists until quite recently was that "at $100 per barrel there is an ocean of oil". That line was written 5 years ago (2010). Until mid-summer 2014 the price of oil hovered around $100 per barrel. This prompted a lot of exploration and investment, especially in Africa, Brazil and Russia. Discoveries continued, especially off-shore, but they are getting smaller on average and more expensive to exploit. No "ocean of oil" has been discovered, nor is such a discovery likely.

However, a technological breakthrough of sorts did occur: hydraulic fracturing ("fracking") of shale, to produce natural gas and "tight" oil (Sect. 10.3). The technology involves horizontal drilling and very high pressure fluid injection to release hydrocarbons attached to the surface of sand particles.[19] Since the 1990s, and especially after 2003 this technology has been successfully applied in Texas and subsequently in other parts of the US, especially North Dakota. It was extraordinarily low interest rates after the "Dot Com" crash in 2000 that financed large-scale fracking. The first outcome was a glut of natural gas that cut the US price to as little as a third of the European price.

A second consequence, resulting from the oversupply of gas, was a shift by the fracking industry from gas to oil, using almost the same technology. Oil production in the US nearly doubled in the years after 2007, making the US a major producer by 2014, second (barely) only to Saudi Arabia (and accounting for virtually all of the increase in total global production during those years). The simultaneous slowdown in the rate of Chinese economic growth, from 11 % p.a. to 7.5 % p.a., has cut demand and triggered another temporary glut. That glut has triggered a 40 % decline in global oil prices from July to December 2014. (As of January 2016 OPEC—meaning Saudi Arabia and the Gulf, has not agreed to cut production, probably to maintain market share.)

The third consequence, happening now, is a severe economic recession for all the oil exporters, whose national budgets are based on $ >100/bbl oil. Russia, Nigeria and Venezuela are particularly hard-hit. A fourth predictable consequence for the near future is that a lot of marginal drilling projects in remote places, and some US "frackers" with junk-bond financing, will have to drop out; capital expenditure on discovery and drilling by the "majors, will slow drastically (for a while) and investment in renewable energy will also slow or stop.

The fifth consequence is that cheap oil, for as long as it lasts, is very good news for the oil importing countries. Economic growth may rebound in Europe and China, thus stimulating demand for all commodities, including oil. The world is probably facing an extended period of oil price volatility, together with alternating periods of growth and recession.

The major point I have tried to make in this section is that the price of oil matters a lot in real-world economics, even if the neo-classical theory of growth pretends otherwise. How did this misunderstanding come about? Read on.

11.9   The Special Case of Petroleum

In the current economic system, the price of energy, at the margin, is essentially the price of oil. This is because oil from different places can be compared easily (e.g. in terms of viscosity, sulfur content and "sweetness") and because it is portable, storable, and therefore traded internationally to a far greater extent than coal or gas. It is also the least substitutable of the fuels because it is a liquid, and almost all

---

[19] The fluid is mainly water, of course, but there are a number of other additives, including methanol, isopropanol, ethylene glycol and a variety of salts, gels and proprietary chemicals.

of the internal combustion engines in the world are designed for liquid fuels. Hence the price of oil is a key economic indicator. Yet, in recent years the price of oil has been very volatile, to say the least, ranging over a factor of more than five from peak to valley, on several occasions: see Fig. 10.9 in Sect. 10.2, which traces oil prices and recessions since 1970.

With every oil price fluctuation, up or down, economic activity, especially automobile sales, has reacted (Hamilton 2003; Hamilton and James 2009). The most immediate response to higher oil prices, passed on to consumers at the filling station, is reduced travel mileage and reduced automobile sales. The immediate effect of oil prices on airlines depends on how much of the fuel price increase is passed on in ticket prices, but either way airline profits suffer and tourism is adversely affected. These sectors are important enough to affect the rest of the economy significantly, because as energy (gasoline) prices rise, households have less discretionary income to spend on other goods and services. In 2007–2008 when some home-owners were struggling to meet mortgage payments for houses they couldn't really afford, the rate of defaults spiked. The prices of houses declined sharply and the whole banking system would have collapsed, like the walls of Jericho, if not for the Troubled Asset Relief Program (TARP) and the bank bailouts that followed.

One longer-term effect after the first "oil shock" in 1973–1974 and again later after the second shock (1979–1980) was to shift home heating demand in the US and Europe away from oil to gas. There was also a modest shift away from big "muscle" cars with V-8 engines, toward 4-cylinder "compact" cars. Most of those were imported from Germany and Japan, resulting in significant job losses for the US auto industry.

As Fig. 10.9 shows, oil price spikes tend to be followed by recessions (gray bars on the chart). And, conversely, when oil prices fall, all of those economic responses go in the other direction. There is a "recovery". People have more money to spend. They buy more cars, drive them further, fly more often to distant resort destinations, buy yachts and so on. Economists tend to attribute recessions and recoveries to "boom and bust" business cycles resulting from over-production during booms (recoveries), followed by "inventory adjustments" (recessions). But that explanation does not explain either the great depression starting in 1930 or the "great contraction" after 2008.

If energy (oil) were not very important to the economy—as standard neo-classical theory asserts—then these boom and bust cycles would not be strongly correlated with energy price spikes. But the correlation is obvious in Fig. 10.9. It is true that the great depression that started in 1930 and the deep recession of 2009–2011 were financial in origin and not directly related to energy prices. And, as already suggested, it is possible that the very high oil prices of 2007–2008 contributed to the spike in mortgage default rates that started the financial collapse; (e.g. Hamilton and James 2009). As they say, what goes around comes around.

This pattern of energy-induced boom followed by "bust" has been operating, on a larger canvas, since the industrial revolution and especially since the major oil discoveries in the US starting in Texas (1901). There were brief periods after the US automobile industry started to grow rapidly c. 1910, when demand for gasoline threatened to outrun supply, but never for long. From 1920 or so through the early 1970s the supply of oil products was effectively unlimited, and prices were very low. No attractive new product or service was shut out of the market—apart from supersonic passenger planes—because of the high price of oil during those years. There was no incentive to economize on fuel. There was no incentive to invest in greater efficiency.

In fact, cheap oil was an economic stimulant in those years, and may be again for a few more years. Oil burners rapidly replaced coal-fired furnaces in homes and office buildings after WW II (before being replaced, in turn, by natural gas.) All kinds of motorized toys and appliances, from toy planes to motorboats, leaf blowers and golf carts were popularized. The first "oil shock" of 1973–1974 was a loud warning signal, but it was not heard by many. Geo-political events and "speculators" were blamed, as again in 1979–1980. All the experts expected cheap oil to be back soon. (And, for a number of years, it was.)

As mentioned in Sect. 10.1, the supersonic Concorde aircraft (and a Russian counterpart, the Tupolev) were designed, starting in 1954 (just after the huge oil discoveries in Saudi Arabia). Development continued into the late 1960s and 1970s. Service started in 1976 and continued until 2003. But when oil prices stopped declining in 1986 and started a long steady climb, the Concorde project (and supersonic transports in general) were dead ducks still flying. In fact, I argued in Sect. 10.4 that the whole automobile-based US consumer economy (and its offshore imitators) are unsustainable as presently configured, albeit the decline will not be quite as dramatic as that of the Concorde. The success of the Toyota Prius hybrid and the plug-in electric Tesla are just the first sounds of distant drums.

The European economic expansion from 1800 to 1910, led by Britain, France and Germany, was similar in many ways to the US experience after 1870, albeit more focused on empire-building and less on consumer goods, apart from textiles. Output was essentially final demand limited, rather than resource supply limited, as was the case before coal became widely available and cheap. After that transition (Sect. 9.2) fuel prices were declining and lack of useful energy (exergy) was never the limiting factor in any domestic economic enterprise or foreign venture.

Moreover, most of the products and industries that drove nineteenth century economic growth—the steam engine and the railroad, iron and steel, gas-light, coal-tar chemicals and electric power generation from steam turbines– were direct consequences of the use of coal. Similarly, the main driver of twentieth century growth, except for electrification, has been the internal combustion engine (and motor vehicles), direct consequences of the availability of liquid fuels from petroleum. All were due to overcoming various technological challenges.[20] In short, they all illustrate Schumpeter's idea of "creative destruction".

---

[20] Electric power, chemistry and telecommunication are partial exceptions, due more to scientific progress and inventive genius than natural resource availability or scarcity. Having said that, much of the history of the chemical industry can be traced to (a) the availability of by-products of coking and (b) ammonia-synthesis technology developed to meet the agricultural and military need for nitrogen-based chemicals.

The idea that economic growth may be limited by physical resource scarcity has had no traction in policy circles until very recently. (And it will be disregarded again in the coming year or so). But it actually has quite a long history, dating back to the eighteenth century. It goes back, at least, to the French "physiocrats"(Quesnay 1766) and especially Thomas Malthus, who saw arable land as the limiting factor (Malthus 1798 [1946]). In recent years "Malthusian" concerns have been expressed about future supplies of food (e.g. "to feed China") (Brown and Lester 1973; Brown 1995), about fresh water availability (Postel 1999) as well as about future supplies of certain metals that are now important for electronics (e.g. Graedel and Erdmann 2011).[21]

It is perfectly true that temporary localized shortages of commodities are usually relieved fairly quickly, helped by price adjustments. But it is also true that global reserve capacity for temporarily increasing the output of many resources, including oil and fresh water, is declining. In some cases, such as water, it is close to zero. However, Malthus' gloomy forecasts turned out to be wrong (at least premature) and the adjective "Malthusian" has become a derogatory term, as used by economists, especially in regard to the famous *"Limits to Growth"* book (Meadows et al. 1972).

Since the 1950s a few economists have explored the idea that economic growth may be limited by the buildup of industrial waste products in the environment (e.g. Boulding 1966; Kneese et al. 1970). Growth-limiting pollution is not a Malthusian problem, *per se*. But the Earth's capacity for diluting and detoxifying pollutants is a resource, and it is finite. Moreover, acidification and eutrophication of the environment, deterioration of soil, destruction of the ozone layer and climate change due to the buildup of greenhouse gases (GHGs) are all consequences of waste emissions. The costs of pollution abatement measures, or damage repair, whatever they may be, are also subtractions from the net surplus produced by the economy, in much the same way as medical expenses are deductions from household disposable income.

Unless there is another global depression comparable to the one in the 1930s, the cost of oil to the global economy is virtually certain to rise in coming decades. This follows from several lines of reasoning. One is the fact that the average exergy return on exergy invested (EROI) has been declining since the 1930s. In fact, if we had earlier data, we would probably see that it has been declining most of the time since 1860, but with interruptions for momentary "spikes" due to occasional large oil discoveries e.g. in Azerbaijan, Texas, Kuwait, Iran, Mexico, Venezuela, California, Texas again, Saudi Arabia, and more since then. However, really big giant and "super-giant" discoveries, such as Prudhoe Bay (Alaska) and the North Sea have been getting very scarce.

In 1931, the number of barrels of oil produced for each barrel (energy equivalent) needed to dig the wells and transport the product to a refinery (EROI) was over

---

[21] For example lithium, platinum, cobalt, tantalum, indium, tellurium and several of the rare Earths.

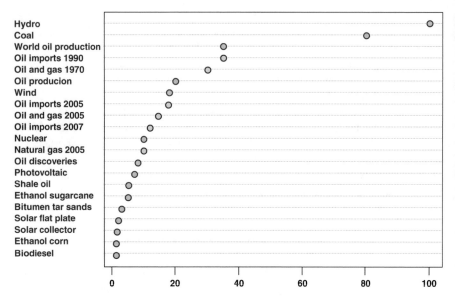

Fig. 11.7 Ratio of Energy Return On Energy Invested (EROI)—USA 2010. *Source*: Murphy and Hall (2010) Ann NY Acad Sci 1185:102–118

100:1. By 1980 the global average EROI was down to 30 or so. See Fig. 11.7. According to a recent estimate, from 1999 to 2006 the EROI for global oil and gas production declined by nearly 50 % from 35 to 18. From 1993 to 2009 the EROI for natural gas in Canada fell from 38 to 20. The EROEI of nuclear power ranges from 5 to 15. The EROI for conventional hydropower is still greater than 100; for wind turbines it is about 18; for PV it ranges from 6 up to 12 (much better than earlier estimates). But EROI for ethanol from sugar cane ranges from 0.8 to an upper limit of 10. Ethanol from corn is 0.8–1.6 while biodiesel (from rape or sunflower oil) is about 1.3.

Many economists and "peak oil" skeptics, and even some physicists, have argued that everything, including petroleum, has a substitute (e.g. Goeller and Weinberg 1976). But the substitute is not necessarily an equivalent (Graedel and Erdmann 2011). Canadian tar sands and ethanol from corn are supposed to be substitutes for Saudi oil, but they use more energy and cost a lot more to produce. "Clean coal" is sometimes suggested as another alternative. But coal is not a viable substitute for liquid fuels except insofar as it is used to make electricity for electrified trains or vehicles. It is true that coal can be converted into a liquid fuel by a hydrogenation process (Nazi Germany did that in the 1940s), but from a third to half of the exergy content of the coal is lost (Kenney 1984).

Coal is the dirtiest of the fossil fuels, so the sobriquet "clean coal" is only an advertising slogan (and an oxymoron). All coal contains contaminants of sulfur, nitrogen, mercury, ash and other pollutants. Most old coal-burning power plants do not do a very good job in capturing and treating all those pollutants before they get into the atmosphere. Improvements in that regard—including carbon capture and sequestration (known as CCS)—are possible (even probable) but not at zero cost. In fact a CCS system for a coal-fired power plant will probably consume 20–30 % of the electricity output, to separate, compress and transport the carbon-dioxide to a safe underground storage place. And, as a skeptic might ask, "who wants to live over an underground cavern full of compressed carbon dioxide? What if there is an Earth tremor? What if it leaks?" However, carbon dioxide is not captured and removed from the exhaust gas of any power plant anywhere in the world, up to now.

As pointed out in Sect. 11.5, most economists seem to believe that there is really no such thing as geological scarcity, because better technology always overcomes declining resource grades. They cite the historical record. History confirms that most natural resource commodity prices have declined more or less monotonically since the beginning of records (Barnett and Morse 1963). However, the evidence since 1970 is much less clear-cut, and since 2002 the story for oil looks different. The fact is that petroleum is obtained from underground reservoirs that have finite lifetimes, while the rate of discovery of new oil has been declining since the peak year (1963). All of the six major Saudi Arabian fields[22] accounting for about 95 % of Saudi output were discovered between 1939 and 1967. The biggest of all, Ghawar, was discovered in 1948. They are all located near each other in a tiny corner of Saudi Arabia, just over 17,000 km^2 in area (Simmons 2004).[23]

Saudi Arabia has not published detailed production data by oil field since 1987, so the true reserve situation is obscure, to say the least. But some things are known. One is that official claims of proved reserves in all of the Persian Gulf states were increased sharply, one after the other, in the late 1980s, without any discoveries. The reason for this series of competitive revisions probably have to do with market share in the OPEC cartel. But what is certain is that those "proven reserve" figures are not reliable (Campbell and Laherrère 1998). The awkward truth is that nobody outside Aramco and the Saudi government really knows how big the remaining reserves are and how long they will last (Laherrere 2014). However, see Figs. 10.3, 10.4, 10.5 in Sect. 10.2 showing a large increase in reserves not based on new discoveries.

Another fact is that the peculiar geological conditions that make those Saudi wells so extraordinarily productive have not been found anywhere else in the world, despite intensive worldwide search. The territory of Saudi Arabia itself has been fairly thoroughly explored by Aramco (except for a few remote areas), as one would expect, given the importance of oil to the Saudi economy and the age of its producing wells. Most of those six "super-giant" fields have been in production since the 1950s. In Simmons words, written in 2004:

> the chances of finding a great oil giant in Saudi Arabia that has eluded discovery so far must now be deemed remote. . ..Unless some great series of exploration miracles occurs soon, the

---

[22] Ghawar, Safaniya, Abqaiq, Berri, Zuluf and Marjan. Ghawar is by far the biggest, still accounting for around half of the total flow.

[23] Matthew Simmons is an investment banker specializing in oil.

only certainty about Saudi Arabia's oil industry is that once its five or six great oilfields go into steep decline, there is nothing remotely resembling them to take their place. (Simmons 2004)

Those "legacy" wells are now 10 years older than when Simmons wrote those words, and indications of decline are increasingly ominous. The very secrecy of the Saudi government may well be one such indication. Moreover, as early as 1974 there were strong indications that Ghawar and the others were over-producing (thus cutting lifetime output), probably to maintain dominance in OPEC and to establish the notion that Saudi Arabia has the world's only reserve pumping capacity. It must be acknowledged that the Saudi situation, and that of Ghawar in particular, is critical to the near-term future of global oil supply. Official pronouncements on that subject should be taken with a tablespoon of salt.

Regarding the future price of oil, Matthew Simmons had this to say in 2004 (Simmons 2004)

> ... the uncertainty ... relates to the poor correlation between Generally Accepted Accounting Principles (GAAP) and the real cost to create added oil supply. GAAP methodology as it relates to oil and gas is flawed. The numbers it creates are heavily tied to the practice of capitalizing all costs of developing oil and gas supplies and then expensing these costs by an estimated volume of proven barrels over the life of an oilfield. Over the past decade or two these numbers created an impression that the cost of finding and developing oil and gas was around five to six dollars per barrel. These numbers, while widely used as guidelines ... are close to meaningless as a basis of determining what oil should cost ... (Simmons op cit. p. 345)

In an interview with the Association for the Study of Peak Oil (ASPO) in March 2014, ex-Aramco geologist Dr. Sadad-Al-Husseini predicted that oil prices would spike at $140/bbl by 2016–2017. (In March 2014 the price of Brent crude was about $108/bbl.) In the interview his 2009 output projections were compared to current 2014 actuality. North American production was above projection by 2.631 Mb/d and the Middle East was above trend by 1.183 Mb/d. Asia was slightly above projection by 0.32 Mb/d. All other regions were lagging below forecast outputs, notably Africa (−1.922 Mb/d) and South America (−1.069 Mb/d) with Europe at −0.461 Mb/d.

The volatility of historical oil prices, and the complexity of the exploration, development and marketing systems is daunting. Most economists, even today, sidestep the difficulties by simply assuming that the future oil supply will be determined by future demand, which will be determined by a GDP forecast. The price issue is almost never explicitly addressed.

In early 2011 the IMF's global outlook forecast global economic growth of more than 4.5 % p.a. for the years 2011 through 2016 (actually accelerating from 4.4 % in 2011 to 4.7 % in 2016). It assumed constant prices—hence no price effect—and an average income elasticity of 0.685, which means that 1 % increase in global GDP would call forth a new supply of 0.685 %. Thus a global growth rate of 4.5 % implied an increase in global oil supply of about 3 % p.a. In terms of cumulative totals, this implied an increased global output of 17 Mb/d by 2016, whereas there had been very little increase (3–4 Mb/d) since 2004 despite higher prices. Where was this new oil supposed to come from?

It has been pointed out that if the constant price assumption were relaxed, and prices were allowed to increase just enough to cut consumption growth from 3 % p.a. to 2 % p.a. using the IMF's assumed price elasticity (-0.019), total new supply of "only" 11 Mb/d would be needed in 2016 (Stanisford 2011). But, that, in turn, implied an annual price increase of 53 % for the next 5 years. This did not happen, of course, because a much smaller increase would have stopped economic growth and brought on another recession or depression. If Stanisford (or the IMF) slipped a decimal point and the assumed price elasticity is -0.19 % (much more likely) the annual oil price increase would still have had to be 5.3 %. To keep global oil consumption constant (no price increase) the annual price increase would have had to be several times higher still—and the economic consequences would have been severe. OK, we have a glut, as I write, and the price of oil is down, rather than up. But that is not a recommendation of the standard IMF model, either.

As Reiner Kuemmel and I have argued in previous publications,[24] and as this chapter argues, the real demand for oil (and energy) in future—after 2020 or so—is likely to be constrained by the supply. In that case, the direction of causality will be reversed: GDP growth will be determined by oil supply (and price). Just after the IMF publication mentioned in the previous two paragraphs, members of the IMF research staff "led by Michael Kumhoff" modified the big computable general equilibrium model (CGE) to take oil prices into account *for the first time*. The IMF's main annual "outlook" reports still do not reflect these modifications to the model, but staffers have been able to use it for research. Figure 11.8 is the result of the first runs (Kumhoff 2012). Note that, in fact, there was no price dip after 2011, whereas 2014–15 saw a 50 % decline. It would seem that the decline was late. Economic models are still far from reliable forecasting tools.

The model forecast showed a slight price decline in 2011, followed by a steady increase. In reality, there was no decline until mid-2014, after which the decline was much sharper. However, the model is no worse than many and I suppose that it could be improved after re-calibration based on actual data.

11.10   The Role of Resources in Standard Economic Theory

It is high time to confront the single major problem with neoclassical economics, which is the dominant paradigm today. Many criticisms can be made of neoclassical economics. They include the assumption of growth-in-equilibrium, utility (as consumption) maximization, perfect knowledge, declining returns, and the "representative agent" idea. This is not the place to discuss all the problems with

---

[24] See Kuemmel (1980), Kuemmel et al. (1985, 2010), and Ayres and Warr (2009).

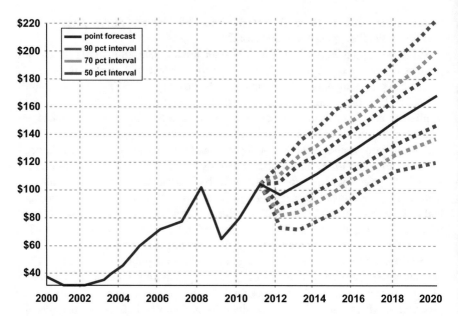

Fig. 11.8 Oil price forecast made in 2012 by IMF staffers. *Source*: IMF Report. Benes et al. (2012). The future of oil: Geology versus technology

contemporary theory. For a recent and very readable book that discusses many of them, see *Debunking Economics* (Keen 2001). See also Arthur (1994).

The Big Idea of this section (and this chapter) is simple: economic theory, especially as applied to economic growth, has grossly under-estimated the importance of "useful" energy (exergy) and "useful" work. As already pointed out in several places, most of the useful energy consumed in the world economy today comes from fossil fuels. Not only that, most of the industrial technology that supported economic growth in the nineteenth century and most of the twentieth century was "invented" and developed to utilize natural exergy resources, notably coal and petroleum.

Yet the standard theory of economic growth assumes that growth is essentially automatic ("the economy wants to grow") and that it happens smoothly thanks to the accumulation of capital per worker, although the precise mechanism is unclear. The role of increasing returns, complexity and path-dependence (Sect. 11.5) is not widely understood. Most economists agree that increasing knowledge or "social learning" play a part. Surprisingly, they also agree that growth does not require (much) energy, notwithstanding the fact that energy (oil) price "spikes" since the 1970s have been repeatedly followed by damaging recessions (Hamilton 2003, 2005). *The standard economic theory since the 1950s says (in effect) that natural resources are not essential because there is always a way to a substitute a little more capital or a little more labor) for any scarce physical resource* (Solow 1974a; Stiglitz 1974; Dasgupta and Heal 1974).

The mid-1970s were a time when economists were so impressed with the achievements of technology (and scientists were so impressed by themselves) that a simple story of the current and former uses of the element mercury, and their substitutes, was published with great acclaim as *"The Age of Substitution"* in *Science* and then reprinted in the *Journal of Economic Literature* (Goeller and Weinberg 1976). The element mercury has, indeed, had a long history of uses (and misuses) most of which have subsequently been replaced because of toxicity (like its nineteenth century use in making felt hats).[25] But *exergy* is not (like mercury) just one of 100 chemical elements whose properties are mostly substitutable by others. Exergy, by contrast, has no substitutes.[26]

In this context, it seems worthwhile to mention the work of the most extreme anti-Malthusian economist of all, Julian Simon (Simon 1977, 1981, 1996). Simon saw no problems at all from population growth or resource consumption. He argued forcefully that, in the absence of government regulation or restrictions, human ingenuity would overcome any geological scarcity. In 1980 he famously bet (against Paul Ehrlich, John Harte and Jon Holdren) that the prices of a basket of five metals (aluminum, copper, nickel, chromium and tin) would decline during the following decade. His opponents thought the prices of those metals would increase. Simon won that bet in 1990.[27] The libertarians, who oppose government regulation of all kinds, loved his ideas, and he was lionized by them in his later years. When he died of a heart attack in February 1998, the *Economist* devoted two full pages to his obituary.

So much for personalities. I need to explain now why most mainstream economic theorists think (assume) that exergy is unimportant. It starts with a model, described in most macro-economic textbooks. The model assumes that a large number of small producers ("representative agents") compete, making a single "good" (call it "GDP") from rented capital and rented labor. The capital and labor are independent of each other, and can be used in arbitrary amounts. (The rental agent is on Mars.) Since the small producers are rational, it follows that they will utilize capital and labor in proportion to their respective productivities. The "proof" of this theorem is found in every textbook of macro-economics. See, for example (Mankiw 1997). (N. Gregory Mankiw is currently the Chairman of the economics department at Harvard University.) In fairness, I need to say that the proof is OK if you accept the assumptions. That is where the problem arises.

It follows that, in the same model economy that was assumed for that proof, the *output elasticities* (logarithmic derivatives of output with respect to inputs) with

---

[25] The "mad hatter" in Lewis Carroll's story was based on a real health problem from mercury poisoning in the hat industry.

[26] Nor are there any realistic substitutes in practice, for fresh air, fresh water, wood, concrete, steel (as a construction material), copper (as an electrical conductor) or platinum group metals as catalysts.

[27] On the other hand he lost a later bet with David South, a professor at Auburn, that timber prices would fall. They rose. Simon attributed that loss to government interference (regulation) in the Pacific Northwest.

respect to capital and labor must be equal to their respective *cost shares* in the economy. (Remember that the national accounts define GDP as the sum of all payments to labor plus all payments to capital). All the expenditures in the economy are traditionally allocated between those two categories. Payments to labor consist of wages and salaries, Payments to capital include interest, dividends, rents, and royalties. There are no payments in the national accounts to energy, as such, only to energy companies employing capital and labor. Thus, by assumption, energy can be left out of the production function because it is (by assumption) "produced" by some combination of capital and labor.

This "cost-share theorem" is accepted, without question, by most academic economists as a fact applicable to the real economy. See, for example, Denison (1979). Few have challenged this assumption, at least in the journals that most economists read. However, as has been pointed out, if some of the payments to labor and capital are segregated as payments to energy-producing activities, or just payments to energy, it can be shown easily that the "cost share" theorem is not applicable when there are three or more "factors" that are interdependent. In other words, the cost-share theorem is invalid if the three factors of production depend on each other or are subject to exogenous constraints (Kuemmel et al. 2010). A more detailed mathematical exposition can be found in Kuemmel's work (Kuemmel 2011). By "inter-dependent" in the first sentence, I simply mean that the capital stock requires some labor and energy input to function, while the labor force requires metabolic energy (food) as well as tools (capital) and the energy supply depends on both capital and labor inputs.

The cost share of energy, excluding food and animal feed—counted as payments to energy activities (companies)—in the OECD countries for most of the past 30 years or so, has been around 5 %. However, while fossil energy has been cheap, mainly because the fossil fuel resources have been so plentiful and easily discovered, a small direct share of GDP does *not* imply unimportance as a factor of production if the other factors depend on energy. On the contrary, energy (as useful work) is a direct input to every economic process.

More to the point of this chapter, energy (as work) is also an invisible but crucial input to both of the other factors of production: capital goods and labor. In other words, the assumption that the factors of production are independent of each other is wrong and, frankly, absurd. Unfortunately, the prices of capital and labor services do not explicitly reflect that fact, although it is obvious after a moment's thought.

Yet, as was argued in Sect. 11.6, rapid economic growth since the industrial revolution has been largely due to the technological inventions and innovations that followed the discoveries of coal and oil resources. Nuclear power might possibly have opened a new door and triggered another set of powerful new technologies, but nuclear power seems to have reached a dead end, at least for now.[28] The new

---

[28] In principle, there is a much safer form of nuclear power, based on thorium, using liquid reactants rather than solid fuel rods (Hargraves and Moir 2010). However the development of practical thorium-based plants appears to be far in the future.

technology that is driving growth in the industrialized countries today is centered on the internet. But when everybody has a smart-phone that enables one to surf the internet, trade stocks and make bank deposits or withdrawals while sitting in a café (or a car), or by the pool, what then?

Innovation in the "real" economy that produces goods seems to be slowing, along with economic growth. The great inventions that drove economic growth in the past, like the steam engine or the internal combustion engine, electrification and digital information processing, were "one time" events in the sense that their influence on productivity is now largely spent (Gordon 2012). A new resource or discovery that can incubate a new set of productive technologies is needed. Is there such a resource? Are there any new production technologies waiting to be discovered?

Since energy from fossil fuels has been so central to the modern history of economic growth it should be no surprise that some economists eventually woke up to the phenomenon of resource depletion and diminishing returns. (The surprise is how long it took.) The dismal forecast of Thomas **Malthus** in 1798 was, in a sense, the mother of all catastrophist theories (Malthus 1798 [1946]). Malthus, **David Ricardo** and **John Stuart Mill**, the successors of **Adam Smith**, all predicted that natural resource scarcity—they were thinking of fertile agricultural land—would result in "diminishing social returns to economic effort". Economic effort could be either human (and animal) labor, capital investment, or both. The end result, they thought, would necessarily be a slowdown and cessation of economic growth resulting in a static and stationary society. From those considerations, the "classical" view of scarcity evolved, namely that with depletion of finite resources, and diminishing returns, the cost of extraction must increase. Kenneth Boulding and Herman Daly are the most recent economists to take that scenario seriously (Daly 1973, 1991).

The classical view has been increasingly challenged since the landmark paper by **Harold Hotelling** (Hotelling 1931), based on his earlier work on the mathematical theory of depreciation. Hotelling's "rule" is that the exhaustible resource prices should rise at the rate of interest (or the social discount rate). That rate is taken to be the rate of interest (returns on capital) for the economy as a whole. However, the model was based on a number of very unrealistic assumptions, e.g. that extraction costs are independent of extraction rate, and that they remain constant over time.

Hotelling's model was ignored for three decades, due to its unfamiliar mathematics and the competing concerns of the Great Depression and WW II. Moreover, its central prediction of rising prices for exhaustible resources (the "Hotelling Rule") was contradicted by the empirical results of Barnett and Morse, discussed in Sect. 11.5 (Barnett and Morse 1963). Nevertheless, the so-called optimal control methodology employed by Hotelling, became central to the furious response of economic theorists to the publication of the Club of Rome report "Limits to Growth" in 1972 (Meadows et al. 1972).

Many economists, from Solow to Hayek and especially those of the Julian Simon-Herman Kahn "cornucopian school", have observed that apparent scarcity in the material domain often encourages innovation leading to a substitution for the scarce resource. I don't disagree, of course: that was the whole point of Sect. 11.6. It can even happen that a radical innovation (or substitution) actually reverses the process of decreasing returns. This has happened several times in the history of technology when the scarcity of one resource triggered technological innovations that not only enabled the shift to a more available (if not "better") resource for the original purpose, but also opened completely new markets.

The substitution of coal for charcoal (due to deforestation) was one such example. It led to gas light, aniline dyes, nitrogen fertilizers and—finally—to the stationary internal combustion engine. That, in turn, led to the automobile and all its cousins. The substitution of petroleum for whale oil was another example: It led to powerful drilling technologies, fractional distillation technology, naphtha "cracking" (to make gasoline) and applications of all of the refinery by-products, from asphalt to lubricating oil. To those I would add (with natural gas) the whole petrochemical spectrum: synthetic fibers, synthetic rubber, plastics, aspirin, vitamins and sulfa-drugs.

In the mechanical sphere, the substitution of steam turbines for reciprocating steam engines in the 1890s was another example. The electric power industry and the aircraft industry could not exist in anything like their present form without steam turbines. Finally, the scarcity (unavailability) of natural rubber during World War II was the inducement for synthetic rubber development.

The last few pages certainly acknowledge the creative power of human ingenuity, especially when confronted by barriers. The term "breakthrough" is apposite. (I have written about a "barrier-breakthrough" model of technological development (Ayres 1988)). But human ingenuity cannot overcome every barrier. Time travel, faster than light travel, anti-gravity and eternal life are fairly obvious examples. (The fact that "science fiction" writers routinely ignore these barriers is not proof that they can be overcome in the real world).

A fundamental question at issue in executive suites is whether geological scarcity is nullified, for practical purposes, by free-market economics, i.e. market incentives, appropriate public policies and new technology. Another book from RFF, "*Scarcity and Growth Reconsidered*" was based on a Forum in San Francisco in the fall of 1976 (Smith and Krutilla 1979). It is obvious from the timing that the then-recent 1973–1974 Arab Oil boycott and the resulting turmoil, was fresh in a lot of minds. The link to thermodynamics was particularly emphasized by Herman Daly (Daly 1979). His point is made by citing a statement from the 1963 book that provided the sub-text for the 1976 RFF Forum (Barnett and Morse 1963, p. 11).

> ... Science, by making the resource base more homogeneous, erases the restrictions once thought to reside in lack of homogeneity. In a neo-Ricardian world, it seems, the particular resources with which one starts increasingly become a matter of indifference. (Cited by Daly, op cit, p. 69)

In effect, Barnett and Morse were conceding that the high quality resources are being used up, but arguing that it didn't really matter, since—thanks to endogenous technological progress—resources could always be inter-converted or extracted even from the homogeneous stuff of the Earth's crust. That conclusion is a sign of ignorance about geology and chemistry, and it is dangerously wrong.

Let me explain with an example. A growing economy needs increasing amounts of copper. Copper is currently mined from so-called porphyry deposits of decreasing grade, ranging from 2 % in a few mines to 0.5 % in others. The copper minerals of interest are chalcopyrite, chalcocite, bornite, azurite, covellite and malachite. They are all individually rare. Chalcopyrite is the most common among them (0.0066 % of the crust) but they are found together in veins near "the ring of fire". The first stage of extraction is grinding to separate the copper minerals from the others. The lower the grade of ore the more grinding (and more exergy for grinding) is needed per unit of copper content. The next stage, to separate copper-containing minerals from the other minerals, can be based on relative solubility in strong acids, relative density (which requires centrifuges) or surface characteristics requiring additives such as flotation agents.

These unit operations invariably yield two output streams, one of which has a higher concentration than the other. The low concentration stream is either fed back into the main feedstock (thus diluting it) or thrown out. The higher concentration stream often has to go through the same process again—perhaps multiple times—to obtain a smelter feedstock with a reasonably high percentage of copper-containing minerals. The lower the grade of ore, the more repetitions it takes to produce a concentrate. More repetitions also consume more exergy.

To some extent, improved technology can overcome the progressively larger exergy costs of separation as grades decline. But beyond some point the multi-stage separation methodology itself becomes inapplicable. Suppose all the mineable copper minerals (chalcopyrite, etc.) have been mined and used up. Suppose all that is left is average rocks with a few atoms of copper hiding in among the atoms of crystal structures of quartz, albite, oligoclase, orthoclase, andesine, paragonite, biotite, hydromuscovite/ illite, augite, and hornblende. You don't recognize any of those names (except the first one on the list)? In fact, they are the ten most common minerals in the Earth's crust, in order of frequency.

The reason you probably don't recognize any of their names is that they are all metallurgically uninteresting. They are compounds of eleven elements: oxygen, hydrogen, silicon, aluminum, magnesium, sodium, potassium, calcium, iron, a trace of titanium and trace of fluorine. Every one of those 11 minerals (except quartz) contains at least 4 different elements and several of them contain 8. The next 20 minerals (in order) are almost equally uninteresting. (The first mineral containing a new element (P) is phosphate rock, with only three elements Ca $(PO_4)_2$ which is number 32 on the list. It has a concentration of 0.03 % of the Earth's crust.)

But the question was: Can we extract copper from those minerals? If we grind them up we get what miners call "gangue". Think of it as dirt. It does not dissolve in water. There is no acid that will dissolve "dirt", so chemical processing is no longer possible. (Any soluble minerals like chlorides or nitrates are already dissolved in the oceans). If you really want to separate a very rare element that is atomically dispersed in "dirt" the only thing to do is to heat it into an ionized "plasma"—at a temperature of at least 3000 °C or in an electric arc—consisting of individual ionized atoms. Then one can use a mass spectrometer to separate the atoms in the ionized plasma by weight. This is possible in principle, but extremely far beyond practicality for industrial metallurgical operations.

Bottom line: most of the world's copper is permanently unavailable.

What Barnett and Morse failed to mention or understand (among other things) is what might be called a "double whammy": The first whammy is that the exergy requirements of separating desired materials from a mixed mineral mass rises without limit, as the concentration (grade) of the ore or resource falls. The second "whammy", as you have probably already guessed, is that the fuels and electric power needed for copper ore treatment are themselves produced from increasingly low-grade sources, and this applies not only to fossil fuels but also to renewables. It follows that essentially all extractive resources will be increasingly exergy-intensive as time goes on. This contradicts one of the fondest assumptions of neoclassical economists, especially Julian Simon and his followers, who believe that human ingenuity will always triumph and there is a substitute for everything.

One of the discussants at the 1976 RFF Forum was Nicolas Georgescu-Roegen. He was scathing in his criticism of the mainstream economist's view, as presented on that occasion by Stiglitz and others. His criticism was partly focused on the standard emphasis on exponential economic growth, partly on utility maximization as a strategy (at least in theory) and partly on the standard treatment of inter-generational equity and discounting the future, with all of which he strongly disagreed.

But G-R's other disagreement with mainstream economists was with their failure to recognize the physical existence and "finitude" of resources. The standard approach, exemplified by Stiglitz and Solow (as well as others), treats resources and capital as abstractions and assumes that man-made capital can be substituted for natural resources, essentially without limit. Solow has been explicit about this (Stiglitz 1974; Solow 1974a, b).

Georgescu-Roegen says that treating resources as abstractions, without mass or real properties, is a "conjuring trick". Again and again, his writings point out that capital, in the sense of equipment and infrastructure, is not an abstraction, and that real physical resources are required to create the capital assets that on which the economy depends. G-R could have added that abstract transactions between abstract "agents" need not have any consequences for third parties (externalities) but that when real physical and energy transformations are involved the likelihood of externalities (e.g. pollution emissions) is very high.

Mainstream economics has still not absorbed any of these arguments. The Solow-Stiglitz response to G-R amounts to arguing that the neoclassical assumptions are OK for the foreseeable future and after that nobody cares (Stiglitz 1997). According to the standard (Solow) theory, economic growth can continue without limit and without useful energy (exergy). This is because the theory allows for only two factors of production, capital and *labor*, which implies that energy is an intermediate good, created (in his model) by abstract capital and abstract labor. But energy is conserved; it cannot be created or destroyed, so it cannot be an intermediate good.

The economic vs. thermodynamic perspectives cannot (and need not) be reconciled at this point. But it is important to recognize that they differ in some truly fundamental respects.

Chapter 12
New Perspectives on Capital, Work, and Wealth

12.1 Active vs. Passive Capital

In the last chapter I pointed out how neo-classical economics evolved and why it neglects resources, including the prime mover of all, which is energy. In this chapter I will attempt to repair those omissions. For a start, it is useful to think of capital in energetic terms by making a distinction between "active capital" and "passive capital". For animals the active components are muscles, the digestive system, the cardio-vascular system and the brain, while the passive components are bones and skin. In the economy the active components are human workers (actually only their eyes, hands and brains) plus information processing equipment (computers, telephones etc.) and power-projection equipment consisting of engines, motors, generators, and machines with moving parts. Passive capital in the economy consists of structures that exist mainly to support, contain and protect the active capital equipment. The latter also includes "infrastructure" (roads, rails, bridges, tunnels, harbors, pipelines, electrical transmission lines). It also includes knowledge and intellectual capital.

Air, food and water are passive carriers of essential nutrients into the human (or any animal) body, while carbon dioxide, urine, perspiration and feces are metabolic wastes. Blood plasma and lymphatic fluid are the passive internal carriers of oxygen via red blood cells, carbon dioxide and other wastes, as well as the protective "white" cells that protect the body from alien bacteria. Blood vessels merely contain the blood. Nerve cells convey electrical signals to and from the brain. The blood vessels, plasma, lymph and nerve fibers are inactive.

Obviously, pipes and wires are intrinsically inert carriers of liquids, gases, electric power and information (signals or messages) throughout the socio-economic system. Money also moves through the economy, much like oxygen in the blood. Money can be active (financing investment) or passive (repayment of debt). More on this topic later.

© Springer International Publishing Switzerland 2016
R. Ayres, *Energy, Complexity and Wealth Maximization*, The Frontiers Collection,
DOI 10.1007/978-3-319-30545-5_12

This distinction between active and passive capital is important, because all of the active capital in the economy (engines, etc.) requires a flow of exergy in the form of fuel or electric power to do "useful work", whereas the passive capital (structures, pipes, etc.) merely exists and requires no exergy input to perform its functions. (Maintenance and repair operations are, of course, active). Passive capital can be highly productive, in the sense of *enabling* production of goods and services by active means. For instance, canals and roads enabled production (before railroads) but only when there was active capital doing the carriage work: mules towing barges or teams of horses pulling carts or carriages. When the railroads came, it was the steam-powered locomotives that did the work. (In fact, one can argue that it is only the engine or electric motor that does the work while the rest of the vehicle is also passive capital. I lean toward this view).

Intellectual capital, such as journal articles, software, chemical formulae, novels or films is essentially passive. The same is true of TV sets to receive news or entertainment, TV stations and cable networks to broadcast it, or distribute the movies. It is only productive—value-creating—when there is also active capital at work, such as computers crunching numbers, or algorithms controlling flows in a chemical plant. (The distinction between active and passive is blurry in this domain. A TV set or lap-top computer may be regarded as "active" when it is turned on, but not processing data, and "passive" when it is turned off.) The important point here is that the active component is essential but quantitatively unimportant in terms of mass or energy embodied.

Industrial capital equipment is passive when it is unutilized and active when it is being utilized (and when human workers are being employed as operators). There is a debate among economists as to whether production (i.e. GDP) is entirely driven by human labor and capital services, or whether exergy (or work) should be regarded as an independent "*factor of production*".

A point of interest to economic growth theorists is that the exergy consumption by an economy may be a good proxy for the current utilization rate of capital stock. This would explain why production functions that do not explicitly reflect exergy (or work) flows may still have a reasonable explanatory power. This is a topic that deserves further consideration.

12.2   Exergy, Useful Work and Production Functions

Evidently there is a very close causal connection between active capital, exergy consumed, work done and the production of goods and services. That work adds information or "order" to raw or processed materials and, in so doing adds economic value. For instance, an alloy that has a very important characteristic (such as hardness, conductivity, or ferromagnetism) may require a very precise combination of elements in an alloy. The precision required is defined by a formula, and the formula can be regarded as information. The information can be regarded as negative entropy (negentropy). The greater the precision, the greater the information content. So it is not unreasonable to assert that economic value is closely related to useful information (negentropy) added. However, they are not quite the same thing, because not all information has value, and information is not a substitute for exergy. But, hopefully you get the idea.

Adding structural information (desired form, desired characteristics, surface finish) to physical materials is what manufacturing is all about. The sequence of steps can sometimes be done in more than one way. After extraction from nature, the first step is cleaning (or purification), to remove unwanted chemical elements or contaminants. For wood chips the next step will be reaction with caustic soda and pulping. For paper, the pulp is bleached, rolled, dried and coated. It is then ready for printing.

For metal ores, grinding may be followed by sorting by size and weight, followed by flotation or magnetic separation, or dissolution in acid followed by precipitation, to separate valuable minerals from gangue. This is likely to be followed by chemical reactions (e.g. electrolysis to remove the chlorine from chlorides or the fluorine from fluorides, or smelting with coke) to remove oxygen from oxides. The next step may be further refining or recombination (alloying or chemical synthesis) to improve the properties of the materials. Heat treating for hardening of metals would fall into that category. Next (for metals) might be forming into useful or beautiful shapes, ranging from films to flat plates, bars, wires and tubes to complex shapes like crankshafts or Michelangelo's statue of David.

The thermodynamic work done to add information can be measured in terms of electrical work kilowatt-hours (or horsepower-hours) consumed in the various stages of mechanical work, chemical work or thermal work. These are easily measured in any given case, but there is no general theory to determine how much exergy is required per gigabyte of useful information added, nor is there any general theory that quantifies the information (in gigabytes) corresponding to a complex shape or a chemical combination.[1] Clearly, the exergy required to add information depends on the specific technology. It happens that the process technology in some cases has been improving at a spectacular rate. It also happens that every stage in the manufacturing process entails losses (entropy creation). But nobody really knows what the limits are, or how near (or far) we are now from those limits.

However, from a macro-economic perspective, information technology can be regarded as an artificial extension of the human brain. Human labor, nowadays, is mostly brainwork (decisions), assisted by eyes, ears, and sense of touch, and implemented by muscular motions of fingers, hands, arms and (rarely) legs. The days of John Henry, "the steel driving man", and others like him are long gone in the industrial world. Computers do not do muscle work, so they do not directly replace horses and oxen. But computers and computer-controlled machines are replacing humans in a variety of computational and word processing tasks, with

---

[1] For a discussion of these problems, see (Ayres 1994). Chapters 9–11, pp. 215–270.

many more applications to come. Driving cars may be one of the next task descriptions to be taken over by computers. Hence information technology should probably be regarded as a form of labor-augmentation.

Thermal work is the work done by endothermic chemical processes (including smelting of ores) or to melt or soften metals for forming. Electric furnaces are examples of machines to do thermal work. Electrolytic work is work done to reduce metals from oxide, chloride or other salts. In both of these cases the work done can be equated to the electrical energy consumed. Thermal work done by fuel combustion, as in a blast furnace or ammonia synthesis, can be measured in exergy terms since mechanical work is not involved and the Carnot limit is not relevant.

Mechanical work done by stationary machines is directly proportional to the electric power consumed, whatever the downstream use of the electricity (for lighting, heating, electrolysis, or electric motors). In fact, since we have no way to measure the exergy "content" of information, *per se*, it may be reasonable to include computers with other machines, measuring (in effect) work done by heat created.

Non-electric mechanical work is the work done by internal or external combustion engines to propel vehicle or agricultural or construction machinery. Standard measures are vehicle-km traveled or ton-km traveled.[2] But the common measure for all of them is fuel consumed per unit distance travelled: liters per km in metric countries.[3] Whether the engine is an Otto cycle (spark ignition), diesel cycle (pressure ignition) or Brayton cycle (gas turbine), the fuels are liquid fractions of petroleum, labeled as gasoline, gas-oil (distillate), or kerosene. These fuels differ slightly in density (and price), but they are all storehouses of chemical energy that can be utilized in one of the three main types of heat engine.[4]

The work performed by any of the engines is equal to the thermal efficiency of the engine times the exergy (enthalpy or heat) consumption of the engine in the vehicle or other machinery. Otto cycle engines burn gasoline and they are used in most private cars, although diesels have become increasingly competitive in recent years. The difference between the two is that, for the same power output, diesel engines are self-igniting (no need for spark-plugs), more efficient in terms of fuel consumption because of higher compression, but heavier to withstand the higher

---

[2] The units are confusing even in the metric system used by physicists, where energy (and exergy) is variously measure in calories, ergs, joules, or watt-hours, preceded by kilo, mega, giga, tera, or penta as appropriate. Energy is also sometimes expressed in quantities of fuels, such as tonnes of coal, tonnes of oil or cubic meters of gas. In the "Imperial" system used in the Anglo-countries, one has to cope with "British thermal units" (Btus), gallons or barrels of oil, horsepower-hrs, etc. The conversions are tedious but straightforward and if you care about these things, you already know where to look them up. This variety of units seems to have arisen from the early days of thermodynamics (Chap. 1) when the equivalence of mechanical, thermal, electrical and chemical energy was a major subject of study.

[3] The Anglo-Saxon countries use an inverse ratio, miles per gallon (mpg).

[4] Alcohols, such as methanol or ethanol, can also be burned, but being also partly oxidated they have only about half the energy-content of hydrocarbons.

pressures. Diesels are also more polluting, at present, because they generate quite a lot of ultra-fine unburned carbon particles that take the form of haze or soot. (Those ultra-fine particles are also a major health hazard, a fact that has resulted in strict emissions requirements, and a major scandal when Volkswagen was recently found to have systematically cheated on emissions testing.)

Brayton cycle engines (gas turbines) are as efficient as diesels, but they have much higher power-to-weight ratios, meaning that they are much lighter for a given power output. However, the light weight is achieved by operating at very high temperatures, requiring the use of costly high-strength alloys that are difficult to cast or machine. Thus, of the three, large gas turbines are ideal for continuous operation (as in aircraft or stand-by power), but very unsuitable for variable-load ("stop-start") applications in vehicles. Hence, the efficiency of heat engines, in practice, depends on the particular application. Thus ground transportation (automobiles, buses and trucks) that must operate mostly in "stop start" environments depend on Otto-cycle or diesel engines. During recent decades diesel cars have been taking a larger share (outside North America) but the Volkswagen scandal may change that.

The efficiency of converting chemical energy (enthalpy) to useful work in an engine, under ideal conditions, depends on the temperatures of the hot fuel mix before ignition, and the combustion products after ignition. That depends, in turn, on the compression ratio of the engine. Otto-cycle (spark-ignition) engines currently have compression ratios in the range of 8–10, while diesel engines go higher (13–15 in smaller engines, 16–20 in larger ones). The so-called research octane number (RON)—or just "octane"—determines how high the compression can go before self-ignition. The latter (called "knocking") cuts efficiency and can damage the engine if it occurs before the completion of the compression stroke.[5]

The liquid-fuel burning heat engines in the modern economy are dominant in the transport sector (cars, trucks, buses, aircraft) plus farm machinery, mining and construction equipment. The work done by these engines is mechanical. The work done is the product of net exergy consumption (exajoules, EJ) times the thermodynamic efficiency of conversion. The efficiency calculations for different applications are based on engineering information on fuel economy.

N.B. in the transport sector it is common to measure work in terms of payload moved, whether of persons (passenger-km) or goods (ton-km). Both of these measures depend on vehicle weight, speed, rolling friction, air friction, and route conformation. Hence the passenger transport measures used for buses, trains and airliners are not directly comparable to engine (or motor) efficiencies. More important, the usual transport statistics refer to the efficiency of moving vehicles from one place to another. But, of course, what matters more is the efficiency of moving payload (passengers and goods) rather than moving specific vehicles.

---

[5] For historical reasons, pure octane (C_8H_8) was chosen as the standard for 100-octane. Since gasoline is usually a mixture of lighter C_nH_n hydrocarbons (pentane, hexane, heptane), an aromatic such as benzene or an alcohol is usually added to increase the octane number.

Electricity can be regarded as almost "pure" work, since it can be converted to mechanical or thermal energy with nearly 100 % efficiency. In the case of electricity generation, the fuel inputs—whether coal, gas, nuclear heat or hydro-power—are individually known, so the efficiency of generation (output divided by input in energy units) is a calculated quantity. (This number rose rapidly from 1890 to 1930 or so but has stagnated around 33 % or so since the 1960s.)

To be sure, electric power is often used to do mechanical work by means of motors, or to produce illumination, or even to drive endothermic chemical reactions. Those "secondary" uses cut the end-use efficiency of electricity. The same can be said of gasoline or diesel engines powering vehicles. The efficiency of the engine, *per se*, may be 25–30 % running continuously on a test bench before allowing for stop-start inefficiencies, drive-train losses, parasitic equipment, air resistance, rolling resistance and so forth. The efficiency of an automobile powered by an engine is not much over 10 % when those things are taken into account.

My point here is that it is possible to measure work done (subject to the above conventions) by fossil fuels, with some accuracy. It is also possible, with a little more work (the kind done by engineers at desks) to calculate the thermodynamic efficiency of some of the other energy conversion processes in the economy. From this one can estimate the thermodynamic efficiency with which energy (exergy) is converted to work in various industrial and economic activities, and by nations as a whole. Indeed, exergy efficiency, as applied to a nation, is also a good—perhaps the only—quantitative measure of the state of technology of that nation. The efficiency of a sector or a process is simply a ratio of exergy output (work done) to exergy input. If you are interested in details, see the *Global Energy Assessment* (Johansson et al. 2012). The difference between exergy input and output, due to losses and irreversibilities, is usually called *entropy*.[6]

Unfortunately the term *efficiency* used in the above sentence, is also used inconsistently (and often incorrectly) in other contexts. For one thing, economists speak of "economic efficiency" with another meaning entirely. In economics the term "efficient" is really a comparative adjective without any quantitative meaning. In physics and engineering, however, it is a number between zero and unity and that number is the ratio between two flow quantities: output divided by input.

As I have also explained, not all energy is potentially useful. The ocean is much warmer than outer space; but its heat is not useful or potentially useful to us who live on Earth because the temperature gradient between land and sea surface is too small to make practical use of.[7] Steam is more useful in practical applications

---

[6] Technically, the lost exergy is "anergy", which is measured in energy units, whereas entropy is measured in units of energy (exergy) divided by absolute temperature.

[7] There is a scheme for ocean thermal energy conversion (OTEC). It would extract useful energy via a heat engine from the difference in temperature between the ocean surface and the deeper water, which can be as great as 25 °C in some regions. There are several versions under development e.g. by Lockheed-Martin, and one small operating OTEC plant in Japan. However the maximum Carnot efficiency of an OTEC plant would be 6 or 7 %, and the capital required to build it would be very large. So it is hard to see how OTEC could be economically justified unless capital were much cheaper or other benefits attached.

because it is a lot hotter than the ocean, so there is a greater temperature difference. From a technical thermodynamic perspective, work means exerting force to overcome inertia or some kind of resistance. It may be lifting a weight, winding a spring, pumping a fluid through a pipe, propelling a projectile (or a vehicle) through air or water, driving an endothermic chemical reaction or driving an electric current against resistance.

The difference between work, in the technical sense, and *useful work* arises from the existence of human intention and control. The expansion of the universe constitutes work against the force of gravity, but it is not "useful" in an economic sense. Similarly, the evaporation of water from the oceans, the eruption of volcanoes, the rise of mountains and the carving of valleys are other examples of work done by natural forces. But those forms of work are not *useful* in the economic sense. Work is useful to us humans only insofar as it is *productive* (which economists take to mean contributing to GDP) and/or contributes to human welfare. Those are not equivalent, but an explanation of the differences would be quite long and technical. However, there is significant literature on why GNP or GDP is not a particularly good measure of welfare. See, for example (Daly and Cobb 1989). Other measures have been suggested, but none has been adopted by government statisticians.

To extract useful work (or any kind of work) from a reservoir of exergy there must also be a thermodynamic disequilibrium that can be exploited. The existence of disequilibrium is usually marked by a *gradient* of temperature, pressure, voltage, chemical composition or gravitational field. Natural forces always act in such a way as to oppose gradients and approach thermodynamic equilibrium.

Energy *per se* is conserved in every process. That is the First Law of Thermodynamics. But exergy is not conserved; in fact, exergy is lost in every action or process. Yet, when we speak of "consuming energy" in everyday language (as in the context of heating a house or driving a car) it is really exergy that is being consumed (destroyed). The chemical energy embodied in a fuel such as natural gas is mostly *exergy*, while the heat energy in the ocean is mostly *anergy*. Unfortunately exergy, anergy and work are all measured in the same units as energy (e.g. joules, BTUs, calories or kilowatt-hours) and it is easy to confuse energy in the form of useful work with primary energy in the form of wind, flowing water or fuel.

Since inert structures and inactive muscles produce nothing, it is a short step to realize that all economic production is really comprised of products made from materials transformed by potential *useful* work, plus services generated by those products. This is even true of information production, processing and transmission. Moreover, useful work is an essential input to every economic activity, just as capital (in some form) and labor (in some form) are supposed to be essential. It can be argued that useful work is obtained from raw material or primary energy by the application of capital and labor, but useful energy (exergy) is also essential for that purpose. Contrary to neoclassical economic assumptions, there is no way to produce useful work from inert capital and inactive labor alone.

It follows that exergy (or useful work) could be regarded as explicit factors of production in the economic sense, along with capital and labor. Yet it is a fact that most economic models today still consider only capital and labor as factors of production, even though the increasing productivity of labor is largely due to inputs of energy (exergy) that drive the machines that continue to replace human labor. Yet in traditional economics the benefits of increasing productivity are nevertheless allocated to labor, even though they result from capital investment in machines! Traditional economics still regards energy (or useful work) as an intermediate product of the application of capital and labor. This tradition continues, even though capital and labor, without activating exergy, cannot produce anything.

In view of the standard neo-classical assumption that energy is not a factor of production, the high correlation between primary energy consumption and GDP suggests to most neoclassical economists that global energy production is simply driven by global demand. (I discussed this assumption from another perspective in Chap. 11, especially Sect. 11.6). Demand, in turn, is driven by GDP (Y), which is assumed to be determined by capital stock (K), labor supply (L) and a function of time A(t). This exogenous multiplier is usually interpreted as technological progress. It is also convenient to assume that those three variables have great inertia: i.e. they change quite slowly, they change independently of each other, and they change independently of GDP.

This is the main justification for introducing a so-called "production function" of the form $Y = F(K,L,t)$ where t is time. This formulation implies that there are no energy supply constraints and that neither capital K, or labor L, or A(t) depends on useful energy supply. There are further mathematical assumptions, important but rarely noticed, namely that the variables are continuous, with continuous first derivatives, that returns to scale are constant (not necessarily true), and that the economy is in equilibrium and that it grows while in equilibrium. I will have more to say about those assumptions in Appendix A.

The growth equation is the output maximization condition (where the first derivative vanishes) namely $dY/dt = 0$. Written out in full, this is an expression involving the partial logarithmic derivatives or marginal productivities of Y with respect to the factors K, L and A(t). The logarithmic derivatives of the factors are the so-called *output elasticities*, which reflect the relative contributions of each factor to overall growth. See Appendix A for more detail.

For purposes of forecasting future energy needs, and to carry out benefit cost calculations for various policy options, it seems to follow that a GDP growth forecast, together with a forecast of price and income elasticities, should be sufficient. This is exactly the methodology employed by the IEA, the OECD, the US EIA and the IMF. However, continuous growth-in-equilibrium does not occur in practice, and is virtually a logical self-contradiction, since growth only occurs when there is a dis-equilibrium of some sort, resulting in Schumpeterian "creative destruction". With apologies to Robert Solow, Ken Arrow and others, I think it is high time to challenge and rethink that standard forecasting methodology.

12.3 Wealth as "Condensed" Work and Useful Complexity

As a starting point, it is useful to distinguish two kinds of wealth. One is created by humans: it is economic wealth. Part of it is passive, although it may be replicated: literature, music and art—or pure knowledge—that contributes to quality of life or happiness. The rest is productive "capital", both passive and active. Financial wealth is "active" if it can be exchanged for "real" wealth in a marketplace, or if it enables the production of "real" goods. The qualifier "if" is important, because the possibility of exchange is not a certainty. Active wealth is autocatalytic: given exergy inputs it generates material consumables and can also create more wealth. This process also creates waste that destroys exergy and degrades material resources.

The other kind of wealth can be characterized as "gifts of nature". Some of it is passive, but exchangeable and monetizable, like minerals in the ground, or rights to exclusive access (e.g. to water rights, salmon fishing rights on Scottish rivers, hunting licenses for scarce animals, or access to beaches). Another part is cultural and social. These things can be enjoyed but they do not produce more of themselves. There is an important category of wealth that is non-exchangeable and monetizable (no markets) but that contributes to health and productivity: clean air, clean water, and the waste assimilation (and detoxification) capacity of natural systems. To that might be added our culture, our "rights" (such as free speech, equal opportunity), our safety and security provided by the State, public health, public education, and infrastructure, ranging from roads and streets to parks and schools.

Financial wealth consists partly of "liquid" assets—money and money equivalents—and partly of non-liquid assets. Liquid assets are cash or assets that can be sold for cash quickly in an existing market. Again, the existence of the market is fundamental. When there is no market, there is no financial wealth. Non-liquid assets for which there is no immediate market include undeveloped land, buildings, partnerships, club memberships, art, first editions of (some) books, or an item of clothing worn by a famous person. Such items can be bought and sold, depending on the coincidence of a willing seller and a willing buyer. Even then, other conditions may apply. For instance, a Spanish billionaire was recently arrested for trying to smuggle a Picasso painting, classed as a "national treasure", out of Spain without an export license. He was outraged that a designation such as "national treasure", imposed by the government, could reduce the value of "his" possession.

Passing over the complexities of money and banking, there is a relationship between *money* and *capital* in the economic sense. Karl Marx wrote a major tome *Das Kapital* on this subject (Marx 1867). But the important point is that some capital, as *exergy*, is necessarily "embodied" in capital equipment, while some "circulating" capital (as money) must be available for covering current expenses: i.e. purchasing raw materials, paying for utilities, paying for labor, paying interest on debts and reserve against losses. The income stream available to cover these expenses (excluding reserves) is called "cash flow". It is one of the standard measures used to evaluate a business. Many a business has failed due to insufficient free cash flow.

Economists use the term "net worth" to account for both categories—liquid and non-liquid—of wealth. The biggest component of net worth, for most people who are not extremely wealthy is "home equity". If you are a homeowner, you probably have a mortgage, which is a debt to the bank. Home equity is the difference between the price it can hypothetically be sold for (based on the recent sale prices of similar houses), and the unpaid amount of the mortgage. Millions of homeowners discovered in 2008 that this difference can disappear in a flash, if the market for homes collapses suddenly. Under present rules, the bank doesn't share the loss: mortgage debt remains when equity disappears. This is why so many victims of the 2008 "sub-prime mortgage" collapse were left "under water" with debts greater than the market value of their properties. (Yet the "bailouts" helped the banks, not the homeowners.)

Other components of wealth include partnerships in law firms or accounting firms. These normally have monetary value. When a partner leaves, the firm buys back the equity of the partner. But partner shares are not liquid. Moreover the value of a partnership can also disappear in a flash if, for instance, the other partners decide to reduce the number of partners, or if one of the partners—or the whole firm—is accused of a fraud. (You may recall that after the collapse of Enron the accounting firm Arthur Anderson was accused of fraud, and its partners all lost their equity).

Another way to think about money, and what it can buy, is credit. With credit you can buy goods without cash, in exchange for debt. Mortgages enable people to buy real property, using the property as collateral. The borrower can enjoy a current benefit—a comfortable, convenient place to live in that will eventually be owned debt-free—in exchange for a share of future income during a fixed period, such as 20 years. Virtually the entire construction industry depends on this mortgage creation mechanism.

So called "private equity" firms borrow from banks to purchase other firms, usually for purposes of "financial engineering". (This activity can be creative, although it is more often simply destructive of jobs and long-term assets. I will say more about this in the context of debt). Auto loans and credit cards are the other obvious example of a way to consume goods and services before earning the money to pay for them. Of course, both mortgage payments and credit card payments include interest payments on the debt. These provide an income to the bank and enable it to finance future lending.

It is also helpful to distinguish between what we humans personally own—those things created either by ourselves, or by nature—vs. what is only "owned" by society as a whole. Ownership is another key concept in economics that could be discussed at some length. However, in law it amounts to an exclusive right to use *and dispose of* a property, while excluding other claimants. Everyone may be entitled to use a park or beach without owning it, but that does not include the right to sell it or prevent others from using it. The rights conveyed by ownership of land can be regarded as a further extension of human skin, beyond the "skin" provided by clothes and the house (Sect. 8.1). At the national level, the non-monetizable social and environmental assets are extremely important to welfare, but they are not owned, hence not exchangeable, except insofar as they are tied to land ownership.

Having said all that, and caveats aside, the monetizable wealth (net worth) of an individual is the sum total of his or her money, financial assets and other potentially salable property, at market prices (if there is a market) less debts. The same formula applies to nations. However, all debts are financial assets of others. My mortgage is the bank's asset. Hence debts and assets *in the same currency* simply cancel out at the national level. Debts and assets in different currencies don't necessarily cancel, however, because currency exchange rates can (and do) vary. Small countries have been bankrupted by sharp depreciations of their currency, sometimes due to events beyond their control.

The hyper-inflation in Germany after WW I is a good example: The Treaty of Versailles required Germany to pay reparations in gold to the victorious nations, especially to France and Britain. But Germany had very little gold in reserve, and was also being prevented from exporting manufactured goods to finance imports. The Weimar government tried to square the circle by borrowing money from the United States, using "paper" Deutschmarks, in order to buy gold to repay France and Germany. The D-mark plummeted in value during 1920–1921, leaving the American holders of those D-marks with a loss, and much of the German middle class in destitution. This episode wiped out the German domestic debt but not the foreign debt. The rise of Hitler and Nazi-ism followed, and many see a causal connection.

So much for the evils of hyper-inflation. But what most people have not thought about, unless they have studied economics intensively, is the role of money itself. I refer to the macro-economic effects of "money supply" (Friedman and Schwartz 1963; Friedman 1968). Friedman saw money-supply—the national version of Marxian "circulating capital"—as a tool for central banks to smooth out economic fluctuations (boom-bust cycles) without political oversight. His anti-Keynesian ideas on monetary policy were adopted and put to the test by Margaret Thatcher (Conservative leader of the U.K. from 1975 to 1990). When she was elected as Prime Minister in 1979 her priority was to reduce inflation, and her proposed policy was to control the money supply and raise taxes to balance the budget by cutting consumption. In 1981 she did raise taxes while unemployment was rising, contrary to the Keynesian consensus at the time. But there was no dramatic recovery and her monetarist experiment failed.

Here is where it gets tricky because of the special role of banks in "creating" money. According to standard economics textbooks, written long ago, banks do not create money. The standard textbook model is the "savings bank" that can only make loans based on some fixed multiple of capital (plus reserves) of that has been already deposited by savers. The bank manager has a certain annual outlay (interest paid to depositors) that needs to be covered by income from borrowers. This picture does apply to local savings banks and credit unions. However it has nothing to do with Citibank or JPMorgan-Chase, and even less than nothing to do with Goldman-Sachs.

The reality in the money-centers is different. Big commercial bankers have realized that when they make a mortgage loan for a house or a car loan or finance a credit card purchase, the money circulates through the economy. As it circulates, virtually all of it is soon re-deposited in a bank. Thus loans actually generate bank deposits, multiplying the aggregate lending power of all banks. *So the banks actually have an incentive to lend as much as possible*, as long as they can find borrowers with assets that can be used as collateral or insurance. The result is the system known as "fractional reserve" banking. Lending is only limited by the regulatory rules on the ratio of total loans to permanent ("core") capital. That ratio is called *leverage*.

Unlike the savings banks that were once restricted to making home loans, the commercial banks in the US (since 1998) have been free to use depositor's money to gamble in the financial markets—they don't call it gambling, of course—instead of investing in the "real" economy (Werner 2005). There are several kinds of gambles for sophisticated investors. They include buyouts, takeovers, "financial engineering", and a variety of games (bets) involving "futures" or derivatives, that need not be described here.

For analytic purposes (in models) capital is now generally assumed to be cumulative savings or investments, depreciated by some assumed discount rate (Maddison 1993). Of course, some investments depreciate much faster than others. Roads and tunnels depreciate slowly (50+ year life), ships, bridges and buildings a little faster (30+ year life), trucks and buses much faster (15 year life) and software much faster (3–5 years). This makes it hard to estimate real capital stock. Or, capital can be defined as whatever generates profits, as measured by monetary "returns" (payments), as classified in the National Accounts. These payments include interest on bank accounts, dividend payments, rents and royalties.

But there are problems with either definition of capital. Neither definition adequately accounts for intellectual capital such as skills, patents, copyrights, art or old movies (increasingly monetizable). Nor does it properly take into account empty buildings, undistributed corporate profits sitting in overseas banks, or underground reserves of coal, oil, gas or other potentially valuable minerals.

As an interesting example of the monetizing problem, consider the element thorium. At present nuclear power plants and nuclear weapons use from uranium. Hence uranium—which is also the source of plutonium—is a valuable commodity and uranium mining and processing is a big business. Meanwhile, thorium is considered a toxic pollutant in the rare-earth industry (because it is a minor component of some rare-earth ores, such as monazite). But what if uranium were banned and the nuclear power industry switched to thorium? This change is technologically feasible and environmentally very desirable, although it would require some R&D and several decades to accomplish (Hargraves and Moir 2010; Hargraves 2012). In that hypothetical—but possible—future world, thorium would become a valuable by-product of monazite. How should monazite be valued today?

Now reconsider the idea of wealth from another perspective: not only as a product of work performed (exergy expended) but as a kind of condensed form of exergy, a little bit analogous to matter itself as the solid "condensate" of high-energy radiation (Chap. 4). After all, much of the non-liquid, not easily monetizable, forms of human wealth nowadays are invisible and intangible. This is true of skills, music, artwork, software and scientific knowledge. All of those forms of wealth are examples of long-lived order, created by humans from disorder by "feeding on negative entropy" (Schrödinger's words).

That new order, closely related to "knowledge", had to be created and learned initially by human brains using sensory organs and hands, as well as physical tools, writing implements, means of recording, and so on. My point is that energy (exergy) is required by the brain and the body to enable it to function (metabolize) as well as to create order from disorder. Active brains need food.

What does this mean for economic valuation? I suppose it means that education and basic research are likely to be profitable to society. But the return on investment is not immediate. The people who run "private equity"—and the libertarian followers of Ayn Rand—would fire most of the teachers of history, civics, art and humanities and sell off the buildings.

Wealth can be partly measured in terms of money, but wealth is not money. It consists of assets. A productive asset—like a living organism—can be characterized as one having the means to replicate itself, given an exergy input. Intellectual property (IP) is a special case: it has private value only as long as it is owned and not replicated, but it may have public value after replication. Most ownable assets have physical representations: these include long-lived productive capital, goods (infrastructure, machines) as well as passive consumptive assets, like houses, cars, clothes, works of art and books. These are also forms of wealth.

The stock of both material and intellectual wealth in the world has grown much faster since 1500 CE than it did earlier, and it has accelerated with every advance in replication technology (Sect. 8.3). As I have noted before, everything produced or built before 1500 CE was hand-made by one of a small group of skilled craftsmen. After 1500 CE factories especially designed for the replication of one product— such as muskets—began to appear, gradually at first. But by the end of the nineteenth century, hand-craftsmanship for most goods was becoming obsolescent and mass production of material goods was virtually the only game in town. This change, above all, accounts for the acceleration of economic growth in the past two centuries.

Again, human wealth (not including some kinds of natural wealth) constitutes a store of "condensed work" (exergy). Some of this exergy is captured in the form of physical structures and products while some is disembodied and intangible but no less real. Of course most of the food energy consumed by the population as a whole is used for metabolism: circulation of the blood, breathing, digestion of food, walking around. A small part is also used for reproduction of the species. But when an author writes, an artist draws, an engineer performs calculations (even with a computer), or when a worker performs any job, the final product is at least partly a bit of "created order" (knowledge) made possible by capturing a stream of exergy, mainly from the sun.

Incidentally, information (knowledge) storage is becoming a challenge. Big Data is only the beginning. Google is trying to make a digital recording of every book and manuscript ever published or printed, partly because traditional libraries, consisting of paper copies of books, are subject to various decay or loss processes, from fire (the Alexandrian library) to water damage or termite attack. New technologies for data storage such as nano-graphics (my word) are under development.

Knowledge, as such, may (or may not) be useful to others, e.g. for increasing economic productivity. Of course, most acquired knowledge is later also "used up" so to speak, in the metabolism of the economy as a whole. People need to acquire skills and knowledge just to keep the machine going. The postman needs to learn his route. The farmer needs to know when (and how) to plant his crops and milk the cow. But a small part of all that brain-work may be embodied in something new and permanent: a song, a book, a jewelry design, a new gourmet dish, a new mousetrap, or a clever new business plan (e.g. Twitter or Facebook). Some of it may increase the productivity of the existing production system, thus creating new wealth for society as a whole.

To summarize, I have said a good deal about what is, or isn't, wealth. The question arises: Is there a "wealth function" i.e. a functional relationship between measurable inputs and societal wealth? The essential features of such a function, as applied to an economic agent, can be stated formally (Ayres and Martinás 1996, 2006). Subject to reasonable assumptions, it can be shown that a unique wealth function Z exists for each firm or economic actor. It can also be derived from the theory of games (von Neumann and Morgenstern 1944).

However, societal wealth is not the simple sum of individual net worth, as normally calculated (based on individual ownership of exchangeable assets). The reason is that citizens of a country, or a city, do not "own" a share of social assets, because—unlike shareholders in a company—they cannot dispose of those assets or exclude others from access to them. As a resident of Paris I have access to parks, vistas, excellent restaurants, a superb transportation network (so good that it makes no sense to keep a car) and a very good health service. Moreover, the air is still breathable, and it is safe to walk the streets or take the metro at any hour. These services are available to me as a resident and taxpayer, *but access to them cannot be bought or sold in a market, except insofar as they are rights associated with physical presence.*

I could move to a small town in the countryside, or in another country, where the prices of houses are lower. But I would not be wealthier as a consequence. Quite the contrary. Or I could move to a country where the personal income tax is smaller, leaving me a greater share of my earnings to spend. *But I can only spend my money on the goods and services that are available at the shopping mall or on the internet.* Many of the services I receive as a resident and taxpayer in Paris are not available from Amazon or at Macy's, or in "low tax" countries. My imputed share of societal capital supporting these urban social services in Paris amount to a large percentage of my effective personal wealth. Londoners and New Yorkers will say similar things about their cities.

There is an implicit relationship between physical capital, as embodied exergy, and "circulating capital" as money. Several authors—have suggested that the historical correlation between energy (exergy) consumption, and GDP, is a kind of thermodynamic law. The idea is that cumulative energy (power) consumption is proportional to cumulative GDP.

In the first place, GDP is a measure of economic activity. It bears some relationship to monetary national (or global) income, but income is not wealth. Wealth is essentially assets, some ownable, exchangeable and monetizable and some not. Natural resources are a component of wealth, and much of that component has been used up. Some wealth is produced from natural resources and some is not. Some material objects have monetary value and those can be a part of wealth. But wealth isn't money, *per se*, even though if you have more money you may acquire more assets. (The ultra-wealthy today seem to lust after paintings by Pablo Picasso, which they hide away in vaults.) But many social and environmental services contribute to wealth without any direct monetary cost to those who enjoy those services, although there is a social cost.

My point is that economic wealth in the real world is based on productive assets (active capital) and is not closely related to economic activity or income. The cathedrals constructed at great cost in the Middle Ages consumed wealth but created GNP. They are tourist assets today. The religious wars after the Reformation consumed immense wealth and produced none, but they did constitute economic activity.

Note that active capital is a tiny percentage of the total, measured by cost or by mass. Most physical capital is passive—mainly roads, bridges, tunnels and buildings—and the lifetimes of those things can be very long if they are maintained. Roman roads and aqueducts are still in use. The best measure of economic wealth, in this sense, is probably something like output per kg. of energy conversion equipment (engines, machines) and information processing equipment (computers, telecom devices). In short, miniaturization of active, exergy-consuming capital equipment is probably the best measure of economic wealth today (even though an important social and environmental component of real wealth is not taken into account by such a measure). By that measure of productivity, the human race is still getting richer very rapidly, while simultaneously running out of cheap petroleum.

My point here is that there is no a simple money equivalent of energy because the reality of wealth is very different and far subtler. Given the unlimited availability of solar energy, the end of the age of hydrocarbons is not necessarily the end of civilization.

12.4 Debt: The Downside of Financial Wealth Creation

Economists and politicians today bemoan the fact that US savings rates have been declining since the 1970s and have been nearly zero or negative for the past decade. Except for startups financed by venture capital, investment in recent decades has been financed largely by debt. Moreover, the magnitude of the debt has been growing faster than the economy.

In fact, the original function of banks was to capitalize expected future income streams from depositors by leveraged lending. The leverage arises from the fact that banks loaned (and are now officially allowed to lend) many times more than their total reserves. This scheme is called "fractional reserve" banking. The idea is to make a small profit on each loan but a large profit on the total. This is called "financial leverage".[8]

A 100 years ago local banks mainly provided working capital for farms and small companies, allowing them to finance growth. Lending policy was largely based on trust (which is why that word appears in the names of many banks, even now.) That situation, and the policy of lending on the basis of trust, has changed. In fact, most of those small "savings banks" in the US disappeared in the 1970s and 80s due to the squeeze between inflation and legal restrictions on interest on deposits (Regulation Q) in the 1933 US banking reform law. Small US commercial banks either failed outright or were swallowed up by larger commercial banks. (Europe never had many savings banks in the first place.)

The reality today is that the large money-center banks are not even interested in savers who deposit their money in branch offices. Branch offices are regarded as unnecessary overhead. What the big banks do now is to create credit for large potential borrowers with collateral. This shortly becomes debt (and income) to the banks (Minsky 1982; Werner 2005; Keen 2011b). When borrowers fail, as some do, the collateral is taken over by the banks.

Of course credit and debt are two sides of the same coin: all debts are also assets for a lender. But they are not symmetric. Both parties take risks: every loan is a "bet". However the lender typically demands repayment in full, as a "moral obligation" subject to penalties for lateness. The lender accepts no risk of loss (although losses do occur when the banking system as a whole fails, as it almost did in 1929 and 2008.) Quite often it is the borrower who accepts the whole risk of loss without understanding the risk. Loan documents tend to be legalistic, and some would argue that they are often intentionally obscure.

It is true that most people who borrowed money to buy homes in the US and Europe during the 60 years from 1945 to 2005 enjoyed increasing home equity and

---

[8] The leverage being employed by most major banks at the time of the collapse of Lehman Brothers in 2008 was above 30 to 1 (Ayres 2014a, b) p. 175. Today the limit has been set by the Bank of International Settlements (BIS), in Basel, at around 10:1, slightly lower for the biggest banks. In consequence, banks are reluctant to lend to the riskiest borrowers, namely small business. This is one of several reasons for the slow economic recovery after 2008.

net worth. This happened because suburban real estate prices rose faster than the cost of living, due to urbanization and "automobilization". But a lot of people who bought homes during the later years of that period, 2004–2007, ended up as losers. House prices fell sharply in 2008–2009 and have not increased much since 2009. (Renters have done better recently).

Investment bankers promoted "securitization" of mortgage-based bonds and actively encouraged unqualified people to buy houses they couldn't afford, by offering adjustable rate mortgages (ARMs) "sub-prime" mortgages that cost the buyer almost nothing in interest payments during the first year or 2, after which the rate would rise to a higher level. The banks took minimal risks themselves, because they packaged the loans into bonds known as "collateralized debt obligations" (CDOs) that they sold to other institutions seeking higher returns. Those institutions were mainly endowments, insurance companies and pension funds.

Yet the banks still insist on repayment in full from borrowers, as a "moral imperative", even though they passed most of the risks to others and kept very little "skin in the game". This fact has arguably reduced their traditional risk aversion. Not only that, the large US banks, being "too big to (be allowed to) fail"—because failure would threaten the whole financial system—were bailed out by the taxpayers via the Troubled Asset Relief Program (TARP). This bailout has encouraged banks to make loans they should not make.

The fact that large banks in recent years have (mostly) succeeded in shifting the risk of loss from themselves to other institutions has not eliminated bankruptcy risks *per se*. Large numbers of small companies have failed as a result of slowing (or negative) economic growth. Moreover, many developing countries dependent on borrowing from money-center banks in Frankfurt, London or New York have suffered severe crises, resulting from increases in interest rates controlled by others. The peripheral countries of Europe, especially Ireland, Cyprus, Portugal and Greece, have also suffered. In fact, recent experience of financial failures is merely a natural continuation of eight centuries of financial folly ("This Time Is Different"), as documented by Reinhart and Rogoff (2009).

In one sense, the US mortgage debt crisis of 2008 was "caused" by the greed of unscrupulous bankers and real estate promoters. But it was possible because many large institutional investors with long-term commitments eagerly bought those CDOs. The buyers were seeking higher yields because the triple-A rated government bonds they traditionally bought were producing so little income. Those pension funds and endowment funds had made commitments to provide future retirement income to savers based on the assumption of 6 or 7 % returns. But by 2003 they were faced with rock-bottom interest rates on US, German, British and Scandinavian government bonds and slow global economic growth.

That fact, in turn, reflects the absence of any powerful "engine" of economic growth since the 1970s, when the post-war demand by "baby-boomers" for cars, houses, TVs, washing machines and other electrical appliances was approaching saturation. The post 1980 markets for second (or third) cars, vacation homes, and computer games have also showed signs of saturation. In any case, it has not provided a driver of demand growth comparable to the post-war baby-boomer" period.

The search for higher yields have led those same institutional investors to invest heavily in the peripheral countries of Europe and "emerging markets" abroad, especially the BRIC countries (Brazil, Russia, India, China) as well as Turkey, South Africa, and a few others. The so-called "sovereign debt" crisis in Europe followed, as the night follows the day. Between November 2011 and April 2012, interest rates on government bonds of peripheral states in Europe, led by Greece, Cyprus, Ireland, Portugal and Spain, and even Italy, soared to extraordinary levels as investors in those countries pulled their money out and bought "safer" German or US government bonds. See Fig. 12.1. The immediate causes varied from case to case, but the situation put enormous pressure on the European Central Bank (ECB) and the IMF.

The modest "recovery" of the Eurozone in 2012–2013 was partly due to ECB Chairman Mario Draghi promising to do "whatever is necessary" to save the Euro (including a trillion Euros in liquidity and lower interest rates), and partly due to massive investments in those currencies by US-based hedge funds. See Fig. 12.1. Ireland and Portugal were able to exit from their IMF programs in 2014, but Greece needed a second bailout in September 2015 with poor prospects of success, as seen by financial experts.

Fig. 12.1 Sovereign bond interest rates between 2009 and 2015

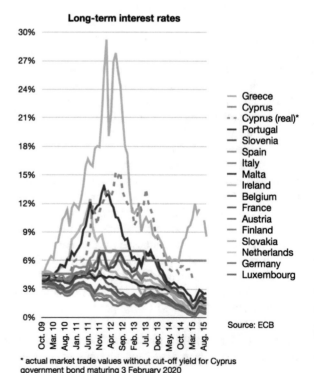

* actual market trade values without cut-off yield for Cyprus government bond maturing 3 February 2020

The Euro is still in danger because the ECB-IMF intervention in Greece was ineffective. While the Greek GDP has fallen 25 % from its 2007 peak, and unemployment in Greece has reached 27 %, there is no sign that renewed growth can ever produce enough surplus to repay its debt. Indeed, the brief period of positive growth in 2014 has ended. The IMF has acknowledged that the Greek debt is unpayable. But the German and other conservative governments have insisted that the debt must be repaid in full because "rules are rules". Recently, the conservative minority government of Portugal has also been kicked out of office by a left-wing coalition that includes the communists. The conservative government of Spain is also in trouble.

A third leg of the latest financial crisis—due to the search for higher yields—is now (2015) in prospect as foreign investors are pulling funds out of emerging markets, especially Brazil. The Brazilian debacle was triggered by a scandal involving the debt of the Brazilian oil company, Petrobras. The Russian seizure of Crimea and support for pro-Russian rebels in the Ukraine has resulted in sanctions from the NATO countries that have weakened the Russian economy significantly. The Chinese devaluation in the summer of 2015 due to its growth slowdown (from 14 % p.a. in 2007) is now expected to be significantly less than the promised 7 % p.a. This has worried the financial markets. The real problem—apparently not well known to the investors—is the sheer magnitude of the overhang caused by the construction boom of 2003–2011 (see Fig. 12.2). A large number of

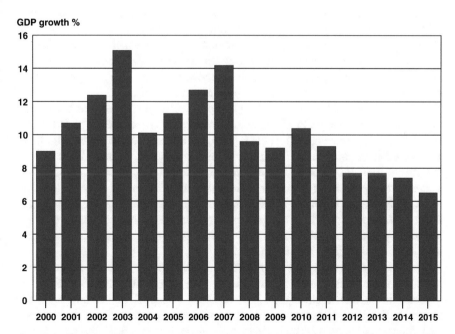

Fig. 12.2 Chinese GDP growth 2000–2015. *Data Source: The Conference Board. 2015. The Conference Board Total Economy Database™, May 2015* http://www.conference-board.org/data/economydatabase/

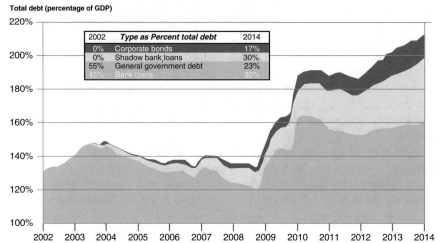

China's shadow bank loans have increased threefold since 2009

Fig. 12.3   Chinese bank debt 2002–2014

unfinished buildings with unsold apartments means that a large loan portfolio held by Chinese banks and "shadow banks" is non-performing and much of it will never be paid.

Less well-known, but more important, Chinese bank debt has tripled since 2009 (Fig. 12.3). Luckily, most of the debt is in Chinese currency. Even so, this leaves the Chinese government a major problem of maintaining employment by financing new urban infrastructure (especially for public transport), newly created pensions and social safety-net schemes and large infrastructure projects, such as high-speed rail lines and pipelines. Chinese debt is sure to increase further.

The magnitude of the global debt buildup in recent years has been hyped by a variety of interest groups, but it is still not sufficiently appreciated. The US case is not the worst in percentage terms, but it is still the largest in dollar terms (Fig. 12.4). That fact has consequences. With some justification, the former Greek finance minister, Yanis Varoufakis, has called the US debt the "global minotaur"—a reference to the legendary monster in the palace of King Minos of Crete, that had to be fed by annual sacrifices of subscriptions young people from the mainland of Greece (Varoufakis 2011). He argues that the federal government debt, which is financed by foreign savings (surplus), has paradoxically enabled the US to maintain its position as the global financial hegemon.

That thesis deserves serious consideration, although the author's suggestion— that it came about as a result of a deliberate plan by a small number of senior US officials during the Bretton-Woods Conference (Chap. 3 of his book)—is too paranoid. The US debt is mostly the result of declining international competitiveness thanks to fighting wars rather than investing in domestic productivity. It is also (like the Greek debt) partly a consequence of unrealistic exchange rates ("strong dollar") favoring military allies in the anti-communist crusade. According to

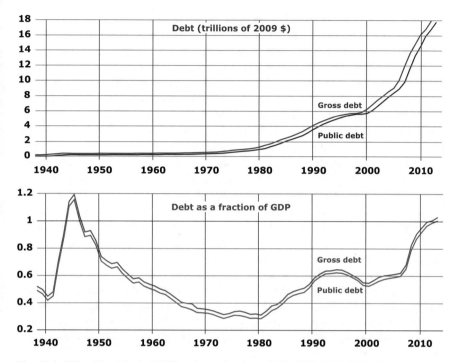

Fig. 12.4 US public debt in 2009$ and as a fraction of US GDP 1939–2014. *Source*: https://research.stlouised.org/fred2/series GDPCA, FYGFD, and GFDEBTN

Varoufakis, the idea was to enable Germany, Japan and Taiwan to recover rapidly by giving them free access to the US market without reciprocal access to their internal markets. Japan and Taiwan used their large dollar surpluses to buy US treasury securities rather than US products.

In the European case the dominant exporter is Germany, which has large trade surpluses with respect to all the other members of the EU. The peripheral countries need to finance imports with loans from Germany, since they all use the same currency. The Germans used their currency surplus to finance those loans. This strategy supports employment in German factories and also increases overall German national wealth at the expense of all those borrowers. The slow but steady shift of wealth from the other countries of the EEC into Germany is probably the major destabilizing force in the European Union today.

The rise of China after 1980 has not been due to any US *quid-pro-quo*. It began thanks to massive Japanese, Taiwanese and German investments in Chinese factories to make labor-intensive consumer goods that—thanks to low labor costs—dramatically underpriced all competitors. Those goods flowed especially to the US. (European countries were more protective of domestic suppliers.) The strong dollar policy coupled with an artificially weak Renminbi policy helped Chinese exports, of course.

But China did not need to import much (apart from commodities, airliners and heavy construction equipment), so it used its dollars to buy US government bonds. That kept US interest rates from rising and kept the export boom going. The situation in China from 2004 through 2008, depicted in Fig. 12.3, is much more complicated. The trend toward increasing debt, prior to 2004, was apparently interrupted for 5 years, for reasons that are unclear. But during the single year 2009 a huge amount of new debt was created also for purposes that are unclear. (It is hard not to wonder about the reliability of the data.) The pre-2004 trends seem to have continued after 2010 as if there had been no interruption.

US federal debt at the end of 2007 was about 1.3 times GDP, but when state, local, corporate and private debts are counted, the total then was 3.6 times GDP. Including interbank debts (usually not included) the total reached 4.75 times GDP at the end of 2007. It peaked in that year and fell in 2008 and 2009, but federal debt has risen to compensate in the past few years after 2008 as the Federal Reserve Bank undertook responsibility for a lot of private bank debt (quantitative easing, or QE), leaving the totals relatively unchanged.

Leaving out the interbank debt, US federal government debt/GDP ratio in 2013 was 103.5 (including external debts) according to the IMF. IMF projections suggest that it will remain close to that level for the next decade or so. US citizens and institutions hold two thirds of the US government debt, while one third is held by foreign countries, mainly Saudi Arabia, Japan and China. (Meanwhile, the Japanese debt to GDP ratio in 2013 is much larger, 2.43:1.)

It is important to note that the US and Japanese debts are denominated in dollars and Yen, respectively. The Chinese debts are also denominated mostly in Renminbi. One of the overhanging problems confronting developing countries and emerging markets is that their debts to banks in New York, London or Frankfurt are also denominated in hard currency: dollars, pounds sterling or euros. Hence if the balance of trade is negative and the currency is weak, the consequence of a demand for repayment in dollars (because the banks themselves are being forced to repair their balance sheets) can be catastrophic. This can mean bankruptcy to a small country. This has been, and still remains a major problem for parts of Latin America, Africa and Eastern Europe. It is also a problem for the peripheral countries within the Eurozone, as already noted.

In the post WW II era, governments of the industrialized countries needed more revenue to finance reconstruction and to pay for a growing range of social programs. Taxation of corporate profits was one previously untapped source of revenue. However taxation of corporate profits resulted in an unexpected and unintended consequence. It created incentives for firms to borrow (usually by issuing bonds) rather than to maximize taxable earnings (and dividends). This happens because tax authorities treat interest payments on corporate debt as a cost of doing business, not as a profit. Not surprisingly, corporate debt in the western world has multiplied faster than government debt. Financial debt, between financial institutions has grown even faster, as Fig. 12.5 (for the US) shows. The solid green line shows how debt between financial institutions has increased from

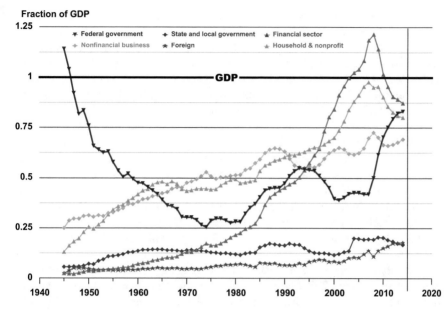

Fig. 12.5 US debt/GDP ratio since 1945 by category. *Source*: Calculated from US Federal Reserve Board "Credit Market Debt Outstanding by Sector", Series D.3 http://www.federalreserve.gov/datadownload

almost nothing in 1945 to roughly the size of the GDP by 2010, even after significant deleveraging.

What we do not know and cannot easily estimate is how much of the US debt added since 2007 or so has been invested in "hard" assets to increase standard (GDP) productivity, how much has been wasted in projects that will never pay for themselves (like unfinished housing stock) and how much has been invested in building up the stock of physical capital that provides social services, ranging from roads, buses and trains, electric power plants and water treatment to education and health services.

There is much more that could be said about the origins and the quantity of debt by sub-national entities. There is a lot being written about the forthcoming bankruptcies of US states, territories, cities, counties, hospitals and universities e.g. (Edstrom 2015). What usually happens at the sub-national level is that unpayable debt is converted to some kind of equity, which subsequently loses value. The banks that made unpayable "sub-prime" home loans to unqualified borrowers ended up with repossessed houses that subsequently declined in value. When a municipality is bankrupt, a court allocates the marketable assets (such as land and buildings) to unpaid creditors, in order of "seniority." But bond-holders can end up unpaid. Something of the kind occurs when a corporation fails. The creditors take over the assets, whatever they are, and swallow the losses under another name.

When nations like Greece can't pay, the situation is more complex, because it becomes a negotiation between parties, without rules. The creditors (banks and bond-holders) demand "guarantees" of future payment-in-full, in exchange for additional loans. But additional loans may make the aggregate debt even more unpayable. The usual outcome of a crisis in the modern world is to renegotiate the terms, usually in secret, because there is a lot of behind-the scenes jockeying between different creditors. But the outcome of the process is usually that the interest rate on the debt is set lower than the expected inflation rate, so that the unpayable debt is eventually "inflated" away.

But this strategy only works if the debtor country can grow its economy (and tax receipts) at a rate faster than the interest rate on the debt. Increasing debt interferes with growth by diverting the economic surplus into amortization rather than investment. Debt also has a significant negative impact on business confidence, without which investment does not take place. If the debt is too great—again, Greece—the interest burden stops economic growth altogether and turns it negative.

The only possible—and inevitable—outcome is for the debtor country to refuse to pay at some point. This has happened many times in the past; in China (1911), Bolshevik Russia (1917), Nazi Germany (1933), and more recently Argentina (2010). The consequence is that the defaulting country is cut off from future borrowing. Eventually, if the defaults mount up, the banks and institutions with money available to lend (or invest) find themselves without customers able to pay high interest rates. The world seems now to be slipping gradually into this situation. Yet, oddly enough, economic growth models have not, hitherto, incorporated this relationship. I will discuss a specific approach later. (See Appendix A.)

12.5 The Direct Costs of Economic Growth

An issue that is frequently brought up in connection with investment in energy efficiency is the so-called 'rebound effect', otherwise known as the 'Jevons Paradox' (Jevons 1865 [1974]; Khazzoom 1980, 1987; Saunders 1992). The underlying point, very simply, is that the demand for any good or service tends to increase as the price falls, while (conversely) demand falls as the price rises. This phenomenon is called the *price elasticity of demand*. It is generally agreed that if the cost (price) of energy should decrease as result of increasing energy efficiency, demand would also increase. In some cases (e.g. iron early in the nineteenth century, steel later, electric power, telephones and cars after that and computers and semi-conductors in modern times) the increased demand can overwhelm come any energy savings from efficiency gains.

As a matter of fact, this phenomenon is essentially the same mechanism that has driven overall economic growth for the past two centuries or more (Ayres and Warr 2009). However, it does not follow that energy savings from increased efficiency in a given industry will necessarily be compensated by increased demand in that sector. It depends, in the first place, on the fraction of total cost of the product or service attributable to energy and, in the second place, on the degree of saturation of the market. (These factors determine the price elasticity of demand). Beyond that, there is a question as to how the monetary savings—in contrast to the energy savings—are spent. If monetary savings from cheaper consumer goods are spent on air travel or larger houses energy consumption will still rise

For instance, the direct energy component of housing costs for most consumers is now very small, and the expenditure share that goes for illumination is smaller still. Hence, savings from energy conserving lighting will not induce consumers to buy larger houses, or even to buy more lighting fixtures, though some lights may be left on longer. Savings from insulation will not induce most consumers to raise the thermostat temperature and consume more fuel for heating. Much the same holds for automobiles: better fuel economy will not induce most drivers to drive many more miles (Schipper and Grubb 1998). On the other hand, it is a fact that kilometers driven per year increased at just about the same rate that US fuel economy was increasing during the 1980s (Herring and Sorrell 2009). This point needs to be understood.

The fact that energy savings in one sector may "rebound" (generate more expenditure in that sector)—hence greater energy consumption—also needs to be understood. In the case of producers, it can safely be assumed that energy savings per unit of output go to the "bottom line", i.e. to profits. Of course profits get spent somehow, either to increase wages or dividends or investment, and hence GDP. In the case of consumers, any savings in energy use will be treated like an increase in income. How will it be spent? In developing countries, especially, it can be argued that a marginal increase in income will be used for goods and services that are more energy intensive than the average for the economy. If this is so, the "rebound effect" must be taken very seriously and policy instruments may have to be created to compensate for the effect. However, it can also be argued that the increased income available for other goods and services is precisely what economic growth is all about.

However, although the rebound phenomenon is real, there is no theoretical basis for assuming that efficiency *per se* causes increased consumption. The linkage between higher efficiency and greater consumption must be through lower prices, via the price elasticity of demand. In a number of historical cases, often cited, efficiency improvements were, indeed accompanied by lower prices. This was true of the iron-smelting case cited by Jevons (not to mention the case of Bessemer and open hearth steel a few years later). It has been true of several twentieth century examples, especially cars, computers, cameras, cell-phones and information technology.

The situation we face in the coming decades is very different from the past. Energy prices are likely to rise in the long run (beyond the next decade), not fall. Greater efficiency of energy use may abate the price rise; that seems to be happening right now. But in the OECD countries it will not give rise to significant increases in absolute energy consumption, as some of the "rebounders" have suggested. The situation in China, India, Africa, and elsewhere in the developing world, the situation is different. Nevertheless it is important for analysts to quantify the effect as far as possible and to take the phenomenon into account where policy choices are indicated.

In fact, virtually all studies, by economists, of the long-term consequences of climate change start from the presumption that there will be significant economic costs (in terms of lower growth) of breaking the current dependence on—or addiction to—fossil fuels. The argument is that it will be costly as compared to "business-as-usual", but less costly than inaction. The best example is the famous "Stern Report" by Sir Nicholas Stern and colleagues (Stern 2006, 2008; Stern et al. 2006; Dietz and Stern 2008).

The Stern Report has been rather harshly criticized by a number of mainstream economists on the basis of its assumption about discounting. In brief, economists normally "discount" benefits that will be received in the future (as compared to benefits enjoyed today) on several grounds. One is the fact that the future benefits will not be enjoyed by us, personally, but only by our children or grandchildren. Assuming the future benefit is monetizable (as economists always assume) the question is: how much should we pay today for that future benefit for our descendants? People vary in how they answer that question, but economists tend to assume for benefit-cost studies that the "natural" discount rate is around 7 % per annum, based on data from a variety of experiments and contemporary situations e.g. (Weitzman 2007).

Obviously, if we assume a 7 % discount rate, future benefits 50 or a 100 years from now from today's actions aren't worth very much today. (This happens to be the rate that most pension funds and insurance companies have been expecting to return to pensioners. Stern knew this, of course, but he argued that long-term habitability of the Earth is not like a pension.) The Stern Report assumed a much lower discount rate for essentially moral reasons.

Stern et al were not without support. There is a substantial literature on "hyperbolic discounting", which means discounting faster than exponential in the short run, and slower in the long run (Ayres and Axtell 1996; Weitzman 2001; Axtell and McCrae 2007). In a 1998 paper, Weitzman himself argued for the "lowest possible" discount rates for the long term (Weitzman 1998). Talbot Page argued for zero discount rates for restoring environmental damages (Page 1977). Several other well-known economists have said that they are uncomfortable with non-zero discount rates.

The real argument against the Stern Report (and all others using the same methodology) is quite different. The real objection is that the underlying business-as-usual assumption—that the economy would grow steadily at 3 % p.a. from now until the year 2100 if no actions were taken to counter the climate change threat—is unjustified. There have been very few periods in history when any country grew steadily like that, and the next 85 years is extremely unlikely to be such a period. In fact to say "extremely unlikely" implies that there is a finite chance of steady growth. I think there is no chance whatever, because "growth" is inherently subject to cycles and "bubbles" that are getting bigger and more dangerous. I will explain what I mean in some detail later.

For the moment, the point of the last several paragraphs is that the assumption that economic growth would occur at historical rates in the absence of radical change-inducing policies is analytically convenient but dangerously mistaken. In fact, a serious program to "decarbonize" the world economy, in the shortest possible time, may well be the only viable long-term growth strategy. In other words, I am saying that kicking the fossil-fuel habit may be profitable as well as environmentally beneficial, whereas the policy of resisting any change (as currently advocated by Republicans in the US and much of Wall Street) is probably the road to economic as well as environmental disaster. This statement needs further explanation, of course. Read on.

12.6 More on Economic Growth: Cycles and Bubbles

Cycles come in all sizes and lifetimes, as do organisms, and like organisms, they are ubiquitous. All organisms have life-cycles, of course, but there are larger cycles in nature. There are cycles in ecology, as pointed out in Sect. 3.8, and in biochemistry, Chap. 6. These so-called "nutrient recovery cycles" conserve information (negative entropy) and prevent the loss of biologically important compounds, such as amines and phosphates. This is accomplished by passing them along a chain of organisms for re-use before being lost to the environmental "recycle bin" (from which little or nothing is actually recycled). The life-cycle is crucial to biological growth. The life-cycle in economics is an analog of the life cycle of an organism: It is born, grows, matures, ages and dies (unless it is merged into, or swallowed by, another larger firm). In that case the DNA of the "prey" may survive as part of the "predator's DNA" (Margulis 1970, 1981). This, of course, is the mechanism by means of which multi-cell organisms and multi-national corporations came about.

The economy as a whole can be regarded as a collection of profit-making firms, plus some not-for-profit (but not to be ignored) government agencies and social institutions of various kinds, all under a governmental umbrella. There are many varieties of firms, of institutions and of governments, but the similarities and differences do not matter for the purpose of this discussion. The point is that aggregate economic growth is a summation over this collection of entities (albeit not including some of the social service capital mentioned earlier). Among the economic sectors, some are growing rapidly, others not at all.

This has been called the "granularity of growth" by stock-market professionals who constantly seek to identify (and buy) the young "growth" companies while disinvesting from the old "dogs" (Viguerie et al. 2007). For each of the profit-making firms in the economy, growth is a highly discontinuous process, rarely a monotonic one. It is often accompanied (if not triggered) by "creative destruction" of some sort. Yet growth of the economy as a whole is typically presented as a monotonic increase while the system is in equilibrium. This is misleading, both because growth in history has never been smooth or monotonic, and because growth-in-equilibrium is essentially impossible, given that growth is fundamentally a disequilibrium process.

Business cycle studies began long ago. The speculative manias of the early eighteenth century, as well as the commercial crises of 1763, 1772, 1783 and 1793 induced several serious efforts to understand the phenomena, but those efforts were mainly descriptive e.g. (Mackay 1841 [1996]). The Napoleonic wars and the economic changes that followed, gave rise to a new set of concerns, prompting different responses. Ricardo's formulation of economic theory was partly a response to these changes. But the classical theorists, from Adam Smith to Alfred Marshall throughout the nineteenth century, generally focused on the "normal" state of the economy and paid little attention to the perturbations of trade. The first economists to focus attention prominently on the booms and busts were critics of the system.

The year after Waterloo was a long crisis. British export goods overstocked the markets in 1816, forcing many of the consignees into bankruptcy. The people of Europe had no money to buy. There was a recovery in 1817 and 1818 but another recession started in 1819. Sustained recovery did not begin until 1821. This caused some observers, notably J.C.L. Simone de Sismondi, to question the "laissez faire" philosophy of Adam Smith and his followers (Sismondi 1819 [1827]). Why, he wondered, if everyone was free to produce as much as he wanted, was there so much unemployment and misery among the workers?

Sismondi made several suggestions, not thoroughly examined. One was the fact that when producer's decisions are guided by prices, there is a danger that all of the producers will react the same way at the same time, resulting in a glut or a shortage. Another one of his suggestions was that consumers might have too little money to spend left, due to income fluctuations, or that "over-saving" might play a role. Finally, Sismondi suggested that mal-distribution of income was a major part of the problem. He complained that economists assumed that demand (as in "wants') is effectively unlimited, whereas the income of the "laboring man" is narrowly limited.

Kindleberger and Aliber have noted that there were financial crises almost exactly every 10 years in the early part of the nineteenth century: 1816, 1826, 1837, 1847, 1857, 1866, 1973 (Kindleberger and Aliber 1978 [1989]; Kindleberger 1989). Thereafter the frequency slowed down: major crises occurred in 1907, 1921, and 1929. All of these mechanisms, and others, were discussed at length in subsequent years, as summarized by a major effort by the National Bureau of Economic Research (NBER) (Mitchell 1927 [1955]). The most familiar economic cycles are nowadays named after their discoverers or proposers. They include:

- 3–5 year "inventory cycle" of Joseph Kitchin (1923). These "minor cycles" span on the average about 31/3 years or 40 months; Kitchin concluded that two or three 'minor cycles' fit within one 'major cycle' (that is nowadays referred to as the Juglar or Business Cycle).
- 7–11 year fixed investment cycle of Clement Juglar. The Juglar (1862) cycle reflects a three-phase cycle in the world economy with a periodicity of about a

decade. His ideas were elaborated by Joseph Schumpeter in his landmark book "*Business Cycles*" (Schumpeter 1939).

- 15–25 year "swing" of Simon Kuznets (1930). He investigated secondary secular movements and concluded that the secondary secular movements in prices (estimated at 23 years) are similar to those in production (estimated at 22 years). He explained the swing in terms of immigration and emigration movements, and construction demand. A critic has argued that the swing is an artifact of the filter he used to distinguish these waves from combinations of the others. I doubt that it is worth discussing further in this chapter.

- 45–60 year "long-wave" of Nikolai Kondratieff (1925). This "K-wave" is highly controversial, but has been widely discussed since the 1925 publication of the Soviet economist Nikolai Kondratieff in the "*Voprosy konyunktury*" or "Long Economic Cycles". A German translation of Kondratieff's article was published in the *Archive für Sozialwissenschaft und Sozialpolitik* (1926) and an English summary of the German article was published by Stolper (1935). There has been an avalanche of articles and theories about the K-wave since it was first noted.

The short Kitchin cycle is generally explained as a temporary imbalance between supply and demand. When there is economic growth, demand rises and inventories are depleted because there is a lag between orders and delivery. But when the consumers have bought all they want, growth slows down and inventories rise again as goods on the shelves or still in the pipeline are delivered to the shops. Then comes a period of slowdown as the shops sell from excess inventory before ordering more goods from the factory. During this pause, there is an uptick in unemployment. When the excess inventory is used up, the cycle starts again.

Schumpeter developed Juglar's asymmetric three-phase model into a sinusoidal four-phase model consisting of prosperity, recession, depression and recovery (Schumpeter 1936, Vol. 1, p. 167), of which the depression phase is not necessarily included in the cycle. The Schumpeterian wave is indicated in Fig. 12.6. Schumpeter's student, Hyman Minsky took this argument further. Basically, the Minsky model asserts that the flow of credit constitutes a feedback system. During the rising (recovery) part of the cycle, borrowers and lenders are both optimistic. Increased credit may cause an increase in demand faster than the increase in supply for the "good" in question. This can cause a price spike. The price spike, in turn,

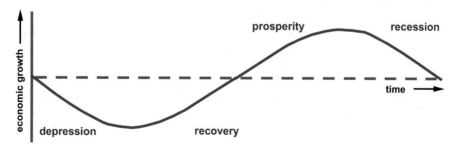

Fig. 12.6 Four phases of a Schumpeterian Business Cycle

attracts further investment from speculators using borrowed funds and operating on margin. There is always a theory as to "why it is different this time" to explain why the rising asset price is not a bubble e.g. (Reinhart and Rogoff 2010).

A similar 4-phase cycle, with different labels, has been recognized in the ecological literature, focusing on *resilience* (Holling 1973, 1986). The ecological version was originally applied to explaining the succession of vegetation types following a forest fire, a major storm, a pest outbreak or clear-cutting by lumber companies. The sequence is "renewal" followed by "exploitation" and eventually "conservation" or "climax". The latter is usually followed—sooner or later—by "creative destruction" due to the lack of resilience in the aging system.

During the "renewal" phase of rapid growth, (colonization at first by fast growing species such as succulents or "weeds") carbon is captured and embodied in biomass, while surplus oxygen is produced. During this stage many species compete and diversity is great. The system is very resilient. In the "exploitation phase" the ecosystem matures (perennial shrubs, deeper-rooted small trees), net growth slows down, diversity declines. Longer-lived and larger species drive out small competitors. In its "conservation" or "climax" phase (domination by large, deep-rooted forest giants that capture and store most of the nutrients and water in the system), the ecosystem becomes a net consumer of oxygen and producer of carbon-dioxide. In this phase, biomass (stored capital) is maximized but old trees begin to die and decay. The "climax" ecosystem is ripe for destruction. It is during that time that biological innovations can occur, usually as some new invasive species gets its chance.

The business cycle, as described by Schumpeter, in his early work, differs from the ecological cycle in one crucial detail: the end of an established industry is often brought about not by old age or natural disaster, but by competition from a newer and more dynamic innovator. Steam-powered railways put many canals out of the shipping business. Steamships put sailing ships out of business. Gas lights superseded candles and incandescent electric lights superseded gas lights only to be replaced by fluorescent lights and yet again by LEDs. Automobiles replaced horse-drawn carriages, TV has replaced radio, Cable TV replaced broadcast TV, transistors replaced vacuum tubes, e-mail has replaced postal communication. Similar innovations have occurred in business organization, e.g. interchangeable parts, mass production, mail-order, franchising of name brands, providing free internet services paid for by advertising, etc. Otherwise, the cycle of technological substitution is much the same as the eco-cycle described above.

The term "creative destruction" is a phrase first used in Schumpeter's Ph.D. thesis work in Austria (Schumpeter 1912). He was making the argument that in a competitive capitalist marketplace, creative innovations (new inventions or new organizational systems) fight for market share against established monopolies by offering better or cheaper alternatives to the existing product or service. But, by "winning" they also improve the overall system and increase the overall welfare of society. When a new product provides a cheaper or better product to its clients, it saves them money previously used to buy the competing product, or it helps in other ways (cutting maintenance or repair costs, for instance). Either way, the innovation increases the wealth of both the producer and the consumer and enables them to increase their demand for other goods and services. The "destruction" applies to the previous incumbent, of course, but Schumpeter argued that the gains of the winners, and the public, in such cases, generally exceed the losses of the losers. (*This is not true for externalities, as I pointed out in* Sect. 11.3).

"Creative-destruction" has undoubtedly been a major mechanism behind global economic growth. This is especially important when successful innovators create— or enable—completely new businesses based on technologies that didn't exist before. Printing books on paper using movable type was an example. The printing press started a paper manufacturing industry and a machinery industry, not to mention the Protestant Reformation. Coking saved the forests and created a coal industry. Steam engines enabled railways, steamships and gas lights. Electric power generation enabled light for all purposes, as well as electric furnaces, electric motors, electrolytic processes and telecommunication. Programmable electronic computers enabled calculations and capabilities—such as airline scheduling, chemical process simulation and language translation—that were previously impossible. Computers allied with high-speed telecommunications have also enabled "browsers" and "search engines" that enabled internet shopping, internet match-making, self-publishing, "crowd-funding" and medical diagnosis from a distance, among other things.

Nikolai Kondratieff's original work on long cycles was essentially empirical. However it should be noted that the field now has a life of its own, due to the existence of several different competing theoretical partial-explanations, based on different interpretations of the empirical data. On the other hand, several Nobel laureate economists have dismissed the entire idea as "*implausible*" (Leontief) or "*science-fiction*" (Samuelson). Nevertheless, I think a brief summary of the evidence is in order.

Kondratieff (1926) covered a period of 2½ waves of which the first two were 60 and 48 years. Based on this analysis he indicated an average length of about 50–55 years. Table 12.1 provides an overview of the various Kondratieff waves that have been suggested since the beginning of the Industrial Revolution (IR). Kondratieff originally postulated just two phases, expansion and depression, or recession and recovery. He later expanded this to three phases with "stagnation" between expansion and recession. He identified "turning points" at the peak of the expansion or recovery. The asterisks in the table are the turning points in his formulation.

Kondratieff broke his "waves" into two phases, "depression" and "recovery", based on observed trends in economic growth (including recessions) in a few European countries plus the US for which economic data were available. Since

Table 12.1 Historical overview of past Kondratieff waves

Kondratieff	First wave	Second wave	Third wave	Fourth wave	Fifth wave
Depression	1764–1773	1825–1836	1872–1883	1929–1937	1973–1980
Recovery	1773–1782	1836–1845	1883–1892	1937–1948	1980–1992

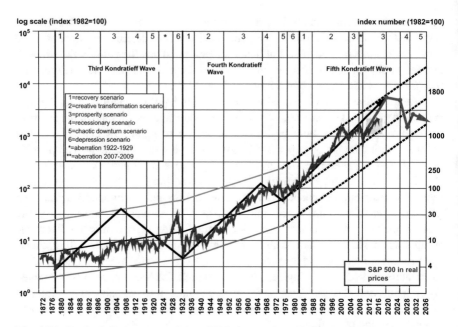

Fig. 12.7 Kondratieff chart expanded to 2035. *Source*: http://beleggenopdegolven.blogspot.fr/2010/11/de-kondratieffgolf-volgens-ter-veer-en.html 2011–2015 real data added by Ayres

his original work, a number of authors (including myself) have attempted to explain the K-wave in terms of technological transitions e.g. (Ayres 1989):

The chart reproduced in Fig. 12.7 shows the K-wave superimposed on economic data from Standard and Poor (S&P) from 1875 extrapolated to 2035. The long K-wave departs radically from the data during the 1920s and 1930s, but matches fairly well until around 1997 when another boom-bust episode "took over". Kondratieff made several empirical observations concerning the main features of long wave cycles in the capitalist world, including the following (paraphrased) from (Stolper 1988):

- During the recession of long waves, agriculture—as a rule—suffers an especially pronounced and long depression. This was what happened after the Napoleonic Wars; it happened again from the beginning of the 1870s onward; and the same can be observed in the years after World War I.
- During the recession of long waves, an especially large number of important discoveries and inventions in technique of production and communication are made. However, these advances are usually applied on a large scale only at the beginning of the next long upswing.
- It is during the period of the rise of the long waves, i.e. during the period of high tension in the expansion of economic forces, that—as a rule—the most disastrous and extensive wars and revolutions occur.

The first observation can be generalized by stating that during the recession of long waves the older sectors will face a more severe recession than the newer industries because they are less resilient. The second observation is supported by Gerhard Mensch's book *Stalemate in Technology* (Mensch 1979) and the *Schumpeter Clock* (Haag et al. 1988). Grübler and Nakićenović (1991, p. 337) wrote that "*It is the disruptive crisis of the old that provides the fertile ground for new systems to develop*". In other words: creative destruction drives the economy.

Finally, Kondratieff's third observation fits the historical record remarkably well—except where it doesn't. On the positive side, the 1802–1815 period of the first wave refers to the Napoleonic Wars that followed the French revolution (1789–1799). The American civil war took place from 1861–1865 which is at the end of the prosperous period of the second wave. World War I (1914–1918) involved many of the world's great powers and took place at the end of the prosperous period of the third cycle. The 6-day-war (1966) and Yom Kippur War (1973) took place in the Middle-East at the end of the fourth wave. (That wave was arguably based, in part, on competition for oil as a main energy source.)

However, Fig. 12.7, and other factors, suggest (to me) that the K-wave—if it really was a cycle—has recently broken down. This is partly because purely economic cycles are increasing in relative importance as compared to radical technological innovations. At the end of the fifth hypothetical K-wave, there is much turbulence in the world. This time the social unrest is caused by a combination of long-term trends, including the digital economy, changes in sexual mores, urbanization, climate change, globalization, the "clash of civilizations" (especially Islamic fundamentalism), the rise of China, the stagnation of Europe and the decline of American power, among other things.

Major technological "breakthroughs" occur at times that are hard to predict, although the technological "gaps" and "needs" that inspire research and invention are usually obvious to many people. This fact accounts for the phenomenon of "simultaneous discovery" by different inventors in different places. Edison's incandescent light was actually preceded by a similar one in England by Joseph Swan. Alexander Graham Bell's patent of the telephone was virtually simultaneous with a patent application by Elisha Gray (which led to years of legal dispute over priority). The invention of the electrolytic reduction process for aluminum occurred virtually simultaneously in the US (by Charles M. Hall) and in France (by Paul Heroult), in 1886, which is why the process—still in use—is now known as the Hall-Heroult process.[9] There are quite a few other examples.

In these economic cycles, the common element is imbalance. Imbalances can occur in demand for different products, in labor supply, in savings or investment. The economic system has some homeostatic (self-correcting) tendencies, emphasized in economics 101 or soon after. Imbalances between supply and demand of

---

[9] There are hundreds of books about the history of technology. I mention only three: two edited multi-volume histories available in libraries (Singer et al. 1958; Kranzberg and Pursell 1967)and a more recent and more readable survey from a systems perspective (Gruebler 1998).

products and labor in an ideal free market, are normally self-correcting. The price system is the principal agent of self-correction. Prices work to correct imbalances, in "normal" conditions. Admirers of "free markets" often use characterizations like "magic" to describe its workings. However, a free market does not determine some key factors. There are times when the self-correction mechanism in the real world fails.

In fact, markets do not exist for everything, and when they do exist, they are rarely "free" of external influence or intervention. For example, the price of money (the interest rate) is fixed by a small group of economists at the central banks of the world, particularly the US Federal Reserve. Admittedly they take advice from many sources, *but their decisions are not based on supply and demand*. They are based on observed trends in "core" inflation data and unemployment statistics.

It is increasingly obvious that US and European central bank policy on reducing interest rates during a recession has caused major imbalances in the form of bad real estate (and other) investments. (This was the chief complaint of the Austrian school of anti-Keynesians in the 1930s.) In many cases, not least the "sub-prime" disaster of 2008, the bad real-estate investments were encouraged by poorly executed public policies favoring home ownership. The mechanism was an administrative lowering of standards for home loans in certain areas, by government sponsored housing institutions. This tinkering was well-intentioned, but sent a bad message to markets and led to a bad result in practice (Ayres 2014a, b).

Similarly, fiscal policy, formed by governments (democratic or not), does not prevent unbalanced budgets, and unbalanced trade flows. In the fifteenth or sixteenth centuries, when the royal treasury of a European country was depleted (due to military expenditure or over-lavish royal weddings) the King would demand that his officers collect more taxes. In early seventeenth century England, the Parliament resisted the royal demands and the end result was the English civil war. In Spain, the government met the shortfall (from military expenditures) by borrowing from Italian and German bankers. When the bankers refused to lend more (or couldn't) the Empire came apart. In France, Louis XIV finally resorted to "tax farming". When the rising middle class refused to pay (or couldn't pay) the end result was the French Revolution. Similar problems in China and elsewhere have led to analogous consequences.

From the beginning of long-distance trade, there was a steady outflow of gold, moving to India (and silver to China) to pay for spices, tea, silk and other goods desired by the English (and other) rich.[10] Those countries had no interest in British products like wool. This resulted in a persistent trade deficit, compensated at first by the Spanish looting of Aztec and Inca gold, and mining of silver by slave labor. But that inflow of bullion diminished, while the outflow of gold and silver increased. In the eighteenth century the British adopted a "mercantilist" policy demanding that British exports of manufactured goods (and taxes) to its colonies be paid for in gold or silver specie. The "Boston Tea Party" and the American Revolution was one

---

[10] For a detailed history, see (Graeber 2011).

consequence of that ill-judged policy. (The British should have taxed imported silk, tea and spices.)

The British East India Company was another mechanism to do the same thing (i.e. to cut the outflow of precious metals) by more effective means. Those means included direct taxation and deliberate destruction of the local Indian textile industry. The "opium war", sponsored by British interests, was a crude attempt to create a Chinese market for a British product (opium from Afghanistan) that the Chinese didn't want or need. Again, the underlying purpose was to stop the persistent outflow of gold and silver.

Speaking of imbalances, in the nineteenth century the "gold standard" was adopted by most of the trading countries, led by the British. This kept the global money supply (outside of China) under control, to a degree. But the supply of new gold from discoveries in America and South Africa was insufficient to compensate for the outflow of British, Dutch and French gold to India and China. US gold from the California discoveries moved to Europe in exchange for European investments in railroads and other things. The result was a shortage of money that restricted bank lending and made life very hard for US farmers. That shortage dominated a presidential election. (In 1895 President Cleveland had to go to JP Morgan to replenish the gold in the US treasury, which had been depleted by a law setting a fixed price for silver. That bizarre law, which had nothing to do with supply and demand, helped to make the silver-mining Guggenheim family among the richest people in the world).

The point of the last few paragraphs of potted economic history is that *some imbalances are not self-correcting. Moreover, in the past, they often ended in violence.* But the violence often leads to a new paradigm—a new symmetry-breaking. The Versailles Treaty, intended to de-industrialize and impoverish Germany, led to the hyper-inflation of 1921–1922 that destroyed the savings of the German middle class. The end-result was Hitler and Nazism. The collapse of a giant US stock-market bubble in October 1929 was turned into the Great Depression by stupid budget-balancing policies (raising taxes during a recession) on the part of the Hoover administration. Keynes' proposals to support demand (and employment) by deficit spending were opposed bitterly by powerful "laissez-faire" advocates, such as US Treasury Secretary (and banker) Andrew Mellon, who expected the imbalance to fix itself when all the "bad investments" were closed out and off the books. (He didn't say what would happen to the unpaid debts.)

As things worked out, the US economic recovery, under Franklin Roosevelt, was very slow, mainly because there was very little stimulus from the government until the start of World War II. The German recovery, under Hitler, was much faster, due to heavy German spending on armaments and highways.

Sometimes government interference is necessary, as when regulation is needed for health and safety reasons. At other times, government intervention can have negative consequences, as for example by raising taxes ("to balance the budget") during a recession or when central banks hold interest rates too low for too long, resulting in another bubble. The seemingly endless dispute between Keynesians and anti-Keynesians boils down to an argument between those who trust governments despite plenty of evidence of poor decisions by politicians, and those who don't trust governments but trust "the market", despite plenty of evidence of market failures and even criminal behavior (e.g. Bernie Madoff).

At any rate, there does seem to be a "regulatory cycle", at work. A period of weak regulatory enforcement and rapid growth is likely to be followed by a disaster of some sort that prompts a new and tighter regulatory regime. This, in turn, can dampen the enthusiasm of investors and slow down the rate of economic expansion. There may also be a credit-debt cycle, after a war or revolution in which a lot of debt is converted to equity and subsequently wiped out. Starting fresh, reconstruction is likely to result in very fast growth for relatively little outlay of new funds because much of the needed capital investment is already in place. Only the ownership changes. But after the "recovery" is over and a period of expansion has gone on for some time, all sorts of interest-groups get into the act. I am reminded of some hedge funds that can be described as "vultures" who feed on distress. (Some US hedge funds bought Argentinian government bonds at low prices, betting that US courts would force Argentina to pay the face value of the bonds "in-full", thus making a large profit for themselves from the taxpayers of Argentina.)

A fairly complete theory of this phenomenon, entitled *The Logic of Collective Action*, followed by *The Rise and Decline of Nations* was published a few years ago by Mancur Olson (Olson 1965 [1971], 1982). His theory basically argues that political stability contains the seeds of its own destruction, very much like climax forests. Borrowing another metaphor from biological evolution, political stability is like oxygen photosynthesis: the longer photosynthesis goes on, the more it poisons the air with toxic oxygen. In the economy, it permits the formation of destabilizing networks of narrowly focused special-interested groups, such as cartels, lobbying organizations, political splinter groups and ideologies. Those narrow interest groups working against each other gradually impede decision-making, and choke off growth. Olson's theory has explained a number of historical phenomena, starting with the differential growth rates of different countries in the post-war period, as well as earlier periods, as well as the British class structure and the Indian caste system.

A non-cyclic consequence of financial imbalances is the growth and collapse of asset bubbles. When there is "free money" floating around—money not otherwise committed either to de-leveraging (debt repayment) or capital investment—there is an opportunity for an asset bubble. There have been a number of such bubbles in history, starting with the Dutch Tulip bubble and subsequently the Mississippi Bubble in France and the South Sea Bubble in England (Mackay (1841) [1996]). Mackay put most of the emphasis on irrationality. Kindleberger and Aliber put the phenomenon in its macro-economic context, and provide a generic "anatomy" of a typical crisis. I lean toward the latter explanation.

A formal theory that might explain at least some of the K-waves, along with the financial crisis of 1929–1930, is the "debt-deflation theory" of Irving Fisher (Fisher 1933), later modified as the "financial instability theory" of Hyman Minsky and Stephen Keen (Minsky 1982; Keen 1995, 2011b). The core point of the Minsky-

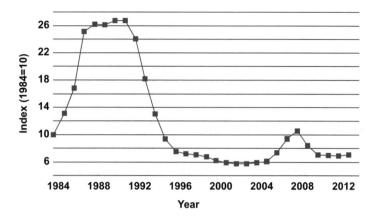

Fig. 12.8 Commercial real estate prices in Ginze 7-chome Tokyo, 1984–2013

Keen thesis is that banks do not just lend the money of their depositors. They create money and debt simultaneously, as pointed out earlier (Sect. 12.4). Debt tends to grow faster than output. *This means the banks also create imbalances.* In fact, I think the Minsky-Keen theory is on target, and I hope to help develop it further.

In every bubble, there comes a time when the "free money" runs out and prices stop rising. At that point, speculators start to cash in, the asset price starts to decline, and the feedback cycle reverses. The three oil price spikes (1972–1974, 1979–1980 and again in 2007–2008) are almost certainly examples of speculative bubbles, or deliberate manipulation, divorced from actual supply-demand conditions (see Fig. 10.9). The Japanese stock market and real-estate bubble in the late 1980s is another example, as was the US "dot com" bubble in 1999–2000. Lesser examples of bubbles in gold, silver and rare earths are also of interest.

The Japanese land-and-stock bubble1984–1989 has almost been forgotten, despite the fact that it was the primary cause of the Japanese "malaise" after 1990. In the Japanese case the bubble was based on the supposed marvels of "just-in-time" manufacturing prowess and other social innovations such as life-time employment e.g. (Kahn 1971). Figure 12.8 shows the price history of a section of Tokyo, which happens to correlate with that bubble of 1984–1992.

There have been a series of bubbles in the rare metals markets, generally based on spurious theories of huge increases in demand, against supply limitations. Consider the history of silver prices since 1960. There have been two huge price spikes (Fig. 12.9). The first of them (c. 1980) was misinterpreted by "gold bugs" but was actually caused by a deliberate attempt, by the notorious billionaire Hunt brothers, to "corner the market" (Streeter 1984). The recent spike (c. 2012) has no such obvious explanation, except that the silver price is always somewhat correlated with the price of gold. In both cases, supply and demand had nothing to do with the prices.

Figure 12.10 shows the prices of three mining company stocks associated with "rare earth" mining. This bubble was kicked off in 2011 when China (which

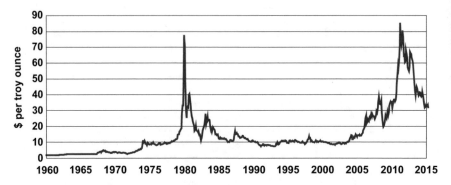

Fig. 12.9 US monthly silver prices 1960–2015

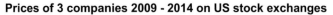

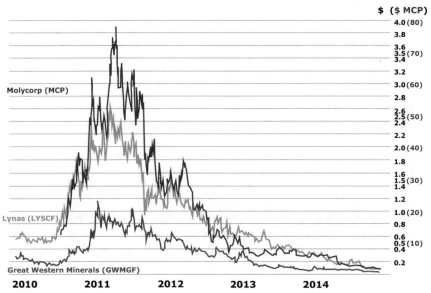

Fig. 12.10 Rare earth stock prices—a classic bubble

dominates the market for rare earth metals), announced plans to cut back on exports of ores or concentrates, and to concentrate on products utilizing those metals. A lot of people (including me) thought that a mining boom would ensue. It didn't. There was no supply shortage because the major users of rare earth metals found substitutes or paid the Chinese for refined metals, or products, rather than purchasing ores.

The US stock-market "dot com" bubble of 1991–2002 (Fig. 12.11) was similar to the Japanese bubble, in that it was based on a spurious theory that a new kind of wealth had been discovered. In the US case it was based on the curious notion

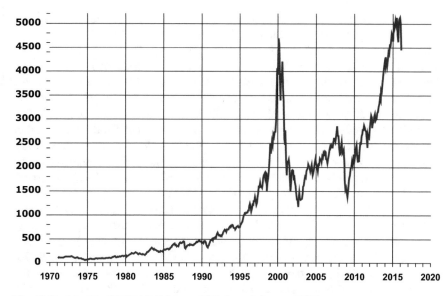

Fig. 12.11   Monthly NASDAQ February 1971 through January 2016

(based on the growth of the Internet), that fast-growing digital "dot.com" companies without visible earnings but lots of "followers" were far more valuable than
"old fashioned" companies that actually made things. This idea had merit, though
the vast majority of the dot-coms popped (as bubbles do). The more recent success
of Google, Amazon, Netflix and Facebook (and others) reflects that fact that ways
of monetizing electronic media marketing have evolved (at the expense of the print
media).

What we are seeing today in the marketplace is also an ever-increasing trend
toward oligopoly and away from "perfect" competition. This is happening in almost
every country and industry. The evidence is not just anecdotal: quantitative measures of market power have been developed, and they indicate that business
consolidation is continuing. The days when 10,000 flowers might (metaphorically)
bloom and grow independently are long gone.

Qualitative changes are generally categorized as innovations and all innovations
are regarded (by most economists) as good. Yet many of them (like diet pills, cancer
cures, perpetual motion devices and formulae for manufacturing gasoline from
water) have been "invented" (and failed) many times. Most innovations are mere
novelties, not improvements to our well-being. But radical changes do occur from
time to time. The steam engine was one. The electric light was another. The radio
was another. The Internet was another. The smart phone was yet another. There is
an evolutionary process going on, and the process itself is complex. It is qualitatively and locally the same as the evolutionary process in biology. It can be
characterized as a search algorithm: variation, selection and replication
(Beinhocker 2006).

Most new business ideas, like new designs for mousetraps, exercise machines, sailboats or bicycles, fail immediately. They are not as good (cost-effective) as the original for whatever task the original was performing. But sometimes a variant performs better in competition with rivals, at least in a test. Then it is selected, developed, tested again, manufactured in prototype quantities, and exposed to some segment of the marketplace. Changes may be indicated. If it is a physical object that finally succeeds in the marketplace, it will be replicated (mass-produced), like Henry Ford's Model T. It is worth mentioning that the Model T followed 18 earlier models of which 8 were actually produced (Models A, B, C, F, K, N, R, S). Most new products fail somewhere along the line. But if it succeeds it is perceived to be "fitter" than its rivals, at least in some domain, until the next challenger comes along. Hence, there is a tendency to think innovation is always good, Q.E.D.

I'm not saying that innovation and increasing complexity is good or bad; merely that it drives evolution. It is argued by some scientists that increasing complexity corresponds to increasing information content (negentropy) which is *ipso facto* good. But that argument is somewhat circular. On the contrary, I argue that increasing complexity (in terms of goods and choices) is not *ipso facto* a measure of progress or well-being, although it is correlated with GDP. I suspect it is a consequence of economic growth as we understand it, but not a driver.

In biology, the "R&D" process is based on random mutations probably triggered by cosmic rays. It is unguided, unsupervised, accidental, and unintentional. In human society, things are very different. That is a major reason why technological change is so much faster than any natural mutagenic evolutionary change in the biosphere. But behind the difference there is also similarity. Like a lot of physicists and economists, I suspect that there is a "force law" or an extremum (maximization or minimization) principle at work. In the evolution of atoms, chemicals and living organisms it could be characterized as maximization of "binding energy". In economics the maximand is probably something like disposable wealth.

But be careful. At first sight, the measure of binding energy in the biosphere is probably the stock of the protein chlorophyll. If that is so, the "Gaia hypothesis" would suggest that the animal kingdom exists simply to consume the toxic oxygen, produced by oxygen photosynthesis, thus preventing life from destroying itself. The maximization of chlorophyll occurs when there is a homeostatic balance between photosynthesis and respiration. In other words, in the "Gaia world" the balance would be an animal kingdom with just enough biomass to keep the atmospheric oxygen level from increasing further.

Of course that doesn't prevent human biomass from increasing at the expense of other species of oxygen breathers. But the biological analogy does suggest a question: Is there a parallel case of balancing two contrary tendencies in economics? The apparent conflict between competition and cooperation might be such an example. The conflict between wealth maximization and current income (consumption) maximization might be another. At present there is very little discussion of this sort of question in the literature.

It is qualitatively clear that economic growth cannot continue without limit for a variety of reasons, already touched upon. What is lacking is an endogenous anti-growth mechanism that can be incorporated in growth models. Currently, profit maximization or cost minimization and utility maximization are found virtually everywhere in economic models, large and small. In most models, since Samuelson and Solow, utility is taken to be a function of aggregate consumption (Samuelson and Solow 1956). Until now, nobody has argued much with this assumption i.e. $U = U(C)$. In other words, it is widely assumed in mainstream economics that people (consumers) want to maximize their quantitative consumption of goods and services. That fits, more or less, with the assumption that they want to maximize monetary income. It also fits with the idea that governments want to maximize GDP.[11]

There is no mechanism in the current models to reflect the anti-growth forces (such as debt accumulation, oligopoly, and externalities) that also exist. In fact, I suggest later that a realistic production function must include a measure of the current investable economic surplus, taking into account the current burden of (total) debt and the current cost of compensation for externalities, such as global climate change.

Another anti-growth trend has received almost no attention from economists. That is the declining "ore grade" of all natural resources. Since every extractable natural resource, except fresh water, can be measured in terms of the exergy that would be required to re-create it in its natural form, the exergy return on exergy invested (EROI) measure becomes pertinent. Global EROI is gradually increasing as high quality natural resources are extracted and dissipated. What this means is that the potential economic surplus will inevitably be reduced by the amount of additional effort needed to extract essential resources, as the best quality sources are used up. However, apart from some almost anecdotal studies cited previously, it is not measured systematically and consequently not (yet) usable in economic models.

The issue here can be characterized in terms of imbalances. What we can say for sure is that increasing GDP, in the short run, means increasing energy (exergy) inputs. Economic activity of all kinds requires exergy inputs, even activity directed toward increasing the efficiency with which exergy is used in production and in consumption. But as pointed out in the last, a society without any fossil fuels would be much less active than the society we live in: it would be a very quiet place. And a society with no source of exergy at all would itself be a fossil.

But if exergy is too freely available (cheap) with no other resource constraint, a lot of exergy will be used inefficiently because there is no need to minimize it. For example, farmers will pump water from deep wells for irrigation; builders will put electric heat in apartment houses and homes. They will build very large "McMansions" or large yachts for smuggling works of art or drugs. Aircraft designers will create supersonic jets like the Concorde. Auto companies will sell

---

[11] To be sure, there are ecological economists, notably Herman Daly, who say that GDP is not a good measure of human welfare (Daly and Cobb 1989). Their critique of GDP as a measure is hard to gainsay. But the vast majority (myself included), use GDP for analytic purposes simply because there is no good alternative.

gas guzzlers like the V-8 "muscle cars" of the 1960s. What gets maximized in such cases is convenience, or speed, or ostentation, or even cost, but certainly not economy. Whatever is being maximized it is not necessarily profits or income.

By the way, according to established wisdom, there are no "free lunches", because the busy and rational maximizers with perfect information have eaten them all. As the economics professor of legend said to his young graduate assistant as they crossed the Harvard Yard: "That $100 bill you see on the sidewalk is a counterfeit; if it were real, somebody would have picked it up." In more conventional phraseology, economic models tend to assume that rational cost minimizing business owners (or managers) would always take every opportunity to save energy at a profit. Yet the UN Industrial Development Organization (UNIDO) once hired me to do a study under their title: "If energy conservation pays, why don't companies do it?"—or words to that effect. I won't tell you what happened to that report.

Yet, here again, reality is quite different. Actually there are lots of uneaten "free lunches" out there, such as combined heat and power (CHP), LED lighting, car-sharing, riding bicycles and so on. A lot of $100 bills are still lying on the floor, partly because markets and people are imperfect, and partly because the business bosses think they must be counterfeit, but mainly because most managers are not cost minimizers at all. Not yet, at least. I think most entrepreneurs are growth maximizers: if they have money to invest, they would rather use it develop a new product, open an office in China, or take over a rival. If they only use it to become more energy-efficient, they may be making themselves more attractive targets for takeover. But if growth were the only imperative the world would soon starve or choke on its waste. *There must be a growth-limiting imperative "out there" that needs to be conceptualized and quantified and incorporated in the models.*

Examples of all kinds of maximization behavior in market economies can be found without difficulty. At the individual level, we can see many different maximization drives. Some want to maximize power, or fame or reputation. Some scholars (not the famous ones) want knowledge for its own sake. Some men (like Don Giovanni) want to make love to as many women as possible. Some men and women want to talk to God. Many hate change and want to be told what to believe and what to do. They are stability maximizers. Others crave change and hate boredom. Some crave risk. Others (like Ferdinand the Bull) want to smell the flowers and enjoy life. The point is that no single or simple formula describes the motivations of all humans.

But, as a second-best solution to the problem, maximizing money (income) turns out to be a reasonable approximation for many of the short-term wants and needs of most people, because money can buy many—but by no means all—of the other things that people want. It does not buy "happiness" as has been said and proved many times, but its absence is very likely to promote unhappiness. In short, maximization of consumption (which corresponds roughly to income) is an imperfect representation of what drives humans to behave as they do, in the aggregate, but it is used by most economists on the fragile grounds that it is the best we have.

However, there is an alternative. Consumption is not what is maximized by real people, or firms, most of the time: a consumption maximizer with perfect foresight (as assumed in many theorems) will die destitute, leaving nothing to his or her heirs. This is not how most real humans behave. Humans do not just save for old age: we are *wealth-maximizers* and *legacy maximizers*. As history since the pyramids shows clearly, humans crave dynasties (for biological as well as economic reasons). In macro-economic terms, what this means is that when exergy is plentiful, we try to maximize consumption.

Only when exergy gets scarce or expensive will there be a large market for CHP, insulation, double-glazed windows, LED lights, electric cars and electric bicycles. The problem is that when exergy gets very expensive, the economic surplus will shrink. Shall we start to maximize exergy efficiency while there is still a large surplus? During an oil glut? *I suspect that the answer is that increased exergy efficiency will only attract investment when it can be shown unambiguously that it also increases resource productivity and societal wealth.* But for the sake of the survival of the human race, economics must address that issue sooner than later. My small contribution along those lines is outlined Appendix A.

12.7 Planetary Limits: The Downside of Material Wealth Creation

It is evident that the production system creates wastes as well as wealth, and that some of the wastes are toxic or harmful to the planet. Production based on the extraction, processing and transformation of materials into "goods" also means pollution (Ayres and Kneese 1969). There are no "zero-emissions" processes and no perpetual motion machines. The laws of thermodynamics say so, and there is no getting around that fact of life. The questions we need to address are: (1) How bad (inefficient) is our present system of production? If we do nothing but "business as usual" how bad will the situation get by, say, 2050 or 2100? (2) What can be done soon, at zero net cost or at a profit? (3) If the human race got its act together, what could be done in the long run?

Far-sighted people have been thinking along these lines for a number of years. Arguably the whole field of ecological economics is about these questions. I cannot do justice to all the environmental issues, still less the others, in a few pages. The best summary of the current situation is probably a series of articles and reports from a group of 28 planetary scientists, led by Johan Rockström, Director of the Resilience Center at Stockholm University, for the Club of Rome. Their first report, entitled "Planetary Boundaries: Exploring the safe operating space for Humanity" was published initially in the journal *Ecology and Society*, Vol. 14(2) and reprinted in edited form in *Nature* (Rockström et al. 2009). Since publication, there have been a great many commentaries and supplementary studies, far too many to cite.

The group identified nine "planetary boundaries", of which they were able to quantify seven. I have split one of their categories (bio-geochemical flows) into two separate parts: nitrogen (N) and phosphorus (P) making eight quantifiable and two non-quantifiable limits. Those ten (of which three had been exceeded by 2009 – the time of publication) are as follows:

- Radiative forcing should not exceed 1 W/m^2 on average. But, since 1750 radiative forcing has averaged 1.66 W/m^2. Almost certainly this will cause Earth temperature to increase above +2 °C. The safe temperature limit has not yet been passed, but very likely soon will be. See Sects. 6.3 and 6.4.
- Ocean acidification, in terms of saturation of aragonite. (This is defined as the level at which sea-creatures can form shells). The safe lower limit is estimated to be 80 % of pre-industrial level, which was 3.44). The current ocean acidification level is 2.9, which is close to (84 % of) the limit. This limit will probably be exceeded soon.
- Ozone concentration reduction in the stratosphere. The safe lower limit is estimated to be 5 % below the pre-industrial level. The ozone level is affected by chloro-fluorocarbons. Fortunately that source has been controlled by international agreements.
- N-cycle limit. It is estimated that the upper limit for anthropogenic emissions is about 35 Tg (N) per year; partly fertilizer, partly NOx from fossil fuel combustion. Nitrogen fertilizers are synthesized from ammonia, produced from natural gas. The safe limit has already been exceeded.
- P-cycle limit. It is estimated that the safe upper limit is ten times the natural background level in runoff due to weathering (the current level is three); but when freshwater P-runoff is considered, limits have already been exceeded. Phosphate rock (the only source) will run out in 200 years, or less according to some estimates.
- Freshwater use for consumption should not exceed 4000 km^3 per year
- Land use: No more than 15 % of ice-free land can safely be used for crops
- Biodiversity loss: should not exceed ten species extinctions per million living species, per year
- Chemical pollution: undetermined
- Atmospheric aerosol loading: undetermined

Those limits may or may not be "hard", but human society has already exceeded several (see Fig. 12.12) and that means there is a significant risk of a "runaway" situation. It also means that the total of planetary natural wealth (including resiliency) has been depleted.

The next question, of course, is: what can be done to ameliorate the situation? There are three parts to the answer. One part refers to technological and scientific potential. The second part concerns economics. The third part concerns politics and behavior. As regards the first, I am moderately optimistic—as Chap. 10 indicates—although there are some open questions, as noted. As regards the second, it is what the rest of this book is about. The political question is beyond my competence to

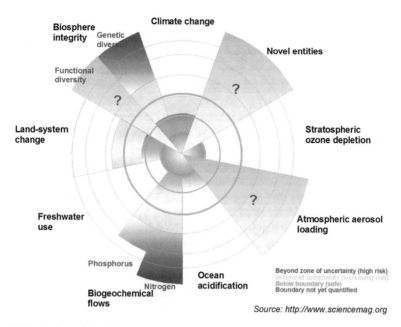

Fig. 12.12 A view of the planetary boundaries

comment. Only the elected leaders of the nations of the world, and spiritual leaders like Pope Francis, can make a difference.

N.B. of the ten boundaries (treating N and P separately), at least six (the first four and the last two) are directly related to the excessive consumption of fossil fuels by human industrial civilization. That, in turn, is directly related to the problem of what drives economic growth, and whether economic growth can (or should) continue without the use of fossil fuels.

There is no doubt that economic growth *per se* has a downside, starting with the resulting demand for raw materials that have to be extracted from the Earth's crust. That extraction process causes consequent damage to the landscape and depletion of scarce resources. The environmental problems associated with pollution and waste have only grown worse. Not surprisingly "de-growth" is back in the news. Indeed, since Pope Francis's recent encyclical, *Laudato Si*, the rationale for economic growth itself is being questioned. Below I make a few arguments for discussing economic growth in a more positive light:

- Redistribution? De-growthers insist on it. The reason is that if everybody is on Earth is to have a decent standard of living, some will have to consume less so that others can consume more. Yes, that is true if the pie is fixed. But the "pie" may grow. (Also, is there any conceivable policy to make radical redistribution happen in the real world?)
- Exit from capitalism? "To each in accordance to his need, from each in accordance with his ability . . . ". Great slogan, almost as good as the Golden Rule. But

it has been tried in any number of idealistic communities and it doesn't work in practice. Unless there is a personal incentive to take risks (i.e. to increase personal wealth), nobody works more than the minimum because everybody hates "free riders". The Russian Bolsheviks never figured this out, but the Chinese communists did, and China is now growing faster than any country in history.

- Job guarantees with minimum income (negative income tax?) financed by the State. But how? By printing money? Somebody has to be producing a surplus in order to have anything left over to redistribute. Again, we need some growth.
- Debts. Everybody's debt is somebody's asset. It is hard to love banks, but debts can't be just wiped off by the stroke of a pen, except in a bankruptcy court or in a revolution. But there are—as yet—no bankruptcy courts (or rules) for sovereign nations. Unpayable debts need to be wiped out by a legal process. Revolutions don't work well. What happens is usually bloody.
- In summary, de-growth is what will happen sometime in the future if we run out of something essential—like fresh water—but otherwise, it is really contrary to human nature. What we need to do is to figure out how to have economic growth without environmental destruction. I am optimistic enough to think that it should be possible.

So what do we mean by "growth", and is there such a thing as "green growth"? Is it sustainable? To what extent are qualitative changes equivalent to growth? Are they making us better off? Do they improve our quality of life? From a very high level perspective, what economic (GDP) growth offers now is two things: more and more choice, and more and more material "stuff" that has shorter and shorter useful lifetimes. The existing measures of wealth do not reflect capital investments generating social services. But they should.

Moreover, capital invested in material products is currently lost as soon as those products are discarded,—which is much too soon. Most products could be redesigned as service providers (as telephones and IBM computers once were). Material products should be designed to last as long as physically possible, with allowance for repairs and upgrades. The current industrial practice of designed obsolescence to maximize current consumption will unnecessarily denude the Earth of many scarce metals in a few generations.

Most of us—not only economists—tend to assume that more choice and more stuff to choose from is *ipso facto* better than less choice and less stuff. From this high-level perspective, more GDP means both more variety and greater quantity of goods produced (and discarded). The biological analogy of product variety, complexity and quantity is bio-diversity and biomass. Both of these attributes (see Chap. 5) were very important in biological evolution. But in biological evolution, the virtue of diversity is resilience. Extinctions occur, but they are followed by explosions of diversity. In technological society, that argument holds to a point. But what happens when the liquid oil runs out? More precisely, what happens when the exergy return on exergy invested in oil discovery falls below unity? What happens when the last remaining copper mine on Earth is extracting ore with a grade of 0.01 %? What happens when the last phosphate rock deposit is exhausted?

Most scientists and economists since Darwin assume that biological evolution is all about competition and "survival of the fittest". Is that the case in economics? At first sight, you may think that economic competition is like nature, "red of tooth and claw". In fact, that is not a good description of nature. Biological evolution is *not* all about species competition for food and space. There are other modes of getting along within a society or an ecosystem, ranging from inter-dependence between complementary species to parasites, symbiotes and outright mergers. Yes, I mean mergers. Multi-cellular organisms are now understood to be evolved from symbiotes that were originally independent organisms that learned how to combine their specialized functions into a common DNA. We humans are the result of a large number of such bio-mergers, and we (like cows) depend on symbiotic bacteria in our digestive systems.

The resemblance to large firms that were created by mergers of many small ones is irresistible. I deplore the fact that large firms are able to force me to buy things— like "smart phones" that have capabilities I don't need or want, and that break or wear out in a couple of years and can't be repaired. On the other hand, the industrial giants give me capabilities I enjoy at prices I can afford.

However, these behemoths currently behave as though raw materials are not scarce. They do so simply because these materials have not been scarce in the past, and because scarcity is supposed to be overcome by substitution. Economists, as a profession, have consciously supported this view e.g. "the age of substitutability" (Goeller and Weinberg 1976). However Kenneth Boulding—among others from Henry George to Adlai Stevenson—invoked the metaphor of the "cowboy economy", which our grandfathers or their grandfathers enjoyed, in contrast to the "space-ship economy" where our children and grand-children will actually need to live (Boulding 1966). In a spaceship economy, products that deliver a service need to be designed for very long—virtually infinite—lifetimes. Re-use, refurbishing, renovation, and remanufacturing must be central features of the economic system of spaceship Earth. Recycling, in the usual sense of re-smelting mixed trash like ore, is a last resort. Old landfills will become the last mines.

Here I should mention a proposal for an environmental indicator, **M**aterials **I**nput per unit of **S**ervice (MIPS) that is gaining credence in Europe. MIPS reflects the fact that all services are performed by people, or by machines, that involve material inputs. Those inputs may be direct (like the shampoo used by a barber) or indirect (like food for the barber, or fuel for the electrical generator that powers his electric razor, or steel in the razor itself). The idea was formulated initially by Friedrich Schmidt-Bleek at IIASA and has been developed primarily at the Wuppertal Institute in Germany (Schmidt-Bleek 1993; Hinterberger and Schmidt-Bleek 1999).

A few more words about phosphorus need to be added here. In the first place, biological processes tend to concentrate this element. As pointed out in Sect. 6.9, phosphorus is actually concentrated by the food chain. Higher organisms have more phosphorus in their bodies, in percentage terms, than lower ones. On the other hand, calcium phosphate, the main constituent of bones and teeth, is comparatively insoluble. It is more likely to be buried in sediments and lost for long periods of geological time, than to be recycled biologically (as far as we know).

According to one calculation, P_2O_5 now constitutes about 0.045 % of the mass of river flows, but less than 0.005 % of mass of the oceans. This is noteworthy because the concentration of dissolved salts (mainly sodium, potassium and magnesium chlorides) in the oceans is much higher than the concentration of those salts in the rivers. That is because those chlorides are accumulating in the oceans and are not utilized extensively in biomass, nor are they transferred back to the land by natural processes. Yet the concentration of P_2O_5 in the oceans is much *lower* than the concentration in rivers: P_2O_5 accounts for less than 0.005 % of the ocean (0.18 % of the dissolved salts in the ocean).

The most plausible explanation is that most of the phosphorus that moved into the early oceans was quickly taken up by living organisms and remains there. If it had been subsequently deposited in sediments, and mineralized as apatite, this would result in an accumulation on the ocean floor of 0.2 mm per year (Blackwelder 1916) cited by Lotka (1925). But even at the low deposition rate of 0.2 mm/y the accumulation in a billion years would be 200 km, which is clearly not what happened. Hence the true deposition rate must have been much lower, probably because there is more recycling within the ocean than geologists thought. This is a question for future oceanographers and biochemists.

12.8 The "Circular Economy" and the Limits to Growth

This book has mentioned a number of different examples of cycles, so it seems appropriate to end by discussing a cycle that does not exist today, but without which the human race cannot survive on Earth for much longer. I refer to the entropy minimizing "circular economy." It is an old idea, recently given a new lease on life by the publication of a widely circulated report by the Ellen MacArthur Foundation (2013) with Stiftungfonds für Umweltsökonomie und Nachtshaltigkeit (SUN) and the McKinsey Center for Business and the Environment. The Report collects a number of ideas about the state of the environment that have been around for decades, but puts them into a broad societal context and makes a powerful case for business pro-activity.

The Ellen MacArthur Foundation Report is available on the Internet. It argues for a radical increase in material productivity, both to protect the environment by minimizing waste, and to provide a model for enabling developing countries to achieve rising standards of living without bankrupting the Earth. A follow-up report by McKinsey & Co. makes the case that this radical transition is not only techno-logically and economically feasible, but can be very attractive to businesses with long-term growth perspectives (Ellen MacArthur Foundation and McKinsey Center for Business and Environment 2016).

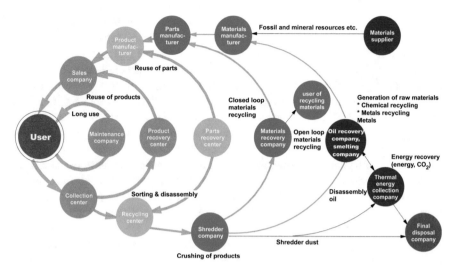

Fig. 12.13 The Ricoh "Comet Circle". *Source*: https://www.ricoh.com/environment/manage
ment/img/co_img_s.gif *Copyright 1994*

It is important to acknowledge that the circular economy is not literally circular.
It consists of a set of loops, and (thanks to the second law of thermodynamics) not
all the loops are closed. See Fig. 12.13. The "name of the game" is to minimize
those losses and thereby minimize the entropy generated by the system. As the
McKinsey report emphasizes, the reduction of losses corresponds to large potential
economic gains. The problem is how to govern the overall system so as to achieve
those gains and allocate them appropriately. Capitalism "as usual" rarely leads to
that ideal result.

Consider the problems for a widget manufacturer. For example, in order for the
scheme illustrated below to work it is important to provide incentives for people to
return their used and broken widgets to collection centers, rather than discarding
them. Moreover, to encourage refurbishment and remanufacturing it is important to
design the widget to make it easy to dismantle. Yet widget manufacturers usually
design their products to make them as hard as possible to take apart, so as to force
people to buy new ones rather than repairing old ones. Moreover, companies don't
like to have to compete with remanufactured versions of their widgets, because that
reduces sales of the new versions.

This is not the place for an extended discussion on how to make a circular
economy happen in the real world. Nor am I competent to do so. It will be a step-by-
step process if it happens. In the rest of this penultimate section, I will focus on a
few nutrient elements, and once again on "limits to growth". The rationale for the
shocking "limits" argument by the authors of that notorious report to the Club of
Rome, back in 1972 (Meadows et al. 1972), boils down to expectation of exhaustion
of low entropy physical resources. The thesis of that report happened to

be—intentionally or not—in perfect agreement with the entropy-law thesis put forward by Nicolas Georgescu-Roegen a year earlier (Georgescu-Roegen 1971).

Mainstream academic economists ignored Georgescu-Roegen. But in 1974 they took issue with Meadows et al., probably due to the very wide circulation of "Limits to Growth".[12] They argued that resource exhaustion is no problem as long as the *elasticity of substitution* is sufficiently large (Solow 1973, 1974b; Stiglitz 1974, 1979). A different group of economists focused on declining resource prices, arguing that resource exhaustion is not happening thanks to technological progress in exploration and extraction (Simon 1980, 1981; Nordhaus 1992). Both groups dismissed Malthusian "de-growthers" (such as the Meadows) as naive amateurs who don't understand how "reserves" and "resources" are measured, while underestimating the power of technology. (I must say in response that some economists, like Julian Simon, tend to over-estimate what technology can do to overcome scarcity).

But a few economists such as Boulding—even before Georgescu-Roegen—realized that recoverable material resources in the Earth's crust, with concentrations significantly above the average, are strictly limited (Boulding 1966). Stockpiles of low-entropy resources, including but not limited to fossil fuels, are limited. Inputs of exergy from "outside" are necessary to maintain the Earth as a viable system. Clearly the exergy cost of resource extraction must rise in the long run, as ore grades decline, other factors remaining constant. The fact that Meadows et al. misinterpreted the figures on "reserves" as published in *Minerals Yearbooks* and other publications of the US Bureau of Mines (which they did) does not totally undermine their qualitative conclusions. It only alters the time-scale.

But, contrary to Georgescu-Roegens's pessimism, there is no physical upper limit on the availability of solar power, especially given that means of harvesting it are not restricted to the Earth's surface. We could, for instance, harvest solar power from the moon or from solar satellites (Criswell 2000, 2002). Does this mean there is no limit to future economic growth? That seems to be what mainstream economists (e.g. Stiglitz, Solow, Nordhaus) were implying in their argument against Meadows et al. I think that implication is wrong.

I agree that solar exergy resources are not free. They require physical devices, from space ships to solar cells to computer chips to magnets for turbo-generators—and a lot more. These devices increasingly depend on geologically scarce metals that I have called "hitch-hiker metals" because they are not mined for themselves (Ayres and Talens Peiro 2013; Ayres et al. 2013, 2014). Most are not recycled, and even if some recycling—by conventional means—occurs in the future it will be insufficient. Taken as a group, I think that these scarce metals constitute a future limit to economic growth. The limit may not be "hard", like hitting a brick wall. But it is a real barrier nonetheless.

---

[12] It was a huge best-seller. As I recall, well over ten million copies were sold in a number of languages.

Consider the energy (actually exergy) return on energy invested, or EROI. The point is that EROI for petroleum has been declining rapidly; it was over 100:1 in 1930 and (globally) is now down to about 20:1 (Murphy and Hall 2010; Hall et al. 2014). For deep-water wells, oil sands and shale fracking, the EROI is much lower. That is physics, not economics. But a lower EROI (10:1 to 5:1) means the magnitude of energy expenditure needed to keep the overall level of consumption steady doubles, and doubles again. When EROI for hydrocarbons gets close to 1:1, the fossil fuel game is over. OK, wind and solar power may have EROI's somewhere in the 10–20 range. But what about the scarce metals needed to manufacture the solar cells (indium? tellurium?) and the wind turbines (neodymium?) and the "smart grids" (gallium?). Nobody thinks about that yet, but I think somebody should.

The first step would be to calculate the economic cost of scarce metals as a fraction of total GDP for the past half century or so. This can be done for the US, at least. The US Geological Survey recently compiled historical prices of all commercially available metals (USGS 2012a). The product of price times consumption, for each metal, gives the "share" of that individual metal in the gross economy. Of course, the sum for all the scarce "hitch-hiker" metals—those not mined for themselves—plus the "precious" metals (excluding gold) will provide an index of relative importance that is increasing over time. Currently it is fairly small, but this index will almost certainly increase in coming decades. Recycling is currently negligible, but it must increase in the future. The economic implications of these trends for technology and for economic growth must be considered.

In Sect. 6.9 I pointed out that most of the phosphorus minerals in the Earth's crust are of biological origin. Moreover, at current consumption rates, they will be exhausted in a few hundred years at most. Phosphorus in fertilizers that is not taken up by plants (currently 95 %) goes into surface runoff and ends up in the deep oceans. There it is effectively lost to the terrestrial world, except via natural upwelling. For this reason, I think phosphorus availability is probably the real near-term limit to global economic growth. The question arises: can phosphorus be recycled? And, if so, how? If not, is there another source?

One possibility is undersea mining. I mentioned earlier that phosphorus is more highly concentrated in the deeper waters of the ocean. The sedimentary fluorapatites, from which phosphate rock was formed long ago, are formed by a regeneration process in anoxic sediments on the sea floor of continental shelves. There may be "sweet spots" where phosphate nodules could be mined commercially. This possibility is already under consideration for a future undersea mining activity.

Nitrogen and sulfur are also essential to all forms of life. But neither is scarce and currently both of these elements are being mobilized by the combustion of fossil fuels. Nitrogen from the atmosphere is mobilized as NOx (by lightning or forest fires) or as synthetic ammonia (made from natural gas). Sulfur is mobilized as SO_2 (from sulfur in the fuels), or from sulfide metal ores, mainly of copper, zinc and lead. Sulfur, in the form of sulfuric acid, is one of the most important industrial chemicals, needed as a solvent in a vast number of chemical processes from which it is essentially never recycled, because there is no need. The sulfur generally ends up as an insoluble calcium sulfate. Recovery from that would be costly, but feasible given enough exergy.

What if combustion of fossil fuels were to cease? The reduction in NOx emissions would be very beneficial, but the cost of synthetic ammonia without natural gas would sky-rocket. To be sure, there is plenty of gas for ammonia synthesis for a long time, especially if the gas is no longer being utilized for heating and power, and if the nitrates in fertilizers are used less wastefully. The situation for sulfur is more critical, because the only source (other than as contaminants of fossil fuels) would be sulfide metal ore (mainly copper) reduction, currently utilizing coke as the reducing agent. That process would also be infeasible without coke.

Of course, as in the case of ammonia, natural gas can be used instead of coke, if enough gas is available. In both cases, the alternative chemistry in the long run would depend on hydrogen, presumably obtained by electrolysis of water (using solar, wind or nuclear power). This suggests a long, carefully planned transition. But that—in turn—is difficult to assure in a capitalist world where growth in the short run, not decline in the long run, is the mantra.

Then, there is the problem of fresh water, also identified as a planetary limit by the Resilience Center (Rockström et al. 2009). Certainly there is no substitute for fresh water. But desalination is already practiced on a large scale in the Middle East, using natural gas for heat. Solar desalination is more capital-intensive, but it works. There are a number of water-saving and water recycling technologies now in use in large cities throughout the world and in dry countries, like Israel. The most important ones are drip irrigation and "greenhouses". Drip irrigation is an established technology that delivers water directly to the roots of the plants. It saves 35–65 % of the water compared to sprinklers, which are already an improvement over old style irrigation by ditches. Simple drip systems can be significantly upgraded by adding humidity sensors and controllers. Most producers of vegetables and some grain farmers in western countries already use some version of this method.

The next step for agriculture may be "greenhouses" which condense fresh water from humid air or solar desalination of brackish water (Fig. 10.20). Greenhouses not only conserve water, but they allow multi-cropping (strawberries in October, flowers in January). More important in the long run, greenhouses also allow the possibility of recovering and re-using excess nitrates and phosphates. Policies that would make water and fertilizer recycling more profitable than it already is, would have a huge economic payoff for the whole world. The fact that the Netherlands is a net exporter of flowers and food products today is largely due to its extensive use of greenhouses.

Vertical farms, an idea getting attention from urban planners nowadays, are a natural extension of the horizontal greenhouse. In fact, the idea is to stack horizontal greenhouses on top of each other, partly to conserve land and partly to facilitate water and fertilizer recycling. To be sure, land near large cities is expensive and scarce, which is why high-rise buildings are built. As it happens, China currently has a large oversupply of steel structures—originally intended for housing—that

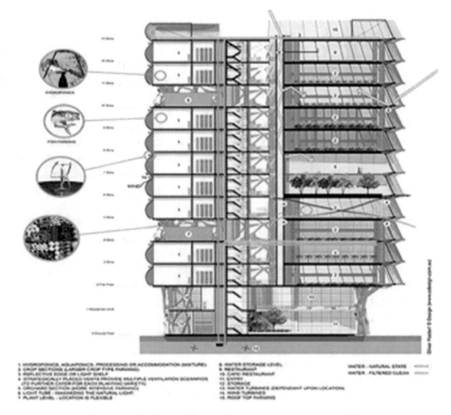

Fig. 12.14   Diagram of vertical farming system

might be adapted for a combination of housing and vertical agriculture. Suppose some of those structures were enclosed by glass. See Fig. 12.14.

Rainwater would be collected on the roof (or from a humidity condensation system) and would be distributed to lower floors by a gravity system. If the structure were combined with residential quarters, the sewage would also be recycled as organic fertilizer, and the "gray" water would be recycled to the plants. Most important in the long run, the nitrogen, phosphorus, and potassium fertilizers could also be recovered and recycled. This would cut out one of the major pollution problems in agricultural soils and marine estuaries, as well as conserving essential phosphorus. The electric power for lighting (to accelerate plant growth), for pumping water and recycling it, would be provided by solar PV or wind (with new storage technology). Intermittent light may be OK for plants.

Admittedly, if the vertical farm is producing food for consumption outside the farm itself, there will be some loss of nutrients that must be replaced. However, if the vertical farm is located in or near a city, a secondary sewage recovery facility can increase the recovery and recycling of nutrients. As urbanization increases, urban sewage plants can become increasingly important sources of agricultural fertilizers.

In fact, the vertical farm suggests the possibility of a nearly self-sufficient vertical village, in which most of the transportation takes place via electrically powered elevators and (possibly) moving walkways. Obviously the vertical village, as a "circular economy" sketched above—including total elimination of fossil fuels as a source of exergy—requires a lot of new investment in technology. Electrification of the transport sector is under way, but far too slowly.

Energy storage technology to compensate for the intermittency of solar and wind power has a high priority. Pumped storage and batteries are not the only option: in fact, advanced heat pumps and flywheels look increasingly competitive. Solar desalination can be a major tool for enabling the agricultural use of dry land. Vertical farms consisting of stacks of greenhouses, interspersed with human habitations, would recycle water, sewage and chemical fertilizers while sharply reducing the need for horizontal distribution of goods (and wastes) within localities.

In fact, during the next century I think clusters of vertical farms can easily evolve into *vertical villages* that export food, fiber and other agricultural products to surrounding territories, being linked to other villages and large cities by an electrified rail network. The vertical villages will be almost completely self-sufficient in terms of energy, food and water as well as for most social services like education and health. The remaining need for transport between villages and central cities could then be accomplished by electrified rail networks. This, in turn would solve the major problem now preventing remanufacturing of complex manufactured products, namely the need to collect used but reparable items and transport them to specialized rehabilitation centers. Some of the villages will become specialized centers for recovery and recycling of usable components and metals.

Being electrified, the system will eventually be integrated into a combined communication-cum-transportation system, i.e. "e-mail for people and goods" (EMPAG). To use it within a country or within the Schengen zone of Europe, you will electronically enter your personal address and the address of the place where you want yourself or your goods to be "mailed". Just as e-mail today moves messages by the most efficient rout between computers, the combined system will figure out both the least cost and the fastest route to take you (or it) to the desired address. Crossing national borders and crossing oceans will be more complex, but only marginally so. I bet Google or UPS is working on how to implement EMPAG right now.

The electrified e-mail system for people and goods will not need many internal combustion engines. The companies making those things will fade away, along with the market for liquid hydrocarbons. But the manufacturers of electric motors and motor-controls, especially elevators and moving walkways, will prosper.

Conversion of nuclear power from uranium to thorium is another long-term option deserving of serious attention (Hargraves 2012). I hope that the political will to make it happen does not require another kind of crisis (a nuclear one). In any case, the technical problems that would be encountered in making this change are beyond the scope of this chapter, or this book.

12.9 A Trifecta?

It is easy to end a book like this with pious generalities about wealth maximization in theory and ideal government policy changes that are unlikely to be carried out in this corrupt world. But here is an idea that might—just might—be worthy of serious consideration by policy-makers. Let me start with several uncontroversial facts.

- Since the end of the "gold standard" in 1933 and the end of the "gold window" in 1971 (the rule that permitted other countries to exchange dollars for US gold at a $35/oz as part of the so-called Bretton Woods system), the market price of gold rose dramatically, although not monotonically. In 1980 it hit $850 per Troy oz in current dollars ($2458 in 2014 dollars) It hovers nowadays around $1100/oz, which looks very much like a continuing downward trend.
- As a "store of value", gold still has many faithful followers (known as "gold bugs") including some national banks. The Swiss and French (and more recently, the Chinese) are especially devoted to gold. The world Gold Council estimates that 32,400 ton are held in bank vaults around the world, or about 19 % of the total amount of gold estimated to have been mined to date, which amounts to 170,000 tonnes. A large but unknown fraction resides on the arms of women in India, where gold is the established form of dowry (and mode of intergenerational wealth transfer). The purchases of banks and on behalf of Indian brides probably keep the price well above the level it would command based only on industrial and commercial uses. The same is true to a lesser extent for silver.
- The gold standard was never a determinant of US domestic money supply; its role was mainly in international money transfers. The domestic money supply in the US and Europe is effectively controlled by the lending decisions of private banks (now including "shadow banks"), not central banks (Keen 1995). The central banks have some control, e.g. over interbank interest rates and capital requirements (leverage). The FRB and the ECB are said to "print" money by "quantitative easing" (QE), but that is merely an asset swap that keeps the overall bank liabilities unchanged. It does not increase the money supply.
- As is well-known, the total amount of money in circulation has grown rapidly since the 1970s. Many conservative economists and business leaders have worried that this flood of "cheap money" must inevitably result in renewed inflation in terms of the prices of consumer goods. (There are "warnings" about this from gold bugs and others, even now.) But inflation has not happened; what has actually occurred is more like the opposite, as Paul Krugman keeps pointing out. The excess money supply has been mostly used, thus far, to finance private, corporate and government debt.
- As the amount of money in circulation has grown, so has the total debt, recorded as assets held by the banks. But (as pointed out in Sect. 12.4) lenders— bankers—operate on the principle that all risks are to be borne by the borrowers. Bankers invariably try to enforce repayment in full, as a *moral* obligation, even when payment is impossible (as in the case of Greek public debt). As you may

guess, I believe that lenders in future will need to accept a share of the risk of every loan. This will make them less inclined to gamble with other people's money.

- Bank lending to small and medium-sized enterprises (SMEs) has slowed drastically after 2008. This was partly because the major banks first had to be recapitalized and "deleveraged"—a continuing process. It was also partly because newly formalized "Basel rules" (set by the Bank for International Settlements, BIS, in Basel, Switzerland) classified loans in terms of supposed "riskiness". In practice, this meant that banks need more reserves to back loans to businesses than to governments, government-sponsored entities (GSEs) or mortgages secured by property. *BIS rules thus favored loans to governments, despite budget deficits, rather than loans to profitable businesses.* Hence, despite the ultra-low interest rates, economic recovery has been very slow.

- I mentioned above that the price of gold is now much higher than it would have been if it were determined only by demand for industrial and commercial uses. Speaking of industrial and commercial uses, the majority of the non-radioactive elements in the periodic table (as well as uranium and plutonium) are now employed for specialized purposes somewhere in modern technology. There are only a few non-radioactive elements without significant uses, and several of those might find uses if they were reliably available at stable prices.

- The output of the scarce by-product metals is not driven by price, and the price of rare metals is not driven by demand as economists normally assume (Ayres and Talens Peiro 2013). Yet the investment in additional process technology needed to extract those by-products is obviously determined by expected future market prices.

- The loss of rare by-product metals that are not now being recovered from mining of major industrial metals is a serious and unnecessary depletion problem for the planet Earth. Moreover, the waste is quite unnecessary. The technology to recover most of the rare by-product metals is well established. The problem is that prices are not high enough, at present, to justify the additional investment.

I now focus on the first and last points, above. For argument's sake, I wonder if a stockpile of scarce by-product metals could become an actual currency base, in parallel with conventional paper money and the new "bitcoins". Today there is really no base at all for the primary currency, which is one reason why bitcoins have thrived. In 1924 Chancellor Gustav Streseman and Banker Hjalmar Schacht created a new monetary system for Germany (*rentenmarks*, based on land) to replace the depreciated gold reichsmark. The exchange rate was set at one trillion to one. It ended Germany's hyper-inflation virtually overnight (Ayres 2014a, b p. 88). Simple. Could it be done again?

The main requirement for a monetary base is that it be both stable and tangible. If there were an actual *physical* stockpile of those metals and a fixed buying price for each of them, it would begin to reconnect marginal supply and demand, thus providing a floor on prices, reducing the wild swings that have occurred in the past. It would also set a ceiling on prices, by providing an emergency source of supply in case of extra-ordinary need (e.g. in time of war).

Suppose, for argument's sake, that the physical (but otherwise undefined) stockpile were financed by a depletion tax—call it an "entropy tax". It needn't be too onerous. Or it could be financed by one of the billionaires who would like to find an environmentally beneficial use for surplus money. The existence of a stable market would slow down depletion in two ways. First, it would reduce unnecessary losses and wastage in the metals extraction process by inducing metal refiners to invest in recovering by-product metals that are currently not recovered. And second, higher prices would encourage efficient end-uses of the metals. In fact, when the stockpile is large enough it could initiate a policy of *rental* of the rare metals rather than outright sale. That, in turn, would be a huge step toward achieving the idealized circular economy discussed in Sect. 12.8.

The physical stockpile, a quantity of scarce but valuable and recoverable metals, still in the ground, would have a different function. It could serve as the backing of the money supply for the global—or any national—economy if every monetary instrument were convertible to metal, in principle, at a fixed price. That is exactly how the gold standard worked until 1933: paper money (legal tender) was legally convertible to gold, although the actual gold sat untouched in Fort Knox, Kentucky (recall the James Bond movie "Goldfinger"). In reality, once the stockpile existed, it was no longer needed. The inflation-adjusted price of gold doubled between 1931 and 1934 and remained at the higher level until 1942 when it began a long downward slide, hitting bottom in 1971.

The size of the monetary base could be based on a physical stockpile, or determined each year by the quantity of "reserves" as determined by the US Geological Survey, or some international version or extension thereof.[13] (The exact mechanism for this annual assessment can be overlooked for the present). In the latter case, most of the by-product metals would remain in the Earth's crust, in the copper, zinc, lead and other ores where they now reside, until needed. They would serve the same function as the invisible gold formerly held at Fort Knox that once served as backing for paper dollars.

Which metals? I suggest that the stockpile would be most valuable for metals that are not mined for themselves and are only available from the ores of major industrial metals or—in the case of PGMs and REEs—from each other. Details on these metals are presented in Appendix C. The major industrial metals (iron, aluminum, copper, nickel, zinc, and lead) are all found in concentrated deposits on the right side of Fig. 12.15. See (Skinner 1976). There are a few cases of rare metals, such as mercury, molybdenum, niobium, tantalum, tin, tungsten, uranium and zirconium that are also found in concentrated deposits that are—or were—

---

[13] Reserves, for geologists, are based on quantities that are economically recoverable given existing technology and prices. Reserves have been increasing fairly steadily since the nineteenth century when data started to be kept. Unfortunately there is no published reserve data for gallium or germanium.

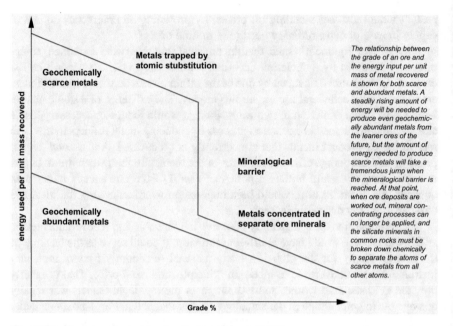

Fig. 12.15   The mineralogical barrier. *Source*: Skinner (1976)

mined for themselves. Tantalum and niobium are often found together (in tantalite, also known as the conflict mineral "coltan"), as are the platinum group metals (PGMs). Rare earth elements (REEs) are always associated with each other, and occasionally with iron ore; gallium is found in bauxite (aluminum ore), indium is mostly found in zinc ore. Rhenium is found with molybdenum (which is often found with copper). Germanium is obtained from zinc ore or coal. Many other rare metals, such as selenium, silver, gold and tellurium are also found in copper ores. The major point worthy of emphasis is that the amount of a rare metal obtainable from a mineral ore (reserves) is typically far less than the amount in the earth's crust, because the latter is largely dispersed as atomic substitutions in ordinary rock, from which extraction by pyrolytic or hydrolytic means is not possible. The exergy requirements for extraction of rare metals from ordinary rock would be so large as to make it essentially impossible.

The dots in the log-log graph (Fig. 12.16) indicate crustal abundance and price, as of 2009 or so, according to one source.[14] There is an inverse relationship, but with a lot of variability. The industrial metals are clustered in the lower right, while the rare elements are clustered in the upper left. The red dots are the ones identified as critical to the future renewable energy supply, by a joint study sponsored by the

---

[14] The numbers are not carved in granite. An industry expert in the rare-earth business has commented that Sc is not more abundant than Sm or Dy, while Nd is not more abundant than La (Rollat 2015).

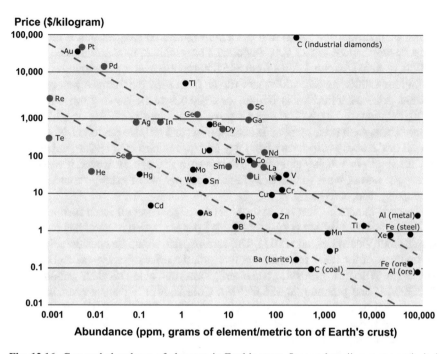

Fig. 12.16 Cost and abundance of elements in Earth's crust. *Source*: http://www.aps.org/units/fps/newsletters/201107/images/elementalprices.jpg

American Physical Society (APS) and the Materials Research Association (MRS) (Jaffe et al. 2011). A number of other reports in recent years have suggested sets of "critical" elements, but there is a good deal of overlap, viz. (European Commission 2010; Graedel and Erdmann 2011). A short discussion of potential metals stockpiles is given in Appendix C.

Of course, a "virtual basket" of different metals cannot be quantified simply in terms of weight. (Well, it could, but that would make very little sense). There is just one measure that is applicable to all metals—indeed all products—namely "embodied energy", or *exergy* content. The exergy embodied in a metal is the thermodynamic minimum exergy required to extract and purify it, which is also the same as the heat (enthalpy) that would be released if it were to be oxidized (burned) or somehow recombined into the original mineral form.

The exergy content of platinum and palladium is almost the same per mole; 146.5 kJ for platinum and 145.7 kJ for palladium Valero Capilla and Valero Delgado 2015). Those figures correspond to 750 and 1375 kJ/kg respectively. The exergy in kJ per mole figures for the other metals: rhodium (183), iridium (256), ruthenium (315) and osmium (371) are considerably higher than for palladium and platinum because they are much rarer and harder to separate. For the PGM group as a whole, averaging palladium and platinum, and disregarding the others, the exergy "content" is approximately 1100 kJ/kg or 1.1 GJ/tonne.

The dollar value of a tonne of PGMs varies, of course, but current (Nov. 2015) prices of platinum ($850/oz) and palladium ($530/oz) suggest that a metric ton must be worth close to $22,400,000/tonne. For selenium, the current price is about $10/lb, or $22,000/tonne, down from $46,000/tonne a year ago. Indium is currently selling for $300/kg or $300,000/tonne (down from $700,000/tonne a year ago). The current price of tellurium is $50/lb, or $100,000/tonne down from $400/lb or $800,000/tonne in 2012. The current price of rhenium is $87/oz or $2,784,000/tonne. The current price of silver is about $16/oz or $790,000 per tonne. Needless to say all of the above prices vary from day to day, and rare metal prices today (Nov. 15, 2015) may be roughly half of what they were a year ago. However, if we convert prices per unit of mass (oz, lb or kg) into prices per MJ of exergy "content" some interesting facts emerge.

The first interesting fact is that the cost (price) per EJ of all fossil fuels extracted in 2007 was $6.13 billion per EJ, or $0.0061 per MJ, based on numbers in (Valero Capilla and Valero Delgado 2015). The current price would be considerably lower, around half of the 2007 price. By comparison, the price of exergy as embodied in silver (in 2015) is $790/MJ. Rhenium works out at $1156/MJ, palladium at $12,300/MJ and platinum at $36,600/MJ. Gold is even higher, at present. At the opposite extreme, the exergy cost of selenium is only $0.005/MJ, slightly less than the cost of fossil exergy in 2007. Tellurium comes to $0.043/MJ and even indium works out at only $0.079/MJ.

One can scarcely avoid the implication that, in some sense, the metals used in electronics, especially for PV—solar energy production—are considerably undervalued in dollar terms, while the "noble" metals are over-valued. It follows that there is a strong incentive to recycle silver, rhenium, palladium, and platinum and little or no incentive to recycle the PV metals—or even to invest in recovering them from copper or zinc ore for possible future use. *A global wealth maximization strategy would undeniably call for the creation of incentives to recover and recycle these elements.*

Wealth maximization implies minimizing unnecessary depletion of the Earth's original resource base. We should adopt measures to maximize the recovery of by-product metals that would otherwise be lost during mining and refining operations. How can that be done? The simple answer is almost too simple: tax the unrecovered exergy fractions. I would suggest a tax of $1 per MJ of unrecovered (wasted) exergy in the form of rare by-product metals. The tax revenues would then finance the stockpile of those metals. That stockpile, in turn, could be the basis of the money supply for a nation or a group of nations such as the Eurozone.

You say that this tax rate would be arbitrary? There are two parts to the answer: (1) it would be no more arbitrary than what the Central Banks of the world are charging now. And (2) it would attach a time-dependent global price to exergy embodied in rare metals. Unfortunately, markets for those metals are by no means "free"; they fluctuate wildly and the spot prices are currently determined by factors other than supply/demand balance. It is doubtful that conventional econometric studies could winkle out a reasonably accurate "shadow prices" for the wasted rare metals from published data on market prices, over time.

If my "crazy" scheme were adopted, losses of the scarce metals, whether in processing or final disposal, would marginally decrease the money supply. New resource discoveries, and improvements in recovery and recycling technology, on the other hand, would increase the monetary base. That, in turn would allow an increase economic activity (GDP). Other factors being equal, this would also stimulate research and discovery, increasing global wealth. And it would certainly prevent monetary inflation. A trifecta? We could use one.

Epilog

"In the very beginning, there was only pure energy and the laws of physics".

To sober scientific types, the opening sentence of the Thesis (Chap. 2) of this book may seem, to be "over the top". I couldn't resist the idea of pure knowledge (the laws of physics) being the starting point, as well as the end point. The end-point? Whether the ultimate destiny of Man is to navigate starships through "wormholes" among other galaxies, or even to mine asteroids, is far beyond my ken. For my purpose, it suffices to say that the only possible antidote to global natural resource exhaustion in the coming decades and centuries is substitution of new, non-material resources, such as knowledge applied to doing a lot more with a lot less: recycling scarce metals, dematerialization of the active means of production or increasing the usable bandwidth in the IT domain.

Passive material goods, such as roads and buildings, are not the long-term problem. The problems for human civilization in the future are the production of necessities—especially clean air, clean water and food—that cannot be dematerialized. The single most critical resource on the planet may be phosphorus, because it is geologically scarce and every living organism needs it. But that is a subject for another book.

The lifetime of the human race on this planet could come to a shocking end in a few decades (as some predict) if nothing changes in the board rooms and legislative bodies of the world. But it can be extended for a very long time, I think, by the application of human intelligence and scientific knowledge. Indeed, the most intelligent thing we can do, probably, is to invest more of our remaining economic surplus in science. Some of the knowledge we humans need to survive (longer than a few decades) already exists in the pages of academic publications and the brains of makers and doers. The rest is available, like hanging fruit, for the grasping. The key to long term survival is the implementation of what we know how to do already, or how to find out what we don't yet know. COP 21 may have been the start.

That is the "good news". Not very good, you may think, and I agree if the past history of human stupidity is considered. So, let us adopt IBM's old slogan: *THINK*.

© Springer International Publishing Switzerland 2016
R. Ayres, *Energy, Complexity and Wealth Maximization*, The Frontiers Collection,
DOI 10.1007/978-3-319-30545-5

Appendix A: Energy in Growth Theory

A.1 Economic Modeling

The purpose of this Appendix is to summarize, in the language of economics, the hypothesis that energy delivered as *useful work* (in the thermodynamic sense) should be one of the three major factors of production in modern industrialized economies. At one level, this seems obvious: machines have largely displaced unskilled labor as a factor of production in the industrialized countries, since the beginning of the twentieth century. John Henry, the "steel driving man", is dead. More extended treatments of these issues were dealt with in prior chapters and other publications (Kuemmel et al. 1985, 2010; Kuemmel 1989; Ayres and Warr 2005, 2009). Incidentally, in this Appendix the word "we" is used because everything from here on is based on joint work with others in cited publications.

A theorem dating from a century ago (discussed in Sect. 11.5) postulates an ultra-simple economy consisting of many small competing firms producing the same product (call it bread) from rented capital and labor. (The "rentier" is a fat fellow who eats all the hypothetical bread.) The two factors, capital and labor, are assumed to be freely substitutable. However, the profit maximization condition is that total labor requirements will be determined by the marginal productivity of labor, and ditto for capital.

The marginal productivity of labor, in money terms, is equal to the wage rate. The marginal productivity of capital is the profit rate. (In this model labor and capital are homogeneous, so there is only one kind of labor and one kind of capital.) Hence the total payments to labor must be equal to the wage rate times the quantity of labor employed. Similarly the total payments to capital must be determined by the marginal productivity of capital (the profit rate), times the quantity of capital employed. It follows that, in this model economy, the output elasticity of each factor must be equal to its cost share in the GDP (which is equal to the sum total of wages and profits, respectively.)

© Springer International Publishing Switzerland 2016
R. Ayres, *Energy, Complexity and Wealth Maximization*, The Frontiers Collection,
DOI 10.1007/978-3-319-30545-5

Neo-classical economists assume that this logic also applies to heterogeneous capital and heterogeneous labor in the real world, taking wage and profit rates as averages. As it happens, the cost shares of capital and labor in the US GDP have been relatively constant over the years. This seems to fit the requirements of the parametric Cobb-Douglas production function; more about that later. In 1956 and 1957, Robert Solow noted that increasing capital per worker but keeping the cost shares constant did not explain US economic growth from 1909 to 1949 (the years he analyzed) (Solow 1956, 1957). But instead of introducing a third factor of production he attributed most of the unexplained historical growth to the residual multiplier A(t) that he characterized as *technical progress* (Now it is called *total factor productivity* or TFP.)

The fundamental problem with the big economic models is their assumption that energy is not a factor of production in the first place. If that assumption is modified, one has a production function of three variables, rather than two, and the three variables cannot be regarded as independent of each other. The third factor could be primary energy (exergy) E or useful work U. Useful work is the product of primary exergy E times conversion (to work) efficiency f, viz. $U = fE$. One important result of the work reported in this Appendix, and in the journals cited above, is that the average exergy efficiency f turns out to be a very good measure of technical progress (or factor productivity). In other words, $A(t) \cong f(t)$.

An enormous amount of effort has been expended on trying to eliminate the unexplained "Solow residual" by introducing *quality adjustments* to conventional labor and capital, while continuing to treat labor productivity as the ratio between total output and labor input. Certainly education and increasing literacy (and numeracy) would be a quality adjustment for labor, while technical performance improvements should also be taken into account for capital. Unfortunately none of the proposed adjustments are satisfactorily quantifiable. Hence the usual approach is to assume a convenient exponential form for $A(t)$, with a constant annual growth rate, around 1.6 % per annum for the US during the twentieth century. It follows that economic growth is assumed to continue, as in the past, *with no limits on energy availability* (as an intermediate).

Perfect substitutability of capital or labor for each other, or for useful work, would allow for the possibility of producing output from capital alone (and no labor or useful work), or from labor alone with no capital or useful work, or from useful work alone, with no labor or capital. Clearly, none of the factors is a perfect substitute for either of the others. In fact, the extent of substitutability between the factors is quite limited, and those limits can be expressed as mathematical constraints on the utility or profit maximization conditions (in equilibrium). If we think there is an optimization process in the economic system, and if the production function has an explicit mathematical form (such as the Cobb-Douglas), the next step will be to carry out a form of Euler-Lagrange optimization.[1]

---

[1]Either as utility-of-consumption maximization or profit maximization. See Sect. 4.4.2 in Kuemmel's book for a detailed explanation and derivations (Kuemmel 2011) pp. 186 et seq.

Given the postulated constraints on the possible relationships among the factors introduces new terms called *Lagrange multipliers*. These are interpreted as *shadow prices* of the constraints. You ask how a constraint can have a price? A shadow price is what gets added to—or subtracted from—the "real" price on account of the existence of that constraint. If the constraints are very weak or very far away they may not be "binding". In that case the shadow prices will be zero. But suppose that, thanks to interdependence, the short-term substitutability among the factors is much more limited than most economists assume? With interdependencies and binding constraints, the shadow prices will be non-zero. In that case the output elasticity of each variable (labor, capital and exergy or work) will be larger (or smaller) than its cost share. Since the three variables K, L and E or U (and any others that might be introduced) are definitely inter-dependent, the standard cost share theorem is not valid even in the absence of binding constraints.

But if the output elasticity of exergy (or work) is larger than its cost share, it follows that energy (exergy) is really more productive than its price suggests. That would imply in turn that the optimum (profit-maximizing) consumption of energy (work) in the economy is actually larger than current consumption. Can this possibly be true? Economists doubt it. They point out (correctly) that there is plenty of empirical, if anecdotal, evidence that firms use too much—not too little— commercial energy. The evidence from many empirical studies says that firms consistently neglect cost-saving investments in energy conservation, with very high returns, in favor of investments in marketing, new products or in capacity expansion. Governments and consumers behave in roughly the same way, preferring to spend money on current consumption goods or services rather than investment in future cost savings.

The evidence cited above does imply that the economy as a whole, and most firms, have been behaving (until very recently) as though there is no energy constraint. In other words, they behave as though growth is driven only by aggregate demand, while mergers are driven by cost minimization when there is little or no growth.

So what is the contrary evidence that firms should use more energy and less labor? The answer is simple: It pays to automate, i.e. to replace human workers by machines wherever possible. Machines keep getting cheaper and smarter and workers (in the industrialized economies) keep getting more expensive but not necessarily smarter. So businesses continue to replace workers by machines. The same tendency can be seen in consumer behavior. It is exemplified by the numerous home appliances that reduce household manual labor, from washing machines to electric mixers, knife grinders, vacuum cleaners, hedge clippers, leaf blowers and lawn mowers.

Moving on, we can now postulate parametric functional forms for the output elasticities, with appropriate asymptotic behavior, such that realistic physical constraint requirements, plus the constant returns to scale condition are satisfied. This was done some years ago by Kuemmel (1986). Bear in mind that there is no law that says the production function, or its derivatives, need to be parametric. Any *a priori* choice of a parametric function introduces uncertainties and potential sources of error. More recent work has shown how non-parametric functions of the variables can be derived from the data itself (Ayres and Voudouris 2014; Voudouris et al. 2015). However in this Appendix only parametric functions are considered.

A.2   Growth Equations

One can conceptualize the economic system as a multi-sector chain of linked processing stages, starting with resource extraction, reduction, refining, conversion, production of finished goods and services (including capital goods), final consumption (and disposal of wastes). Each stage has physical inputs and physical outputs that pass to the next stage. At each stage of processing value is added and useful information is embodied in the products, while low value, high entropy, low information wastes are separated and discarded.[2]

Global entropy increases at every step, of course, but the value-added process tends to reduce the entropy of useful products, while increasing the entropy of the wastes. An adequate description of the economic system, viewed in this way, must include all materials and energy flows, and information flows, as well as money flows. These flows and conversion processes between them, are governed by the first and second laws of thermodynamics, as well as by monetary accounting balances.

Suppose we now postulate a twice differentiable, two sector, three-factor parametric production function $Y(K, L, E, t)$ where K is a measure of capital stock, L is labor supply, E is energy (or exergy) and t is time. Suppose all variables are indexed to unity in the starting year (1900). The growth equation is the total (logarithmic) time derivative of the production function Y multiplied by dt (Kuemmel et al. 2002), viz

$$\frac{dY}{Y} = \alpha\frac{\partial K}{K} + \beta\frac{\partial L}{L} + \gamma\frac{\partial E}{E} + \delta\frac{\partial t}{t}. \tag{A.1}$$

The last term reflects the possibility that some part of the growth cannot be explained in terms of K, L, E and is therefore an explicit function of time alone. This term, originally identified by Solow with technical progress, is now usually called *total factor productivity* or TFP. It cannot be accounted for within a single sector model (because a single sector model produces only a single composite product), whence it must be regarded as exogenous. The same argument applies to two-sector models.

---

[2]The language here is suggestive of an energy (or information) theory of value. Unfortunately, perhaps, the term "value added" is so thoroughly established in economics that it cannot reasonably be avoided. In any case, I am not espousing the discredited energy theory of value. For a more thorough discussion of the economy as a self-organized system of concentrating "useful information" see Ayres (1994) Chap. 8.

The four output elasticities are α, β, γ and δ. The first three are defined below. Here, δ can be thought of as the elasticity of "technological change" (in a broad sense). It can be neglected hereafter for simplicity. Equation (A.4) follows from the convenient but unrealistic assumption of constant returns to scale:

$$\alpha(K,L,E) = \frac{K}{Y}\frac{\partial Y}{\partial K} \tag{A.2}$$

$$\beta(K,L,E) = \frac{L}{Y}\frac{\partial Y}{\partial L} \tag{A.3}$$

$$\gamma(K,L,E) = 1 - \alpha - \beta \tag{A.4}$$

To assure that the variables are continuous with continuous derivatives there are integrability conditions for parametric functions. These conditions say that the second-order mixed derivatives of the production function Y with respect to all factors K, L, E must be equal (Kuemmel 1980). It follows that

$$K\frac{\partial\alpha}{\partial K} + L\frac{\partial\alpha}{\partial L} + E\frac{\partial\alpha}{\partial E}\alpha = 0 \tag{A.5}$$

$$K\frac{\partial\beta}{\partial K} + L\frac{\partial\beta}{\partial L} + E\frac{\partial\beta}{\partial E} = 0 \tag{A.6}$$

$$L\frac{\partial\alpha}{\partial L} = K\frac{\partial\beta}{\partial K} \tag{A.7}$$

These differential equations can be solved. The most general solutions to these three equations are:

$$\alpha = a\left(\frac{L}{K},\frac{E}{K}\right) \tag{A.8}$$

$$\beta = \int\left(\frac{L}{K}\frac{\partial\alpha}{\partial L}\right) + J\left(\frac{L}{E}\right) \tag{A.9}$$

where J is any function of the ratio L/E.[3] The simplest (trivial) solutions are constants: $\alpha=\alpha_0$, $\beta=\beta_0$ and $\gamma=1-\alpha-\beta$. Other possible solutions of the above equations are considered later. The most commonly used production function of capital and labor (without energy) takes the so-called Cobb-Douglas form, mentioned earlier:

$$Y = A(t)K(t)^a L(t)^b \tag{A.10}$$

where Y is output, K is the stock of capital and L is the stock of labor, while a and b are constants and $a + b = 1$ (to assure constant returns to scale). This form has the

---

[3]For a derivation, see Kuemmel (1980) or Kuemmel (2011).

convenient mathematical property that output elasticities (logarithmic derivatives) of the two factors are just the constants a and b. That convenience has nothing to do with reality. If reality has any resemblance to the C-D function it is pure coincidence.

Yet the conventional growth model does assume that output elasticities α, β, γ are equal to cost (payment) shares for capital and labor, respectively, in the national accounts, and that they are constant over time. The usual parametric choices are $\alpha_0 = 0.3$ and $\beta_0 = 0.7$, respectively, where $\alpha_0 + \beta_0 = 1$. The time dependent term $A(t)$ is the "Solow residual', i.e. the growth component that is not explained by either capital accumulation or increased labor supply. It is traditionally approximated as

$$A(t) = e^{\lambda(t-1900)} \qquad \qquad (A.11)$$

where t is the year and $\lambda = 0.016$ for the US. In other words, throughout the twentieth century growth attributable to exogenous technical progress or TFP in the US has averaged 1.6 % per annum, although growth was significantly slower in the 1930s and somewhat higher in the early postwar period.

The next logical step is to choose the simplest *non-trivial* solutions of the growth equation (A.8) and (A.9). Since we now disallow constants, we need to select plausible mathematical expressions for the output elasticities, α, β, γ based on asymptotic boundary conditions[4] (Kuemmel 1980; Kuemmel et al. 1985). To satisfy the constant returns-to-scale (**Euler**) condition these elasticities must be homogeneous zeroth order functions of the independent variables. The assumption of constant returns-to-scale implies that, at every moment in time, equation (A.4) is satisfied.

The first of Kuemmel's proposed choices can be thought of as a form of the law of diminishing returns to capital. It is an asymptotic boundary condition conveying the notion that even in a hypothetical capital intensive future state in which all products are produced by machines, some irreducible need for labor L and exergy E will remain, *viz*

$$\alpha = a\frac{L+E}{K} \qquad \qquad (A.12)$$

Kuemmel's second equation reflects the continuing substitution of labor by capital and exergy as capital intensity (automation) increases:

$$\beta = a\left(b\frac{L}{E} - \frac{L}{K}\right) \qquad \qquad (A.13)$$

---

[4]The derivation of (A.14) is from Kuemmel's papers: (Kuemmel 1980, 2011; Kuemmel et al. 1985, 2002).

Both of these relationships are intuitively reasonable. However, it must be noted that when considered as *constraints* on an Euler-Lagrange maximization equation, the question arises: are those constraints "binding" throughout the whole range of the variables, or not? Some economists have expressed doubts on this point. If the Euler-Lagrange optimization procedure were linear, this would be an important question, because one could be left with zero shadow prices. However, for a non-linear optimization, the constraints are indeed binding. This has been shown in Kuemmel et al. (2010).

Moving on, the corresponding production function is then obtained by partial integration of the growth equation, where the output elasticities are given by (A.12) and (A.13) and (A.4), between proper initial and final points in (Y; K,L,E)-space. This yields the linear-exponential (LINEX) form:

$$Y = A(t)E\exp\left[a(t)\left(2 - \left(\frac{L+E}{K}\right) + b(t)\left(\frac{L}{E} - 1\right)\right)\right] \qquad (A.14)$$

The functions (of time) $a(t)$ and $b(t)$ that now appear in the integration have been characterized by Kuemmel as "capital efficiency" and "energy demand" respectively. The resulting time-dependent elasticities of output are then determined by fitting the above functional form to real GDP from 1960 to 2004 for the UK, Germany, Japan and the US. Introducing time dependent parameters $a(t)$ and $b(t)$ the GDP fits are extremely good (Kümmel et al. 2010).

The output elasticities are then obtainable by inserting the econometrically obtained technology parameters $a(t)$ and $b(t)$ and the empirical K, L, E into equations (A.12) and (A.13); for details see ibid. The resulting (calculated) time-dependent elasticities show a significant increase in exergy output elasticity and a decline in labor output elasticity, over time, as might be expected.

Apart from the assumption that Y is a parametric function, it would be desirable if $a(t)$ and $b(t)$ had a straightforward economic interpretation. For the moment, this is still a problem.

A.3   A New Variable: Useful Work U

As a first approximation, it is now convenient to assume that the economy is a two sector system with a single intermediate product, to be defined in a moment. (This assumption is not unreasonable, bearing in mind that most of economic growth theory to date postulates a single sector, composite product model.) The product of the primary sector (Sector 1) can be thought of as "energy services". For our purposes, energy services can be defined more rigorously as "useful work" U in the thermodynamic sense. Useful work includes muscle work by animals or humans, electricity, mobile engine power as delivered (e.g. to the rear wheels of a vehicle) and useful heat, as delivered in a space in a building or in a process. A full explanation of how these terms are defined and quantified can be found in another publication (Ayres and Warr 2003).

By definition U is the product of resource (exergy) inputs E times a conversion efficiency f. U has the same dimensions as E, since f is dimensionless. Capital, labor and useful work are not mutually substitutable; indeed there is some complementarity and interdependence between capital and work, and between labor and work. Thus the third factor U is not truly independent of capital and labor. This means that *not* all combinations of the three factors K, L, U are possible. (Some combinations, including extreme cases like $K = 0$, $L = 0$ or $U = 0$ are impossible.)

Capital and work have a strong synergy. In fact, as pointed out in Chap. 11, capital has two components: a passive component and an active one. The passive component consists of physical buildings, pipelines, roads, bridges and containers. Active capital consists of all exergy-consuming machines and information processing equipment that perform useful work in the economy. (The role of active and passive money cannot be considered here.) Thus capital goods are activated by exergy, while exergy has no economic function in the absence of capital goods. Labor, on the other hand is not passive (by definition), whence labor without exergy does no work.

A.4   The Choice of Production Function

As already mentioned, the Cobb-Douglas (CD) function assumes constant elasticities of output. It is not realistic to assume constant elasticities over the entire century from 1900 to 2007. On the other hand, the model can be used for shorter periods separated by well-defined structural breaks. The essential result that we need to take into account is that energy, and especially useful work (exergy), are important drivers of growth, and that their importance have increased throughout the century. We expect that the imputed marginal elasticity of resource (exergy) inputs will be greater than the factor-payments (cost) share. This has also been suggested by others (e.g. McKibben and Wilcoxen 1994, 1995; Bagnoli et al. 1996).

As already noted, the traditional single-sector, composite-product model should be rejected, if only on the grounds that the very act of isolating commercial energy as an output of the extractive industry (coal, oil, gas) constitutes a *de facto* sectorization. The Cobb-Douglas model using exergy E as a third factor can be implemented by defining a sectoral boundary between exogenous extraction and everything else. However, that version of the model does not explain past economic growth. Our alternative version incorporates "useful work" U as a third factor. The sectoral boundary shifts to include extraction, primary processing of fuels, electricity generation and usage (in motors, lighting, etc.) and work done by mobile internal combustion engines (Ayres et al. 2003).It will be seen that this version of the model has considerable explanatory power.

Obviously, useful work is required for its own production, as trucks carry coal to power plants and diesel fuel and electric power are used in coal mines. It seems reasonable to postulate, as a first approximation, that capital, labor and useful work are used in the same proportions in the production of useful work U as they are in the economy as a whole. In fact, having assumed a parametric production function, Y, it seems reasonable to go a step further and assume that the parametric form of the production functions Y_1, Y_2 and Y are identical, except for a constant multiplier. This being so, it follows that

$$\frac{K - K^*}{K} = \frac{L - L^*}{L} = \frac{U - U^*}{U} = \lambda \qquad (A.15)$$

$$K - K^* = \lambda K \qquad (A.16)$$

$$L - L^* = \lambda L \qquad (A.17)$$

$$U - U = \lambda U \qquad (A.18)$$

It follows that

$$Y_1\left(K^*, L^*, U^*\right) = (1 - \lambda)Y(K, L, U) \qquad (A.19)$$

$$Y_2\left(K - K^*, L - L^*, U - U^*\right) = \lambda Y(K, L, U) \qquad (A.20)$$

And therefore

$$Y_1 + Y_2 = Y \qquad (A.21)$$

It is worth noting that the above logic applies to any parametric production function that is homogeneous and of order unity. The Cobb-Douglas function is a special case, in which the parameters α, β, and γ are set equal to zero and $0 < \alpha$, $\beta < 1$.

The more general LINEX function, adapted from (A.14) by substituting useful work U for exergy E, is equation (A.22). However, I have added an additional term J for an additional variable, namely ICT capital stock, yielding equation (A.23), where A, a, b, c, are fitting parameters.

$$Y = AU\exp\left[a\left(2 - \left(\frac{L + U}{K}\right) + b\left(\frac{L}{U} - 1\right)\right)\right] \qquad (A.22)$$

$$y = AU\ exp\left[2a - b - a\left(\frac{L + U}{K - \delta}\right) + ab\frac{L}{U} + c\frac{J}{L}\right] \qquad (A.23)$$

As noted earlier, the functions (or time) a and b respectively were labelled capital demand and energy demand by Kuemmel. However, for fitting we treat them (with c) as parameters. Here U refers to useful work and Y, U, K and L are functions of time, while a, b and c are constants. In principle, we have time series for U, K, L and δ. The constants a, b and c have to be determined econometrically, i.e. by finding the best fit between Y as calculated and actual historical output (GDP). In practice, K, L and J are available from official sources, but U is not, and it must be estimated,

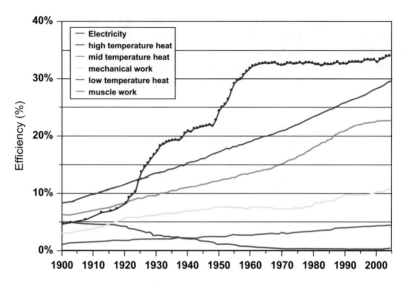

Fig. A.1 Conversion efficiencies for the US

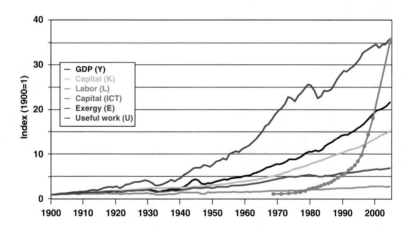

Fig. A.2 GDP and factors of production for the US, 1900–2000

as in Fig. A.1. For the US, this has been done for the case $c = 0$, for the years 1900–1998 (Ayres and Warr 2005, 2009) and subsequently updated and extended to several other countries (Serrenho 2014). All the variables in equations (A.22) and (A.23) are plotted in Fig. A.1 for the US.

Warr and Ayres worked out the three-parameter case (non-zero c) where capital is divided into a "conventional" capital stock (K) and a "new" (information technology) capital stock, (J). (Warr and Ayres 2012). The revised graph of the factors of production from 1900 to 2000 is shown in Fig. A.2.

A.5   Statistical Analysis of the Time Series

Before we can have great confidence in the outcome of calculations with a production function, especially after introducing a new and unfamiliar factor of production, it is desirable to conduct a number of statistical tests. In particular, one must to ascertain the presence of unit-roots, structural discontinuities, co-integrability and Granger-causality. The five economic variables in question are capital K, labor L, energy (actually exergy) E, useful work U and output (GDP) Y.

To do statistical analysis on time-series variables, they must be converted to logarithmic form, to eliminate any exponential time trend. But the logarithmic variables may or may not be *stationary* (trend-free, with finite variance and co-variance) as required for the application of standard statistical procedures such as ordinary least squares (OLS). In general, time-series (of logarithms) are not stationary i.e. they have "*unit roots'*. The first question is whether the unit root is "real" (i.e. due to a missing variable) or whether it is due to an external shock or structural break (discontinuity) in the time series. In general, (but not always), the latter interpretation suffices. Structural breaks can be attributed to wars, hyper-inflations, currency devaluations, depressions, or other major events such as the "oil crisis" of 1973–1974.

Structural breaks can be identified most easily by examination of the residuals of several models. However, different models suggest possible breaks in different places. This is troublesome, because there is an element of circular logic involved: the models, in turn, depend on the locations and magnitudes of the structural breaks. However, when different models (such as Cobb-Douglas and LINEX) are consistent in the sense that they show significant deviations in the same place, and when those deviations are easily explained in terms of exogenous events, it is reasonable to interpret them as structural breaks. To make a rather long story short (based on dozens of combinations), it turns out that the period of WW II (1942–1945) is the only break that needs to be taken into account in both the US case and the Japanese case.[5]

Benjamin Warr has carried out extensive co-integration analysis and Granger causality tests with the LINEX production function (Warr and Ayres 2010). A complete description of the lengthy statistical testing procedures involved, not to mention the underlying rationale, is far too complicated and specialized to reproduce here. The most robust result, however, is simple: both exergy and useful work are causally related to output, and *vice versa*. As intuition suggests, the causality

---

[5]. We note in passing here that for both countries there are several 'mini-breaks' that suggest the possibility of re-calibration of the model parameters. This leads, in practice, to a series of 'mini-models' covering as few as twenty or thirty years. However, an obvious constraint – one that is impossible to incorporate explicitly in the mathematical optimization process – is the need for each variable to be continuous across breaks. Ignoring that condition leads to extremely implausible results.

(in this model) goes both ways. In other words, Y depends on K, L and U, but it is also true that K, L and U depend on Y.

A.6 Results

Figure A.3 shows the results of comparing the best fits for a three-factor Cobb-Douglas production function, assuming constant output elasticities, vs. the best LINEX fit with variable output elasticities. For other results see the Ayres and Kuemmel references already cited. The LINEX fit is obviously better. Moreover the fit does not require an exogenous multiplier $A(t)$ to explain past growth. To put it more bluntly, the LINEX function fits US growth from 1900 to 2000 quite well, except for the years 1949–1946, using only two parameters (constants).

Even so, it must be acknowledged that even the LINEX fit is not very good after the 1980s. There are several possible explanations. One possibility is that an additional variable is needed. The rapid introduction of new information and communications technology (ICT) in the 1980s and 1990s may be one missing link. It suggests a modified production function in which capital is split into "normal" and ICT components. The results of this experiment have been tried by Ben Warr and myself. The results are shown in Fig. A.4. The fit could hardly be closer.

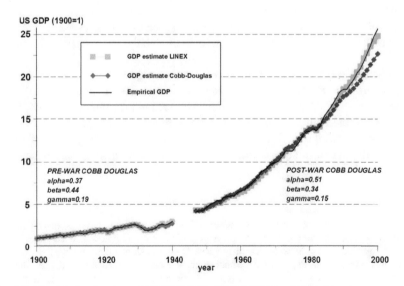

Fig. A.3 Empirical and estimated US GDP 1900–2000; LINEX and Cobb-Douglas

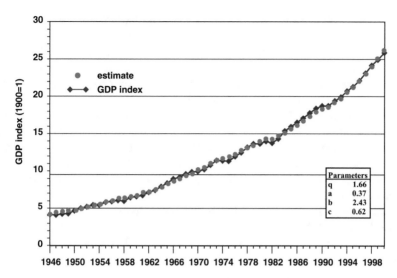

Fig. A.4 Empirical and estimated GDP using LINEX-ICT function:1946–2000

A.7 Conclusion

Despite a large body of literature on the subject, there is still substantial disagreement as regards the econometric evidence as to the direction of causality. (In other words, does energy demand drive growth or vice versa?) See Sects. 12.1 and 12.2. The econometric evidence (much of it not discussed in this Appendix), leans increasingly toward the conclusion that energy consumption is now the driver of growth—"the locomotive"—not the caboose. Using data for 100 years tends to confuse the issue, because energy prices were declining most of the time. The shift toward rising prices is a recent development. (Some economists will surely argue that the days of low and declining energy prices are not yet over. Time will tell.)

There is another problem with the LINEX function, as with the Cobb-Douglas function, the constant elasticity of substitution (CES) function and others: namely *they are parametric functions* chosen, in the first instance, to imitate certain historical trends in the real data. In the case of the U.S., a large country that has been relatively independent and self-sufficient for much of its history, it is not unreasonable to assume that the macro-variables have changed smoothly and, consequently, that two or three parameters might suffice to get a good fit. However most of the world has experienced a more complex history, such that smooth and gradual changes did not happen and could not have been expected. The breakup of the USSR, the merger of East and West Germany, the Cultural Revolution in China, and any number of wars reflect this reality.

Moreover, despite a huge volume of econometric analysis (thousands of statistical analyses—aimed at uncovering critical determinants of economic growth), the "answer" remains elusive. Before 1929, the dominant paradigm was "Laissez-Faire" (and the gold standard). In the 1930s and the early post-war years, a weak version of Keynesianism—called The New Deal—was widely accepted in US and European policy circles. But the stagflation of the late 1960s, the sudden end of the gold standard, the two oil crises in the 70s, and the Iranian hostage situation in 1980 brought Ronald Reagan and some of the avatars of "voodoo economics" (cutting taxes will increase government revenue) into power.

The respectable wing of Reaganism was the "Chicago School", chiefly Friedrich Hayek and Milton Friedman. They espoused a conservative anti-Keynesian ideology known as the "Washington Consensus". Their ideas were widely adopted by policy-makers, notably the World Bank and the IMF. The Washington Consensus promoted "labor market reform" (anti-union), balanced budgets, elimination of subsidies, privatization of government assets, reduced taxes, and austerity. Those policies (especially as implemented by the IMF) have not turned out to be as growth-friendly as the Chicago School claimed. There have been some successes, mainly in developing countries that were being run by left wing governments with socialist ideologies. But the Eurozone situation (2010–2015) illustrates the downside.

The point is that nobody, whether at the University of Chicago or Harvard or MIT or any of the great universities in Europe, has an adequate understanding of the phenomenon of economic growth. It is unreasonable to expect this very complex subject matter to be completely (or even approximately) well reflected in a simple two or three parameter "production function" whose mathematical form was selected *a priori* for convenience.

New advances in statistical methodology have changed the situation somewhat. Now it is possible to postulate non-parametric production functions, utilizing asymmetric, non-Gaussian, distribution functions that satisfy the necessary conditions without any of the unnecessary assumptions that are invisibly incorporated in traditional growth models (including ours). The new approach does not assume constant returns to scale, nor does it assume exponential growth with "normal" (Gaussian) deviations from the assumed path. What it does is to let the historical data explain the growth path (including the future) with far greater realism. This approach has been successfully demonstrated for 4 countries (US, UK, Japan, Austria) in one study (Ayres and Voudouris 2014) and for 15 of the countries of the European Union (EU) in another (Voudouris et al. 2015). Clearly, the new approach is still in its infancy.

Appendix B: Standard Theory of Nuclear Forces

The Standard Theory of Nuclear Forces is mostly based on so-called gauge theories (of which classical electromagnetism is the simplest example). In gauge theories, dynamic force laws are described by "operators" called Lagrangians and extensions of the variational principle of least action. Lagrangians in classical mechanics are expressions for kinetic energy minus potential energy, while Hamiltonians are defined as kinetic energy plus potential energy. Potential energy may include gravitational, electro-magnetic, chemical or nuclear potential. Application of the Euler-Lagrange variational principle yields the equations of motion. (J-L Lagrange demonstrated this method for Newtonian classical mechanics back in the eighteenth century.)

Gauge transformations are coordinate transformations in which "something" is invariant, or more familiarly, "conserved". Simple geometric examples of gauge transformations include translation, rotation and reflection (as in a mirror). Invariance under a gauge transformation is a technical term for a symmetry, while an invariant (unchanging) combination of variables is another word for a conservation law (or situation). Amalie Noether proved an important mathematical result in 1920, using group theory. Noether's theorems showed that every symmetry relationship has a corresponding conservation law. Applied to physics: if there is a symmetry, there must be an associated conservation law, and vice versa. This insight has turned out to be extraordinarily helpful in terms of formulating a viable quantum theory of high-energy physics.

The simplest symmetry of all is time invariance (symmetry) between forward and backward time (t, −t). Since the expressions for kinetic and potential energy in physics—the components of the Lagrangian—are invariant if you interchange t and − t, the conservation law for energy follows. QED. Conservation of angular momentum (spin) follows from rotational symmetry, OK so far. Another symmetry (mirror symmetry) leads to conservation of chirality (or spin direction), which is rather important in biochemistry as well as high-energy physics. But classical intuition only goes so far: what is the symmetry behind conservation of charge or conservation of quarks? Some of these conservation laws that are only applicable to

© Springer International Publishing Switzerland 2016
R. Ayres, *Energy, Complexity and Wealth Maximization*, The Frontiers Collection,
DOI 10.1007/978-3-319-30545-5

interactions between particles are equivalent to "conservation of quantum numbers", such as "iso-spin". (Pauli's exclusion principle is an example.) In high energy physics, several other new symmetries and conservation laws have been discovered in the last half century, as experimental collision energies got higher and higher, taking us experimentally closer and closer to the time just after the BB.

But in some situations it turns out that symmetries (conservation laws) "break". Curiously, "symmetry-breaking" has become a by-word in particle physics, especially since the famous work of Yang and Lee in 1957 (Yang and Lee 1956). They showed that "parity" (mirror symmetry) is violated in the electro-weak interaction of beta-emissions. Actually the idea of symmetry-breaking as an explanation of physical phenomena is much older. The intuition is that, at some critical point in the evolution of a non-linear system (such as the universe) in a "symmetric" situation where all configurations are equally possible, there is a sudden event, a "bifurcation" (or "catastrophe"). This takes the system from a situation consisting of a superposition of several or many possible states to just one of them.

According to the standard model, there are four forces in the universe, carried by four different types of bosons. The weakest force is gravity, which is presumably transmitted by gravitons, (with spin -2) which have yet to be detected for sure. Of course, it is the gravitational "force" that holds the Earth together, keeps us on it, keeps the Earth in an orbit around the Sun, keeps the Sun together and the galaxies together, and so on. Or is it? Einstein would say that what appears to be a gravitational "force" is just curvature of space, and that particles that seem to be moving under the influence of gravity (the force) are really just following the shortest distance between two points (i.e. geodesics) in curved space. This ambiguity is not merely semantic confusion. There is an underlying problem about theories of gravity, discussed in Chap. 3; see EGR vs. YGR.

The electromagnetic force (that keeps atoms together and accounts for electricity and all of the electrical phenomena we exploit) is much stronger. It is transmitted by massless photons. All electromagnetic radiation, including visible light, X-rays, gamma rays and so on consists of photons.

There are two stronger forces (though one of them is stronger than the other). The weaker one, called the "electro-weak force", enables interactions between all fermions with charge. It accounts for radioactive decay and beta-decay, whereby a neutron "decays" into a proton, an electron and an anti-neutrino (to balance the spins). Beta-decay is what converts hydrogen atoms into deuterons that attract electrons and become helium nuclei. That kicks off the chain of fusion reactions that make the stars shine. It also accounts for nuclear fission on Earth. It is transmitted by particles known only by the letters W and Z. The W comes in two charges, $+$ and $-$, while the Z is neutral. They both have mass and spin of -1.

The strong nuclear force is what binds the quarks together (in triplets) inside a nucleon. Its carriers are unimaginatively called gluons, and there is one for each flavor of quark. There are six "flavors" and three "colors". The gluons may have mass, even though QCD says they don't. Composite particles composed of quarks and gluons are called "hadrons". The simplest hadrons are the familiar nucleons (protons and neutrons), but there are other, heavier ones that can pop into existence for a moment, usually after a collision. There are also particles, called "leptons" as a group, that do not feel (interact with) the strong force. The leptons are electrons, muons, tau particles and corresponding kinds of neutrinos.

You probably didn't want to know all this.

Appendix C: Potential Stockpile Metals

The obvious candidates for a stockpile start with the traditional "precious metals": gold (Au) and silver (Ag). They are used mostly (66 % in the case of Au) or partly (30 % in the case of Ag) for jewelry, silverware and coins. Some gold is used in dental alloys and for gold plating e.g. of statues and architectural features. At least 10 % of silver is used for mirrors. Before digital cameras, the major use was in color photography, where there is still some demand but not much. Some silver is used for oxidation catalysts in the chemical industry. A growing use is for silver-oxide and silver zinc batteries in mobile phones and lap-top computers. Gold (5 %) and silver (32 %) are both used in electronic circuitry for their low electrical resistivity (high conductivity). The rest is used in solar panels, water treatment, radio-frequency identification (RFI) tags and so on.

Mine production of gold in 2012 was estimated by the USGS as 2700 tpy and for silver at 24,000 tpy (USGS 2012b). Ultimate recoverable resources of gold are estimated at 34,000 tonnes, as compared with an estimated 170,000 tonnes already mined (and mostly still in jewelry or bank vaults around the world). For silver the estimate of ultimate recoverability is 1,300,000 tonnes (Trudinger and Swaine 1979).

The six platinum group metals (PGM), are used mostly for catalytic converters for vehicles and petroleum refineries (Pd 72 %, Pt 31 %, Rh 69 %). This use is declining, as the converters had used PGMs more efficiently. It will disappear altogether when internal combustion engines and petroleum refining are phased out in the coming decades. On the other hand, the use of platinum in hydrogen fuel-cells might grow in future years. Other major uses of PGMs are in electronics, e.g. for computer hard disks and multilayer capacitors (Pd > 16 %, Pt < 23 %, Ru 62 %) (Rombach and Friedrich 2015). Iridium is used partly (55 %) for crucibles to grow single-crystal sapphire for backlit LED displays. Platinum (7 %) and rhodium (9 %) are also used in the glass industry. About 20 % of PGMs (mainly platinum and rhodium) are used for jewelry and other decorative purposes (op cit). Global production of palladium for 2012 was estimated at 200,000 kg/y (200 tpy); pro-

© Springer International Publishing Switzerland 2016

R. Ayres, *Energy, Complexity and Wealth Maximization*, The Frontiers Collection, DOI 10.1007/978-3-319-30545-5

duction of platinum was 179 tpy, for rhodium 28.2 tpy, for ruthenium 25.1 tpy and for iridium 9.36 tpy (USGS 2012b). There is no data for osmium.

The ultimate recoverable reserve of platinum minerals has been estimated at 84,000 tonnes (Trudinger and Swaine 1979). Based on the ores in the Bushveld complex in South Africa where most of the PGM reserves are located, the palladium, rhodium and other PGM reserves would be roughly in the same proportion to platinum as in current production (Holwell and McDonald 2007).

The rare earth elements (REEs) consist of scandium, yttrium and the 15 lanthanides (atomic numbers 57–71). The lanthanides, in order, are as follows: lanthanum (La), cerium (Ce), praseodymium (Pr), neodymium (Nd), promethium (Pm), samarium (Sm), europium (Eu), gadolinium (Gd), terbium (Tb), dysprosium (Dy), holmium (Ho), erbium (Er), thulium (Tm), ytterbium (Yb) and lutetium (Lu). All but four (Pm, Ho, Er and Tm) have been classified as "energy-critical elements" by the APS-MRS study cited above (Jaffe et al. 2011).

The major uses of REEs start with cerium oxide (Ce), which is widely used for glass polishing. REE catalysts include lanthanum (La), used in oil refineries, and cerium (Ce) used for catalytic converters in vehicles. Lanthanum was also used in nickel-hydride batteries formerly utilized by hybrid electric vehicles, though now it is being replaced by lithium batteries. Cerium, lanthanum and neodymium were all used in some steel alloys, to bind other trace elements. These uses accounted for 59 % of REE consumption in 2011. Cerium and Lanthanum account for about 80 % of these uses. Catalysts accounted for 60 % of end usage in 2015, according to USGS.

The other 41 % of REE usage in 2011 was in fast growing markets, where Dy, Nd and Pr account for 85% of the total. The most notable is for permanent magnets. The strongest known permanent magnet alloy is the neodymium-iron-boron (Nd-Fe-B) compound, widely used for powerful small electric motors (as in hybrid or electric vehicles) and in wind turbines. Dysprosium substitutes for about 6 % of the neodymium, for special applications involving a special magnetic property (curie temperature), but it is even rarer (Rollat 2015). Permanent magnet uses were rapidly growing before 2011, which prompted the APS-MRS group to forecast a 700 % increase in future demand for Nd and a 2600 % increase for dysprosium in the coming years (Jaffe et al. 2011). This rapid growth has not (yet) happened, despite a sharp decline in prices, because of the Chinese export cutback in 2011, which, in turn, prompted manufacturers like Toyota and GE to use substitute magnet alloy materials for the magnets. The substitutes are also REEs, viz. samarium (Sm), terbium (Tb) and praseodymium (Pr).

Dysprosium with iron and terbium are used in Terfenol-D, which is the most magnetostrictive material known at room temperature. It has uses for a variety of transducers, mechanical resonators and high precision mechanical fuel injectors. Dysprosium is also used in metal halide lamps. Phosphors for fluorescent lamps (mostly based on Yttrium) accounted for 3 % of REEs in 2012. This end-use will not grow, as fluorescent lamps are being displaced by LEDs. Unspecified other uses, such as LEDs, yttrium-aluminum garnets, for lasers, sensors, photochemistry, photoluminescence and high-performance spark-plugs accounted for 6 % of REEs in that year. One of the most interesting for the future is the yttrium-barium-copper oxide (YBCO) superconductor, discovered a few years ago, which has a transition temperature of 93 °K, higher than the temperature of liquid nitrogen (77 °K).

World mine production of REEs in 2012 through 2015 was about 110,000 tpy, mostly from China. Dysprosium output in 2011 was about 100 tonnes, virtually all from China. Yttrium output in 2012 was about 8900 t. Global reserves of REES are estimated at 130 million tonnes, of which 55 % are in China. Most Chinese output is used in China itself. Prices of REEs (as metal) spiked in 2011, but they declined rapidly after that, despite limited supply, possibly in order to discourage competitors outside of China.

Other rare metals of possible interest for stockpiling include semi-conductors gallium, germanium, selenium and tellurium, plus rhenium and tantalum. Gallium (Ga) is used in the form of gallium arsenide (GaAs) or gallium nitride (GaN) for integrated circuits and opto-electronic devices including laser diodes, LEDs photo-detectors, and copper, indium, gallium, and selenide solar cells (CIGS). Integrated circuitry accounted for 71 % of US domestic consumption in 2012, while opto-electronics accounted for the rest. Primary production of gallium in 2012 was 273 tonnes, as a by-product of bauxite refining for the aluminum industry. There are no data on reserves.

Germanium (Ge) is used for fiber optics (35 %), infra-red optics (30 %), as a catalyst for polymerization (15 %) and another 15 % for high efficiency solar PV "multi-junction cells" with gallium-indium-arsenide and gallium-indium-phosphide layers. These cells can obtain 28 % conversion efficiency in spacecraft and up to 40 % efficiency in solar concentrators. Germanium is found in sphalerite (zinc ore) so it is a by-product of zinc refining, but some germanium is also recovered from coal combustion (in the flue gas) from some deposits in China and in Russia (Kamchatka). Output in 2011 was 118 tonnes (80 % from China). Reserves in zinc ore are estimated at 10,000 tonnes, but in coal the reserve is 112,000 tonnes. (However, if coal mining is shut down in coming decades, the latter source would disappear.)

Selenium and tellurium are semi-conductors that are both by-products of copper refining. Selenium was originally important in xerography (copying machines), but that market has now been replaced by organics. It is still used for some photo-cells and in the PV compound copper-indium-gallium selenide (CIGS). But use of CIGS is declining as silicon cells are improving.

Global production of selenium was 1510 tonnes (in 2007); recovery efficiency (from ore) in this case is 77 % (Ayres et al. 2014 Table 4.1), but recycling from end-of life products is apparently negligible. The main use (40 %) today is in metallurgy; 25 % is used for giving glass a red color. About 10 % is used in agriculture and for dietary supplements. Electronics (in CIGS PV cells) accounts for about 10 % and chemicals and pigments and other account for 15 %. None of these markets except PV look likely to grow rapidly in the future. There is no immediate need for a stockpile.

Tellurium is a better candidate. Global usage in 2011 was about 500–550 tonnes. It is now used mainly (40 %), as cadmium telluride, for thin film PV. This use was pioneered by First Solar Inc. and it has been successfully scaled up to large scale solar farms for electric utilities. But its scarcity has prompted serious studies of future availability, with somewhat discouraging implications (Woodhouse et al. 2013, Redinger et al. 2013). About 30 % of the element, as bismuth telluride, is used for thermo-electric devices. Other uses are metallurgical (15 %), primarily to increase machinability of steel and copper. It is also a vulcanizing agent and accelerator for rubber (5 %). The other 10 % is for a variety of other uses, including pigments for glass and blasting caps.

Indium is also used for PV in CIGS as well as liquid-crystal-displays (LCDs) and touch-screens. It is a by-product of zinc (670 t/y) with a recovery efficiency of 39 %. It is primarily (74 %) used, as indium-tin-oxide (ITO) for flat panel display devices. Other uses are for architectural glass (10 %), solders for fire control systems (10 %), minor alloys (3 %) and miscellaneous (3 %). Demand for indium looks likely to grow rapidly (Woodhouse et al. 2013). There is now some recycling of ITO.

Rhenium, used in superalloys, is a by-product of molybdenum (56.5 t/y) with a recovery efficiency less than 1 % (op cit). It is mostly (70 %) used in superalloys. There is no recycling of rhenium, as yet.

Glossary

Terms

ACEEE The American Council for an Energy Efficient Economy, in Washington D.C.

adenine One of the five nucleotide bases. It is a **purine**, consisting of five linked copies of hydrogen cyanide (HN)

ADP/ATP Adenosine diphosphate and adenosine triphosphate (ATP) are the primary energy carriers in all living organisms. The transfer mechanism is for ADP to gain, or for ATP to lose, a phosphate (PO_4^-) group. Exergy is needed to add a phosphate group to ADP and exergy is released when ATP gives up the extra phosphate group.

AEC Atomic Energy Commission (US), now part of the US Department of Energy.

aether A hypothetical medium conceived by **Isaac Newton** to explain force at a distance, and re-invented by **James Clerk Maxwell**, as the medium through which electro-magnetic vibrations (e.g. radio waves) were supposed to travel. The aether idea was discredited by **Albert Einstein**'s work. Currently physicists speak of "the void" or "the vacuum". It is now assumed that 'virtual' particles are constantly being created and destroyed in the vacuum.

albedo The technical term for the light reflected from the top of the atmosphere

amino acid A chemical compound containing both an amine ($-NH_2$) and a carboxylic acid (R–COOH) group, plus one or more other side-chains. Proteins are constructed from 20 different amino acids.

anti-matter Matter constructed from anti-particles. There is no anti-matter in this part of the universe, though some cosmologists think there may be another part of the universe where anti-matter dominates.

anti-particles Particles that can be produced in very high energy nuclear reactions but which cannot survive in ordinary space (i.e. this part of the universe) because when a particle and an anti-particle meet, they annihilate each other, producing pure energy.

anergy The component of energy that is unavailable and can do no work. Energy is the sum of anergy and exergy.

anisotropy Lack of 3D symmetry.

ANL Argonne National Laboratory (of the US Department of Energy)

APS American Physical Society

Argo The name for an array of 3600 expendable floating bathythermographs (XBTs) recording ocean temperatures, for the Global Ocean Data Assimilation System (GODAE) part of the Global Ocean Observation Strategy (GOOS). The name, taken from the name of Jason's mythological ship, the Argo, also reflects the relationship between the ocean-based instrumentation and the Jason satellite altimeter mission, which records atmospheric data at the same locations and times, thus providing a 3-dimensional temperature data-base.

ARM Adjustable Rate Mortgage

ATP Adenosine triphosphate: see ADP

available energy energy that can do work; **exergy**

Avogadro's number $N_A = 6.02214129 \times 10^{-23}$ is the number of particles or molecules in a **mole** (the molecular weight in grams).

axon The part of a neuron that links the body of the cell to the synapse. It carries electrical signals as well as protein molecules, signaling molecules, organelles, cyto-skeleton fragments, etc.

baryons Stable nucleons (**protons, neutrons**) that are affected by the strong nuclear force. Baryons each consist of three **quarks**, have non-integral spin (1/2, 3/2, ...) and obey the **Pauli exclusion principle**. There is a conservation law for baryon number (the number of baryons minus the number of anti-baryons in any reaction).

beta decay The spontaneous "decay" of a neutron into a proton and a positron (anti-electron) plus a neutrino.

BIF Banded iron formation, consisting of insoluble ferric iron deposited by **stromatolite** organisms.

bifurcation A branching point of a dynamic trajectory; a type of "catastrophe" in the mathematics of non-linear dynamic systems, so named by the French mathematician **René Thom**.

binding energy Energy that holds atomic nuclei together. When a heavy nucleus (with atomic weight greater than iron) fissions, some of the binding energy is released in the form of smaller nuclei, neutrons and gamma radiation. This is the source of energy in nuclear explosions or nuclear power plants. In fusion processes the situation is reversed: energy as radiation is actually released when light atoms (e.g. hydrogen) to fuse together creating heavier ones (e.g. helium).

biological clock A measure of the aging process.

biopoiesis Term introduced by J.D. Bernal describing the sequence of abiotic evolutionary stages preceding the first living cells. The stages are (1) organic molecule creation and accumulation, (2) polymerization and (3) enclosure in a protective membrane.

BIS Bank of International Settlements, located in Basel, Switzerland

black body, black body radiation A black body is a hypothetical object that absorbs all radiation and emits a continuous spectrum. The term was first introduced by Gustav Kirchhoff.

black hole A prediction of Einstein's general theory of relativity: when a mass is compressed to the point where it's gravitation attraction is so great that light cannot escape. The **Schwarzschild radius** is the radius of a spherical black hole. If Yilmaz' theory is correct, black holes do not exist and gravitational collapse results in "gray holes".

black smoker An undersea hot spring (vent) of volcanic origin, often carrying hydrogen sulfide and other gases in solution

BOF Basic oxygen furnace, the process developed in Austria, now used in all countries to produce iron or steel from iron ore. It is essentially the same as the Bessemer process, except for the oxygen.

Boltzmann constant k $1.3806488 \times 10^{-23}$. The energy in joules added to a particle when the temperature is raised by 1 K.

bosons This is a category of force-carrying particles with zero or unitary spin. Named for **Satyendra Nath Bose**, an Indian physicist who pioneered in the field of quantum statistics (Bose-Einstein statistics). The bosons include **gravitons** (if they exist), **photons** (that carry the electromagnetic force), **W** and **Z mesons** (that carry the electro-weak force) and **gluons** that carry the strong nuclear force. Bosons may have mass. The **Higgs** particle is a massive boson.

BOSS Baryon Oscillation Spectroscopic Survey, uses a new technology for identifying quasars with high red-shift values, $Z > 2.25$.

Brent Brent crude oil, from the North Sea, one of the two grades most used for pricing. The other is **WTI**.

BRIC countries Brazil, Russia, India, China

BRT Bus Rapid Transit. A system for buses with dedicated right-of-way and automated ticketing.

Calvin cycle Named for **Melvin Calvin**. The chemical synthesis used by most plants (built around the enzyme RuBisCO, better known as chlorophyll) that converts carbon dioxide to glucose (and produces oxygen as a waste product).

CAPEX CAPital EXpenditure, business jargon for new investment

carbon-oxygen cycle The capture of carbon from atmospheric carbon dioxide, together with oxygen production, by photosynthesis in plants followed by the reconversion of carbon from biomass to carbon dioxide by respiration for metabolic purposes by plants and animals. See nutrient cycle.

Carnot law The maximum theoretical efficiency of an idealized heat engine operating between a hot and a cold reservoir was determined by Sadi Carnot c. 1830. It states that the maximum efficiency possible is equal to $1 - T_C/T_H$ where T_C is the absolute temperature of the cold reservoir and T_H is the absolute temperature of the hot reservoir. All "real" cycles are less efficient than this idealized one.

cash flow Money available from normal business activities to pay for operational expenditures.

catastrophe In non-linear mathematics, a discontinuity (see **Thom**). In physics, a phase transition, such as condensation or freezing, or the onset of superconductivity or super-fluidity at low temperatures.

CCS Carbon Capture and Storage, a technology proposed for removing carbon from the exhaust gases of a combustion process, such as a power plant.

CD Cobb-Douglas, referring to a mathematical form of production function in economics.

CDAC Carbon Dioxide Assessment Committee (Charney Committee) of the US National Academy of Sciences, which met in 1979 to estimate climate sensitivity.

CDO Collateralized debt obligation—a term for bonds constructed by investment banks from bundles of home mortgages.

CEES Center for Energy and Environmental Studies (at Princeton University).

CERN Acronym for Centre European pour Recherche Nucléare. The laboratory is in France, just across the border from Switzerland near Geneva Airport.

CES Constant Elasticity of Substitution, referring to a parametric production function, in economics.

Charney sensitivity See **ECS**.

chemical potential Another word for potentially **available energy**, or **exergy**, in the form of differences between molecular energy (or excitation) levels.

chemo-litho-autotrophs Organisms that obtain metabolic energy from reactions between hydrogen or hydrogen sulfide from hydrothermal vents and dissolved oxygen or ferrous iron in the water.

chirality The property of not being superposable on a mirror image. Human hands are chiral because the left hand and the right hand are mirror images of each other that cannot be superposed. The combination of a spin and a direction of motion in the case of a particle is chiral, because it must be either "right-handed" (clockwise) or "left-handed "(anti-clockwise).

chlorophyll A specialized protein that carries out the key steps in **photosynthesi**s.

chloroplast A self-sufficient **endosymbiant**—formerly free-living—photosynthesizing unit, living symbiotically inside a larger bacterial or algal cell and containing **chlorophyll**.

CHP Combined Heat and Power. This technology is applicable to decentralized power sources where the otherwise waste heat from electricity generation is utilized as such. In general, the efficiency of electricity production from CHP systems is lower than central power plants achieve, but by beneficially utilizing the heat, the overall efficiency of CHP is higher, often around 60 %.

Chromodynamics The theory of strong force interactions among quarks. See QCD

chromosome A heritable sequence of **genes**, that controls some attribute of a living organism, such as eye color or gender.

CIGS Copper-Indium-Gallium-di-Selenide, a PV compound.

CLASS Cosmic Lens All-Sky Survey, a collaboration since 1994 between UK, Netherlands and the US, looking for cosmic radio sources acting as gravitational lenses. So far about 10,000 lenses have been identified.

climate sensitivity The degree of amplification, by positive feedbacks, of the basic **greenhouse effect** (heating of the Earth due to absorption and re-radiation of atmospheric GHGs.) See **ECS** and **ESS**.

CMIP5 Coupled Model Intercomparison Project, Phase 5. This is a project to compare the predictions of general circulation (climate) models, centered at Lawrence Livermore National Laboratory in California.

COBE Cosmic background explorer, an orbiting satellite for mapping cosmic temperatures, in orbit 1989–1992.

codon A sequence of three **nucleotide bases**. Codons are the simplest components of the genetic code, like letters in an alphabet. A sequence of codons is an **exon**. Exons are like words. **Genes** are composed of a sequence of **exons**. Genes are instructions for making **proteins**.

conservation law A law that states that some physical quantity, or quantum number, is in invariant with respect to some symmetry. Noether's theorem states that there is a conservation law for every symmetry. The most familiar conservation law (conservation of energy, E) reflects time symmetry i.e. $E(t) = E(-t)$,

Coriolis effect The apparent "force" accounting for trade winds and jet streams, due to the rotation of the Earth. It was explained by Gaspard Coriolis in 1835.

CSP Concentrated Solar Power, a term used in reference to a tower structure surrounded by a field of directed mirrors that focus the sun's rays on a heat exchanger, where the heat converts water (or some other fluid) into steam that drives a turbine to generate electric power.

cyanobacteria The first oxygen **photosynthesizer**s also known as blue-green algae.

cytosine One of the five nucleotide bases: it is a **pyrimidine**.

dark energy, dark matter Cosmologists postulate that dark energy and dark matter must exist in order for the universe to appear "flat" (as it does) even though the mass and energy density that we can account for is not nearly enough.

deuteron, deuterium A deuteron is a proton-neutron combination, and is the nucleus of deuterium or "heavy" hydrogen.

dissipative structure An organized pattern that is dependent on—and maintained by—a flow of exergy from "outside". Abiotic examples of such structures include hurricanes, tornados, sunspots, and so-called "Benard cells" among others. Living organisms (and ecosystems) are the most important examples, of course. It is thought that such structures maximize the rate of entropy increase.

divergence theorem in mathematics It expresses (in **vector calculus**) the idea that the sum of all sources, minus sinks, of a vector field—or a fluid—inside a boundary is equal to the fluid flow across the boundary. This theorem also applies to force fields, which is why it is so important in physics.

DEB theory Dynamic Energy Budget theory (biology) which incorporates mass balance and thermodynamics into a generic theory of metabolism applicable to all species. See S.A. Kooijman.

DNA Deoxyribonucleic acid, the chemical name for genetic material.

DOE Department of Energy (US)

ECB European Central Bank, an institution created by the European Union (EU) for the Eurozone.

ECS Earth climate sensitivity (**Charney sensitivity**), defined as the amplification factor (AF) for average surface temperature change of an equilibrium state by doubling the atmospheric CO_2 level, after taking into account all "fast" feedbacks such as surface ice melting, increased atmospheric water vapor, and increased storminess. The IPCC sets the ECS as about 3.1 (between 1.5 and 4.5 with a 66 % confidence level). The sensitivity now appears to somewhat larger, between 4 and 4.5. See CDAC, **ESS**.

EEC European Economic Commission, the executive agency of the European Union (EU) in Brussels.

EET Energy Efficiency Technology

EGR Einstein General Relativity

EIA Energy Information Agency (part of the US Department of Energy)

electric charge One of the conserved quantities in nature. Electrons and protons have opposite charges. Neutrons are electrically neutral. Electro-magnetic fields are produced by electric charges, whether static or in motion.

electricity A flow of electrons, usually through a wire, driven by a voltage (potential) differential.

electrolysis A reduction process energized by an electric current. Aluminum, chlorine, magnesium, potassium, sodium and titanium metal are all produced electrolytically.

electro-magnetism One of the four fundamental forces in nature, along with gravity and the weak and strong nuclear forces.

electron The most common **fermion**, with mass and spin ½. It is classed as a **lepton** because it is unaffected by the strong nuclear force. The anti-particle of an electron is a positron. An electron volt is a unit of quantity in physics.

EMPAG E-mail for People And Goods

endothermic process A chemical process that is not self-energizing but requires **exergy** from outside.

endo-symbiosis The cellular process of engulfing other self-sufficient organisms, but not digesting them. Instead each organisms provides services to the other. **Mitochondria** and **chloroplasts** are evolutionary examples. This is the major difference between **eukaryotes** and **prokaryotes.**

energetics A socio-philosophical movement that was influential in the late nineteenth and early twentieth century, especially associated with the name of **Wilhelm Ostwald.**

energy What this book is about.

ENIAC Electronic Numerical Integrator and Computer, designed by J. Presper Eckert and John Mauchly completed in 1946. The UNIVAC (UNIVersal Automatic Computer) was its successor.

ENSO El Nino Southern Oscillation: the climate subsystem that generates El Nino events, when the water off the west coast of South America is especially warm. La Nina is the reverse situation. The mechanisms are not yet well understood because the first such event was recorded in 1950.

entropy A state of a system (or the universe as a whole) that reflects aggregate disorder. This is often represented in terms of probabilities or uncertainty. Entropy is not a physical substance that can "flow" into or out of a region, even though that metaphor is common.

entropy law The second law of thermodynamics, which says that total entropy increases in every process, even though the entropy of a sub-system can decrease, thanks to **self-organization** of **dissipative structures** if there is an external source of exergy. The existence of life on Earth seems to be an example of self-organization driven by solar radiation.

enzyme A protein molecule that specializes in catalyzing a specific reaction between other organic molecules without being changed itself. Enzymes are essential for digestion (e.g. **glycolysis** and decomposition of proteins into separate **amino acids**), as well as for all aspects of replication and reproduction.

EPICA European Project for Ice Coring in Antarctica, located at Dome C, one of the summits of the Antarctic ice, where the ice depth is 3270 m to bedrock, covering 8 glacial cycles and 700,000 years.

equilibrium In physics or chemistry a state where there are no gradients and nothing changes. In thermodynamics it is a state where free energy G is zero and entropy S is maximum. In economics it means that supply and demand for all products and services are in balance. A *stable* equilibrium is one that recovers from small perturbations where an unstable equilibrium does not.

EROI Energy (exergy) return on energy (exergy) invested, an acronym usually applied to the petroleum and gas sector but applicable to all exhaustible resources.

ESS Earth system sensitivity. As compared to ECS, this is defined as the amplification factor (AF) for a doubling of the atmospheric CO_2 concentration, taking into account all (both fast and slow) feedbacks. It is estimated that ESS is 30–50 % higher than ECS.

EU European Union, now with 29 member states.

Euclid The Greek mathematician who described "flat" space as a space where the shortest distance between two points is a straight line. Euclidean space is normally represented as a 2-dimensional plane with an x-axis and an orthogonal y-axis. One can imagine a set of such planes, parallel to each other, along a vertical z-axis. The distance between two points is the shortest of all possible curves connecting them.

eukaryote A cell with a nucleus containing the DNA. It may also include other mitochondria such as chloroplasts

exergy The part of energy that can do **work** (in contrast to **anergy**). Note that energy is the sum of anergy and exergy. Exergy is also a measure of **distinguishability**. Also called "availability", "available energy" or "essergy".

exergy efficiency The ratio of **exergy** output (or the exergy content of output) to total exergy input, for any process.

exon A short sequence of **codons** in DNA. Codons are like the letters of an alphabet. Exons are like words. Genes are sequences of exons.

exothermic process A process (like combustion) that is self-energizing and produces excess heat.

experience curve A relationship between experience (e.g. the number of objects produced) and the cost per unit. See **PR** (Progress Ratio).

externality In economics, a "third party" effect, i.e. a consequence of some transaction affecting parties not party to the transaction.

factor(s) of production Variables that are essential for production e.g. labor, capital and energy (exergy) or work. These variables should, ideally, be independent of each other. But they may be inter-dependent. The variables need to be well-defined and in the form of statistical time-series.

FAO Food and Agriculture Organization (part of the UN system), located in Rome.

FCC Fluid catalytic cracking, the petroleum refining process currently in use.

fermion A category that applies to all particles that occupy space, and interact with the strong force, named after **Enrico Fermi**, the Italian physicist who (among other things) built the first nuclear reactor at the University of Chicago. They all have half integer **spin.**

force law A law of motion (or action). The original force law of Newton is $F = ma$, where m is mass and a is acceleration. Acceleration is defined as the rate of change of velocity. A mass falling is accelerated by gravity.

forcing (in the context of climate change) Any external perturbation affecting the Earth's energy balance, such as volcanic eruptions, sunspots, or anthropogenic activities. See *radiative forcing*.

formose reaction A generic autocatalytic reaction that reproduces an organic compound, and provides extra copies of the catalyst. Discovered by Butlerow in 1861.

fracking Contraction of "hydraulic fracturing" as applied to releasing gas or oil from "tight" shale rock. This process has been responsible for most of the increase in global oil production since 2005.

fractal A geometric shape, first discovered by Benoit Mandelbrot, that is invariant (i.e. equally "rough") over changes of scale.

FRB Federal Reserve Bank of the United States.

free radical Any compound with a single unpaired electron in an outer shell. Free radicals are extremely reactive, and play an important role in the biological aging process.

GAAP Generally Accepted Accounting Principles.

GAMLSS Generalized Additive Models for Location, Scale and Shape; a computer program for analyzing data without specifying a parametric mathematical model for "fitting".

gauge theory A theory of the electromagnetic, "weak" and "strong" interactions that is invariant under a symmetry transformation. See **symmetry**.

Gaussian (normal) distribution This is a symmetric probability distribution characteristic of random choices around a mean. Named after the mathematician **Gauss**.

GCM General Circulation Model. These are mathematical models involving at least four variables (wind velocity, pressure, temperature, humidity) defined on a 3D grid, to predict atmospheric and oceanic behavior. A typical model may have 10–20 vertical layers and up to 100,000 grid points defined by longitude and latitude. The models use fluid-flow (Navier-Stokes) equations and a variety of parametric corrections to deal with boundary problems and imperfectly understood phenomena such as cloud behavior. All GCMs involve tradeoffs.

GDP, GNP In economics, GDP refers to Gross Domestic Product, which is the sum of all value-added within a country. GNP refers to gross output by citizens (including firms) of a nation, wherever produced.

gene The basic information-carrying unit in a **chromosome.**

genetic code Literally, the code (consisting of sequences of codons) that contains the instructions for producing proteins and all other somatic materials.

geodesic The shortest distance between two points. In flat space a geodesic is a straight line, but on the 2-dimensional surface of a sphere, it is a more complex curve. See Coriolis, Riemann, Einstein.

GEV Giga-electron volt (EV), a unit commonly used in theoretical physics.

GHG Greenhouse gas (one of the gases that contribute to the **greenhouse effect**, namely carbon dioxide, methane, nitrous oxide and water vapor), all of which absorb and re-emit.

GIA Global Isostatic Adjustment

Gibbs free energy The minimum work required to drive a chemical reaction at constant pressure; if negative, the maximum work that can be done by the reaction.

GISS Godard Institute for Space Studies (part of NASA).

gluons The particles postulated to "carry" the strong (attractive) force among quarks. This is the force that keeps nucleons together. They are electrically neutral particles with spin = 1 that mediate the interactions between quarks (inside hadrons).

glycolysis Literally, "sugar splitting", by any of several means.

GMSL Global Mean Sea Level

GODAE Global Ocean Data Assimilation Experiment, an international network initiated in 1997, involving agencies interested in climate change and oceanography from many countries, including Australia, Canada, China, France, Japan, UK and the US.

GOOS Global Ocean Observation System, an international oceanographic data system, supported by the US Navy and USNOAA. Linked to Argo, Jason and GODAE.

GRACE Gravity Recovery and Climate Experiment.

gravity The weakest of the four forces in nature, but the one with the longest range. It is the only force that can be "felt" at galactic distances.

gray holes Results of gravitational collapse if Yilmaz is correct. see black holes.

greenhouse effect A phenomenon whereby outgoing **IR** radiation from the Earth's surface is intercepted and re-emitted (partly back down), thus behaving like a blanket.

GSE Government Sponsored Entity (such as the Federal Housing Authority, FHA)

GT Gigaton (a billion tons)

GTL Gas to Liquid (in reference to natural gas)

guanine One of the five nucleotide bases. It is a purine.

GUT Grand Unified Theory, referring to the standard theory in physics to explain the three forces: strong nuclear, "weak" and electromagnetic.

GW Gigawatt (a billion watts)

GWEC Global Wind Energy Council.

GWP Gross World Product.

hadrons All particles affected by strong nuclear forces, consisting of stable baryons (protons, neutrons) and unstable mesons.

Hamiltonian In classical mechanics the Hamiltonian function is the sum of kinetic plus potential energy. It is a conserved quantity, expressed in partial differential form. Hamilton used the terms "action" and "force". See **Lagrangian**.

Heisenberg uncertainty principle The impossibility of determining velocity and position, or momentum and energy, simultaneously.

Higgs boson (also informally known as the "God particle") Is an electrically neutral, lepton that obtains its mass (125 GeV) from the **Higgs field**. The existence of the Higgs boson has been regarded as the cornerstone of the standard theory of particle physics (even though gravity has not yet been included). Discovery of the predicted Higgs boson was recently (July 2012) announced at CERN.

Higgs field A reformulation of the **aether**, named for Scottish physicist **Peter Higgs**. The key idea of the theory is **gauge invariance**, at ultra-high temperatures (above 10^{34} degrees). The field penetrates all of space with non-zero amplitude, but allows **symmetry-breaking** by means of a "Mexican hat" potential that is not zero at the center. This scheme allows particles to acquire their rest-mass (which is also potential energy) from the field. The magnitude of the mass depends on how tightly the particles in the standard model are coupled to the field. The masses of known particles range from that of the electron (0.0005 GeV) to the top quarks (more than 91 GeV). An important achievement of the Higgs field concept is that is has eliminated some deep inconsistencies in the "electro-weak" theory of the weak force, as well as "explaining" the **Guth** inflation model.

Hubble's constant H_0 is 65 km/s per megaparsec. H_0^{-1} is the age of the universe (13.77 billion years).

Hubble's Law Galaxies recede from us at velocities $v = H_0$ d, where d is the distance in megaparsecs.

hyperons Baryons that contain at least one "strange" **quark** and therefore have a non-zero "strangeness" (quantum number).

IAEA International Atomic Energy Agency (located in Vienna): part of the UN, a regulatory agency charged with monitoring nuclear power projects around the world.

IBM International Business Machines (Corp.)

IBRD International Bank for Reconstruction and Development (World Bank): a UN agency created after WW II for the purposes indicated by the title.

ideal gas law $PV = nRT = NkT$ where P is pressure, V is volume, n is the number of molecules, R is the universal gas constant, k is Boltzmann's constant, and T is the temperature in degrees Kelvin. This law applies to an ideal gas, consisting of point particles.

IEA International Energy Agency (part of the OECD), a statistical agency charged with keeping track of energy sources, reserves, transfers, and uses in all countries.

inflation A theory introduced by MIT physicist Alan Guth to explain how the universe got so big so fast after the Big Bang. The theory itself has been the subject of intense controversy, but is now widely accepted as a fact.

inverse square law A force relationship between objects that depends on the inverse square of the distance between them. Gravity is one example.

IPCC Intergovernmental Panel on Climate Change. A body administered by the UN charged with assessing the state of scientific knowledge with regard to climate change.

IR Infra-red (heat) radiation; alternatively, depending on context, Industrial Revolution.

ITO Indium Tin Oxide, a compound used in Liquid Crystal Displays (LCDs)

JASON-1, also JASON 2 and JASON 3 These are satellites, part of a cooperative project between the French and US space agencies. Began as Journées Altimetrique Satellitaire pour Oceanographie (JASO), The satellites are intended to help assimilate altimetric data into the global circulation models (GCMs) that try to forecast climate change.

JCAE Joint Committee on Atomic Energy (of the US Congress)

JPL Jet Propulsion Laboratory, of NASA

Keynesianism, Keynes A theory in economics associated mainly with the idea that deficit spending is a legitimate and appropriate policy for governments to reduce unemployment.

kinetic energy Energy in the form of motion ($E = \frac{1}{2} mv^2$) where m is mass and v is velocity.

Krebs cycle The citric acid cycle that digests carbohydrates, fats and proteins making acetate and oxidizes acetate for energy metabolism. Named for **Hans Krebs**.

Lagrangian A formalism introduced by Lagrange used to calculate the equations of motion, starting from a definite integral over a function L of potential energy, depending on position $x(t)$ and kinetic energy, depending on velocity (dx/dt). Conditions expressed as equalities. The Lagrangian formalism can be derived from the Hamiltonian formalism, and vice versa.

LBNL (or LBL) Lawrence Berkeley Laboratories of the US Department of Energy (DOE); part of the University of California at Berkeley.

LCD Liquid Crystal Display

LED Light emitting diode. These are rapidly replacing conventional incandescent lights (and some fluorescent lights) in many applications because they use electricity much more efficiently.

leptons Nuclear particles (**fermions**) that occupy space but are not affected by the "strong" force. This group includes **electrons**, **muons** and **neutrinos**. They may carry charge, or not. The total number of leptons (minus the number of anti-leptons) is conserved in every reaction.

LHB "Late heavy bombardment", the Hadean period between 4.1 and 3.85 billion years ago during which Earth was hit by numerous meteorites, and planetoids.

LINEX Linear Exponential, referring to a production function in economics.

LNG Liquefied natural gas

LNL Livermore National Laboratory (US)

LUCA Last universal common ancestor (a hypothetical primitive cell). It supposedly lived about 4 billion years ago.

Lyapunov coefficient In brief, it is a measure of the divergence of trajectories (of dynamical systems) starting from slightly different initial conditions, in phase-space. In effect, it measures unpredictability, or the tendency of a system to become chaotic. It was named after the Russian mathematician Aleksandr Lyapunov (1857–1918).

Maxwell-Boltzmann distribution The distribution of energy states around a mean that would occur if all particles were point particles bouncing off each other. Named for **James Clerk Maxwell** and **Ludwig Boltzmann** who conceived it.

mesons Strongly interacting particles, of which the least massive are **pions**, followed by several heavier varieties (K, ρ, etc.) that are exchanged during strong force interactions. Mesons are composed of a quark with an anti-quark.

MIPS Materials Input Per unit of Service, an environmental indicator proposed by Friedrich Schmidt-Bleek, an eminent German atmospheric chemist and environmentalist.

mitochondria (see organelle) A symbiotic structure within a cell, performing a specialized function, such as photosynthesis (chloroplast). Mitochondria have been identified as proxies for a biological clock.

MRS Materials Research Society

muons Are unstable heavy leptons, with negative charge and mass of 207 times the mass of the electron. They have spin ½.

NADPH Nicotinamide Adenine Dinucleotide Phosphate: an energy carrier (related to ADP and ATP).

NASA National Aeronautics and Space Agency

NBER National Bureau of Economic Research

NDP Neutron Degeneracy Pressure: The effective pressure against multiple occupancy of Planck volumes by fermions with the same quantum state, due to the Pauli exclusion principle.

neuron A component cell of nerves and brain tissue.

neutrino A lepton with spin but no charge, no radius and very little or no mass; they do not constitute any part of matter, as we know it. Predicted by **Wolfgang Pauli** in 1930, because neutrinos had to exist to satisfy the law of conservation of energy. However it is barely possible that there are heavy (tau neutrinos) accounting for "dark matter".

neutron A nucleon with no electric charge; see also proton.

NG Natural Gas

NGL Natural Gas Liquids

NOAA National Oceanographic and Atmospheric Agency

Noether's theorem That every differentiable symmetry in the action of a physical system has a corresponding conservation law. Fundamental for particle physics. See **Amalie Noether**.

NOx Nitrogen oxides

NRDC Natural Resources Defense Council

nuclear force The forces between nuclear particles. There are two kinds: "weak" and "strong" forces. Weak forces affect **leptons** and are responsible for beta-decay of neutrons into protons and electrons. Strong forces affect **baryons** They are based on interactions between **quarks** and **gluons** that hold baryons together. These interactions are known as "**chromo-dynamics**" (because they involve the "colors" of the quarks).

nucleic acid Another name for DNA and RNA

nucleotide base One of the primary molecular components of DNA and RNA, namely: adenine, cytosine, guanine, thymine, uracil. DNA contains thymine but not uracil. RNA contains uracil instead of thymine.

nucleus The massive core of an atom, consisting of protons and neutrons. For an eukaryote cell, the nucleus is an organelle that contains most of the DNA.

OECD Organization for Economic Cooperation and Development, an organization of the industrialized countries charged with collecting economic data and acting as a coordinating body for international negotiations on economic affairs. Located in Paris.

OLS Ordinary Least Squares, referring to a standard methodology for statistical analysis of data by minimizing the deviations from a normal (Gaussian) distribution.

OPEC Organization of Petroleum Exporting Countries, organized in 1960. It is a cartel led by Saudi Arabia and the other Persian Gulf producers, plus Venezuela, Nigeria, Angola, Algeria, Ecuador, Libya, Nigeria and Venezuela. The HQ is in Vienna.

organelle An internal structure in a cell, bounded by a membrane that separates it from the surrounding cytoplasm. Examples include cell **nuclei**, **mitochondria** and **chloroplasts**. All of these internal structures probably evolved from free-living bacteria that were "engulfed" by a larger cell, but not digested and became **endosymbiants** instead.

ORNL Oak Ridge National Laboratory

parity, conservation of A quantum number for reflective symmetry as applied to elementary particles and molecules as well as particles. The interesting thing is that strong interactions and electro-dynamic interactions conserve parity, but the "weak" interactions (e.g. beta decay) do not. The discovery of this violation of parity by Fermi was one of the great milestones of physics.

Pauli exclusion principle A fundamental rule of quantum mechanics, says that no two particles of the same type, with non-integral spin, can occupy the same state. The first application of this rule is that no two electrons (having spin ½) can occupy the same state (think about electron-orbits) of an atom. This has huge importance for chemistry and biochemistry.

PCS Pulse Centered Sun (model)

PGM Platinum Group of Metals, consisting of platinum, palladium, rhodium, ruthenium, iridium and osmium.

pH Measure of acidity (the H refers to positive hydrogen atoms, or protons). For an aqueous solution a pH of 7 is considered neutral, while a higher pH is alkaline (basic) and a lower pH is acid.

PHES Pumped Heat Energy Storage; a system utilizing advanced heat pumps to store electric power as heat and vice versa.

phlogiston theory (of heat) That heat is a kind of fluid that moves from warmer to colder places.

photolysis Molecular breakup due to high energy (ultra violet) radiation.

photon A quantum of light having an energy equal to a multiple (Planck's constant) times the frequency. The "photon" concept was first introduced by Einstein (1905) to explain the photo-electric effect.

photosynthesis The process of carbon fixation from CO_2 to glucose or another sugar molecule, using the energy from photons (light) and producing oxygen as a by-product. A photosynthesizer is a photosynthetic bacteria or alga.

pions Are the mesons with the smallest mass; they come with positive, negative and zero charge. The negative pion is an anti-particle.

Planck's constant The fundamental constant h of quantum mechanics, introduced in **Planck's** theory of **black-body radiation** (1900); $h = 6.624 \times 10^{-34}$ j s.

plasma In medicine it is what remains of blood after the white and red cells are removed. In physics (and in stars) it is matter in a gaseous state consisting of charged particles (electrons, protons and ions).

potential energy In contrast to **kinetic** energy, potential energy is usually inactive, in the form of a force field, such as a gravitational field or an electric or magnetic field. Chemical energy (**chemical potential**) and nuclear energy are examples.

PR Progress Ratio, defined for an **experience curve**.

production function In economics it is a function of "factors of production", which are variables that—in some combination—can explain economic output and growth.

prokaryote A cell without a nucleus (also without **mitochondria** or **chloroplasts**).

protein One of the major constituents of living organisms, constructed from amino acids. The instructions for making proteins are carried by **genes** in a living organism.

Proton A positively charged nucleon. Hydrogen atoms consist of a proton and an electron, bound together by electromagnetic forces.

pulsar (from "pulsating star") A rapidly rotating neutron star left over from a supernova, that emits electromagnetic radiation anywhere from visible to gamma frequencies, along the axis of its powerful magnetic field. Pulsars were the first "atomic clocks" because of their extreme regularity.

purine A class of aromatic (ringed) organic compounds, including the nucleotide bases adenosine and guanine.

PV Photovoltaic (solar) power.

pyrimidine A class of aromatic (ringed) organic compounds that includes cytosine, guanine, thymine and uracil.

QCD Quantum Chromodynamics, the theory of strong force interactions with quarks, formulated by Gell-Mann and Fritzsch. Currently known as the "standard model".

QE Quantitative Easing, a policy of central banks (initiated by the US Federal Reserve Bank in 2010) to purchase commercial bank debts (to each other) and thus (supposedly) free the banks to make more loans to small business.

QED Quantum electrodynamics, the quantized theory of electromagnetic interactions, due mainly to Richard Feynman and Julius Schwinger.

QSO Quasi-stellar object (see quasar).

quantum mechanics (or "wave mechanics") The theory based on Planck's quantization of light asserting that all particles also have wave-like characteristics, and propagate as "wave functions" that obey Schrodinger's modified version of Newton's equations. Quantum mechanics essentially replaces Newtonian mechanics as applied to the domain of extremely small and fast moving objects.

quarks These are fermions that interact with the strong force. They constitute the interior substance of all **hadrons** (a category including **protons and neutrons**). The force itself is said to be "mediated" by so-called **gluons** (like glue) which are massless particles comparable (in that sense) to photons and neutrinos. The interactions between different quarks are called "**chromodynamics**". There are six quarks, each available in three "colors" and degrees of "strangeness". There are conservation laws (and symmetries) involving all of them.

quasar (quasi-stellar radio sources) The brightest objects in the sky (100 times brighter than a galaxy), thought to be radio or light emissions resulting from the absorption of a galaxy into a super-massive **black hole**.

R The universal gas constant $= 8.3145$ J per mole-K.

radiation damping coefficient (for Earth) The change in radiation intensity at the top of the atmosphere (measured in W/m^2) that corresponds to an average temperature change on the ground of 1 °C. It equals 3.7 W/m^2 per degree Celsius.

radiative forcing The equivalent increase in solar radiation power at the Earth's surface (in W/m^2) for each doubling of the atmospheric CO_2 concentration. See **climate sensitivity**.

Rayleigh-Jeans Law The law describing radiation from a black body, in classical physics. It suffered from the so-called "ultra-violet catastrophe" and was replaced by **Max Planck**. That was the starting point of quantum mechanics.

reciprocity relations (see Onsager**)** In near-equilibrium statistical mechanics forces and fluxes are coupled. For instance, the rate of (laminar) fluid flow in a pipe is proportional to pressure, while the pressure is proportional to the flow rate. Other examples: heat is proportional to temperature gradient (Fourier's law); diffusion is proportional to the gradient of concentration (Fick's law) and current is proportional to resistance (**Ohm**'s law). Onsager found that all these (and other) examples are covered by a broader reciprocity relationship where entropy production is minimized in a non-equilibrium steady-state.

red shift Light from a source that is moving rapidly away appears (to an observer) to be "redder" (lower in frequency) than light from a stationary source. **Hubble** discovered that the further the stars, in all directions, the greater the red shift. This means that the universe is expanding.

Redox Portmanteau of *re*duction vs. *ox*idation, applied to reactions that involve changing excitation states or moving electrons from one atom or molecule to another.

REE Rare Earth Element (there are 17 including scandium, yttrium and 15 lanthanides).

RET Renewable Energy Technology

RFF Resources for the Future, Inc. located in Washington DC

RFI Radio Frequency Identifier (identification technology), usually a computer chip, often used to identify the components in a machine, by origin or composition.

ribosome The **enzyme** that "decodes" the DNA, and constructs the **amino acids** that, in turn, create **proteins**.

Riemannian space A space that is non-Euclidean, i.e. the shortest distance between two points is not a straight line, but is curved. The surface of a sphere (like the surface of the Earth) is a curved Riemannian space. When that curved space is projected on a flat Euclidean space (like the Mercator projection) the shortest airline distance between two cities shows as a curve in Euclidean space.

RMI Rocky Mountain Institute (created and directed by Amory Lovins).

RNA Ribonucleic acid (see also DNA).

RON Research Octane Number (usually abbreviated to "octane"), a measure of gasoline quality based on the maximum compression allowable in an Otto-cycle engine before pre-ignition or "knocking" occurs.

RuBisCO Better known as chlorophyll, the key enzyme in the Krebs cycle, that converts carbon dioxide from the atmosphere to glucose.

scalar A simple number, in contrast to a **vector** which typically represents a two-dimensional object or a **tensor** which represents a 3- (or more) dimensional object. A scalar field (like an electric or gravitational field) is directionless, as compared to a **vector field** which has a direction (like spin).

SDSS Sloan Digital Sky Survey, since 2001, a large international project with the intention of mapping the universe; it is particularly important for identifying quasars.

SME Small or Medium-sized Enterprise

SMS Szathmary & Maynard-Smith

SMS transitions The evolutionary sequence of successively more complex reproductive units, postulated by SMS, starting with (1) replicating molecules in compartments, (2) **genes**, (3) **prokaryotes**, (4) **eukaryotes**, (5) sex, (6) cell differentiation in multi-cell organisms (and more).

spin A quantum number. In quantum mechanics, spin is a characteristic of certain particles (**fermions**) lacked by other particles (**bosons**). It is a kind of extrapolation of the notion of rotation to elementary particles. Spin, like charge, is **conserved** in all nuclear reactions between fermions, meaning that both sides of a reaction must have the same spin. This rule helps to determine which reactions are possible and which are not. (In fact other quantum numbers, may be conserved, depending on cases.)

SOX Sulfur oxides (air pollutants)

SSM Standard solar model; the theory of the sun and solar system now assumed by most astronomers.

stromatolite Specialized bacterial **photosynthesizer**s that thrived in the oceans of Earth before the buildup of atmospheric oxygen and left deposits of ferric iron in the sea-bed. Iron ore is mainly based on those deposits.

strong nuclear force The force that holds elementary particles (such as nucleons) together. It is the source of nuclear binding energy, and the source of energy from nuclear reactors (and bombs).

supernova A stellar explosion that occurs when a massive star reaches the end of its supply of hydrogen and helium, at which point gravitational collapse begins. But the gravitational collapse, itself, triggers a violent explosion in which all of the lighter materials are expelled, leaving only a core remnant that may be a neutron star or a black hole.

symmetry and symmetry-breaking Symmetry is defined as invariance of a configuration of elements under a group of automorphic transformations. There are several forms, viz. bilateral, translatory, rotational, and so on. When a law of nature is symmetric with respect to some variable, it applies even if the variable is reversed in sign. Symmetry-breaking means that the law is not independent of the sign of the variable. Examples of such variables include time, charge, spin, parity, iso-spin.

tachyon A (hypothetical) particle that travels faster than light (FTL). Contradicts Einstein's special theory of relativity, but the idea won't go away. There are scientists (Robert Ehrlich for one) who think that neutrinos are tachyons (Ehrlich 2015).

TARP Troubled Asset Relief Program, a bailout program to help large banks after the 2008 financial crisis

TCC "Thermofor" catalytic cracking, a petroleum refining process based on the patents of Eugene Houdry. The "Thermofor" unit is a kiln where the catalyst is regenerated.

TEL Tetra-ethyl lead, an octane enhancing additive to gasoline, now banned in most of the world.

tensor A generalization of the idea of direction (see vector) to higher dimensional spaces. Just as a vector can be represented as three numbers (points on a 3D flat Euclidean space), a 3D tensor can be represented as a matrix of nine numbers (3 vectors) in a 3D flat Euclidean space.

thermal efficiency The ratio of useful output energy to total input energy. There are two different versions in practical use. The so-called *"first law efficiency"* reflects the fact that energy is conserved, and distinguishes between "useful"—often an arbitrary choice—and "rejected" components. The "second-law" efficiency takes into account process irreversibility and measures the actual useful work output as compared to the maximum theoretical output of useful work. *Second law efficiency* is usually much lower than first law efficiency, except in the case of electric power generation, where the two are the same.

thermo-haline circulation Also known as the "conveyor belt" of linked ocean currents that carry warm water toward the poles and brings it back through the cold depths, where it picks up nutrients (especially phosphorus) that are recycled to the surface waters. Places where this upwelling occurs are typically very productive, such as the Peruvian anchovy fishery.

thermosteric sea-level rise Sea-level rise attributable to the thermal expansion of ocean water due to heat absorption.

thymine One of the five nucleotide bases. It is a pyrimidine.

time-series Statistical data that vary over time.

TRC Texas Railroad Commission, the organization created originally to regulate shipping prices that controlled Texas production and, effectively, world prices for petroleum, from 1933 to 1972, when **OPEC** took over.

TVA Tennessee Valley Authority

TW Terawatt, 1000 GW, a million MW or a trillion watts

UN, UNO United Nations Organization.

UNEP UN Environment program (located in Paris, Geneva and Nairobi).

UNIDO UN Industrial Development Organization (located in Vienna).

UPS United Parcel Service.

uracil One of the five nucleotide bases. It is found in RNA but not DNA, where thymine takes its place.

USDA US Department of Agriculture

USEIA US Energy Information Agency, located in Washington DC

USEPA US Environmental Protection Agency, Washington DC

UNESCO United Nations Education, Science and Culture Organization, located in Paris

USGS US Geological Survey, located in Reston, Virginia

USSR Union of Soviet Socialist Republics, led by Russia, now extinct.

utility, utilitarianism In economics, a term used by John Stuart Mill. It encompasses whatever it is that persons or parties engaged in an exchange transaction try to maximize. It is usually used as a synonym for profit (for a business) or for income or consumption in the case of an individual.

UV Ultraviolet radiation, also called "ionizing radiation" because of its ability to detach electrons from atoms or molecules.

vector Basically a direction, with respect to a given frame of reference, usually expressed in mathematical terms as points in a 3D Euclidian space. See *scalar*, *tensor*.

vector field A field with a directional component, such as a magnetic field (in contrast to a **scalar field**)

velocity of light (c) 186,000 miles per second or 299,792.458 km/s.

VLM Vertical Land Motion, the tendency of land formerly covered by glacial ice to rise, after the weight is removed. It is a very slow process.

voltage Measure of electrical potential, named after the Italian physicist Alessandro Volta.

VRLA Valve-regulated lead-acid, a type of battery.

weak nuclear force The force responsible for all nuclear reactions involving **neutrinos,** especially **beta decay**.

WIMP Weakly Interacting Massive Particle (a hypothetical explanation of "dark matter").

WMAT Wilkinson Microwave Anisotropy Probe: A cosmic temperature mapping satellite launched in 2001, in orbit until 2010.

WTI West Texas Intermediate, a standard grade for petroleum. The other major grades are Brent, Dubai, Oman, Urals and OPEC reference grade.

XBT eXpendable BathysThermographs, floating thermometers that can determine the ocean's temperature accurately at any depth up to 2000 m, and transmit the information to a satellite. There are nearly 4000 of them distributed around the oceans today.

YGR Yilmaz General Relativity.

People

Alfvén, Hannes Olof Gösta (1908–1995, Swedish physicist) Father of plasma physics, and the "continuous creation" theory of plasma cosmology. Nobel laureate 1970.

Ampère, André-Marie (1775–1836, French) Physicist. The unit of electric current is named for him.

Archimedes (287–212 BCE, Greek) Mathematician, engineer, inventor, astronomer, proved many theorems in geometry, understood infinitesimals (anticipating calculus), exponents, etc. He discovered pi, invented screw pumps, compound pulleys and defensive war machines.

Aristotle *Aristotélēs* **(384–322 BCE, Greek)** Influential Greek philosopher, second century BC, student of Plato (and teacher of Alexander the Great). His "philosophy" was accepted without question by scholastics throughout the Middle Ages, and was challenged, for the first time (in public) by Galileo, in the famous experiments on gravitational acceleration at the leaning tower of Pisa.

Armstrong, Edwin H. (1890–1954) American inventor He invented the regenerative circuit to make De Forest's triode into an amplifier, then into a transmitter; he invented the super-regenerative circuit and the super-heterodyne radio receiver, but is best known for inventing pioneering frequency modulation (FM) radio.

Arouet, France-Marie *aka Voltaire* **(1694–1778, French)** Writer, historian and philosopher. See du Chatelet.

Arrhenius, Svante Auguste (1859–1927, Swedish) Swedish physical chemist and 1903 Nobel laureate, first to suggest global warming. Author of "pan-spermia" theory that life may have originated in space or another planet.

Arrow, Kenneth (1921-, American economist) Known for his "impossibility theorem" concerning the relationship between individual and social choice, his proof (with Debreu) of the existence of a Walras general equilibrium, pioneering work on exogenous technological changer (learning by doing), and other theorems.

Arthur, W. Brian (1946-) British-American economist known for his work on economic complexity and increasing returns and path dependence in economic systems. Increasing returns introduces positive feedbacks that help to explain economic growth on the one hand, and explain economic downturns on the other.

Bachelier, Louis (1870–1946, French mathematician) Bachelier explained random walks in Brownian motion 5 years before Einstein, but his main contribution was the application of random walk theory to finance, where it was further developed at the Chicago School. His work also inspired Mandelbrot.

Bardeen, John (1908–1991, American physicist) Bardeen was one of the three inventors of the transistor, with Brattain and Shockley at Bell Laboratories, Nobel Prize 1956. Later he shared another Nobel Prize in physics (1972) for the BCS theory of superconductivity with Leon Cooper and Robert Schrieffer (his student).

Bateson, William (1861–1926, English) Biologist who founded and named the science of genetics.

Becher, Johan Joachim (1635–1682, German) Physician, alchemist, precursor of chemistry, scholar and adventurer. Developed the "phlogiston" theory of combustion.

Becquerel, Henri (1852–1908, French) Physicist, discovered spontaneous radio-activity in uranium. This discovery inspired the later discoveries of radioactivity in thorium, polonium and radium by Marie Sklodowska-Curie and Pierre Curie. Nobel laureate.

Bell, Alexander Graham (1847–1922, Scottish-American) Inventor of the first practical telephone, while trying to improve the telegraph. His invention was commercialized (by other) very rapidly, leading to many local Bell Telephone companies, later merged into AT&T. Another inventor, Elisha Gray, at Western Electric Co. in Chicago, submitted a patent for the same invention at the same time. They fought over it for many years, until the courts finally favored Bell.

Benz, Karl Friedrich (1844–1929, German) German engineer-entrepreneur carriage manufacturer who built one of the first successful automobiles. His firm later merged with the Daimler firm to form Daimler-Benz.

Berzelius, Jöns Jakob (1779–1848, Swedish chemist) One of the "fathers" of chemistry, known for determining atomic weights, discovery of new elements, formulating the principles of stoichiometry, electrochemistry and catalysis.

Bessemer, Henry (1813–1898) English inventor, entrepreneur and business-man. He invented the Bessemer process for making steel by blowing air through molten pig iron. This process cut the price of steel to a fraction, thus creating several new markets.

Bethe, Hans Albrecht (1906–2005, German-American) Nuclear physicist (astrophysics, quantum dynamics and solid-state physics). Nobel prize (1967) for his work on the theory of stellar energy production.

Biot, Jean-Baptiste (1774–1862, French) Physicist, astronomer, and mathematician. Established the reality of meteorites and discovered the mineral biotite. Co-author of Biot-Savart law (electro-magnetism).

Black, Joseph (1728–1799, Scottish) Chemist, discoverer of carbon dioxide, measured heat capacity and other key parameters Mentor of James Watt.

Bohr, Niels Henrik David (1885–1962, Danish) Physicist who first explained Rydberg's spectral emissions of the hydrogen atom in terms of quantum mechanics. He formulated the principle of complementarity, to explain why an entity can exhibit apparently contradictory aspects (e.g. wave v. particle).

Boltzmann, Ludwig (1844–1906, Austrian) Physicist, pioneer of statistical, author of the statistical definition of entropy (engraved on his tomb-stone in Vienna).

Born, Max (1882–1970, German) Physicist, best known for the probabilistic interpretation of wave functions. He collaborated with most of the pioneers of quantum mechanics while head of the physics department at Gottingen. Nobel laureate.

Bosch, Karl (1874–1940 German) Chemical engineer, worked with Fritz Haber to commercialize the ammonia synthesis process they developed together. Nobel Laureate.

Bose, Satyendra Nath (1894–1974, Indian) Polyglot (self-taught, no degree). His work on quantum mechanics providing the foundation for Bose-Einstein Statistics.

Boulton, Matthew (1728–1809, English) Entrepreneur and partner of James Watt. Watt and Boulton was very successful selling Watt's condensing steam engines to mines and manufactories of all kinds.

Brattain, Walter (1902–1987 American physicist) Co-inventor of the transistor (with Bardeen and Shockley)

Brillouin, Leon (1889–1969, French) Physicist, pioneer of information theory.

Bruno, Giordano, *aka* **Filippo Bruno, Il Nolano (1548–1600)** He conceived of an infinite universe. Burnt at the stake by the Inquisition for rejecting Earth-centrist Catholic doctrine

Brush, Charles Francis (1849–1929, American) Inventor and entrepreneur. Brush arc light systems and generators.

Budyko, Mikhail Ivanovich (1920–2001, Russian) Climatologist, first to quantify the Earth's heat balance. His theory may explain the "snowball Earth" events. "Budyko's blanket" is a proposal to manage radiation forcing by introducing sulfate particles into the stratosphere.

Calvin, Melvin (1911–1997, American) American chemist who first explained the citric acid cycle that oxidizes sugar and produces adenosine triphosphate (ATP) which provides exergy for all metabolic processes.

Caratheodory, Constantin (1873–1950, German) Mathematician, a more general proof of the Second Law not restricted to equilibrium conditions.

Carlisle, Sir Anthony (1768–1840, English) Surgeon. Discovered electrolysis (with William Nicolson).

Carnot, Nicolas Léonard Sadi (1796–1832, French) French engineer, pioneer of theoretical thermodynamics, the "Carnot limit" for heat engines.

Clausius, Rudolf (1822–1888, German) German physicist born in Poland, known for reformulation of entropy.

Colt, Samuel (1814–1862, American) He was the inventor of the famous Colt 44 revolver, which was mass-produced by his factory. He was possibly the first exemplar of mass production of complex metal products.

Copernicus, Nicolaus (1473–1543, Polish) Polish astronomer and author of sun-centered epicycle theory. He also understood Gresham's Law (that bad money drives out good) before Gresham.

Coriolis, Gustave-Gaspard de Coriolis (1792–1843, French) Mathematician, mechanical engineer and scientist. He is best known for his work on the supplementary forces that are detected in a rotating frame of reference, known as the Coriolis Effect.

Cort, Henry (1741–1800) English iron-smith, invented the "puddling rolling" process for wrought iron, patented 1783. This process made wrought iron a practical industrial material (before steel).

Coulomb, Charles-Augustin de Coulomb (1736–1806, French) Physicist best known for the formulation of Coulomb's law. The measure of quantity of electric charge is named for him.

Crick, Francis (1916–2004, English) Co-discoverer of the "double helix" structure of DNA, with James Watson. Also suggested exogenesis. Nobel laureate.

Croesus (595–547 BCE, King of Lydia) Innovator of standardized coins, for general circulation, made from electrum, an alloy of gold and silver. This innovation made Lydia (and its king) very wealthy.

Curie, Marie Skłodowska (1867–1934, Polish) Physicist, pioneer in study of radioactivity. She and her husband Pierre discovered radium. She also discovered polonium. Nobel laureate (twice).

Daimler, Gottlieb Wilhelm (1834–1900, German) Engineer and industrialist. He was mainly responsible for shrinking and speeding up the stationary internal combustion engine of **Nikolaus Otto**, for use in cars. His firm merged with the **Benz** firm in 1927.

Darby, Abraham (1678–1717 English) Iron-master and pioneer of coking. Adapted coking from brewing to iron smelting, built the first coke-fired blast furnace, made the cylinders for Newcomen's steam engines, provided the iron for the Severn River bridge (first iron bridge).

Darwin, Charles Robert (1809–1882, English) Naturalist, geologist, botanist, and biologist, known for the "natural selection" (for "fitness") theory of evolution.

Davy, Sir Humphrey (1778–1829, English) Physicist. Invented the carbon arc. Many experiments in electro-magnetism. Also known for work on safety procedures vs. methane emissions in coal mines.

De Broglie, Louis-Victor-Pierre Raymond (1892–1987, French) Physicist who postulated that electrons (and all particles) are waves and realized that electron orbits around an atomic nucleus can be explained as spherical standing waves. This accounts for the discrete number of electrons in an orbit.

De Forest, Lee (1873–1961, American) Inventor of the vacuum tube triode, or "Audion", further developed for amplification with Edwin Armstrong. Also invented a way of recording sound on film.

Democritus, Greek philosopher Credited with postulating the atomic theory, with Epicurus, Lucretius, and others. Of course, at the time, there was no experimental evidence of atoms.

Descartes, René (1596–1650, French) Philosopher, believer in power of rationality to understand everything (without depending on empirical evidence).

Dicke, Robert Henry (1916–1997, American) Physicist and cosmologist, one of the first to postulate the "big bang" as the source of the microwave background radiation discovered by Penzias and Wilson.

Diesel, Rudolf Christian Karl Diesel (1858–1913, German) Engineer and inventor of the compression-ignition engine, named after him.

Dirac, Paul Adrien Maurice (1902–1984, English) Physicist. Developed the quantum theory of electrons and predicted the existence of anti-particles (beginning with the positron). Nobel laureate.

Djerassi, Carl (1923–2015, Austro-American) Organic chemist, major contributions to the synthesis of anti-histamines, and progestin norethisterone, the major component of birth-control pills. Also founded Zoecon, a firm specializing in the use of birth control technology for insect pest control.

Dollo, Louis Antoine Marie Joseph (1857–1931, Belgian) Botanist, known for his "law of irreversibility" in genetics, still debated.

du Chatelet, Gabrielle Émilie Le Tonnelier de Breteuil, marquise du Châtelet (1706–1749, French) Mathematician, translator of Leibnitz and Newton. Regarding energy–which nobody yet understood, she favored Leibnitz. Her lover (Voltaire) favored Newton.

du Fay, Charles François de Cisternay (1698–1739, French) Chemist and superintendent of the Jardin du Roi. Discovered the existence of two types of electricity and named them "vitreous" and "resinous" (positive and negative charge).

Eddington, Sir Arthur Stanley (1882–1944, English) Astrophysicist, one of the first to confirm Einstein's general theory of relativity (the bending of light), mass luminosity relation for stars, luminosity limit, composition of stars.

Edison, Thomas Alva (1847–1931, American) The most productive inventor in history (over 2000 patents), known for DC generators, electric power systems, carbon filament light bulbs, microphones, moving pictures, phonographs and many other things.

Ehrlich, Paul (1854–1915, German) Physician and scientist. Studies in hematology, immunology, and chemotherapy.

Eigen, Manfred (1927–) German, primary author of the hyper-cycle theory to explain the self-organization of pre-biotic systems.

Einstein, Albert (1879–1955, German-Swiss) Physicist, Nobel Laureate (for explaining Brownian motion.) He also explained the photo-electric effect and introduced the "photon" concept. He was the author of special and general theories of relativity, asserted the equivalence of mass and energy ($E = mc^2$), predicted the bending of light, tried for a unified theory of all forces, but failed.

Erasmus, Desiderius (1455–1536) Dutch priest, humanist, social critic, writer and translator of the New Testament during the early Renaissance.

Euler, Leonhard (1707–1783, Swiss) Mathematician, many important theorems, including the calculus of variations (with Lagrange) now central to physics and economics.

Faraday, Michael (1791–1867, English) Experimental physicist, discoverer of magnetic induction and several other key phenomena in electro-magnetism.

Fermi, Enrico (1901–1954, Italian) Physicist who helped quantize statistical mechanics (with Dirac and Pauli), leading to Fermi-Dirac statistics for particles with spin and Bose-Einstein statistics for particles without spin. He discovered the so-called "weak force" to explain muon decay processes, predicted the neutrino (Pauli's suggestion), and calculated the strength of that force: the Fermi constant $G = 294$ GeV-2 in energy units. He also built the first controlled nuclear chain reactor at the University of Chicago. Nobel laureate.

Fibonacci, Leonardo (1170–1250) Pisa Mathematician. His book *"Liber Abaci"* helped to promote the use of Arabic numerals in Europe. Known for his sequence of numbers (Fibonacci numbers) each of which is the sum of the previous two.

Fleming, Alexander (1881–1955) Scottish biologist and pharmacist: discoverer of penicillin, the first antibiotic. Nobel Prize in 1945.

Fleming, John Ambrose (1849–1945) British Physicist, inventor of the thermionic valve (vacuum tube diode) called the "Audion", in 1904.

Ford, Henry (1863–1947, American) Engineer-entrepreneur who built Ford Motor Company. He is known for his development of the moving assembly line, the mass-produced "Model T" which sold for close to 20 years, and for his elimination of the labor category of "fitters".

Franklin, Benjamin (1706–1790, American) Writer, magazine publisher, inventor, politician, and diplomat. Known for interpreting "lightning" as electric current, inventor of "lightning rod".

Fresnel, Augustin-Jean (1788–1827, French) Engineer, physicist. Pioneered the wave theory or light. Explained diffraction, polarization and other phenomena. Designed the Fresnel lens to replace mirrors in lighthouses.

Friedman, Milton (1912–2006, American economist) He was a monetarist, (Nobel Prize 1976), advisor to Reagan and Thatcher, anti-Keynesian. Libertarian, free market ideologue; leader of the "Chicago School", known for saying that the only legitimate purpose of business is to make money; known for initiating the idea that business managers must maximize "shareholder value", now a mantra.

Galilei, Galileo (1564–1642, Italian) Astronomer and physicist, known for pioneering work on gravity as well as astronomy (discovery of the moons of Jupiter). His experiments from the Tower of Pisa disproved Aristotle's assertion that the speed of fall of an object must be proportional to its mass. Gallileo also discovered air resistance and friction, as well as the moons of Jupiter.

Galvani, Luigi (1737–1798, Italian) Physicist, discoverer of "galvanic effect" (frogs' legs) through experiments in which he discovered the electro-chemical series. This led to Volta's first cell.

Gamow, George (1904–1968, Russian-American) Founder of the "hot" Big Bang theory, based on Friedmann and Lemaitre's ideas. Worked on galaxy formation with Edward Teller. Co-creator with Alpher of the nucleo-synthesis theory. Predicted the background radiation (with Alpher and Herman). Worked on genetic coding (with Crick).

Gauss, Karl Friedrich (1777–1855, German) Mathematician, best known for the distribution function named for him. Also known as the author of the Divergence Theorem, actually discovered by Lagrange, and much more.

George, Henry (1839–1897) American self-taught economist: He founded the "single tax" movement, advocating a single tax on land as the sole source of government finance; this tax had previously been advocated by John Locke and Baruch Spinoza. His book *"Progress and Poverty"* was a best seller and the most popular book on economics in history

Gibbs, Josiah Willard (1839–1903) American physical chemist and mathematician. Pioneer in thermodynamics (Gibbs energy) and statistical mechanics, provided a statistical formulation of entropy.

Goodenough, John (1922-) American physicist, who discovered a class of layered materials, based on metal oxides, capable of acting as a solid-state electrolyte for rechargeable lithium-ion batteries. These batteries are now in use in a wide range of products, including the Tesla electric car.

Gutenberg, Johannes (1398–1468) German blacksmith, goldsmith, printer and publisher; inventor of movable metal type. Publisher of the Gutenberg Bible.

Guth, Alan Harvey (1947-, American) Physicist, author of the "inflation" theory to explain how the universe got so big, so fast during the first micro-micro second after the Big Bang. This theory is now widely accepted, thanks to the recent discovery of the Higgs boson, but it is still mysterious.

Haber, Fritz (1868–1934, Polish) Chemist worked with Karl Bosch at BASF to develop and commercialize the process to manufacture synthetic ammonia. The Haber-Bosch process is regarded as the most important chemical process in history because it was the basis for the modern fertilizer industry. Nobel laureate.

Haeckel, Ernst Heinrich Philipp August (1834–1919, German) Biologist and chemist. Proponent of scientific monism (in opposition to dualism). He was a zoologist, anatomist, promoter of Darwin's theory of evolution, author of the recapitulation theory ("ontogeny recapitulates phylogeny") and "biogenic theory". He saw monism as the key to the unification of ethics, economics and politics as "applied biology".

Haldane, John Burdon Sanderson (1892–1964, British) Mathematician and scientist, known for his works in physiology, genetics and evolutionary biology. Co-author of the "hot soup" theory of life's origin.

Hall, Charles Martin (1863–1914, American) Simultaneous inventor (with Paul Heroult) of the electrolytic process for making aluminum, using cryolite as the solvent. Hall founded the Pittsburgh Reduction Company, later Alcoa. Thanks to this process the price of aluminum dropped from $2 per pound in the 1880s to $0.30 per pound ten years later, thus creating enormous new markets.

Hamilton, William Rowan (1805–1865, Irish mathematician, physicist, astronomer) Inventor of Hamiltonian dynamics, a re-formulation of Newtonian mechanics, now commonly used in economics; inventor of quaternions.

Hayek, Friedrich August (1899–1992 Austrian) Leader of the so-called Austrian School of economics, opponent of Keynesianism, monetarist, opponent of central bank low interest policies (because they promote "mal-investment"), widely admired by libertarians.

Hegel, Georg Wilhelm Friedrich (1770–1831) German philosopher and a major figure in German idealism.

Heisenberg, Werner (1901–1976, German) Physicist, one of the founders of quantum theory (with Bohr, Born, Dirac, Pauli, Schrödinger and others), best known for the Heisenberg uncertainty principle, which is consistent with the idea that atomic scale "particles" are really more like "wavicles" (smudges) without hard edges.

Helm, George Ferdinand (1851–1923), German) Mathematician and popular writer on energy in the late nineteenth century, a precursor of the energetics movement

Henry, Joseph (1797–1878, American) Physicist, co-discoverer (with Faraday) of magnetic induction, and first director of Smithsonian.

Heroult, Paul Louis-Toussaint (1863–1914, French) Physicist, chemist, engineer; simultaneous inventor of the first successful electric arc furnace for melting steel, and also joint discoverer of the electrolytic process for making aluminum. His aluminum process was adopted by the industry.

Hertz, Heinrich (1857–1894, German) Physicist. He was the first to demonstrate radio waves, as predicted by Maxwell's equations. He engineered the oscillators and receivers which later became the foundation of radio, radio-telegraphy and television. The standard measure of frequency is named after him. See **Marconi**.

Higgs, Peter (1929-, Scottish) Theoretical physicist who predicted the Higgs boson, recently confirmed at CERN. Nobel laureate.

Hinkle, Peter British chemist, author of the chemi-osmotic theory of the formation of ADP and ATP molecules, in the presence of phospho-lipid membrane. Nobel laureate.

Holling, Crawford S. (1930- Canadian ecologist) Foundational theory of adaptive cycles in ecology, with special focus on the relationships between growth, increasing "connectedness", decreasing resiliency and adaptive reorganization.

Holmes, Arthur (1890–1965, English) Theory of geochronology, radiometric dating.

Hotelling, Harold (1895–1973, American economist) Hotelling is regarded as the father of resource economics. He proposed an optimizing model based on the proposition that mining companies would extract exhaustible resources at a rate that maximizes the value of their resource stocks left in the ground. This prediction does not accord with empirical data, but it was the first application, in economics of what is now known as optimal control theory.

Houston, Edwin James (1847–1914, American) Businessman, professor, consulting electrical engineer, inventor and author. Designer (with Elihu Thomson) of an arc light generator. Founder of the Thomson-Houston Electric Company in 1879.

Hoyle, Fred (1915–2001, English) Astro-physicist, completed the Alpher-Gamow nucleosynthesis theory which became the standard model of stars. Also worked with Wikramasingh on the theory of exogenesis.

Hubbert, Marion King 1903–1989, American geologist Author of the "peak oil" theory, which says that production from any given area follows a bell-shaped curve, accelerating after the initial discovery due to infra-structure development and intensive exploitation, and declining after the peak due to depletion. In 1956 he correctly predicted the year of US peak production (1969–1970)

Hubble, Edwin (1889–1953, American) Astronomer, who used earlier results from Slipher (1912-) and Shapley (1920-) regarding the expansion of the universe (based on the **red shift** of light from distant stars). He determined the ratio of distance to velocity and made the first quantified estimate of the rate of expansion.

Huntsman, Benjamin (1704–1776) English watchmaker, toolmaker, oculist and steel maker. He invented the crucible process, never patented, which put Sheffield on the map for cutlery.

Huxley, Julian (1887–1975, English) Biologist and pioneer of "hot dilute soup" theory about the origin of life. See also Oparin.

Huygens, Christiaan (1629–1695, Dutch physicist) Studied the rings of Saturn, made lenses, derived the formula for centripetal force; co-discoverer (with Fresnel) that light consists of waves (whereas Newton thought it was corpuscles); pioneer in optics and inventor of the pendulum clock which was the best time-keeping system until quartz clocks (1927).

Jevons, William Stanley (1835–1882, English Economist) One of the three discoverers (with Menger and Walras) of diminishing marginal utility of consumption, and neo-classical economics. He was the first to recognize the essential importance of energy (coal) to the British economy in the middle of the nineteenth century. It was he who also first noticed the so-called "rebound effect" where energy efficiency savings that cut unit costs can (and sometimes do) drive up demand and increase overall energy consumption.

Joule, James Prescott (1818–1889, English) Physicist and brewer. Studied the nature of heat, and discovered its relationship to mechanical work.

Kelvin, Lord (see William Thomson)

Keynes, John Maynard (1883–1946, English economist), best known for his argument that governments should not try to balance their budgets during times of economic recession but should engage in deficit spending to increase employment (and consumption) while reversing this policy (reducing debt) during boom periods.

Kilby, Jack (1923–2005) American physicist, inventor of the integrated circuit at Texas Instruments (1958). (Credit shared with Robert Noyce at Fairchild electronics.)

Kirchhoff, Gustav (1824–1887, German) Physicist, made fundamental contributions to spectroscopy (black-body radiation), thermochemistry and electrical circuits. Kirchhoff's law of circuitry (the sum of currents meeting at a point is zero and the sum of voltages in a closed circuit is zero).

Klemperer, William A. (1970-, American) Chemist.

Krebs, Hans Adolf (1900–1981, English-German) Chemist, building on the work of Albert Szent-Györgyi de Nagyrápolti who first explained the photosynthesis process. The Krebs cycle is named for him. Nobel laureate.

Kümmel, Reiner (1939-, German) Theoretical physicist. Pioneered the introduction of energy in economics as an explicit factor of production. He (with others) proved that the standard theorem (that the output elasticity of a factor of production must equal its cost share) does not apply in the real world.

Lagrange, Joseph-Louis (1736–1813, Italian) Mathematician and astronomer. Best known for the theory of mathematical optimization (with Leonhard Euler) as applied initially to Newtonian mechanics. He was also the discoverer of the Divergence Theorem which is wrongly credited to Gauss.

Lamarck, Jean-Baptiste Pierre Antoine de Monet, Chevalier de Lamarck (1744–1829, French) Naturalist who developed a theory of inheritance of acquired characteristics, called Lamarckism or use/disuse theory. Since discredited.

Laplace, Pierre-Simon (1749–1827, French) Mathematician who (among other things) helped to mathematize the laws of electro-magnetism.

Lavoisier, Antoine (1743–1794, French) Chemist with many discoveries, who largely discredited the "phlogiston" theory. He was a tax-farmer and was executed during the French Revolution.

Law, John (1671–1729, Scottish economist) He pioneered the scarcity theory of value, the idea that money creation—as shares—could stimulate the economy, and was primarily responsible for the Mississippi Bubble in France and secondarily responsible for the South Sea Bubble in England. In both cases he argued that the sale of shares in government monopolies could help repay the national debt.

Leibnitz, Gottfried Wilhelm (1646–1716, German) Mathematician and physicist, rival of Newton. First to use differential calculus.

Lemaitre, Georges Henri Joseph Edouard (1894–1966, Belgian) Priest and astronomer. He was the first to calculate the rate of expansion of the universe from red shift data, to calculate the so-called "Hubble constant" and postulate the Big Bang.

Liebig, Justus (1803–1873, German) Chemist, pioneer of organic chemistry, especially plant nutrients (N,P,K); he formulated the "law of the minimum" which determines soil fertility.

Linnaeus, Carolus (1707–1778, Swedish) Botanist who developed classification methodology and classified a large number of species.

Lotka, Alfred J. (1880–1949, American) Mathematician, physical chemist, and statistician, famous for his work in population dynamics and bio-physics.

Lovelock, James (1919-, English scientist) Best known for "Gaia" theory, the idea that life on Earth has evolved to create and protect the conditions for its survival. He also was the first to measure the CFC concentration in the atmosphere and call attention to ozone depletion.

Luther, Martin (1483–1546) German theologian and scholar, Posted "95 theses", on the door of the church in Wittenberg attacking Church corruption, such as nepotism, simony, usury and the sale of indulgences. He was told to retract by the Pope and the Emperor Charles V at the Diet of Worms (1524) but he refused saying "Here I stand, I can do no other". Excommunicated, but protected by others. Founder of the Protestant Reformation. Translator of the Bible into German.

Mackay, Charles (1814–1889, English economist) The first to notice and recognize the irrationality in human behavior that enables the creation of "bubbles" such as the "tulip bubble, the South Sea Bubble and the Mississippi Bubble.

Malthus, Thomas Robert (1766–1834 English moral philosopher) His landmark essay *On the Principle of Population* (Malthus 1798a, b [1946]) laid out the argument that population would rise exponentially whereas the area of land available to support agriculture must be constant or grow, at best, arithmetically. This inconsistency must lead to a "Malthusian catastrophe" of universal starvation.

Mandelbrot, Benoit (1924–2010, Polish-French mathematician) Mandelbrot made a number of important contributions in pure mathematics, but he is remembered primarily for fractal geometry, the "Mandelbrot set" of geometric shapes that are equally "rough" at any scale. This led to his theory of roughness.

Marconi, Gugliemo (1974–1937, Italian engineer and entrepreneur) He was the first to recognize practical applications of Herz's discovery of radio-waves. He built an international company based on long-distance radio telegraphy. Known for Marconi's Law. Nobel Laureate in physics.

Margulis, Lynn (1938-, American biologist) Author of the theory of endosymbiotic biogenesis, now proven for mitochondria, centrioles and chloroplasts; also co-developer of "Gaia" theory with Lovelock.

Marx, Karl (1818–1883, German economist and revolutionary) As an economists he was influential, especially in clarifying the role of capital and labor. Author of *Das Kapital* and the *Communist Manifesto*, among other writings. As a revolutionary he argued that progress was due to class struggle, between the "proletariat" and the bourgeoisie (owners of the "means of production").

Maxwell, James Clerk (1831–1879, Scottish) Theoretical physicist, author of Maxwell equations to describe electromagnetism, and co-author of Maxwell-Boltzmann distribution in statistical mechanics. His theory of electromagnetism predicted the existence of electromagnetic radiation, later proved by Hertz.

Maybach, Wilhelm (1846–1929, German engineer and entrepreneur) Junior partner of Daimler, both at Klaus Otto's firm and in the Daimler-Maybach firm that pioneered compact internal combustion engines for cars. Inventor of the carburetor. After Daimler's death in 1900 he left and started his own firm.

Maynard-Smith, John (1920–2004, British) Evolutionary biologist, co-author with Eörs Szathmáry of *The Major Transitions in Evolution* and other works.

Mendel, Gregor (1822–1884, Moravian) Botanist who formulated the laws of inheritance for peas. His name is now attached to the genetic laws of evolution.

Mendeleev, Dmitri (1834–1907: Russian chemist) He was the first to formulate the periodic table of the elements (almost) correctly, in 1869. This was a huge step forward.

Menger, Carl (1840–1921, Austrian economist) He was one of the three economists (with Jevons and Walras) credited with discovering the diminishing marginal returns to consumption, and thus a pioneer of neo-classical economics. He was the first to precisely define the conditions for a market exchange in utility terms.

Midgeley, Thomas (1889–1944, American chemist) He was the main discoverer of tetraethyl lead, the gasoline additive that was used from the late 1920s until 1975, when it was banned because of its toxicity. The compression levels (and efficiency) of US auto engines actually peaked in the 1960s because today's motor fuel has lower octane levels, requiring lower compression ratios. Midgeley was also the discoverer of chlorofluorocarbons (CFCs). CFCs were widely used as refrigerants—among other uses—until the 1970s when they were also banned because of damage to the ozone layer.

Milankovitch, Milutin (1879–1958, Serbian) Geophysicist and mathematician who formulated a mathematical theory of Earth's climate based on three different orbital cycles, eccentricity (98,000 year cycle), obliquity (40,000 year cycle) and precession (19,000–23,000 year cycle).

Mill, John Stuart (1806–1873 English philosopher and political economist) He did not "invent" utilitarianism, as such, but he defined it as "maximizing happiness"—not just profits—and doing "the greatest good for the greatest number". He argued against *laissez faire* and in favor of government intervention to limit population growth (to provide "breathing space") and to protect the environment.

Miller, Stanley Lloyd (1930–2007, American Chemist) Known for the Miller-Urey experiment showing that amino acids and other important bio-chemicals could have been produced in a reducing environment such as that of early Earth.

Moore, Gordon (1929-) American physical chemist, and entrepreneur His forecast of the rate of progress, measured as density of components, in electronic devices (published in 1965) became known as "Moore's Law". This empirical law, now 50 years old, remains a valid guideline. Moore co-founded Intel Corp (with Robert Noyce).

Morgan, Thomas Hunt (1866–1945, American) Biologist who formulated the chromosome theory.

Morse, Samuel Finley Breese (1991–1872, American) He modified the multi-wire telegraph systems developed in England (e.g. Wheatstone) to a single wire system, demonstrated between Washington and Baltimore. He also invented the Morse code that made the system practical.

Murdoch, William (1754–1839 Scottish) Chemist, inventor, employee of Watt & Boulton for 10 years. Improved the Watt & Boulton engine, invented the oscillating cylinder engine, innovated gas lighting, invented the gasometer and many other devices leading to many discoveries in chemistry.

Musk, Elon (1971-) American entrepreneur, best known as the founder of Space-X, the first non-government manufacturer of space launch vehicles, and co-founder of Tesla Motor Co.

Newcomen, Thomas (1663-1729, English) He was an ironmonger, inventor and manufacturer of the first practical engine for pumping water from coal mines. His engine was widely used in England until 1775 when the Watt engine took over. It was based on ideas from Thomas Savery and Dennis Papin.

Newton, Isaac (1643–1727, English) Physicist, who formulated the mechanical laws of motion, the laws of gravity and motion, and co-invented differential calculus. His third law of motion, that force is the time derivative of momentum, is one of the most fundamental insights. Author of *Philiosophiae Naturalis Principia Mathematica*. See **Leibnitz**.

Nicolson, William (1753–1815, English) Chemist, natural philosopher. Discovered electrolysis (with Anthony Carlisle).

Noether, Amalie (1882–1935, German) German mathematician-physicist who proved that any differentiable symmetry has a corresponding conservation law

Noyce, Robert (1927–1990) American physicist and entrepreneur, credited with co-inventing the integrated circuit (microchip) with Jack Kilby (1977) as well as co-founding Intel with Gordon Moore and Andrew Grove.

Oersted, Hans Christian (1777–1851) Danish physicist who discovered that electric currents produce a magnetic field. The unit of magnetic field strength is named for him.

Ohm, Georg Simon (1789–1854, German) Physicist and mathematician. Worked in electrical technology. "Ohm's Law". The unit of electrical resistance is named after him.

Onsager, Lars (1903–1976, Norwegian) Physical-chemist who discovered the so-called "reciprocity relations" between forces and flows, far from (thermodynamic) equilibrium. Nobel laureate.

Oparin, Alexander (1894–1980, Russian) Biochemist. First to suggest the "warm dilute soup" theory of the origin of life. See also Julian Huxley.

Oró, Joan (1923–2004, Spanish) Astrophysicist. Suggested comets as source of primitive self-reproducing organisms.

Ostwald, Wilhelm (1953–1932, Latvian) One of the founders of physical chemistry, first to use the term "mole" (gram molecular weight; worked on catalysis (especially of acids and bases), reaction rates, law of dilution, nitrogen fixation (Ostwald process), Nobel laureate. Promoter of the "energetics" movement.

Otto, Nikolaus August (1832–1891, German) Engineer and entrepreneur. Developed the "silent Otto" internal combustion engine, burning coke-oven gas for fuel. It was widely used. The spark-ignition "Otto cycle" of the internal combustion engine, now used in most automobiles, is named after him.

Pareto, Vilfredo (1848–1923, Italian economist) Pioneer of income distribution analysis, preference ordering, and Pareto optimality—the case where nobody can be better off without somebody else being worse off.

Parsons, Charles (1854–1931 Irish) Scientist, engineer and inventor of a turbo-generator and compound steam turbine that revolutionized electric power production as well as British navy ships.

Pauli, Wolfgang Ernest (1900–1958, Austrian-Swiss) Physicist, introduced the notion that all particles have four quantum numbers (degrees of freedom) and no two particles can have the same quantum numbers. This called the Pauli exclusion principle. He also predicted the existence of neutrinos (to preserve the conservation of energy), later used by Fermi to explain beta decay. Nobel laureate.

Penzias, Arno Allan (1933-, American) Physicist, co-discoverer of background micro-wave radiation, Nobel laureate.

Perkin, William Henry (1838–1907, English) Chemist, accidental discoverer of mauveine, the first aniline dye, later he discovered several other aniline dyes, including alizarin.

Planck, Max Karl Ernst Ludwig (1858–1947, German) Physicist, Nobel laureate (1918), discoverer of the correct form of the law describing radiation of a black body by assuming quantization of radiation. Planck's constant (named for him) is the fundamental constant of quantum physics.

Poisson, Simeon-Denis (1781–1840, French) Mathematician, best known for the distribution function named for him.

Polo, Marco (1254–1324) Venetian traveler who visited China and the court of Genghis Khan (with his uncle) and, after returning to Europe, wrote a book about his travels. His story inspired many followers and new technologies.

Priestley, Joseph (1783–1804, English) Physicist who verified that the electric field obeys an inverse square law of distance. Credited with the discovery of oxygen.

Prigogine, Ilya Romanovich (1917–2003, Russian-Belgian) Physical chemist, pioneer of non-equilibrium thermodynamics and "self-organization" of "dissipative structures far from equilibrium. Nobel Laureate (1977).

Rankine, William John Macquorn (1820–1872, Scottish) Civil engineer, physicist and mathematician. He clearly formulated the laws of thermodynamics and defined *potential energy*. He studied the thermodynamics of condensing heat engines. The Rankine (steam) cycle is named after him.

Revelle, Roger Randall Dougan (1909–1991, American geologist and oceanographer) He was a pioneer in tectonics and quantitative oceanography. With Hans Suess, he was the first to raise the alarm about climate change and focusing attention on the mechanisms for ocean uptake of carbon dioxide.

Ricardo, David (1772–1823, English) Advocated the labor theory of value, advocated free trade ("laissez-faire") and opposed the "corn laws" that kept bread prices high by restricting imports for the benefit of farmers. He saw society approaching a standstill due to declining marginal productivity (of land), resulting in rising rents and class conflict between landowners and workers. His work was the basis of much of Marxism, except that Ricardo supported capitalism.

Riemann, Georg Frederick (1826–1866) German mathematician He created the theory of curved multi-dimensional space (1854), and the mathematical methods for describing curvature in a manifold. His *curvature tensor*, in simplified form, was the basis for Einstein's theory of general relativity.

Rockefeller, John Davison (1839–1937 American industrialist) He founded and ran Standard Oil Company of New Jersey from 1870 until it was broken up in 1911. When he died in 1937 he was the richest person in history, being worth $336 billion (2007 dollars). He founded the Rockefeller University in New York and the University of Chicago, among other philanthropic bequests.

Rumford, Count (see Thompson, Benjamin)

Rutherford, Ernest (1871–1937, New Zealand) Physicist, studied radioactivity as an atomic phenomenon, discoverer of protons and nuclei and much more. Nobel laureate.

Sagan, Carl (1934–1996, American) Astrophysicist, biologist, science writer and pioneer searcher for life in the universe. He saw evolution as a trend toward ever greater complexity, information and intelligence.

Samuelson, Paul Anthony (1915–2008, American economist) Regarded as the most influential economist of the twentieth century, and the pioneer in quantification and the application of mathematical tools. Nobel laureate.

Savart, Felix (1791–1841, French) Physicist primarily known for the Biot-Savart Law of electromagnetism.

Schrödinger, Erwin (1887–1961, Austrian physicist) He generalized Newton's equations (using the Lagrangian formulation) to treat particles like waves (the Schrödinger equation). Nobel laureate. He was also the author of *"What is Life?"* (Schrödinger 1945) describing the implications of quantum theory for genetics.

Schumpeter, Joseph (1883–1950, Austrian-American economist) Best known for his idea of "creative destruction" as a factor in capitalist economic growth; also a pioneer in the theory of economic cycles. He disagreed with conventional theory as regards the bank's role in creating money.

Shannon, Claude (1916–2001, American) Mathematician, Equated uncertainty with negative information. He and Warren Weaver originated information theory.

Shapley, Harlow (1884–1972 American) Astronomer who used variable stars and red shifts to determine the actual distance of galaxies. He is credited with first understanding our "place" in the Milky Way and the universe.

Shockley, William (1910–1989 American physicist and entrepreneur) Co-inventor of the transistor and several advances in the technology of semiconductors including the junction transistor. Author of seminal treatise *"Electrons and Holes in Semi-conductors"*. Founded Shockley Semiconductor Corp. Nobel Laureate.

Siemens, Ernst Werner (1816–1892, German) Inventor and industrialist, founder of the electrical and telecommunications company Siemens.

Silliman, Benjamin (1779–1864 American chemist) He was the first to perform fractional distillation of "rock oil" (petroleum), and he wrote the report for a group of investors that led to the discovery of petroleum in Pennsylvania.

Simon, Herbert (1916–2001 American economist) He developed "bounded rationality" into a useful analytic tool; also artificial intelligence. Nobel laureate.

Slipher, Vesto Melvin, 1875–1969, American Astronomer, discoverer of the "red shift" of distant stars and galaxies. By 1924 41 galaxies had been tracked of which 36 were receding.

Smith, Adam (1723–1790, Scottish political economist and moral philosopher) Author of *"An inquiry into the Nature and Causes of the Wealth of Nations"* (Smith 1776 [2007]), the first significant book on economics as a subject. He thought capitalism and trade could result in increasing prosperity, with no limits and minimal government. He is known for insights, such as division of labor (specialization), economies of scale, and the idea that individual self-interest, as manifest in markets, can have beneficial social effects via the "invisible hand".

Smoot, George Fitzgerald III (1945-, American) Cosmologist and leader of the COBE project that has confirmed the existence of "wrinkles" in space time, to explain observed inhomogeneities in residual microwave radiation all the way back to the origin.

Soddy, Frederick (1877–1956, English) Chemist, Nobel laureate. Studied radio-activity with Ernst Rutherford. Wrote three books on energy in economics, starting with *Cartesian Economics*.

Solow, Robert Merton (1924-, American economist) Best known for originating the current theory of economic growth, utilizing a "production function" with two "factors", capital and labor, with an exogenous driver, representing technological change. Nobel laureate.

Spinoza, Baruch (1632–1677, Dutch Philosopher ethicist) Opposed Descartes' dualism. Argued against medieval philosophy.

Stephenson, George (1781–1848, English) Civil and mechanical engineer, builder (with his son Robert) of the first public railway, the Stockton and Darlington in 1825 and the Liverpool and Manchester in 1830. Robert's firm built the steam locomotives for the first lines.

Szathmáry, Eörs (1959-, Hungarian) Evolutionary biologist, co-author with John Maynard Smith of *The Major Transitions in Evolution*.

Szent-Györgyi, Albert von Nagyrápolti (1893–1986, Hungarian physiologist) Co-discoverer of vitamin C (with J. Svirbely), and the main components of the citric acid cycle (photosynthesis). Nobel laureate, 1937.

Taylor, Frederick Winslow 1856–1915 American Engineer Pioneer of scientific management (subdividing a task into small components), time and motion studies and industrial engineering. His work influenced Henry Ford. "Taylorism" has been criticized as dehumanization.

Tesla, Nikola (1856–1943, Croatian-American inventor) Inventor and patent holder of major elements of the AC power system (40 patents), among other inventions including the Tesla coil, dynamos, X-ray equipment and a radio-telegraphy system in competition with Marconi. He sold his patents to George Westinghouse where they were commercialized at Westinghouse Electric Co. He demonstrated the AC system (for Westinghouse) at the Chicago "Century of Progress" exhibition, and then designed the first Niagara Falls electric power plant.

Thom, René (1923–2002, French) Mathematician who created the so-called "theory of catastrophes", or sudden "flips" from one possible solution of a non-linear equation, to another possible solution, when the system is in a critical state. Condensation is an example. This idea has become central to particle physics and non-equilibrium thermodynamics.

Thompson, Benjamin (1753–1814, English) Engineer who pioneered understanding of heat-as-friction while grinding cannons for Bavaria. Known as Count Rumford.

Thomson, Elihu (1853–1937, English-American inventor) Prolific inventor. Major contributor to the development of electricity as a power and light source. Co-founder of Thomson-Houston Co., absorbed by General Electric.

Thomson, Joseph John (J.J.) (1856–1940, English physicist) He extended Maxwell's equations and discovered the electron from cathode ray experiments (1897). He determined that the atom is not indivisible, that every element contains identical charged particles ("electrons") much smaller than hydrogen ions (protons). He invented the mass spectrometer. In 1919 he and Chadwick discovered and measured the mass of the proton, confirmed by Bohr. Nobel Prize laureate 1906.

Thomson, Sir William (*Lord Kelvin*) (1804–1907, Scottish physicist) He was a pioneer in all aspects of thermodynamics. The absolute temperature scale is named for him.

Trevithick, Richard (1771–1833, English inventor) He built the first successful high pressure steam engines, which were much more powerful than the Watt & Boulton engines. His engines powered the first locomotives.

Urey, Harold Clayton (1893–1981, American physical chemist) Developed isotope separation techniques resulting in the discovery of deuterium. Research on planetary origins. His student Stanley carried out the Miller-Urey experiment on amino acid synthesis in a reducing atmosphere. Nobel laureate.

Veblen, Thorstein (1857–1929, American economist and social critic) Best known for his book *"The Theory of the Leisure Class"* (Veblen 1899 [2007]) (still read in university classes) and many trenchant phrases, such as "conspicuous consumption" and "conspicuous waste".

Vernadsky, Vladimir Ivanovich (1863–1945, Ukrainian geologist) Best known for the "biosphere" concept (a precursor of the "Gaia" theory of Lovelock). It was Vernadsky who first recognized that oxygen, nitrogen and carbon dioxide in the atmosphere are of biological origin.

Volta, Alessandro Giuseppe Antonio Anastasio (1754–1827, Italian) First electrochemical cell (zinc-copper) and the first battery (Voltaic pile). Unit of electrical potential is named for him.

Voltaire see Arouet

Von Helmholtz, Hermann (1821–1894, German physician-physicist) He was the first to state the law of conservation of energy; work on vision, color vision, perception, etc. as well as electro-magnetic theory.

von Neumann, John (1903–1957, Hungarian-American) Mathematician, mathematical foundations of quantum theory, information theory, economics and game theory. He was a consultant on the ENIAC project and got credit for the stored program idea, although it Eckert and Mauchly probably had it first.

Walras, Leon (1834–1910, French economist) One of three discoverers of diminishing marginal returns of consumption. One of three (with Jevons and Menger) pioneers of the marginalist revolution". Creator of general equilibrium theory, the idea of simultaneous supply-demand balances in multiple sectors. He postulated that such a balance is possible in a static economy.

Watson, James (1928-, American) Biochemist. Co-discoverer (with Francis Crick) of the "double helix" structure of DNA. Nobel laureate, 1962 (with Crick).

Watt, James (1736–1819, Scottish) Engineer. Improved on steam engine design by adding a condenser; patented several other mechanical improvements, became a successful entrepreneur, with partner Matthew Boulton.

Wegener, Alfred (1880–1930, German) He was the originator of the idea that all the continents had once been joined (Pangea) and that they had subsequently split and drifted apart. This was the precursor of the modern theory of *plate tectonics*.

Westinghouse, George (1846–1914, American) Inventor-entrepreneur. Founded two successful companies in railway technology (signals, air brakes), followed by Westinghouse Electric Co., using Tesla's AC patents. The AC system was demonstrated at the Chicago Century of Progress Fair. Westinghouse pioneered in long distance transmission (from Niagara Falls to Pittsburgh).

Weyl, Hermann (1885–1955, German-Swiss mathematician) He contributed in many areas of mathematics, including group theory and symmetry, as applied to quantum mechanics. He invented "gauge theory". He also worked on Riemannian surfaces in curved space.

Wheatstone, Charles (1802–1875, English) Scientist, developed the Wheatstone bridge, a device to measure electrical resistance at a distance. He was also one of several co-inventors of the telegraph, in 1837. He helped to make it into a commercial success, first used by the railroads.

Whitney, Eli (1765–1825, American) Inventor best known for inventing the cotton gin (which was pirated) but more important for developing manufacturing systems for making interchangeable parts. To overcome limitations in available machine tools he invented the all-purpose milling machine (1818)

Wilkinson, John (1728–1808, English) Ironmonger, specialized in cast iron; inventor of the first precision boring machine and a hydraulic blowing machine for raising the temperature of a blast furnace. He was the first customer for Watt's steam engine, which he used to drive his boring machine.

Wilson, Robert (1936-, American) Astronomer, co-discoverer of microwave background radiation (with Penzias) that indicated the likelihood of a "big bang". 1978 Nobel laureate in **physics.**

Wolff, Christian (1679–1754, German) Philosopher. A founding father of, among other fields, economics and public administration as academic disciplines.

Yilmaz, Huseyin (1924-2013) Turkish-American physicist primarily known for his theory of gravity, which differs from the Einstein theory of general relativity in strong gravitational fields. It does not predict black holes. It may explain "dark matter".

Zwicky, Fritz (1898–1974, Swiss astronomer) Discoverer of many super-novae. He predicted the existence of neutron stars. Discoverer of gravitational instability of galaxies, giving rise to the need for "dark matter". Known, also for "morphological analysis".

References

Abelson, Philip. 1966. Chemical Events on the Primitive Earth. *Proceedings of the National Academy of Sciences* 55: 1365–1372.

ACEEE, American Council for an Energy Efficient Economy, (NRDC) Natural Resources Defense Council, (UCS) Union of Concerned Scientists, and Alliance to Save Energy (ASE). 1992. *America's energy choices*. Washington, DC: National Research Council, National Academy of Sciences.

Ad hoc Study Group on Carbon Dioxide and Climate. 1979. In *Carbon dioxide and climate: A scientific assessment*, ed. Jule (Chair) Charney. Washington, DC: National Academy of Sciences (NAS).

Adler, Mortimer J., Clifton Fadiman, and Philip W. Goetz (eds.). 1990. *Great Books of the Western World*. 56 vols. Chicago: Encyclopedia Brittanica.

Agriculture Business Week. Carabao models commercial bio-ethanol production. *Agriculture Business Week*, 2008.

Ahrendts, Joachim. 1980. Reference States. *Energy* 5(5): 667–677.

Alderman, A.R., and C.E. Tilley. 1960. Douglas Mawson 1882-1958. *Biographical Memoirs of Fellows of the Royal Society* 5: 119.

Alessio, F. 1981. Energy analysis and the energy theory of value. *The Energy Journal* 2: 61–74.

Alfven, Hannes. 1936. A cosmic cylotron as a cosmic wave generator? *Nature* 138: 76.

Alfven, Hannes. 1966. *Worlds-antiworlds: Antimatter and cosmology*. San Francisco: W. H. Freeman and Company.

Alfven, Hannes. 1981. *Cosmic plasma*. Dordrecht, The Netherlands: D. Reidel Publishing Company.

Alfven, Hannes, and C.-G. Falthammar. 1963. *Cosmic electrodynamics*. Oxford: Clarendon Press.

Allais, M., and O. Hagan (eds.). 1979. *Expected utility hypothesis and the Allais paradox*. Dordrecht, The Netherlands: D. Reidel Publishing Company.

Allan, Richard P., C. Liu, N.G. Loeb, M.D. Palmer, M. Roberts, D. Smith, and P.-D. Vidale. 2014. Changes in global net radiation imbalance. *Geophysical Research Letters* 41: 5588–5597.

Alley, Carroll O. 1994. Investigations with lasers, atomic clocks and computer calculations of curved spacetime and of the differences between the gravitational theories of Yilmaz and Einstein. In *Frontiers of Fundamental Physics*, ed. M. Barone and F. Seleri. New York: Plenum Press.

Alpher, Ralph, Hans Bethe, and George Gamow. 1948. The origin of the chemical elements. *Physical Review* 73(7): 803–804.

Anderson, P.W. 1962. Plasmons, gauge invariance and mass. *Physical Review* 130: 439–442.

© Springer International Publishing Switzerland 2016
R. Ayres, *Energy, Complexity and Wealth Maximization*, The Frontiers Collection,
DOI 10.1007/978-3-319-30545-5

Anderson, Philip W., Kenneth J. Arrow, and David Pines (eds.). 1988. *The economy as an evolving complex system*, vol. 5. New York: Addison-Wesley Publishing Company.

Arp, Halton C. 1987. *Quasars, redshifts and controversies*. Berkeley, CA: Interstellar Media.

Arp, Halton C. 1998. *Seeing red*. Montreal: Aperion.

Arrhenius, Svante. 1896. On the influence of carbonic acid in the air on the temperature on the ground. *Philosophical Transactions of the Royal Society* 41: 237–276.

Arrhenius, Svante. 1903. The distribution of life in space. *Die Umschau*.

Arrhenius, Svante. 1908. *World in the making*. New York: Harper and Row.

Arrow, Kenneth J. 1951. *Social choice and individual values*. New York: John Wiley.

Arrow, Kenneth J. 1962. The economic implications of Learning by Doing. *Review of Economic Studies* 29: 155–173.

Arrow, Kenneth J., and Gerard Debreu. 1954. Existence of an equilibrium for a competitive economy. *Econometrica* 22(3): 265–290.

Arrow, Kenneth J., Burt Bolin, Robert Costanza, Dasgupta Partha, C. Folke, C.S. Holling, B.-O. Jansson, S. Levin, Muler Karl-Gran, C.A. Perrings, and D. Pimental. 1995. Economic growth, carrying capacity, and the environment. *Science* 268: 520–521.

Arthur, W. Brian. 1988. Self-reinforcing mechanisms in economics. In *The Economy as an evolving complex system*, ed. Kenneth Arrow Philip Anderson and David Pines. Santa Fe, NM: Addison-Wesley Publishers.

Arthur, W. Brian. 1994. Increasing returns and path dependence in the economy. In *Economics, Cognition and Society*, ed. Timur Kuran. Ann Arbor, MI: University of Michigan Press.

Arthur, W. Brian, Yu M. Ermoliev, and Yu M. Kaniovski. 1987. Path-dependent processes and the emergence of macro-structure. *European Journal of Operations Research* 30: 294–303.

Atkins, P.W. 1984. *The Second Law: Energy, chaos and form*. San Francisco: W. H. Freeman & Co.

Axtell, Robert, and Gregory McCrae. 2007. *A general theory of discounting*. Santa Fe, NM: Santa Fe Institute.

Ayres, Robert U. 1958. Fermion correlation: A variational approach to quantum statistical mechanics. PhD, Physics, Kings College, University of London.

Ayres, Robert U. 1988. Barriers and breakthroughs: An expanding frontiers model of the technology industry life cycle. *Technovation* 7: 87–115.

Ayres, Robert U. 1989. Technological transformations and long waves. *Journal of Technological Forecasting and Social Change* 36(3).

Ayres, Robert U. 1991. *Computer integrated manufacturing: Revolution in progress*, vol. I. - London: Chapman and Hall.

Ayres, Robert U. 1994. *Information, entropy and progress*. New York: American Institute of Physics.

Ayres, Robert U. 1998. *Turning point: An end to the growth paradigm*. London: Earthscan Publications Ltd.

Ayres, Robert U. 2014a. A Tale of Two Economies: The gap is growing. *Economie Appliquee* LXVII(2): 5–40.

Ayres, Robert U. 2014b. *The bubble economy*. Boston, MA: MIT Press.

Ayres, Eugene E., and Charles A. Scarlott. 1952. *Energy sources the wealth of the world*. New York: McGraw-Hill.

Ayres, Robert U., and Allen V. Kneese. 1969. Production, consumption and externalities. *American Economic Review* 59: 282–297.

Ayres, Robert U., and Richard A. McKenna. 1972. *Alternatives to the internal combustion engine*. Baltimore, MD: Johns Hopkins University Press.

Ayres, Robert U., and Katalin Martinás. 1996. Wealth accumulation and economic progress. *Evolutionary Economics* 6(4): 347–359.

Ayres, Robert U., and Robert Axtell. 1996. Foresight as a survival characteristic: When (if ever) does the long view pay? *Journal of Technological Forecasting and Social Change* 51(1): 209–235.

Ayres, Robert U., and Leslie W. Ayres. 1999. *Accounting for resources 2: The life cycle of materials*. Cheltenham, UK: Edward Elgar.

Ayres, Robert U., and Benjamin S. Warr. 2003. *Useful work and information as drivers of growth*. Fontainebleau, France: INSEAD.

Ayres, Robert U., and Benjamin S. Warr. 2005. Accounting for growth: The role of physical work. *Structural Change & Economic Dynamics* 16(2): 181–209.

Ayres, Robert U., and Katalin Martinás. 2006. *On the reappraisal of microeconomics: Economic growth and change in a material world*. Cheltenham, UK: Edward Elgar.

Ayres, Robert U., and Benjamin S. Warr. 2009. *The economic growth engine: How energy and work drive material prosperity*. Cheltenham, UK: Edward Elgar Publishing.

Ayres, Robert U., and Edward H. Ayres. 2010. *Crossing the energy divide: Moving from fossil fuel dependence to a clean-energy future*. Upper Saddle River, NJ: Wharton School Publishing.

Ayres, Robert U., and Laura Talens Peiro. 2013. Material efficiency: Rare and critical metals. *Philosophical Transactions of the Royal Society A* 371 (1986):20110563. doi:10.1098/rsta.2011.0563.

Ayres, Robert U., and Vlasios Voudouris. 2014. The economic growth enigma: Capital, labor and useful energy? *Energy Policy* 64(C): 16–28. doi:10.1016/j.enpol.2013.06.001.

Ayres, Robert U., Leslie W. Ayres, and Benjamin S. Warr. 2003. Exergy, power and work in the US economy, 1900-1998. *Energy* 28(3): 219–273.

Ayres, Robert U., Laura Talens Peiro, and Gara Villalba Mendez. 2013b. Material flow analysis of scarce by-product metals: Sources, end-uses and aspects for future supply. *Environmental Science & Technology* 47(6): 2939–2947. doi:10.1021/es301519c.

Ayres, Robert U., Laura Talens Peiro, and Gara Villalba Mendez. 2014. Recycling rare metals. In *Handbook of Recycling* (Chapter 4), ed. Ernst Worrell et al. New York: Elsevier.

Azar, Christian. 2008, 2012. *Makten over Klimatet (Swedish) Solving the climate challenge*. Translated by Paulina Essunger, 2012 ed. Stockholm: Albert Bonniers Publishing Co.

Bagnoli, Philip, Warwick J. McKibben, and Peter J. Wilcoxen. 1996. Future projections and structural change. In *Climate Change: Integrating Science, Economics and Policy*, ed. N. Nakicenovic, W.D. Nordhaus, R. Richels and F.L. Toth. Laxenburg, Austria: International Institute for Applied Systems.

Barnett, Harold J., and Chandler Morse. 1963. In *Scarcity and growth: The economics of resource scarcity*, ed. Henry Jarrett. Baltimore, MD: Johns Hopkins University Press.

Barrell, Joseph. 1917. Rhythms and the measurement of geological time. *Bulletin of the Geological Society of America* 28: 745–904.

Barrow, John D. 1991. *Theories of everything: The quest for ultimate explanation*. New York: Ballantine Books.

Beer, Jeroen G. de. 1998. Potential for industrial energy efficiency improvements in the long term. PhD thesis, University of Utrecht.

Beer, Jeroen G. de, E. Wees, and Kornelis Blok. 1994. *ICARUS, the potential for energy conservation in the Netherlands up till the year 2000 and 2015*. Utrecht, the Netherlands: University of Utrecht.

de Beer, Jeroen G., Ernst Worrell, and Kornelis Blok. 1996. Sectoral potentials for energy efficiency improvements in the Netherlands. *International Journal of Global Energy Issues* 8: 476–491.

Beinhocker, E. 2006. *The origin of wealth: Evolution, complexity and the radical remaking of economics*. Cambridge, MA: Harvard Business Review Press.

Belousov, V.V. 1969. Continental rifts. In *The Earth's Crust and Upper Mantle*, ed. Pembroke J. Hart. Washington, DC: American Geophysical Union.

Bernal, John Desmond. 1951. *The physical basis of life*. London: Routledge and Kegan Paul.

Berner, R.A. 1990. unknown. *Geochimica et Cosmochimica Acta* 54: 2889–2890.

Berner, R.A., A.C. Lasaga, and R.M. Garrels. 1983. The carbonate-silicate geochemical cycle and its effect on atmospheric carbon dioxide over the past 100 million years. *American Journal of Science* 283: 641–683.

Bethe, Hans. 1939. Energy production in stars. *Physical Review* 55(1,5):103, 434–456.

Bieler, Andre, et al. 2015. Abundant molecular oxygen in the coma of comet 67L/Churyumov-Gerasimenko. *Nature* 526: 678–681.

Bjornson, Adrian. 2000. *A universe that we can believe*. Woburn, MA: Addison Press.

Blackwelder, E. 1916. The geological role of phosphorus. *American Journal of Sciences* 62: 285–298.

Blaug, Mark. 1997. *Economic theory in retrospect*, 5th ed. Cambridge, UK: Cambridge University Press.

Blok, Kornelis, Worrell Ernst, R. Cuelenaere, et al. 1993. The cost effectiveness of CO_2 emission reduction achieved by energy conservation. *Energy Policy* 5: 355–371.

Bolin, Burt, Bo R. Doeoes, and Jill Jaeger. 1986. *The greenhouse effect, climatic change, and ecosystems*. Chichester, UK: John Wiley and Sons.

Boltzmann, Ludwig. 1872. Weitere Studien ueber das Warmegleichgewicht unter Gasmolekulen. *K. Academy. Wissenschaft. (Vienna)* II(66):275.

Bondi, Hermann. 1960. The steady-state theory of the universe. In *Rival Theories of Cosmology*, ed. W.B. Bonnor, H. Bondi, R.A. Lytleton, and J.G. Whitrow. Oxford: Oxford University Press.

Bondi, Hermann, and J. Gold. 1948. The steady-state theory of the expanding universe. *Monthly Notices of the Royal Astronomical Society* 108: 252–270.

Borgwardt, Robert H. 1999. Transport fuel from cellulosic biomass: A comparative assessment of ethanol and methanol options. *Proceedings of the Institute of Mechanical Engineers Part A: Journal of Power and Energy* 213(5): 399–407.

Boulding, Kenneth E. 1966. The economics of the coming spaceship Earth. In *Environmental Quality in a Growing Economy: Essays from the Sixth RFF Forum*, ed. Henry Jarrett. Baltimore, MD: Johns Hopkins University Press.

Breslow, R. 1959. On the mechanism of the formose reaction. *Tetrahedron Letters* 1(21): 22–26.

Briggs, Asa. 1982. *The power of steam*. Chicago: University of Chicago Press.

Briggs, Michael. 2004. *Widescale biodiesel production from algae*. UNH Biodiesel Group. http://www.unh.edu/p2/biodiesel/research_index.html

Brillouin, Leon. 1950. Thermodynamics and information theory. *American Scientist* 38: 594–599.

Brillouin, Leon. 1953. Negentropy principle of information. *Journal of Applied Physics* 24(9): 1152–1163.

Brillouin, Leon. 1962. *Science and information theory*, 2nd ed. New York: Academic Press.

Britton, Francis E. K. 2009. Car-sharing database. New Mobility. http://www.carshare.newmobility.org

Britton, Francis E. K. 2015. World Streets (accessed November 7).

Broadhurst, T.J., R.S. Ellis, D.C. Koo, and A.S. Szalay. 1990. Large-scale distribution of galaxies at the Galactic poles. *Nature* 343(6260): 726–728.

Broder, Jacqueline D., and J. Wayne Barrier. 1990. Producing fuels and chemicals from cellulosic crops. In *Advances in New Crops*, ed. J. Janick and J.E. Simon, 257–259. Portland, OR: Timber Press.

Brooks, Daniel R., and E.O. Wiley. 1986–1988. Evolution as entropy: Towards a unified theory of biology. In *Science and its conceptual foundations*, ed. David L. Hull. Chicago: University of Chicago Press.

Brown, Lester R. 1973-74. The next crisis? Food. *Foreign Policy* 13: 3–33.

Brown, Lester R. 1995. *Who will feed China? Wake-up call for a small planet*. New York: W. W. Norton.

Brown, Lester R. 2006a. *PLAN B 2.0. Rescuing a planet under stress and a civilization in trouble*. New York and London: WW Norton & Co.

Brown, Lester R. 2006b. Water tables falling and rivers running dry. In *Plan B 2.0: Rescuing a Planet Under Stress amd a Civilization in Trouble*, ed. Lester R. Brown. New York and London: W. W. Norton and Co.

Bryner, Jeanna. Car sharing skyrockets as gas prices soar. *US News and World Report*, July 11, 2008.

Bryson, Reid. 1974. A perspective on climate change. *Science* 184: 753–760.

Buchanan, J.W., and G. Tullock. 1962. Externality. *Economica* 29: 371–384.

Budyko, Mikhail I. 1956. *The energy balance of the Earth's surface (in Russian)*.

Budyko, Mikhail I. 1969. The effect of solar radiation variations on the climate of the earth. *Tellus* 21: 611–619.

Budyko, Mikhail I. 1974. *Climate and life*. New York: Academic Press.

Budyko, Mikhail I. 1977. *Climate change*. Washington, DC: American Geophysical Union.

Budyko, Mikhail I. 1988. *Global climate catastrophes*. New York: Springer-Verlag.

Bullis, Kevin. 2007a. Battery breakthrough. *Technology Review*, 26 February.

Bullis, Kevin. 2007b. Creating ethanol from trash. http://www.technologyreview.com/energy/18084/ (accessed February).

Burbidge, E. Margaret, Geoffrey Burbidge, Fred Hoyle, and William Howler. 1957. Synthesis of the elements in stars. *Reviews of Modern Physics* 29(4): 547–650.

Burke, James. 1978. *Connections*. Boston, MA: Little Brown and Company.

Butlerow, A. 1861. Formation synthetique d'une substance sucree. *Compte rendus de l'Academie des Sciences, Paris* 53: 145–147.

Cairns-Smith, A.G. 1984. *Genetic takeover and the mineral origin of life*. Cambridge, UK: Cambridge University Press.

Cairns-Smith, A.G. 1985. The first organisms. *Scientific American* 252(6): 90–100.

Calvin, Melvin. 1956. The photosynthetic cycle. *Bulletin de la Société de Chimie Biologique* 38 (11): 1233–1244.

Calvin, Melvin. 1969. *Chemical evolution: Molecular evolution toward the origin of living systems on the Earth and elsewhere*. Oxford: Clarendon Press.

Campbell, Colin J. 1997. *The coming oil crisis*. Brentwood, UK: Multi-Science Publishing and Petroconsultants.

Campbell, Colin J. 2003. *The essence of oil and gas depletion: Collected papers and excerpts*. Bretwood, UK: Multi-Science Publishing Co.

Campbell, Colin J., and Jean H. Laherrère. 1998. The end of cheap oil. *Scientific American* 278(3): 60–65.

Caratheodory, Constantin. 1976. Investigations into the foundations of thermodynamics. In *The Second Law of Thermodynamics*, ed. J. Kestin, 229–256. Dowden: Hutchinson and Ross.

Carhart, Steven C. 1979. *The least-cost energy strategy-technical appendix*. Pittsburgh, PA: Carnegie-Mellon University Press.

Carnahan, Walter, Kenneth W. Ford, Andrea Prosperetti, Gene I. Rochlin, Arthur H. Rosenfeld, Marc H. Ross, Joseph E. Rothberg, George M. Seidel, and Robert H. Socolow. 1975. *Efficient use of energy: A physics perspective*. New York: American Physical Society.

Carnot, Nicholas L. Sadi. 1826 [1878]. *Reflexions sur la puissance motrice du feu: et sur les machines propres à développer cette puissance*. Reprint ed. Paris: Gauthier-Villars, Imprimeur-Libraire, 1878. Miscellaneous. Original edition, 1826. Reprint, available as google ebook.

Carroll, Sean. 2012. *The particle at the end of the universe*. London: Oneworld Publications.

Carson, Rachel. 1962. *Silent spring*. Boston, MA: Houghton Mifflin Company.

Casten, Thomas R. 2009. Energy recovery. Personal communication Paris, August.

Casten, Thomas R., and Martin J. Collins. 2002. Co-generation and on-site power production: Optimizing future heat and power generation. *Technological Forecasting and Social Change* 6 (3): 71–77.

Casten, Thomas R., and Marty Collins. 2006. WADE DE economic model. In *World Survey of Decentralized Electricity*. Edinburgh, Scotland: The World Alliance for Decentralized Energy.

Casten, Thomas R., and Robert U. Ayres. 2007. Energy myth #8: The US energy system is environmentally and economically optimal. In *Energy and American Society: Thirteen Myths*, ed. B. Sovacool and M. Brown. New York: Springer.

Casten, Thomas R., and Philip F. Schewe. 2009. Getting the most from energy. *American Scientist* 97: 26–33.

Casti, John L. 1989. *Paradigms lost: Images of man in the mirror of science*. New York: W. Morrow and Company.

CDAC, Carbon Dioxide Assessment Committee. 1983. *Changing Climate: Report of the Carbon Dioxide Assessment Committee of the National Academy of Sciences*. Washington, DC: National Academy Press.

Chambers, D.P., J. Wahr, and R.S. Nerem. 2004. Preliminary observations of global ocean mass variations with GRACE. *Geophysical Research Letters* 31: Li3310.

Chandrasekhar, S. 1964. The dynamic instability of gaseous masses approaching the Schwartzschild limit in general relativity. *Astrophysical Journal* 140(2): 417–433.

Church, J.A., and N.J. White. 2011. Sea-level rise from the late 19th to the early 21st century. *Surveys in Geophysics* 32: 585–602.

Cleveland, Cutler J. 1999. Biophysical economics: From physiocracy to ecological economics to industrial ecology. In *Bioeconomics and Sustainability: Essays in honor of Nicholas Georgescu-Roegen*, ed. J. Gowdy and K. Mayumi, 125–154. Cheltenham: Edward Elgar.

Cleveland, Cutler J. 2010. *An assessment of the energy return on investment (EROI) of oil shale*. Boulder: Western Resource Advocates.

Cleveland, Cutler J., Costanza Robert, C.A.S. Hall, and Robert K. Kaufmann. 1984. Energy and the US economy: A biophysical perspective. *Science* 255: 890–897.

Coase, R.H. 1960. The problem of social costs. *Journal of Law and Economics* 3: 1–44.

Collier, J. 1986. Entropy in evolution. *Biology and Philosophy* 1: 5–24.

Comay, Eliyahu. 1984. Axiomatic deduction of equations of motion in classical electrodynamics. *Nuovo Cimento* 80B: 159.

Comay, Offer. 2014. *Science or fiction? The phony side of particle physics*. Tel Aviv, Israel: Dekel Academic Press.

Commissioner of Labor. 1898. *Hand and Machine Labor Annual Reports*. Washington, DC: United States Government Printing Office.

Conlisk, John. 1996. Why bounded rationality? *Journal of Economic Literature* XXXIV: 669–700.

Corliss, James B., and R.D. Ballard. 1978. Oases of life in the cold abyss. *National Geographic* 152: 441–453.

Costanza, Robert, and R.A. Herendeen. 1984. Embodied energy and economic value in the US economy. *Resources and Energy* 6: 129–163.

Costanza, Robert, and Herman E. Daly. 1992. Natural capital and sustainable development. *Conservation Biology* 6: 37–46.

Costanza, Robert, Ralph d'Arge, Rudolf de Groot, Stephen Farber, Monica Grasso, Bruce Hannon, Karin Limburg, Shahid Naeem, Robert V. O'Neill, Jose Paruelo, Robert G. Raskin, Paul Sutton, and Marjan van den Belt. 1997. The value of the world's ecosystem services and natural capital. *Nature* 387: 253–260.

Council for Agricultural Science and Technology; 1976. *Effect of increased nitrogen fixation on stratospheric ozone*. Ames, IO: Council for Agricultural Science and Technology.

Cournot, Antoine Augustin. 1838. *Recherches sur les principes mathematiques de la theorie des richesses*. Paris: Calmann-Lévy.

Criswell, David R. 2000. Lunar solar power: Review of the technology readiness base of an LSP system. *Acta Astronautica* 46(8): 531–540.

Criswell, David R. 2002. Energy prosperity within the 21st century and beyond: Options and the unique roles of the sun and the moon. In *Innovative Energy Strategies for CO_2 Stabilization*, ed. by Robert G. Watts. Cambridge, UK: Cambridge University Press.

Crooks, Gavin E. 1999. Entropy production fluctuation theorem and non-equilibrium work relation for free energy differences. *Physical Review E* 60: 2721.

Crutzen, Paul J. 1970. The influence of nitrogen oxides on the atmospheric ozone content. *Quarterly Journal of the Royal Meteorological Society* 96: 320–325.

Crutzen, Paul J. 1974. Estimates of possible variations in total ozone due to natural causes and human activities. *Ambio* 3: 201–210.

Crutzen, Paul, and E. Stoermer. 2000. The Anthropocene. *Global Climate Change Newsletter* 41: 17–18.

Csendes, T. 1984. A simulation study of the chemoton. *Kybernetes* 13(2): 79–85.

D'Errico, Emilio, Pierluigi Martini, and Pietro Tarquini. 1984. *Interventi di risparmio energetico nell'industria*. Italy: ENEA.

Dalgamo, A., and J.H. Black. 1976. Molecular formation in the interstellar gas. *Reports on Progress in Physics* 39(6).

Daly, Herman E. 1973. *Toward a steady state economy*. San Francisco: W. H. Freeman and Company.

Daly, Herman E. 1979. Entropy, growth and the political economy. In *Scarcity and Growth Reconsidered*, ed. V. Kerry Smith, 67–94. Baltimore: Johns Hopkins University Press.

Daly, Herman E. 1980. The economic thought of Frederick Soddy. In *History of Political Economy*, 469–488. Chapel-Hill, NC: Duke University Press.

Daly, Herman E. 1991. *Steady-state economics*, 2nd ed. Washington, DC: Island Press.

Daly, Herman E., and John Cobb. 1989. *For the common good*. Boston, MA: Beacon Press.

Darwin, Charles. 1859. *On the origin of species*. London: J. Murray.

Dasgupta, Partha, and Geoffrey Heal. 1974. The optimal depletion of exhaustible resources. In *Symposium on the Economics of Exhaustible Resources*.

Davis, Otto A., and A. Whinston. 1962. Externalities, welfare and the theory of games. *Journal of Political Economy* 70: 241–262.

Dawkins, Richard. 1976 [2006]. *The selfish gene*. Oxford: Oxford University Press.

Dawkins, Richard. 1982. *The extended phenotype*. San Francisco: W. H. Freeman and Company.

Dawkins, Richard. 1996. *The blind watchmaker*. New York: W. W. Norton.

de Lapparent, Valerie, Margaret Geller, and John P. Huchra. 1986. A slice of the universe. *Astrophysical Journal* 302: L1–L5 (Part 2—Letters to the Editor).

de Soet, Franz. 1985. Personal communication to R. U. Ayres.

Debreu, G. 1959. *Theory of value*. New York: John Wiley.

Deffeyes, Kenneth S. 2001. *Hubbert's peak*. Hardcover ed. Princeton, NJ: Princeton University Press.

Deffeyes, Kenneth S. 2005. *The view from Hubbert's peak*. Hardcover ed. New York: Hill and Wang.

den Elzen, Michel G.J., Arthur Bensen, and Jan Rotmans. 1995. *Modeling global biogeochemical cycles: An integrated modeling approach*. Bilthoven, The Netherlands: Global Dynamics and Sustainable Development Programme, National Institute of Public Health and Environment (RIVM).

Denison, Edward F. 1979. Explanations of declining productivity growth. *Survey of Current Business* 59(Part II): 1–24.

Descartes, Rene. 1633 [2000]. Le monde (description du corps humain). In *Oeuvres completes*, ed. Vrin. Paris: BTP Vrin. Original edition, 1633.

Descartes, Rene. 1644. *Principles of philosophy*. Dordrecht: D. Reidel.

Dewhurst, J. Frederick. 1947. *America's needs and resources*. New York: Twentieth Century Fund.

Dewhurst, J. F. Associates. 1955. *America's needs and resources: A new survey*. New York: Twentieth Century Fund.

Diamond, Jared. 1991. *The rise and fall of the third chimpanzee*. London: Radius.

Diamond, Jared. 1998. *Guns, germs and steel*. New York: Vintage Books.

Dietz, Simon, and Nicholas Stern. 2008. Why economic analysis supports strong action on climate change: A response to the Stern Review critics. *Review of Environment Economic Policy* 2: 94–113.

Dincer, Ibrahim, and Marc A. Rosen. 2007. *Exergy: Energy, environment and sustainable development*. Oxford, UK: Elsevier.

Dirac, Paul A.M. 1928. The quantum theory of the electron. *Proceedings of the Royal Society of London A: Mathematical, Physical and Engineering Sciences* 117(778): 610–624. doi:10. 1098/rspa.1928.0023.

Dirac, Paul A.M. 1948. The theory of magnetic poles. *Physical Review* 74: 817.

Dollo, L. 1893. Les lois d l'evolution. *Bull. Soc. Belg. Geol.* 7: 164–167.

Durack, Paul J., Susan E. Wijffels, and Richard J. Matear. 2012. Ocean water salinities reveal strong global water cycle intensification during 1950 to 2000. *Science* 336: 455–458.

Dyson, Freeman J. 1999. *Origins of life*. Revised ed. Cambridge, UK: Cambridge University Press.

Eco, Umberto. 1984. *The name of the rose*. Translated by William Weaver. Picador ed. London: Pan Books Ltd.

Eddington, Arthus S. 1920. The internal constitution of stars. *Nature* 106: 106.

Eddington, Arthur. 1926. *The internal constitution of stars*. Cambridge, UK: Cambridge University Press.

Eddington, Arthur. 1928. *The nature of the physical world*. Cambridge, UK: Cambridge University Press.

Edgeworth, F.Y. 1881. *Mathematical physics*. London: Kegan Paul.

Edstrom, William. 2015. Waiting for collapse: USA debt bombs bursting. Counterpunch: Telling facts and naming names (accessed September 28, 2015).

Ehrlich, Robert. 2015. Six observations consistent with the electron neutrino being a tachyon. *Journal of Cosmology and Astro-particle Physics* 66: 11–17.

Ehrlich, Paul R., Richard Holm, and Dennis Parnell. 1983. *The process of evolution*. New York: McGraw-Hill.

Eigen, Manfred. 1971. Self organization of matter and the evolution of biological macromolecules. *Naturwissenschaften* 58: 465–523.

Eigen, Manfred, and P. Schuster. 1979. *The hypercycle: A principle of natural self-organization*. Berlin: Springer-Verlag. Miscellaneous.

Einstein, Albert. 1905. Zur Elekrodynamik bewegter Koerper (On the electrodynamics of moving bodies). *Annalen der Physik* 17: 891.

Einstein, Albert. 1916. The foundation of the general theory of relativity. *Annalen der Physik* 49: 769–822.

Einstein, Albert. 1936. Physics and reality. *Journal of Franklin Institute* 221: 349.

Einstein, Albert. 1939. On a stationary system with spherical symmetry consisting of many gravitating masses. *Annals of Mathematics* 40(4): 922–936.

Einstein, Albert. 1956. *The meaning of relativity*, 5th ed. Princeton, NJ: Princeton University Press.

Eldredge, Niles, and Steven Jay Gould. 1972. Punctuated equilibria: An alternative to phyletic gradualism. In *Models in Paleobiology*, ed. T. J. M. Schopf, 82–115. San Francisco: Freeman, Cooper and Co.

Ellen MacArthur Foundation. 2013. *Towards the Circular Economy*, Cowes, vol. II. Isle of Wight, UK: Ellen MacArthur Foundation.

Ellen MacArthur Foundation, and McKinsey Center for Business and Environment. 2016. Growth within: A circular economy vision for a competitive Europe. Ellen MacArthur Foundation McKinsey Center for Business and Environment.

Elswijk, Marcel, and Henk Kaan. 2008. European embedding of passive houses. PEP project. www.aee-intec.at/0uploads/dateien578.pdf (last modified accessed January 15, 2008).

Energetics. 2004. *Energy loss reduction and recovery in industrial energy systems*. Washington, DC: US Department of Energy (DOE).

Energetics, and E3M. 2004. *Energy use, loss and opportunities analysis: US manufacturing and mining*. Washington, DC: US Department of Energy (DOE).

England, Jeremy. 2013. Statistical physics of self replication. *Journal of Chemical Physics* 139 (12): 121923.

Enos, John Lawrence. 1962. *Petroleum progress and profits, a history of process innovation*. Cambridge, MA: MIT Press.

Epstein, Paul, Jonathan J. Buonocore, Kevin Eckerle, Michael Hendrix, Benjamin M. Stout III, Richard Heinberg, Richard W. Clapp, Beverly May, Nancy L. Reinhart, Melissa M. Ahern, Samir K. Doshi, and Leslie Glustrom. 2012. The full cost of accounting for the life cycle of coal. *Annals of the New York Academy of Sciences* 1219: 73–98 (Ecological Economics Reviews). doi:10.1111/j.1749-6632.2010.05890.

European Commission. 2010. Critical Raw Materials for the European Union. European Commission.

Evans, D.A., R.I. Ripperdan, and Joseph Kirschvink. 1998. Polar wander and the Cambrian. *Science* 279: 9a–9e.

Fisher, I. 1933. The debt-deflation theory of the Great Depressions. *Econometrica* 1(4): 337–357.

Flem-Ath, Rand. 1993. www.flem-ath.com/does-the-earth's-crust-shift?/.

Forbes, R. J., and E. J. Dijksterhuis. 1963. *A history of science and technology 2: The 18th and 19th centuries*. Paperback ed. 2 vols, vol. 2. Harmondsworth, Middlesex, UK: Penguin Books.

Fouquet, Roger. 2011. Divergences in long-run trends in the price of energy and energy services. *Review of Environmental Economics and Policy* 5(2): 196–218.

Fouquet, Roger, and Peter J.G. Pearson. 2003. Five centuries of energy prices. *World Economics* 4 (3): 93–119.

Fox, Sydney. 1980. New missing links. *Science*: 18–21.

Fox, Sydney. 1988. *The emergence of life*. New York: Basic Books.

Frank, Louis A. (with Patrick Huyghe). 1990. *The big splash*. New York: Avon Books.

Frank, Louis A., J.B. Sigwarth, and C.M. Yeates. 1990. A search for small solar-system bodies near the earth using a ground-based telescope: Technique and observations. *Astronomy and Astrophysics* 228: 522.

Friedman, Milton. 1962. *Capitalism and freedom*. Chicago: University of Chicago Press.

Friedman, Milton. 1968. The role of monetary policy. *American Economic Review* LVIII: 1–17.

Friedman, Milton, and Anna Schwartz. 1963. *A monetary history of the United States, 1867-1960*. Princeton, NJ: Princeton University Press.

Friedmann, Alexander. 1922. Uber die Krummung des Raumes. *Zeitschrift zur Physik* 10: 377–386.

Friend, J. Newton. 1926. *Iron in antiquity*. London: Charles Griffin and Company, Ltd.

Fritzsch, Harald. 1984. *Quarks: The stuff of matter*. London: Pelican Books.

Fritzsch, Harald, and Murray Gell-Mann. 1972. Current algebra: Quarks and what else? In *Proceedings of the XVI International Conference on High Energy Physics*, ed. J. D. Jackson, and A. Roberts.

Galbally, I.E. 1985. The emission of nitrogen to the remote atmosphere: Background paper. In *The Biogeochemical Cycling of Sulfur and Nitrogen in the Remote Atmosphere*, ed. J.N. Galloway et al. Dordrecht, The Netherlands: D. Reidel Publishing Company.

Galloway, James N., William H. Schlesinger, H. Levy, A. Michaels, and J.L. Schnoor. 1995. Nitrogen fixation: Anthropogenic enhancement-environmental response. *Global Biogeochemical Cycles* 9(2): 235–252.

Gamow, George. 1947. *One, Two, Three ... Infinity*. New York: Viking.

Ganti, T. 1979. *A theory of biochemical supersystems and its application to problems of natural and artificial biogenesis*. Budapest, Baltimore: Akademiai Klado, University Park Press.

Ganti, Tibor. 2003. *Chemoton theory: Theory of living systems*. Dordrecht, NL: Kluwer Academic/ Plenum.

Garratt, G. R. M. 1958. Telegraphy. In *The Industrial Revolution*, ed. Charles Singer, E. J. Holmyard, A. R. Hall and Trevor I. Williams. London: Oxford University Press.

Garrett, Timothy J. 2011. Are there physical constraints on future anthropogenic emissions of carbon dioxide? *Climate Change* 3: 437–455.

Gates, David M. 1993. *Climate Change and its biological consequences*. Sunderland, MA: Sinauer Associates Inc.

Gell-Mann, Murray. 1964. A schematic model of baryons and mesons. *Physics Letters* 8(3): 214–215.

Geller, Margaret, and John P. Huchra. 1989. Mapping the universe. *Science* 246(4932): 897–903. doi:10.1126/science.246.4932.897.

Georgescu-Roegen, Nicholas. 1971. *The entropy law and the economic process*. Cambridge, MA: Harvard University Press.

Gerlach, T.M. 1991. unknown. *Eos* 72: 249–255.

Gerland, P., A.E. Raftery, H. Elkova, N. Li, T. Gu, D. Spoorenberg, L. Alkema, B.K. Fosdick, J. Chunn, N. Lalic, G. Bay, T. Buettner, G.K. Heilig, and J. Wilmoth. 2014. World population stabilization unlikely this century. *Science* 346: 234–237.

Gibbs, Josiah Willard. 1948. *The collected works of. J. Willard Gibbs*, vol. 1. New Haven, CT: Yale University Press.

Gilbert, Walter. 1986. The RNA world. *Nature* 319: 618.

Glasby, G.P. 2005. Abiogenic origin of hydrocarbons: An historical overview. *Resource Geology* 56(1): 83–96.

Gleick, J. 1987. *Chaos: Making a new science*. New York: Viking Press.

Goeller, Harold, and Alvin Weinberg. 1976. The age of substitutability. *Science* 191: 683–689.

Gold, Thomas. 1993. *The origin of methane (and oil): The origin of energy gases*. Washington: USGS.

Gold, Thomas. 1999. *The deep hot biosphere: The myth of fossil fuels*. New York: Copernicus Books.

Goldanskii, Vitalii I. 1977. Interstellar grains as possible cold seeds of life. *Nature* 269: 583–584. doi:10.1038/269583a0.

Goldanskii, Vitalii I. 1979. Facts and hypotheses of molecular chemical tunneling. *Nature* 279: 109–115. doi:10.1038/279109a0.

Goldanskii, Vitalii I. 1997. *Non-traditional pathways of solid phase astrochemical reactions*, vol. 69. Pure and Applied Chemistry.

Goldanskii, Vitalii I., M.D. Frank-Kamenetskii, and I.M. Barkalov. 1973. Quantum low-temperature limit of a chemical reaction rate. *Science* 182(4119): 1344–1345.

Goldemberg, Jose, Thomas B. Johansson, Amulya K.N. Reddy, and Robert H. Williams. 1988. *Energy for a sustainable world*. New York: John Wiley and Sons.

Gordon, Robert J. 2012. Is US economic growth over? Faltering innovation confronts the six headwinds. CEPR No. 63. Centre for Economic Policy Research. www.cepr.org

Gould, Stephen Jay. 2002. *The structure of evolutionary theory*. Cambridge, MA: Harvard University Press.

Gould, Stephen Jay. 2007. *Punctuated equilibrium*. Cambridge, MA: Harvard University Press.

Gould, Stephen Jay, and Eldredge Niles. 1972. Punctuated equilibrium: The tempo and mode of evolution reconsidered. *Paleobiology* 3: 115–151.

Graeber, David. 2011. *Debt: The first 5000 Years*. New York: Melville House Publishing.

Graedel, Thomas E., and Lorenz Erdmann. 2011. Criticality of non-fuel minerals: A review of major approaches and analyses. *Environmental Science & Technology* 45(18): 7620–7630.

Green, Sheldon. 1981. Interstellar Chemistry: Exotic molecules in space. *Annual Reviews of Physical Chemistry* 32: 103–138.

Greenberg, J. Mayo. 2002. Cosmic dust and our origins. *Surface Science 500*: 793–822.

Gruebler, Arnulf. 1998. *Technology and global change*. Cambridge, UK: Cambridge University Press.

Guth, Alan. 1981. The inflationary universe: A possible solution to the Horizon and Flatness problems. *Physical Review* D23: 347.

Guth, Alan, and J. Steinhard. 1984. The inflationary universe. *Scientific American* 250(5): 116–128.

Haag, Gunter, W. Weidlich, and Gerhart Mensch. 1988. The Schumpeter clock. In *Economic Development and Structural Adjustment*, ed. G. Haag. New York: Springer.

Haeckel, Ernst. 1866. *Generelle Morphologie der Organismen*. Berlin: G. Reimer.

Haldane, J.B.S. 1985. *On being the right size and other essays*. London: Oxford University Press.

Hall, Charles A.S., Jessica Lambert, and Stephen B. Balogh. 2014. EROI of different fuels and the implications for society. *Energy Policy* 64: 141–152.

Hamilton, James D. 2003. What is an oil shock? *Journal of Econometrics* 113: 363–398.

Hamilton, James D. 2005. Oil and the macroeconomy. In *The New Palgrave: A Dictionary of Economics*, ed. J. Eatwell, M. Millgate, and P. Newman. London: Macmillan.

Hamilton, James D. 2009. Causes and consequences of the oil shock of 2007-08. *Brookings Papers on Economic Activity*.

Hannon, Bruce M. 1973. The structure of ecosystems. *Journal of Theoretical Biology* 41: 535–546.

Hannon, Bruce M. 1975. Energy conservation and the consumer. *Science* 189: 95–102.

Hannon, Bruce R. 2010. The role of input-output analysis of energy and ecological sustainability. *New York Academy of Sciences* 1185: 30–38.

Hannon, Bruce M., and John Joyce. 1981. Energy and technical progress. *Energy* 6: 187–195.

Hannon, Bruce R., Robert Costanza, and Robert E. Ulanowicz. 1991. A general accounting framework for ecological systems: A functional taxonomy for connectivist ecology. *Theoretical Population Biology* 40: 78–104.

Hansen, James, A. Lacis, D. Rind, G. Russell, P. Stone, J. Fung, R. Ruedy, and J. Lerner. 1984. Climate sensitivity analysis of feedback mechanisms. In *Geophysical Monograph 29*, ed. Maurice Ewing. New York: American Geophysical Union.

Hansen, James, G. Russell, A. Lacis, I. Fung, D. Rind, and P. Stone. 1985. Climate response times: Dependence on climate sensitivity and ocean mixing. *Science* 229: 582–589.

Hansen, James, M. Sato, P. Kharecha, D. Beerling, R. Berner, and V. Masson-Delmotte. 2008. Target atmospheric CO_2: Where should humanity aim? *Open Atmospheric Science Journal* 2: 217–231.

Hansen, James, M. Sato, G. Russell, and P. Kharecha. 2013. Climate sensitivity, sea level and atmospheric carbon dioxide. *Philosophical Transactions of the Royal Society A* 371: 20210294.

Hapgood, Charles. 1958. *The Earth's shifting crust*. New York: Pantheon Books.

Haramein, Nassim. 2016. Personal communication to R. U. Ayres. January.

Hardin, Garett. 1968. The tragedy of the commons. *Science* 162: 1243–1248.

Hargraves, Robert. 2012. *Thorium: Energy cheaper than coal*. Hanover, NH: Robert Hargraves.

Hargraves, Robert, and Ralph Moir. 2010. Liquid fluoride thorium reactors. *American Scientist* 98: 304–313.

Harman, Denham. 1956. Aging: A theory based on free-radical and radiation chemistry. *Journal of Gerontology* 11(3): 298–300.

Harman, Denham. 1972. A biological clock: The mitochondria? *Journal of the American Geriatrics Society* 20(4): 145–147.

Hart, Pembroke J. (ed.). 1969. *The Earth's crust and upper mantle*, Geophyisical monographs, vol. 15. Washington, DC: American Geophysical Union.

Hatfield, Craig Bond. 1997. *How long can oil supply grow?* Golden, CO: M. King Hubbert Center for Petroleum Supply Studies, Colorado School of Mines.

Hatsopoulos, G., and J. Keenan. 1965. *Principles of general thermodynamics*. New York: John Wiley & Co.

Hawke, David Freeman. 1988. *Nuts and bolts of the past: A history of American technology 1776-1860*. New York: Harper and Row.

Hayek, Friedrich August. 1960. *The constitution of liberty*. Chicago: University of Chicago Press.

Hays, J.D., J. Imbrie, and N.J. Shackleton. 1976. Variations in the earth's orbit: "Pacemaker of the ice ages". *Science* 194(4270): 1121–1132.

Heinberg, Richard. 2009. *Blackout: Coal, climate and the last energy crisis*. Santa Rosa, CA: New Society Publishers.

Helm, Georg. 1887. *Die Lehre von der Energie*. Leipzig: Felix.

Herbst, Eric, and William Klemperer. 1973. The formation and depletion of molecules in dense interstellar clouds. *Astrophysical Journal* 185: 505.

Herendeen, Robert. 1990. System level indicators in dynamic ecosystems: A comparison based on energy and nutrient flows. *Journal of Theoretical Biology* 143: 523–553.

Herring, Horace, and Steve Sorrell. 2009. Energy efficiency and sustainable consumption. In *Energy, Climate and the Environment*, ed. David Elliott. London: Palgrave Macmillan.

Hidalgo, Cesar. 2015. *Why information grows: From atoms to economies*. New York: Basic Books.

Hinkle, Peter C., and Richard E. McCarty. How cells make ATP. *Scientific American*, March 1978.

Hinterberger, Friedrich, and Friedrich B. Schmidt-Bleek. 1999. Dematerialization, MIPS and factor 10: Physical sustainability indicators as a social device. *Ecological Economics* 29(1): 53–56.

Hirata, H., and R.E. Ulanowicz. 1984. Information theoretical analysis of ecological networks. *International Journal of Systems Science* 15: 261–270.

Hoffmann, Roald. 2006. Old gas, new gas. *American Scientist* 94(1): 16–18.

Holland, H.D. 1978. *The chemical evolution of the atmosphere and the oceans*. New York: John Wiley.

Holling, C.S. 1973. Resilience and stability of ecological systems. *Annual Review of Ecology and Systematics* 4: 1–24.

Holling, C. S. 1986. The resilience of terrestrial ecosystems: Local surprise and global change. In *Sustainable Development of the Biosphere*, ed. William C. Clark and R. E. Munn. Cambridge, UK: Cambridge University Press.

Holling, C.S. 1996. Engineering resilience versus ecological resilience. In *Engineering within Ecological Constraints*, ed. P. Schultze. Washington, DC: National Academy Press.

Holmes, Arthur. 1913. *The age of the Earth*. London: Harper Brothers.

Holwell, D.A., and I. McDonald. 2007. Distribution of platinum group elements in the Platreef at Overysel, northern Bushveld Complex: A combined PGM and LA-CP-MS study. *Contributions to Mineralogy and Petrology* 154(2): 171–190.

Hotelling, H. 1931. The economics of exhaustible resource. *Journal of Political Economy* 39: 137–175.

Houghton, J.T., B.A. Callander, and S.K. Varney. 1992. *Climate change 1992: The supplementary report to the IPCC scientific assessment*. Cambridge, UK: Cambridge University Press.

Hounshell, David A. 1984. *From the American system to mass production, 1800-1932: The development of manufacturing technology in the United States*, vol. 4. Baltimore, MD: Johns Hopkins University Press.

Hoyle, Fred. 1947. On the formation of heavy elements in stars. *Journal of the Physical Society (London)* 49: 942–948.

Hoyle, Fred. 1950. *The nature of the universe*. New York: Harper and Brothers.

Hoyle, Fred. 1954. Synthesis of the elements between carbon and nickel. *Astrophysical Journal Supplement* 1: 121.

Hoyle, Fred. 1981. *Ice, the ultimate human catastrophe*. New York: Continuum.

Hoyle, Fred, and N. Chaudra Wickramasinghe. 1978. *Life cloud*. New York: Harper and Row.

Hubbert, M. King. 1956. *Nuclear energy and the fossil fuels*. Houston, TX: Shell Development Corporation.

Hubbert, M. King. 1962. *Energy resources: A report to the Committee on Natural Resources of the National Academy of Sciences—National Research Council*. Washington, DC: National Research Council/National Academy of Sciences.

Hubbert, M. King. 1973. Survey of world energy resources. *The Canadian Mining and Metallurgical Bulletin* 66(735): 37–54.

Hubbert, M. King. Oil and gas supply modeling. In *Proceedings of a symposium held at the Department of Commerce*, Washington DC, May 1980.

Hughes, J. David. 2010. *Hydrocarbons in North America, Post Carbon Read Series*. Santa Rosa, CA: Post Carbon Institute.

Huxley, J.S. 1956. Evolution, cultural and biological. In *Current Anthropology*, ed. Thomas, 3–25. Chicago: University of Chicago Press.

Ibison, Michael. 2006. Cosmological test of the Yilmaz theory of gravity. *Classical and Quantum Gravity* 23: 577–589.

IEA, International Energy Agency. 2007. *Tracking industrial energy efficiency and CO_2 emissions*. Paris: IEA/OECD.

IEA, International Energy Agency. 2009. *Energy technology transitions for industry: Strategies for the next industrial revolution*. Paris: OECD/IEA.

IPCC. 1995. The science of climate change: Contribution of working group I. In *2nd Assessment Report of the Intergovernmental Panel On Climate Change*. Cambridge, UK: Cambridge University Press.

IPCC. 2007. *IPCC Fourth Assessment Report: Climate Change 2007 (AR4)*. 4 vols. Geneva: Intergovernmental Panel on Climate Change.

IPCC. 2013. *Constraints on long-term climate change and the equilibrium climate sensitivity*. Geneva.

IPCC. 2014. *IPCC Fifth Assessment Report: Climate change*. Cambridge: Cambridge University Press.

Ivanhoe, L.F. 1996. Updated Hubbert curves analyze world oil supply. *World Oil* 217(11): 91–94.

Jacob, Daniel. 2002. *Atmospheric Chemistry*. Princeton: Princeton University Press.

Jacquard, Mark. 2005. *Sustainable fossil fuels: The unusual suspect in the quest for clean and enduring energy*. Cambridge UK: Cambridge University Press.

Jaffe, Robert, Jonathan Price, Murray Hitzman, and Francis Slakey. 2011. Securing Materials for Emerging Technologies. Amierican Physical Society (APS) panel on Public Affairs and Materials Research Society (NRS).

Jantsch, Erich. 1980. *The self-organizing universe: Scientific and human implications of the emerging paradigm of evolution*. Oxford, UK: Pergamon Press.

Jarrett, Henry (ed.). 1954. *A nation looks at its resources*. Baltimore, MD: Resources for the Future Inc and Johns Hopkins University Press.

Jarzynski, Christopher. 1997. Equilibrium free energy differences from non-equilibrium measurements—A master equation approach. *Physical Review E* 56: 5018–5035.

Jaynes, Edwin T. 1957a. Information theory and statistical mechanics, I. *Physical Review* 106: 620.

Jaynes, Edwin T. 1957b. Information theory and statistical mechanics, II. *Physical Review* 108: 171.

Jevons, William Stanley. 1865 [1974]. The coal question: Can Britain survive? *Environment and Change* (extracts from 1865 original).

Johansson, Thomas B., Nebojsa Nakicenovic, Anand Patwardha, and Luis Gomez-Echeverri (eds.). 2012. *Global energy assessment: Toward a sustainable future*. Cambridge, UK: Cambridge University Press.

Joly, John. 1895. Radioactivity and earth heating. *Proceedings of the Royal Society?*

Joosten, Louis. The industrial metabolism of plastics: Analysis of material flows, energy consumption and CO_2 emissions in the lifecycle of plastics. PhD thesis, Science, Technology and Society Department, University of Utrecht, 2001.

Kahn, Herman. 1971. *The emerging Japanese superstate*. New York: Prentice-Hall.

Kahneman, Daniel, and Amos Tversky. 1979. Prospect Theory: An analysis of decision under risk. *Econometrica* 47: 263–291.

Kaku, Michio, and Jennifer Thompson. 1997. *Beyond Einstein: The cosmic quest for a theory of the universe*. Oxford: Oxford University Press.

Kander, Astrid, Paolo Malanima, and Paul Warde. 2013. Power to the People. In *Princeton Economic History of the Western World*, ed. Joel Mokyr. Princeton, NJ: Princeton University Press.

Kasting, James F., and James C. G. Walker. 1992. The geophysical carbon cycle and the uptake of fossil fuel CO_2. AIP Conference Proceedings 247(1):175–200. DOI:10.1063/1.41927

Kauppi, P., K. Mielikainen, and K. Kuusula. 1992. Biomass and carbon budget of European forests; 1971 to 1990. *Science* 256: 311–314.

Kavalov, Boyan, and S.D. Peteves. 2007. *The future of coal*. Petten, the Netherlands: Institute for Energy, Joint Research Center, European Commission.

Kay, J.J. 1984. *Self-organization in living systems*. Ontario, Canada: Systems Design Engineering, University of Waterloo.

Keen, Steve. 1995. Finance and economic breakdown; Modeling Minsky's financial instability hypothesis. *Journal of Post-Keynesian Economics* 17(4): 607–635.

Keen, Steven. 2001. *Debunking economics: The naked emperor of the social sciences*. Australia: Pluto Press.

Keen, Steve. 2011a. *Debunking economics*. London: Zed Books.

Keen, Steve. 2011b. A monetary Minsky model of the Great Moderation and the Great Depression. *Journal of Economic Behavior and Organization* 86: 221–235.

Kennedy, Paul. 1989. *The rise and fall of the great powers*. New York: Vintage Books.

Kennett, James P., Kevin G. Cannariato, Ingrid L. Hendy, and Richard J. Behl. 2003. *Methane hydrates in quaternary climate change: The clathrate gun hypothesis*. Washington, DC: American Geophysical Union.

Kenney, W.F. 1984. Energy conservation in the process industries. In *Energy Science and Engineering: Resources, Technology, Management*, ed. Jesse Denton. New York: Academic Press.

Kesler, Stephen E. 1994. *Mineral resources, economics and the environment*. New York: Macmillan Co.

Kestin, J. 1966. *A course in thermodynamics*. Waltham, MA: Blaisdell.

Keynes, John Maynard. 1935. *The general theory of employment, interest and money*. London: MacMillan Company.

Keys, David. 1999. *Catastrophe: An investigation into the origins of the modern world*. London: Century Books, Random House UK.

Khazzoom, J. Daniel. 1980. Economic implications of mandated efficiency standards for household appliances. *Energy Journal* 1(4): 21–39.

Khazzoom, J. Daniel. 1987. Energy savings resulting from the adoption of more efficient appliances. *Energy Journal* 8(4): 85–89.

Kindleberger, Charles P. 1989. *Economic laws and economic history*. Paperback ed. New York: Cambridge University Press.

Kindleberger, Charles P., and Robert Z. Aliber. 1978 [1989]. *Manias, panics and crashes: A history of financial crises*. Second (revised) ed. New York: Basic Books.

Kirschvink, Joseph, and T.D. Raub. 2003. A methane fuse for the Cambrian explosion? *Geoscience* 335: 65–78.

Kirschvink, Joseph L., E.J. Gaidos, L.E. Bertani, N.J. Beukes, J. Gutzmer, L.N. Maepa, and R.E. Steinberger. 2000. Paleoproterozoic snowball Earth: Extreme climatic and geochemical global change and its biological consequences. *Proceedings of the National Academy of Sciences* 97(4): 1400–1405.

Kneese, Allen V. 1977. *Economics and the environment*. New York: Penguin Books.

Kneese, Allen V., Robert U. Ayres, and Ralph C. D'Arge. 1970. *Economics and the environment: A materials balance approach*. Baltimore, MD: RFF Press distributed by Johns Hopkins University Press.

Koene, Coerneille-Jean. 1856, 2004. *The chemical constitution of the atmosphere from Earth's origin to the present, and its implications for protection of industry and ensuring environmental quality*. Lewiston, NY: Edwin Mellen Press.

Kooijman, S.A.L.M. 2000. *Dynamic energy and mass budgets in biological systems*. Cambridge, UK: Cambridge University Press.

Kooijman, S.A.L.M. 2010. *Dynamic energy budget theory for metabolic organization*. Cambridge, UK: Cambridge University Press.

Kopits, Steven. 2013. *Oil and economic growth: A supply constrained view*. New York: Douglas-Westwood Inc.

Koschmieder, E. Lothar. 1993. *Benard cells and Taylor vortices*. Cambridge, UK: Cambridge University Press.

Kotsanas, Kostas. 2011. *Ancient Greek technology: The inventions of the ancient Greeks*. Pyrgos: Lostas Kotsanas.

Kranzberg, Melvin, and Carroll W. Pursell Jr. (eds.). 1967. *Technology in Western Civilization: Volume II Technology in the 20th century*. 2 vols. New York: Oxford University Press.

Krauss, Ulish H., Henning G. Saam, and Helmut W. Schmidt. 1984. Phosphate. In *International Strategic Minerals Inventory: Summary Reports 930 series*. Washington, DC: USGS.

Kudryavtsev, N.A. 1959. Geological proof of the deep origin of petroleum (in Russian). *Trudy Vsesoyuz. Neftyan. Nauch. Issledovatel Geologoraz Vedoch. Inst.* 132: 242–262.

Kuemmel, Reiner. 1980. Growth dynamics in the energy dependent economy. In *Mathematical Systems in Economics*, ed. Wolfgang Eichhorn and R. Henn, vol. 54. Cambridge, MA: Oeigeschlager, Gunn & Hain.

Kuemmel, Reiner. 1982. Energy, environment and industrial growth. In *Economic Theory of Natural Resources*. Würzburg, Germany: Physica-Verlag.

Kuemmel, Reiner. 1986. Energy, technical progress and economic growth. In *Energy Decisions for the Future*, ed. M. Miyata and K. Matsui, 1005–1022. Tokyo: Institute of Energy Economics.

Kuemmel, Reiner. 1989. Energy as a factor of production and entropy as a pollution indicator in macroeconomic modeling. *Ecological Economics* 1: 161–180.

Kuemmel, Reiner. 2011. *The Second Law of Economics: Energy, entropy and the origins of wealth (The Frontiers Collection)*. London: Springer.

Kuemmel, Reiner, Wolfgang Strassl, Alfred Gossner, and Wolfgang Eichhorn. 1985. Technical progress and energy dependent production functions. *Journal of Economics* 45(3): 285–311.

Kuemmel, Reiner, Julian Henn, and Dietmar Lindenberger. 2002. Capital, labor, energy and creativity: Modeling innovation diffusion. *Structural Change and Economic Dynamics* 13: 415–433.

Kuemmel, Reiner, Robert U. Ayres, and Dietmar Lindenberger. 2010. Thermodynamic laws, economic methods and the productive power of energy. *Journal of Non-Equilibrium Thermodynamics* 35: 145–181.

Kumhoff, Michael. 2012. *Modelling challenges for the near future: Income inequality, financial systems and exhaustible resources*. Washington, DC.

Kuuijman, S. A. L. M. 1993 [2009]. *Dynamic energy budget theory*. Cambridge: Cambridge University Press. Original edition, 1993.

Laane, Jaan. 2009. Frontiers of molecular spectroscopy. In *Frontiers of Molecular Spectroscopy*, ed. Jaan Laane. Boston: Elsevier.

Laherrere, Jean. 2014. *Oil and Gas Perspectives in the 21st century*, Europe London Campus. ASPO France.

Landau, Lev Davidovich, and E. M. Lifshitz. 1973. *The classical theory of fields*, vol. 2. Oxford: Butterworth and Heinemann. Original edition, first Russian edition, 1951.

Landes, David S. 1969. *The unbound Prometheus: Technological change and industrial development in western Europe from 1750 to the present*. Cambridge, UK: Cambridge University Press.

Landsberg, Hans H., Leonard L. Fischman, and Joseph L. Fisher. 1962. *Resources in America's future*. Baltimore, MD: Johns Hopkins University Press.

Lavoisier, Antoine Laurent de. 1783. *Réflexions sur le phlogistique, pour servir de développement à la théorie de la combustion & de la calcination*. Paris: Académie royale des sciences (France). Original edition, Le texte de 1783 ne sera publié dans les Mémoires de l'Académie royale des sciences qu'en 1786.

Leakey, Richard, and Roger Lewin. 1995. *The sixth extinction: Patterns of life and the future of humankind*. New York: Doubleday.

Leaton, James, Nicola Ranger, Bob Ward, Luke Sussams, and Meg Brown. 2013. *Unburnable carbon 2013: Wasted capital and stranded assets*. London: Carbon Tracker.org. Grantham Research Institute on Climate Change and the Environment, LSE.

Lederman, Leon. 1993. *The God particle*. New York: Houghton-Mifflin Co.

Lees, Andrew. 2009. The right game: Treating the disease rather than the symptoms. self. http://www.ritholz.com/blog/2009/07/andy-lees/

Lemaitre, Georges. 1950. *The Primeval Atom: An essay on cosmology*. Translated by Betty H and Serge A. Korff. New York: D. Van Nostrand & Co.

Lenton, Timothy M., and Andrew Watson. 2011. *Revolutions that made the Earth*. Oxford: Oxford University Press.

Lenton, Timothy M., Hermann Held, Elmar Kriegler, James W. Hall, Wolfgang Lucht, Stefan Rahmstorf, and Hans Joachim Schellnhuber. 2008. Tipping elements in the Earth's climate system. *Proceedings of the National Academy of Sciences* 105(6): 1786–1793.

Lerner, Eric J. 1991. *The big bang never happened*. New York: Random House.

Levinson, Herbert S. 2003. Bus Rapid Transit on city streets: How does it work? In *Second Urban Street Symposium*, Anaheim, CA.

Lewis, Michael (ed.). 2007. *The real price of everything; Rediscovering the six classics of economics*. New York: Sterling.

Lewis, Michael. 2008. *Panic: The story of modern financial insanity*. New York: W.W. Norton, Penguin.

Lewis, Michael. 2010. *The big short*. New York: W.W. Norton, Penguin.

Lewis, Michael. 2011. *Boomerang: The melt-down tour*. New York: W.W. Norton, Penguin.

Lewis, G.N., and M. Randall. 1923. *Thermodynamics*. New York: McGraw-Hill.

Lindeman, Raymond L. 1942. The trophic-dynamic aspect of ecology. *Ecology* 23: 399–418.

Lindzen, Richard S., and Ming-Dah Chou. 2001. Does the Earth have an adaptive IR "iris"? *Bulletin of the American Meteorological Society* 82(3): 417–432.

Lindzen, Richard S., and Yong-Sang Choi. 2011. On the observational determination of climate sensitivity and its implications. *Asia-Pacific Journal of Atmospheric Science* 47(4): 377–390.

Linnaeus, Carolus. 1735. *Systema Naturae*. Leiden: Haak.

Linnhof, B., and D. R. Vredevelt. 1984. Pinch Technology has come of age. *Chemical Engineering Progress* 80: 33–40.

Linnhof, B., and S. Ahmad. 1989. SUPERTARGETING; Optimal synthesis of energy management systems. *ASME Journal of Energy Resources Technology* 111: 121–130.

Liu, S.C., and R.J. Cicerone. 1984. *Fixed nitrogen cycle*, Global Tropospheric Chemistry, 113–116. Washington, DC: National Academy Press.

Lollar, Barbara Sherwood, Georges Lacrampe-Couloume, et al. 2006. Unravelling biogenic and non-biogenic sources of methane in the Earth's deep subsurface. *Chemical Geology* 226(3–4): 328–339.

Lorentz, Hendrik A. 1904. Electromagnetic phenomena in a system moving with any velocity less than that of light. *Proceedings of the Royal Academy of Sciences at Amsterdam* 4: 669–678.

Lotka, Alfred J. 1910. Contribution to the theory of periodic reaction. *Journal of Physical Chemistry* 14(3): 271–274.

Lotka, Alfred J. 1922. Contribution to the energetics of evolution. *Proceedings of the National Academy of Sciences* 8: 147.

Lotka, Alfred J. 1925. *Elements of mathematical biology*. 2nd Reprint ed. Baltimore: Williams and Wilkins. Original edition, 1924.

Lotka, Alfred J. 1945. The law of evolution as a maximal principle. *Human Biology* 17(3): 167–194.

Lotka, Alfred J. 1956. *Elements of mathematical biology*. New York: Dover.

Loulergue, L., F. Parrenin, T. Bluniert, J-M Barnola, R. Spahni, A. Schilt, G. Raisbeck, and J. Chappellaz. 2007. New constraints on the gas age-ice age difference along the EPICA ice cores, 0–50 kyr. *Climate of the Past* 3: 527–540. doi:10.5194/cp-3-527-2007.

Lovelock, James E. 1972. Gaia as seen through the atmosphere. *Atmospheric Environment* 6: 579–580.

Lovelock, James E. 1979. *Gaia: A new look at life on earth*. London: Oxford University Press.

Lovelock, James E. 1988. *The ages of Gaia: A biography of our living earth*. London: Oxford University Press.

Lovelock, James, and Lynn Margulis. 1974. Atmospheric homeostasis by and for the biosphere: The Gaia hypothesis. *Tellus* 26(1): 2–10.

Lovins, Amory B. 1996. Hypercars: The next industrial revolution. In *13th International Electric Vehicle Symposium (EVS 13)*, Osaka, Japan, October 14.

Lovins, Amory. 2011. *Reinventing fire*. White River Junction, VT: Chelsea Green Publishing Co.

Lovins, Amory B., L. Hunter Lovins, Florentin Krause, and Wilfred Bach. 1981. *Least-cost energy: Solving the CO_2 problem*. Andover, MA: Brickhouse Publication Co.

Lovins, Amory B., Michael M. Brylawski, David R. Cramer, and Timothy C. Moore. 1996. *Hypercars: Materials, manufacturing, and policy implications*. Snowmass, CO: The Hypercar Center, Rocky Mountain Institute.

Lovins, Amory et al. 2014. *The Economics of Grid Defection*. Washington, DC: Rocky Mountain Institute (RMI).

Lucas Jr., Robert E. 1988. On the mechanics of economic development. *Journal of Monetary Economics* 22(1): 2–42.

Luiten, Esther. 2001. Beyond energy efficiency: Actors, networks and government intervention in the development of industrial process technologies. PhD, Department of Science, Technology and Society, Utrecht University.

Lung, Robert Bruce, Aimee McKane, and Mitch Olzewski. 2003. *Industrial motor systems optimization projects in the US: An impact study*. Washington, DC: Lawrence Berkely National Laboratory.

Lunt, Daniel J., Alan M. Haywood, Gavin A. Schmidt, Ulrich Salzmann, Paul J. Valdes, and Harry J. Dowsett. 2010. Earth system sensitivity inferred from Pliocene modeling and data. *Nature Geoscience* 3: 60–64.

Lutz, Wolfgang, Warren Sanderson, and Sergei Scherbov. 2004. *The end of population growth in the 21st Century*. Laxenburg, Austria: IIASA and Earthscan.

Lyell, Charles. 1838. *Elements of geology or the ancient changes of Earth and its inhabitants a illustrated by geological monuments*, 1st ed. 1 vols. London: John Murray. Original edition, 1838.

MacArthur, R.H., and E.O. Wilson. 1963. An equilibrium theory of insular zoo-geography. *Evolution* 17: 373–387.

Mackay, Charles. 1841 [1996]. *Extraordinary popular delusions and the madness of crowds, Wiley Investment Classics*. New York: John Wiley & Sons Inc. Original edition, 1841. Reprint, 1996.

Mackay, David J.C. 2008. *Sustainable Energy—Without the Hot Air*. Cambridge: UIT.

Maddison, Angus. 1993. Standardized estimates of fixed capital stock. In *Essays on Innovation, Natural Resources and the International Economy*, ed. R. Zoboli. Ravenna, Italy: Studio AGR.

Malthus, Thomas Robert. 1798 [1946]. An essay on the principle of population as it affects the future improvement of society. In *Masterworks of Economics: Digest of Ten Great Classics*, ed. Leonard Dalton Abbott, 191–270. New York: Doubleday and Company, Inc. Original edition.

Malthus, Thomas Robert. 1798 [2007]. An essay on the principle of population. In *The Real Price of Everything*, ed. Michael Lewis, 653–659. New York: Sterling. Original edition, 1776.

Mankiw, N. Gregory. 1997. *Macroeconomics*. London: Worth Publishing.

Mann, Michael E., Stefan Rahmstorf, Byron A. Steinman, Martin Tingley, and Sonya K. Miller. 2016. The likelihood of recent record warmth. *Scientific Reports* 6: 19831. doi:10.1038/srep19831. http://www.nature.com/articles/srep19831#supplementary-information.

Margalef, Ramon. 1968. *Perspectives in ecological theory*. Chicago: University of Chicago Press.

Margulis, Lynn. 1970. *Origin of eukaryotic cells*. New Haven, CT: Yale University Press.

Margulis, Lynn. 1981. *Symbiosis in cell evolution*. San Francisco: W. H. Freeman and Company.

Marshall, Alfred. 1930. *Principles of economics*, 8th ed. London: Macmillan.

Martinez-Boti, M.A., Gavin L. Foster, T.B. Chalk, E.J. Rohling, P.F. Sexton, Daniel J. Lunt, R.D. Pancost, M.P.S. Badger, and D.N. Schmidt. 2015. Plio-Pleistocene climate sensitivity evaluation using high resolution CO_2 records. *Nature* 5: 2.

Marx, Karl. 1867. *Das Kapital*. German ed. 2 vols.

Max, Michael D. 2003. *Natural gas hydrate in oceanic and permafrost environments*. Rotterdam: Kluwer.

Maxwell, James Clark. 1871. *Theory of heat*. London: Longmans, Green, and Co.

Maynard-Smith, John, and Eros Szathmary. 1995. *The major transitions in evolution*. Oxford: W. H. Freeman/Spectrum.

Mayr, Ernst. 1991. *One long argument*. Cambridge, MA: Harvard University Press.

Mayr, Ernst. 1997. *This is biology*. Cambridge, MA: Harvard University Press.

Mayr, Ernst. 2001. *What evolution is*. New York: Basic Books.

McElroy, M.B., and S.F. Wofsy. 1986. Tropical forests: Interactions with the atmosphere. In *Tropical Forests and World Atmosphere*, ed. G.F. Prance. Washington, DC: AAAS.

McKibben, W.J., and P.J. Wilcoxen. 1994. *The global costs of policies to reduce greenhouse gas emissions III*. Washington, DC: The Brookings Institution.

McKibben, W. J., and P. J. Wilcoxen. 1995. The theoretical and empirical structure of the G-cubed model. In *Brookings Discussion Papers on International Economics*. Washington, DC.

McMullen, J.T., J. Morgan, and R.B. Murray. 1983. *Energy resources*, 2nd ed. London: Arnold.

Meadows, Dennis L., Donella H. Meadows, Jorgen Randers, and William Behrens III. 1972. *The limits to growth, Club of Rome Reports*. New York: Universe Books.

Meadows, Donella H., Dennis L. Meadows, and Jorgen Randers. 1992. *Beyond the limits: Confronting global collapse, envisioning a sustainable future*. Post Mills, VT: Chelsea Green Publishing Company.

Mellon Institute. 1983. *Industrial energy productivity project*. Washington, DC: Mellon Institute for United States Department of Energy.

Menger, Carl. 1871 [1994]. *Grundsaetze der Volkswirtschaftslehre (Principles of economics)*. hardcover ed, *Institute for Humane Studies Series in Economic Theory*. Vienna: Libertarian Press, Incorporated. Original edition, 1871. Reprint, June

Mensch, Gerhard. 1979. *Stalemate in technology: Innovations overcome the depression*. Cambridge, MA: Ballinger.

Michaelian, Karo, and Oliver Manuel. 2011. Origin and evolution of life constraints on the solar model. *Journal of Modern Physics* 2: 587–594.

Milankovitch, Milutin. 1941 [1998]. *Canon of insolation and the Ice Age problem*. 1998 English translation ed. Belgrade: Zavod za Udz. Original edition, 1941.

Mill, James. 1808 [1965]. *Commerce defended*. Hardcover ed. London: Augustus M Kelley Pubs. Original edition, 1808. Reprint, 1965.

Mill, John Stuart. 1848. *Principles of political economy with some of their applications to social philosophy*. London: C.C. Little and J. Brown.

Miller, Stanley. 1953. Production of amino acids under possible primitive earth conditions. *Science* 117: 528–529.

Miller, Stanley, and Harold C. Urey. 1959. Organic compound synthesis on the primitive earth. *Science* 130: 245–251.

Miller, L.M., F. Gans, and A. Kleidon. 2011. Jet stream wind power as a renewable resource. *Earth Systems Dynamics* 2: 201–212.

Minsky, Hyman. 1982. *Can "it" happen again? Essays on instability and finance*. Armonk, NY: M. E. Sharpe.

Mirowski, Philip. 1984. Physics and the marginalist revolution. *Cambridge Journal of Economics* 8(4): 361–379.

Mirowski, Philip. 1989. More heat than light: Economics as social physics; physics as nature's economics. In *Historical Perspectives on Modern Economics*, ed. Craufurd D. Goodwin. Paperback ed. Cambridge, UK: Cambridge University Press. Reprint, 1999.

Mitchell, Wesley. 1927 [1955]. *Business cycles: The problem and its setting, Studies in Business Cycles*. New York: National Bureau of Economic Research (NBER). Original edition, 1927.

Mitchell, Peter. 1961. Coupling of phosphorylation to electron and hydrogen transfer by a chemi-osmotic type of mechanism. *Nature* 191: 144–148.

Morgan, Thomas Hunt. 1913. *Heredity and sex*. New York: Columbia University Press.

Morgan, J.P. 2015. Brave New World. In *Eye on the Market*, ed. Michael Cembalest. New York: J.P. Morgan Asset Management.

Morgan, Thomas Hunt, Alfred H. Sturtevant, H.J. Muller, and C.B. Bridges. 1915. *The mechanism of Mendelian heredity*. New York: Henry Holt.

Morowitz, H.J. 1992. *The beginnings of cellular life: Metabolism recapitulates biogenesis*. New Haven: Yale University Press.

Morowitz, H.J. 2002. *The emergence of everything: How the world became complex*. New York: Oxford University Press.

Morozov, L.L., and Vitalii I. Goldanskii. 1984. Violation of symmetry and self-organization in pre-biological evolution. In *Self-Organization: Autowaves and Structures Far from Equilibrium*, ed. V.I. Krinsky, 224–232. New York: Springer-Verlag.

Morrison, Elting. 1966. *Men, machines and modern times*. Cambridge, MA: MIT Press.

Muller, Richard A., and Gordon J.F. MacDonald. 1997. Glacial cycles and astronomical forcing. *Science* 277(5323): 215–218.

Murphy, David J., and Charles A.S. Hall. 2010. EROI, or energy return on (energy) invested. *Annals of the New York Academy of Sciences* 1185: 102–118.

NAS/NRC. 2005. *The Hidden Costs of Energy*. Washington, DC: National Academy of Sciences, National Research Council.

Natural Resources Defense Council (NRDC). 2006. *Ethanol: Energy well spent: A survey of studies published since 1990*. New York: Natural Resources Defense Council.

New Scientist. Jojoba oil could fuel cars and trucks. *New Scientist*, March 6, 2003.

Nicolis, Gregoire, and Ilya Prigogine. 1971. *Thermodynamics of structure, stability and fluctuation*. New York: Interscience.

Nicolis, Gregoire, and Ilya Prigogine. 1977. *Self-organization in non-equilibrium systems*. New York: Wiley-Interscience.

Niessert, V. 1987. How many genes to start with? A computer simulation about the origin of life. *Origins of Life* 17: 155–169.

Nordhaus, William. 1992. Lethal Model 2: The Limits to Growth revisited. *Brookings Papers on Economic Activity* 23(2): 1–60.

Nordhaus, William D. 1998. Do real-output and real-wage measures capture reality? The history of lighting suggests not. In *Cowles Foundation papers*. New Haven, CN: Yale University.

Odell, Peter R. 1983. *Oil and world power*, 7th ed. New York: Penguin Books.

Odum, Eugene P. 1953, 1971. *Fundamentals of ecology*, 2nd ed. Philadelphia, PA: Sanders.

Odum, Howard T. 1971. *Environment, power and society*. New York: Wiley.

Odum, Howard T. 1983. *Systems ecology: An introduction*. New York: John Wiley and Sons.

Odum, Howard T. 1986. Energy in ecosystems. In *Environmental Monographs and Symposia*, ed. N. Polunin, 337–369. New York: John Wiley.

Odum, Howard T., and E.C. Odum. 1976. *Energy basis for man and nature*. New York: McGraw-Hill.

Olah, George A., Alain Goeppert, and G. K. Surya Prakesh. 2009. *Beyond oil and gas: The methanol economy*, 2nd ed. New York: Wiley-VCH.

Olson, Mancur. 1965 [1971]. *The logic of collective action*. Cambridge: Harvard University Press. Original edition, 1965.

Olson, Mancur. 1982. *The rise and decline of nations: Economic growth, stagflation and social rigidities*. New Haven, CT: Yale University Press.

Onsager, Lars. 1931a. Reciprocal relations in irreversible processes, part 1. *Physical Review* 37: 405.

Onsager, Lars. 1931b. Reciprocal relations in irreversible processes, part 2. *Physical Review* 38: 2268.

Oparin, Alexander I. 1938. *Origin of life*. Translated by S. Margulis. First English translation ed. New York: MacMillan Company. Original edition, 1936 Moscow in Russian.

Oppenheimer, J. Robert, and H. Snyder. 1939. On continued gravitational attraction. *Physical Review* 56: 455–459.

Orgel, Leslie. 1983. The evolution of life and the evolution of macromolecules. *Folia Biologica* 29: 65–77.

Oro, J. 1961. Comets and the formation of biochemical compounds on the primitive earth. *Nature* 190: 389.

Orsato, Renato J. 2009. *Sustainability strategies: When does it pay to be green?* Houndmills, UK: Palgrave Macmillan.

Ostwald, Wilhelm. 1907. The modern theory of energetics. *The Monist* 17: 481–515.

Ostwald, Wilhelm. 1909. *Energetische Grundlagen der Kulturwissenschaften*. Leipzig: Duncker.

Pacini, Franco. 1967. Energy emission from a neutron star. *Nature* 216: 567.

Page, Talbot. 1977. *Conservation and economic efficiency*. Baltimore, MD: Johns Hopkins University Press.

Paley, William (Chairman). 1952. Resources for freedom. Washington DC: President's Materials Policy Commission.

Pareto, Vilfredo. 1916. *Trattato di Sociologica Generale*. Florence: Barbera.

Pareto, Vilfredo. 1971. *Manual of political economy*. Translated by Ann Schwier. New York: Kelly.

Parker, Geoffrey, and Lesley M. Smith (eds.). 1997. *The general crisis of the 17th century*. Hove, UK: Psychology Press.

Pasek, Matthew A. 2008. Rethinking the early Earth phosphorus geochemistry. *Proceedings of the National Academy of Sciences (PNAS)* 105(3): 853–858.

Passer, Harold C. 1953. *Electrical manufacturers 1875-1900*. Cambridge, MA: Harvard University Press.

Peltier, W.R. 2004. Global glacial isostasy and the surface of the ice-age Earth: The ice-5G model and GRACE. *Annual Review of Earth and Planetary Sciences* 32: 111–149.

Peng, T.-H., W.S. Broecker, H.-D. Freyer, and S. Trumbore. 1983. A deconvolution of the tree-ring based BC record. *Journal of Geophysical Research* 88: 3609–3620.

Penzias, A.A., and T.W. Wilson. 1965. A measurement of excess antenna temperature at 4080 mc/s. *Nature* 142(1): 419–421.

Peratt, Anthony L. 1986. Evolution of the plasma universe. *IEEE Transactions on Plasma Science* 14(639–660): 763–778.

Peratt, Anthony L., and James Green. 1983. On the evolution of interacting, magnetized galactic plasmas. *Astrophysics and Space Science* 91: 19–33.

Petit, J.R., J. Jouzel, D. Raynaud, N.I. Barkov, J.-M. Barnola, I. Basile, M. Bender, J. Chappellaz, M. Davis, G. Delaygue, M. Delmotte, V.M. Kotlyakov, M. Legrand, M.Y. Lipenkov, C. Lorius, L. PÉpin, C. Ritz, E. Saltzman, and M. Stievenard. 1999. Climate and atmospheric history of the past 420,000 years from the Vostok ice core, Antarctica. *Nature* 399: 429–436. doi:10.1038/20859.

PetroSun Inc. 2008. 30 mgy Algal biodiesel refinery to be built in Arizona. The Energy Blog. http://thefraserdomain.typepad.com/energy/2008/01/30-mgy-biodiese.html

Pigou, A.C. 1920. *The economics of welfare*, 1st ed. London: MacMillan Company.

Postel, Sandra. 1989. *Water for agriculture: Facing the limits*. Washington, DC: Worldwatch Institute.

Postel, Sandra. 1999. *Pillar of Sand: Can the irrigation miracle last?* New York: W.W.Norton & Co.

Potter, Neal, and Francis T. Christy Jr. 1968. *Trends in natural resource commodities*. Baltimore, MD: Johns Hopkins University Press.

Powers, William. 2013. *Cold, hungry and in the dark: The popping of the shale gas bubble*. Gabriola: New Society Publishers.

Previdi, M., B.G. Liepert, D. Peteet, Hansen James, D.J. Beeerling, A.J. Broccoli, S. Frolking, J.N. Galloway, M. Heimann, C. Le Qu'er'e, S. Levitus, and V. Ramaswamy. 2013. Climate sensitivity in the Anthropocene. *Quarterly Journal of the Royal Meteorological Society* 139: 1121–1131.

Price-Williams, David. 2015. South Africa and the search for human origins. Capetown South Africa.

Prigogine, Ilya. 1955. *Introduction to the thermodynamics of irreversible processes*. Springfield IL: Charles C. Thomas. Reprint, Third edition 1968 Wiley Intersceince.

Prigogine, Ilya. 1976. Order through fluctuation: Self-organization and social system. In *Evolution and Consciousness*, ed. Erich Jantsch and Waddington. New York: Addison-Wesley Publishing Company.

Prigogine, Ilya. 1980. *From being to becoming*. San Francisco: W. H. Freeman.

Prigogine, Ilya. 1991. Foreword. In *The New Evolutionary Paradigm*, ed. Ervinn Laszlo. Vienna: Gordon and Breach Science Publishers.

Prigogine, Ilya, and I. Stengers. 1984. *Order out of chaos: Man's new dialogue with nature*. London: Bantam Books.

Prigogine, Ilya, Gregoire Nicolis, and A. Babloyantz. 1972. Thermodynamics of evolution. *Physics Today* 23(11/12): 23–28(N) and 38–44(D).

Prothero, Donald. 1993. *The Eocene-Oligocene transition: Paradise lost?* New York: Columbia University Press.

Quesnay, Francois. 1766. *Tableau Economique*.

Ramirez, Andrea. 2005. Monitoring energy efficiency in the food industry. PhD, Chemistry, Universiteit Utrecht.

Ramsey, Frank P. 1928. A mathematical theory of saving. *Economic Journal* 38(152): 543–559.

Rankine, William John MacQuorn. 1881. *Miscellaneous scientific papers*. London: Charles Griffin and Company.

Rashevsky, Nicholas. 1948. *Mathematical biophysics*. Chicago: University of Chicago Press.

Ray, Michael W., Emmi Ruokokski, Saugat Kandel, Mikko Motloken, and Hall David S. 2014. Observation of Dirac monopoles in a synthetic magnetic field. *Nature* 505: 657–660.

Raymo, R.E., and W.F. Ruddiman. 1992. Tectonic forcing of late Cenozoic climate. *Nature* 359 (6391): 117–122.

Redinger, Mike, Martin Lokanc, Roderick Eggert, Michael Woodhouse, and Alan Goodrich. 2013. *The present, mid-term and long-term supply curves for tellurium and updates in the results from NREL's CdTe PV module manufacturing cost model*. Boulder, CO: National Renewable Energy Laboratory (NREL).

Reinhart, Carmen M., and Kenneth S. Rogoff. 2009. *This time is different; Eight centuries of financial folly*. Princeton, NJ: Princeton University Press.

Reinhart, Carmen M., and Kenneth S. Rogoff. 2010. *Growth in a time of debt*. Washington: National Bureau of Economic Research.

Revelle, Roger, and Hans Suess. 1957. Carbon dioxide exchange between atmosphere and ocean and the question of an increase of atmospheric CO_2 during the past decades. *Tellus* IX: 1–27.

Ricardo, David. 1817 [2007]. Principles of political economy and taxation. In *The Real Price of Everything*, ed. Michael Lewis, 764–974. New York: Sterling. Original edition, 1776.

Ricci, G., and T. Levi-Civita. 1901. Methods de calcul differential absolu et leurs applications. *Mathematische Annalen* 54: 125–201.

Righelato, Renton, and Dominick Spracklen. 2007. Carbon mitigation by biofuels or by saving and restoring forests? *Science* 317: 902.

Ringwood, A. E. 1969. Pyrolite model and mantle composition. In *The Earth's Crust and Upper Mantle*, ed. Pembroke J. Hart, 1–17. Washington, DC: American Geophysical Union; NAS-NRC.

Robinson, Joan. 1953-54. The production function and the theory of capital. *Review of Economic Studies* 21(1): 81–106.

Rockström, Johan, Will Steffen, Kevin Noone, Asa Perrsson, F. Stuart Chapin III, Eric F. Lambin, Timothy M. Lenton, Marten Scheffer, Carl Folke, Hans Joachim Schnellhuber, Björn Nykvist, Cynthia A. de Wit, Terry Hughes, Sander van der Leeuw, Henning Rodhe, Sverker Sörlin, Peter K. Snyder, Robert Costanza, Uno Svedin, Malin Falkenmark, Louise Karlberg, Robert W. Corell, Victoria J. Fabry, James Hansen, Brian Walker, Diane Liverman, Katherine Richardson, Paul Crutzen, and Jonathan A. Foley. 2009. A safe operating space for humanity. *Nature* 461: 472–475.

Rollat, Alain. Comment on by-product metals. Paris, 30 November 2015.

Rombach, Elinor, and Bernd Friedrich. 2015. Recycling of rare metals. In *Handbook of Recycling*, ed. Ernst Worrell and Markus Reuter, 125–150. Amsterdam: Elsevier.

Romer, Paul M. 1986. Increasing returns and long-run growth. *Journal of Political Economy* 94 (5): 1002–1037.

Romer, Paul M. 1987. Growth based on increasing returns due to specialization. *American Economic Review* 77(2): 56–62.

Rosenberg, Nathan (ed.). 1969. *The American system of manufacturing*. Edinburgh, Scotland: Edinburgh University Press.

Roy, N. K., M. J. Murtha, and G. Burnet. 1979. Industrial applications of magnetic separation. *IEEE*.

Rubey, W. W. 1955. Development of the hydrosphere and atmosphere with special reference to the probable composition of the early atmosphere. In *Crust of the Earth (Special Paper)*, ed. A. Poldervaart, 631–650. Geological Soc. Am.

Rubin, Vera. 1997. *Bright galaxies, dark matters*. Woodbury, NY: AIP Press.

Rudnick, R. L., and S. Gao 2004. Composition of the continental crust. In *The Crust: Treatise on Geochemistry*, 1–64. Dordrecht, Netherlands: Elsevier Pergamon.

Rundell Jr., Walter. 2000. *Early Texas oil: A photographic history*. College Station, TX: Texas A & M University Press.

Rutledge, David. 2007. The coal question and climate change. The Oil Drum. http://www.theoildrum.com/node/2697 (last modified June 20, 2007).

Sagan, Carl. 1977. *The dragons of Eden: Speculations on the evolution of human intelligence*. New York: Random House.

Salzman, Daniela. 2013. *Stranded Carbon Assets*. London: The Generation Foundation.

Samuelson, Paul, and Robert Solow. 1956. A complete capital model involving heterogeneous capital goods. *Quarterly Journal of Economics* 70: 537–562.

Sandel, Michael J. 2012. *What money can't buy: The moral limits of markets*. New York: Farrar, Straus & Giroux.

Sant, Roger W. 1979. *The least-cost energy strategy: Minimizing consumer costs through competition*. Virginia: Mellon Institute Energy Productivity Center.

Sant, Roger W., and Steven C. Carhart. 1981. *8 great energy myths: The least cost energy strategy, 1978-2000*. Pittsburgh PA: Carnegie-Mellon University Press.

Saunders, Harry. 1992. The Khazzoom-Brookes Postulate and neoclassical growth. *Energy Journal* 13(4): 131–148.

Schimel, D., I. Enting, M. Heimann, T.M.L. Wigley, D. Raynaud, D. Alves, and U. Siegenthaler. 1994. *The carbon cycle*. Cambridge, UK: Radiative Forcing of Climate.

Schipper, Lee, and Michael Grubb. 1998. On the rebound? Using energy indicators to measure the feedback between energy intensities and energy uses. In *IAEE Conference 1998*, Quebec.

Schlesinger, William H. 1991. *Biogeochemistry: An analysis of global change*. New York: Academic Press.

Schmidt-Bleek, Friedrich B. 1993. MIPS—A universal ecological measure? *Fresenius Environmental Bulletin* 2(6): 306–311.

Schneider, E.D., and J.J. Kay. 1994. Life as a manifestation of the Second Law of Thermodynamics. *Mathematical Computational Modeling* 19(6–8): 25–48.

Schneider, Eric, and Dorion Sagan. 2005. *Into the Cool: Energy flow, thermodynamics and life.* Chicago: University of Chicago Press.

Schrödinger, Erwin. 1945. *What is life? The physical aspects of the living cell.* London: Cambridge University Press.

Schumpeter, Joseph A. 1912. *Theorie der Wirtschaftlichen Entwicklungen.* Leipzig, Germany: Duncker and Humboldt.

Schumpeter, Joseph A. 1939. *Business cycles: A theoretical, historical and statistical analysis of the capitalist process.* 2 vols. New York: McGraw-Hill.

Schurr, Sam H., and Bruce C. Netschert. 1960. *Energy in the American economy, 1850-1975.* Baltimore, MD: Johns Hopkins University Press.

Schwarzschild, Karl. 1916. Über das Gravitationsfeld eines Massenpunktes nach der Einstein'schen Theorie. *Sitzungsberichte der Königlich-Preussischen Akademie der Wissenschaften* 189: 424.

Schwinger, Julian. 1969. A magnetic model of matter. *Science* 165: 757–761.

Scitovsky, T. 1954. Two concepts of external economies. *Journal of Political Economy* 62: 143–151.

Sedjo, R.A. 1992. Temperate forest ecosystems in the global carbon cycle. *Ambio* 21: 274–277.

Serrenho, Andre Cabrera. 2014. Decomposition of useful work intensity: The EU-15 countries from 1960 to 2009. *Energy* 76: 704–715.

Shakhova, N., I. Semiletov, A. Salyuk, D. Kosmach, and N. Bel'cheva. 2007. Methane release from the Arctic East Siberian shelf. *Geophysical Research Abstracts* 9.

Shakhova, N., I. Semiletov, A. Salyuk, and D. Kosmach. 2008. Anomalies of methane in the atmosphere over the East Siberian shelf: Is here any sign of methane leakage from shallow shelf hydrates? *Geophysical Research Abstracts* 10.

Shakhova, N., I. Semiletov, I. Leifer, A. Salyuk, P. Rekant, and D. Kosmach. 2010a. Geochemical and geophysical evidence of methane release over the East Siberian Arctic Shelf. *Journal of Geophysical Research* 115: C08007. doi:10.1029/2009JC005602.

Shakhova, N., I. Semiletov, A. Salyuk, V. Yusupov, D. Kosmach, and O. Gustafsson. 2010b. Extensive methane venting to the atmosphere from sediments of the East Siberian Arctic Shelf. *Science* 327(5970): 1246–1250.

Shannon, Claude E. 1948. A mathematical theory of communication. *Bell System Technical Journal* 27: 379–423.

Shannon, Claude E., and Warren Weaver. 1949. *The mathematical theory of communication.* Urbana, IL: University of Illinois Press.

Shapiro, Robert. 1986. *Origins: A skeptics guide to the creation of life on earth.* New York: Summit.

Sharlin, Harold I. From Faraday to the dynamo. *Scientific American,* May 1961.

Shiklomanov, Igor. 1993. World fresh water resources. In *Water in Crisis: A Guide to the World's Fresh Water Resources,* ed. Peter H. Gleick. New York: Oxford University Press.

Sibley, C.G., and J.E. Ahlquist. 1984. The phylogeny of the hominoid primates, as indicated by DNA-DNA hybridization. *Journal of Molecular Evolution* 20: 2–15.

Sibley, C.G., and J.E. Ahlquist. 1987. DNA hybridization evidence of hominoid phylogeny: Results from an expanded data set. *Journal of Molecular Evolution* 26: 99–121.

Sibley, C.G., J.A. Comstock, and J.E. Ahlquist. 1990. DNA hybridization evidence of hominoid phylogeny: A re-anaysis of the data. *Journal of Molecular Evolution* 30: 202–236.

Sillèn, L.G. 1967. The ocean as a chemical system. *Science* 156: 1189–1197.

Simmons, Matthew R. 2004. *Twilight in the desert: The coming Saudi oil shock and the world economy.* New York: Wiley.

Simon, Herbert A. 1955. A behavioral model of rational choice. *Quarterly Journal of Economics* 69: 99–118.

Simon, Herbert A. 1959. Theories of decision-making in economics. *American Economic Review* 49: 253–283.

Simon, Julian L. 1977. *The economics of population growth*. Princeton, NJ: Princeton University Press.

Simon, Julian L. 1980. Resources, population, environment: An oversupply of false bad news. *Science* 208: 1431–1437.

Simon, Julian L. 1981. *The ultimate resource*. Princeton, NJ: Princeton University Press.

Simon, Herbert A. 1982. *Models of bounded rationality*. Cambridge, MA: MIT Press.

Simon, Julian L. 1996. *The ultimate resource 2*. Princeton, NJ: Princeton University Press.

Singer, Charles, E. J. Holmyard, A. R. Hall, and Trevor I. Williams (eds.). 1958. *A history of technology*. 5 vols. New York: Oxford University Press.

Sismondi, J. C. L. Simonde de. 1819 [1827]. *Nouveaux Principes d'Economie Politique*. Original edition, 1819.

Skarke, A., C. Ruppel, M. Kodis, D. Brothers, and E. Lobecker. 2014. Widespread leakage from the sea floor of the East Siberian ocean and along the northern US Atlantic margin. *Nature Geoscience* 7: 657–661.

Skinner, Brian J. 1976. *Earth resources*. Englewood Cliffs, NJ: Prentice-Hall.

Skinner, Brian J. 1987. Supplies of geochemically scarce metals. In *Resources and World Development*, ed. D.J. Maclaren and Brian J. Skinner, 305–325. Chichester, UK: John Wiley and Sons.

Smil, Vaclav. 1997. *Cycles of life: Civilization and the biosphere*. New York: Scientific American Library.

Smil, Vaclav. 2000. *Transforming the world: Synthesis of ammonia and its consequences*. Cambridge, MA: MIT Press.

Smil, Vaclav. 2001. *Cycles of life: Civilisation and the biosphere*. Cambridge, MA: MIT Press.

Smil, Vaclav. 2003. *Energy at the crossroads: Global perspectives and uncertainties*. Cambridge, MA: MIT Press.

Smil, Vaclav. 2004. *Enriching the Earth: Fritz Haber, Carl Bosch and the transformation of world food production*. Cambridge, MA: MIT Press.

Smil, Vaclav. 2008. *Energy in nature and society: General energetics of complex systems*. Cambridge, MA: MIT Press.

Smit, Rob, Jeroen G. de Beer, Ernst Worrell, and Kornelis Blok. 1994. *Long term Industrial energy efficiency improvements: Technology descriptions*. Utrecht: Universiteit Utrecht.

Smith, Adam. 1759 [2013]. *The theory of moral sentiments*. Paperback ed. Empire Books. Original edition, 1790.

Smith, Adam. 1776 [2007]. An inquiry into the nature and causes of the wealth of nations. In *The Real Price of Everything*, ed. Michael Lewis, 22–652. New York: Sterling. Original edition, 1776.

Smith, Cyril Stanley. 1967. Metallurgy in the 17th and 18th centuries. In *Technology in Western Civiliztion*, ed. Melvin Kranzberg and Carroll Pursell. New York: London.

Smith, George David. 1985. *The anatomy of a business strategy*, ed. Galambas. Baltimore MD: Johns Hopkins University Press.

Smith, V. Kerry, and John Krutilla (eds.). 1979. *Scarcity and growth reconsidered*. Baltimore, MD: Johns Hopkins University Press.

Smoot, George, and Keay Davidson. 1993. *Wrinkles in time*. New York: William Morrow & Co., Inc.

Soddy, Frederick. 1920. *Science and life*. New York: Dutton.

Soddy, Frederick. 1922. *Cartesian economics*. London: Hendersons.

Soddy, Frederick. 1933. Wealth, virtual wealth and debt. In *Masterworks of Economics: Digests of 10 Classics*. New York: Dutton.

Soddy, Frederick. 1935. *The role of money*. New York: Harcourt.

Solomon, P. M., and William Klemperer. 1972. The formation of diatomic molecules in interstellar clouds. *Astrophysical Journal* 138: 389.

Solow, Robert M. 1956. A contribution to the theory of economic growth. *Quarterly Journal of Economics* 70: 65–94.

Solow, Robert M. 1957. Technical change and the aggregate production function. *Review of Economics and Statistics* 39: 312–320.

Solow, Robert M. 1973. Is the end of the world at hand? In *The Economic Growth Controversy*, ed. Andrew Weintraub, Eli Schwartz, and J. Richard Aronson. White Plains, NY: International Arts and Science Press.

Solow, Robert M. 1974a. The economics of resources or the resources of economics. *American Economic Review* 64(2): 1–14.

Solow, Robert M. 1974b. Intergenerational equity and exhaustible resources. *Review of Economic Studies* 41: 29–45.

Sorokin, Pitirim A. 1937 [1957]. *Social and cultural dynamics*, 2nd (revised and abridged) ed. 4 vols. Cambridge, MA: Harvard University Press. Original edition, 1937.

Sorrell, Steve, John Dimitropoulos, and Matt Sommerville. 2009. Empirical estimates of the direct rebound effect: A review. *Energy Policy* 37(4): 1356–1371.

Sousa, Tania, Tiago Domingos, and S.A.L.M. Kooijman. 2008. From empirical patterns to theory: A formal metabolic theory of life. *Philosophical Transactions of the Royal Society B* 363: 2453–2464.

Spath, P.L., and D.C. Dayton. 2003. *Preliminary Screening -Technical and Economic Assessment of synthesis gas to fuels and chemicals with emphasis on the potential for biomass derived syn-gas*. Golden, CO: National Renewable Energy Laboratory.

Stanisford, Stuart. 2011. Early Warning: The Gross World Product will not grow at 4% + for the next five years. http://earlywarn.blogspot.com/2011/04/global-world-product-will-not-grow-at-4.html

Steinberg, M. A., and Yuanji Dong. 1993. Hynol: An economic process for methanol production from biomass and natural gas.

Stephens, Graeme, Juilin Li, Martin Wild, Carol Anne Clayson, Norman Loeb, Seiji Kato, Tristan L'Ecuyer, Paul W. Stackhouse Jr., Matthew Lebsock, and Timothy Andrews. 2012. An update on Earth's energy balance in light of latest global observations. *Nature Geoscience* 5: 691–696.

Stern, Nicholas. 2006. *The economics of climate change*. London: Cambridge University Press for H. M. Treasury U.K.

Stern, Nicholas. 2008. The economics of climate change. *American Economic Review: Papers and Proceedings* 98(2): 1–37.

Stern, P.C., O.R. Young, and D. Druckman (eds.). 1992. *Global environmental change: Understanding the human dimension*. Washington, DC: National Academy Press.

Stern, Nicholas, S. Peters, A. Bakhshi, C. Bowen, S. Cameron, D. Catovsky, S. Crane, S. Cruikshank, N. Dietz, S.-L. Edmondson, I. Garbett, G. Hamid, D. Hoffman, B. Ingram, N. Jones, H. Patmore, R. Radcliffe, M. Sathiyarajah, C. Stock, T. Taylor, H. Wanjie Vernon, and D. Zeghelis. 2006. *Stern review: The economics of climate change*. London: HM Treasury.

Stewart, Ian, and Martin Golubitscky. 1992. *Fearful Symmetry: Is God a Geometer?* Oxford: Blackwell Publications.

Stiglitz, Joseph. 1974. Growth with exhaustible natural resources: Efficient and optimal growth paths. *Review of Economic Studies* 41(5): 123 (Symposium on the Economics of Exhaustible Resources).

Stiglitz, Joseph. 1979. A neoclassical analysis of the economics of natural resources. In *Scarcity and Growth Reconsidered*, ed. V. Kerry Smith. Washington, DC: Resources for the Future.

Stiglitz, Joseph E. 1997. Reply: Georgescu-Roegen versus Solow/Stiglitz. *Ecological Economics* 22(3): 269–270.

Stiglitz, Joseph. 2002. *Globalization and its discontents*. San Francisco: W.W.Norton & Co.

Stiglitz, Joseph. 2012. *The price of inequality*. London: Allen Lane.

Stolper, Wolfgang F. 1988. Development: Theory and empirical evidence. In *Evolutionary Economics: Applications of Schumpeter's Ideas*, ed. Horst Hanusch, 9–22. New York: Cambridge University Press.

Strahan, David. 2007. *The last oil shock*. London: John Murray Ltd.

Streeter, W. J. 1984. *The silver mania*. Dordrecht: D. Reidel (Kluwer).

Sundquist, Eric T. 1993. The global carbon dioxide budget. *Science* 259: 934–941.

Szargut, Jan, David R. Morris II, and Frank R. Steward. 1988. *Exergy analysis of thermal, chemical, and metallurgical processes*. New York: Hemisphere Publishing Corporation.

Tainter, Joseph. 1988. *The collapse of complex societies*. New York: Cambridge University Press.

Taleb, Nassim. 2007. *The Black Swan: The impact of the highly improbable*. New York: Random House.

Tans, P.P., I.Y. Fung, and T. Takahashi. 1990. The global atmospheric CO_2 budget. *Science* 247: 1431–1438.

Tawney, R.H. 1926. *Religion and the rise of capitalism*. New York: Harcourt Brace.

Taylor, F. Sherwood. 1942. *The century of science: 1840-1940*. London: Heinemann.

Taylor, John G. 1973. *Black Holes*. New York: Avon Books.

Taylor, K.E., and J.E. Penner. 1994. Anthropogenic aerosols and climate change. *Nature* 369: 734–737.

Taylor, Robert P., Chandrasekar Govindarajalu, Jeremy Levin, Anke S. Meyer, and William A. Ward. 2008. *Financing energy efficiency: Lessons from Brazil, China, India, and beyond*. Washington, DC: The World Bank.

Thiemens, Mark H., and William C. Trogler. 1991. Nylon production: An unknown source of atmospheric nitrous oxide. *Science* 251: 932–934.

Thompson, Benjamin (Count Rumford). 1798. An experimental enquiry concerning the source of the heat which is excited by friction. *Philosophical Transactions of the Royal Society*.

Tomizuka, A. 2009. Is a box model effective for understanding the carbon cycle? *American Journal of Physics* 77(2): 150–163.

Trail, Dustin, E. Bruce Watson and Nicholas Tailby. 2011. The oxidation state of Hadean magmas and implications for the early Earth's atmosphere, Nature 480: 79–82.

Trenberth, K.E., L. Smith, T.T. Qian, A.G. Dai, and J. Fasullo. 2007. Estimates of the global water budget and its annual cycle using observational and model data. *Journal of Hydrometeorology* 8(4): 758–769.

Tribus, Myron, and Edward C. McIrvine. 1971. Energy and information. *Scientific American* 225 (3): 179–188.

Trudinger, P.A., and D.J. Swaine (eds.). 1979. *Biochemical recycling of mineral-forming elements*. Amsterdam: Elsevier.

Tully, R. Brent, and J.R. Fischer. 1987. *Atlas of nearby galaxies*. Cambridge, UK: Cambridge University Press.

Turchin, Peter, and Sergey A. Nefedov. 2009. *Secular cycles*. Princeton, NJ: Princeton University Press.

Tversky, A., and D. Kahneman. 1974. Judgment under uncertainty: Heuristics and biases. *Science* 185: 1124–1131.

Ulanowicz, Robert E. 1986. *Growth and development: Ecosystems phenomenology*. New York: Springer-Verlag.

Ulanowicz, Robert E. 1997. *Ecology: The ascendant perspective*. New York: Columbia University Press.

United Nations Industrial Development Organization, (UNIDO). 2011. *Industrial Development Report 2011: Industrial Energy Efficiency for Sustainable Wealth Creation*. Vienna: United Nations Industrial Development Organization (UNIDO).

United States Bureau of the Census. 1975. *Historical statistics of the United States, Colonial times to 1970*. Bicentennial ed. 2 vols. Washington, DC: United States Government Printing Office.

United States Department of Energy. 2010. *Save energy now. Industrial technology program*. Washington, DC: US Department of Energy (DOE).

Urey, Harold. 1952. *The Planets: Their origin and development*. New Haven: Yale University Press.

USGS, United States Geological Survey. 2012a. *Metal prices in the United States through 2010*. Reston, VA: USGS.

USGS, United States Geological Survey. 2012b. USGS mineral commodity summaries. In *Mineral Commodities*. Washington, DC: United States Department of the Interior.

Valero Capilla, Antonio, and Alicia Valero Delgado. 2015. *Thanatia: The destiny of the Earth's mineral resources*, 1st ed. London: World Scientific.

Valero, Alicia Delgado, Antonio Capilla Valero, and J.B. Gomez. 2011. The crepuscular planet: A model for the exhausted continental crust. *Energy* 36(1): 694–707.

van Gool, W. 1987. The value of energy carriers. *Energy* 12(509).

Varoufakis, Yanis. 2011. *The global minotaur: America, Europe and the future of the global economy*, 2013 ed. *Economic Controversies Series*. London: Zed Books.

Veblen, Thorsten. 1899 [2007]. Theory of the leisure class: An economic study of institutions. In *The Real Price of Everything*, ed. Michael Lewis, 1048–1227. New York: Sterling. Original edition, 1776.

Vernadsky, Vladimir I. 1945. The biosphere and the noosphere. *American Scientist* 33: 1–12.

Vernadsky, Vladimir I. 1986. *The biosphere*. London: Synergetic Press.

Viguerie, Patrick, Sven Smit, and Mehrdad Baghai. 2007. *The granularity of growth*. London: Marshall Cavendish Ltd and Cyan Communications Ltd.

Vincent, William, and Lisa Callaghan Jeram. 2006. The potential for bus rapid transit to reduce transportation-related CO_2 emissions. *Journal of Public Transportation* (BRT Special Edition).

Volterra, Vito. 1931. *Lecons sur la theorie mathematique de la lutte pour la vie*. Paris: Gauthier-Villars et cie.

von Neumann, John. 1961-63. *Collected works*. New York: MacMillan Company.

von Neumann, John, and Oskar Morgenstern. 1944. *Theory of games and economic behavior*. Princeton, NJ: Princeton University Press.

von Weizsaecker, Ernst Ulrich, Amory B. Lovins, and L. Hunter Lovins. 1998. *Factor four: Doubling wealth, halving resource use*. London: Earthscan Publications Ltd.

Voudouris, Vlasios, Robert U. Ayres, André Cabrera Serrenho, and Daniil Kiose. 2015. The economic growth enigma revisited: The EU 15 since the 1970s. *Energy Policy*.

Wachterhauser, Gunter. 1988. Before enzymes and templates: Theory of surface metabolism. *Microbiological Reviews* 52: 452–484.

Wachterhauser, Gunter. 1992. Groundworks for an evolutionary biochemistry: The iron-sulfur world. *Progress in Biophysics and Molecular Biology* 58: 85–201.

Wadhams, P., V. Pavlov, E. Hansen, and G. Budeus. 2003. Long-lived convective chimneys in the Greenland Sea and their climactic role. *Geophysical Research Abstracts* 5 (05572).

Wald, Robert. 1984. *General Relativity*. Chicago: University of Chicago Press.

Walker, Gabrielle, and David King. 2009. *The hot topic: How to tackle global warming and still keep the lights on*. Revised paperback ed. London: Bloomsbury. Original edition, 2008 Bloomsbury London.

Walker, James C.G., P.B. Hays, and James F. Kasting. 1981. unknown. *Journal of Geophysical Research* 86: 9776–9782.

Wall, Goran. 1986. Exergy conversion in the Swedish society. *Energy* 11: 435–444.

Walras, Leon. 1874. *Elements d'economie politique pure*. Lausanne, Switzerland: Corbaz.

Ward, Peter. 2009. *The Medea hypothesis: Is life on Earth ultimately self-destructive?* Princeton, NJ: Princeton University Press.

Warr, Benjamin S., and Robert U. Ayres. 2010. Evidence of causality between the quantity and quality of energy consumption and economic growth. *Energy: The International Journal* 35: 1688–1693.

Warr, Benjamin, and Robert U. Ayres. 2012. Useful work and information as drivers of economic growth. *Ecological Economics* 73: 93–102.

Wasdell, David. 2014. Climate sensitivity and the carbon budget. Apollo-Gaia project.

Watson, William D. 1978. Gas phase reactions in astrophysics. *Annual Reviews of Astronomy and Astrophysics* 16: 585–615.

Watson, James, and Francis Crick. 1953. A structure for deoxyribose nucleic acid. *Nature* 171: 4356.

Watson, R.T., L.G. Filho, E. Sanhueza, and A. Janetos. 1992. Greenhouse gases: Sources and sinks. In *The Supplementary Report to the IPCC Scientific Assessment*, ed. J.T. Houghton, B.A. Callender, and S.K. Varney. Cambridge, UK: Cambridge University Press.

Weatherford, Jack. 1997. *The history of money*. Paperback ed. New York: Three Rivers Press.

Webb, Peter-Noel, and David Harwood. 1991. Late Cenozoic glacial history of the Ross embayment, Antarctica. *Quaternary Science Reviews* 10: 215–223.

Weber, Max. 1904-05. Die Protestantische Ethik und der Geist des Kapitalismus. *Archiv fuer Sozialwissenschaft und Sozialpolitk* XX,XXI: 1–54.

Wegener, Alfred. 1966. *The origin of continents and oceans*. Translated by John Biram. 4th (German) ed. New York: Dover Editions.

Weinberg, Steven. 1977. *The first three minutes*. New York: Andre Deutsch Ltd.

Weinert, Jonathan, Andrew Burke, and Xuezhe Wei. 2007. Lead-acid and lithium-ion batteries for the Chinese electric bike market and implications on future technology advancement. *Journal of Power Sources* 172(2): 938–945.

Weiss, R.F. 1981. The temporal and spatial distribution of nitrous oxide. *Journal of Geophysical Research* 86: 7185–7195.

Weitzman, Martin L. 1998. Why the far-distant future should be discounted at the lowest possible rate. *Journal of Environmental Economics and Management* 36(3): 201–208.

Weitzman, Martin L. 2001. Gamma discounting. *American Economic Review* 91(1): 260–271.

Weitzman, Martin L. 2007. A review of the Stern Review on the economics of climate change. *Journal of Economic Literature* 45(3): 703–724.

Werner, Richard A. 2005. *New paradigm in macro-economics*. Basingstoke: Palgrave-MacMillan.

Wertime, Theodore A. 1962. *The coming of the age of steel*. Chicago: The University of Chicago Press.

Weyl, Hermann. 1952. *Symmetry*. Princeton, NJ: Princeton University Press.

Whitehead, Alfred North. 1926. *Science and the Modern World*. Cambridge: Cambridge University Press.

Whittet, Douglas, et al. 2011. Observational constraints on methanol production in interstellar and preplanetary ices. *Astrophysical Journal* 742: 28.

Wicken, Jeffrey S. 1987. *Evolution, thermodynamics and information: Extending the Darwinian program*. New York: Oxford University Press.

Wigley, T.M.L. 1989. Possible climate change due to SO2-derived cloud condensation nuclei. *Nature* 339: 365–367.

Wikipedia. 2008. Velib. Wikipedia. http://

Williamson, Harold F., and Arnold R. Daum. 1959. *The American petroleum industry: 1859-1899*. Evanston, IL: Northwestern University Press.

Wingren, Anders, Mats Galbe, and Guido Zacchi. 2008. Energy considerations for a SSF-based softwood ethanol plant. *Bioresource Technology* 99(7): 2121–2131.

Woodhouse, Michael, Alan Goodrich, Robert Margolis, Ted L. James, Martin Lokanc, and Roderick Eggert. 2013. Supply chain dynamics of tellurium, indium and gallium within the context of PV manufacturing costs. *IEEE Journal of Photovoltaics* 3(2): 833–837.

Worrell, Ernst, and Lynn Price. 2000. *Barriers and opportunities: A review of selected successful energy-efficiency programs*. Berkeley, CA: Lawrence Berkeley National Laboratory.

Wyman, Charles E. 2004. Ethanol fuel. In *Encyclopedia of Energy*, ed. Cutler J. Cleveland, 541–555. London: Elsevier.

Wyman, Charles E., Richard L. Bain, Norman D. Hinman, and Don J. Stevens. 1993. Ethanol and methanol from cellulosic biomass. In *Renewable Energy*, ed. Johansson et al. Washington, DC: Island Press.

Yang, Chen Ning, and Tsung Dao Lee. 1956. Question of parity conservation in weak interactions. *Physical Review* 104: 254–258.

Yergin, Daniel. 1991. *The Prize: The epic quest for oil, money and power.* New York: Simon and Schuster.

Yilmaz, Huseyin. 1958. A new approach to general relativity. *Physical Review* 111(5): 1417–1426.

Yilmaz, Huseyin. 1973. A new approach to relativity and gravitation. *Annals of Physics (NY)* 101: 413–432.

Yilmaz, Huseyin. 1986. *Quantum mechanics and general relativity.* New York: New Techniques and Ideas in Quantum Measurement Theory.

Yilmaz, Huseyin. 1995. Gravity and quantum field theory: A modern synthesis. *Annals of the New York Academy of Sciences* 755: 476–499.

Zittel, Werner, and Joerg Schindler. 2007. *Coal: Resources and future production.* Ottobrunn: Energy Watch Group.

Zoravcic, Zorana, and Michael Brenner. 2015. Models of self-replication of microstructures. *Proceedings of the National Academy of Sciences* 111(5): 1748–1753.

Zotin, A.I. 1984. Bioenergetic trends of evolutionary progress of organisms. In *Thermodynamics and Regulation of Biological Processes,* ed. I. Lamprecht and A.J. Zotin, 451–458. Berlin: Walter de Gruyter & Co.

Zweig, George. 1964. *An SU(3) Model for Strong Interaction Symmetry and its Breaking I.* Meyrin, Switzerland: CERN.

Index

© Springer International Publishing Switzerland 2016   577
R. Ayres, *Energy, Complexity and Wealth Maximization*, The Frontiers Collection,
DOI 10.1007/978-3-319-30545-5

Titles in This Series

Particle Metaphysics
A Critical Account of Subatomic Reality
By Brigitte Falkenburg

The Physical Basis of The Direction of Time
By H. Dieter Zeh

Asymmetry: The Foundation of Information
By Scott J. Muller

Decoherence and the Quantum-To-Classical Transition
By Maximilian A. Schlosshauer

The Nonlinear Universe
Chaos, Emergence, Life
By Alwyn C. Scott

Quantum Superposition
Counterintuitive Consequences of Coherence, Entanglement, and Interference
By Mark P. Silverman

Symmetry Rules
How Science and Nature Are Founded on Symmetry
By Joseph Rosen

Mind, Matter and Quantum Mechanics
By Henry P. Stapp

Entanglement, Information, and the Interpretation of Quantum Mechanics
By Gregg Jaeger

Relativity and the Nature of Spacetime
By Vesselin Petkov

The Biological Evolution of Religious Mind and Behavior
Ed. by Eckart Voland and Wulf Schiefenhövel

Homo Novus - A Human Without Illusions
Ed. by Ulrich J. Frey; Charlotte Störmer and Kai P. Willführ

Brain-Computer Interfaces
Revolutionizing Human-Computer Interaction
Ed. by Bernhard Graimann, Brendan Z. Allison and Gert Pfurtscheller

Searching for Extraterrestrial Intelligence
SETI Past, Present, and Future
By H. Paul Shuch

Essential Building Blocks of Human Nature
Ed. by Ulrich J. Frey, Charlotte Störmer and Kai P. Willführ

Mindful Universe
Quantum Mechanics and the Participating Observer
By Henry P. Stapp

Principles of Evolution
From the Planck Epoch to Complex Multicellular Life
Ed. by Hildegard Meyer-Ortmanns and Stefan Thurner

The Second Law of Economics
Energy, Entropy, and the Origins of Wealth
By Reiner Kümmel

States of Consciousness
Experimental Insights into Meditation, Waking, Sleep and Dreams
Ed. by Dean Cvetkovic and Irena Cosic

Elegance and Enigma
The Quantum Interviews
Ed. by Maximilian Schlosshauer

Humans on Earth
From Origins to Possible Futures
By Filipe Duarte Santos

Evolution 2.0
Implications of Darwinism in Philosophy and the Social and Natural Sciences
Ed. by Martin Brinkworth and Friedel Weinert

Chips 2020
A Guide to the Future of Nanoelectronics
Ed. by Bernd Höfflinger

Probability in Physics
Ed. by Yemima Ben-Menahem and Meir Hemmo

From the Web to the Grid and Beyond
Computing Paradigms Driven by High-Energy Physics
Ed. by René Brun, Federico Carminati and Giuliana Galli Carminati

The Dual Nature of Life
Interplay of the Individual and the Genome
By Gennadiy Zhegunov

Natural Fabrications
Science, Emergence and Consciousness
By William Seager

Singularity Hypotheses
A Scientific and Philosophical Assessment
Ed. by Amnon H. Eden, James H. Moor, Johnny H. Soraker and Eric Steinhart

Trick or Truth?
The Mysterious Connection Between Physics and Mathematics
Ed. by Anthony Aguirre, Brendan Foster and Zeeya Merali

How Can Physics Underlie the Mind?
Top-Down Causation in the Human Context
By George Ellis

The Challenge of Chance
A Multidisciplinary Approach from Science and the Humanities
Ed. by Klaas Landsman and Ellen van Wolde

Energy, Complexity and Wealth Maximization
By Robert Ayres

Quantum [Un]Speakables II
Half a Century of Bell's Theorem
Ed. by Reinhold Bertlmann and Anton Zeilinger

Ancestors, Territoriality, and Gods
A Natural History of Religion
By Ina Wunn and Davina Grojnowski

Printed in the United States
By Bookmasters